Giant Planets of Our Solar System

Atmospheres, Composition, and Structure (Second Edition)

Patrick G. J. Irwin

Giant Planets of Our Solar System

Atmospheres, Composition, and Structure
(Second Edition)

Published in association with
Praxis Publishing
Chichester, UK

Dr Patrick G. J. Irwin
Atmospheric, Oceanic and Planetary Physics
Clarendon Laboratory
Oxford
UK

SPRINGER–PRAXIS BOOKS IN ASTRONOMY AND PLANETARY SCIENCES
SUBJECT *ADVISORY EDITORS*: Philippe Blondel, C.Geol., F.G.S., Ph.D., M.Sc., Senior Scientist, Department of Physics, University of Bath, UK; John Mason, M.Sc., B.Sc., Ph.D.

ISBN 978-3-540-85157-8 Springer Berlin Heidelberg New York

Springer is part of Springer-Science + Business Media (springer.com)

Library of Congress Control Number: 2008940285

Second edition published 2009
Abridged paperback edition published 2006
First edition published 2003
Printed in Germany

Cover design: Jim Wilkie
Project management: OPS Ltd, Gt Yarmouth, Norfolk, UK

Printed on acid-free paper

Contents

Preface

I can remember it vividly. I was eleven and standing on a cold, windy hill with my mother and one of her friends who was an amateur astronomer and had offered to let me look though his telescope. It was only a small telescope, but through it I was able to see the disk of Jupiter with the two dark strips of the equatorial belts accompanied by Jupiter's Galilean moons. As for Saturn, the sight of it hanging there in space surrounded by its fabulous ring system quite took my breath away. This experience, among others, fostered a life-long interest and enthusiasm in physics and in the planets of our solar system, especially the giant planets. Ten years later I found myself as a research student in Oxford studying the atmosphere of Mars and twenty years later I found myself on a beach in Florida, watching the launch of the *Cassini/Huygens* mission to Saturn, carrying with it the CIRS instrument that I had helped to design and build. I am now fortunate enough to be involved in the ongoing *Cassini/Huygens* mission and several other space missions to the planets, and also to be involved with ground-based observations of the giant planets using modern large ground-based telescopes such as the UK Infrared Telescope on Mauna Kea, Hawaii. I'm glad to say that the study of planetary atmospheres continues to fascinate and inspire me.

The first edition of this book, written in 2002 and published in 2003, anticipated the arrival of the *Cassini/Huygens* mission to Saturn. As was hoped, this mission has proven to be a huge success and has greatly improved our understanding of Saturn's (and Titan's) atmospheres. In addition, continually improving ground-based observations of the giant planets have recorded a number of new phenomena such as the reddening of Jupiter's White Oval, and Uranus' changing cloud structure during its northern spring equinox in 2007. All these new observations have substantially revised our knowledge and understanding of the giant planet atmospheres and have thus led to this second edition.

This book is aimed at third-year to fourth-year undergraduates of physics and astronomy and first-year postgraduate students of planetary physics. I hope it may also serve as a handy reference for researchers. One of the difficulties I had in compiling the book was in peeling away some of the jargon used in the scientific literature that assumed prior knowledge which was actually sometimes rather arcane. Hence, wherever possible I have tried to approach all of the fields that make up this book from the starting point of an undergraduate with a good grasp of physics but no prior specialist knowledge. Furthermore I have tried to include references to the major books and papers in the various fields, which should allow an interested reader to explore further should they wish to. For the chapters dealing with current and future projects I have also included a number of website addresses, which were very helpful in writing these chapters. In many areas presented in this book the opinion of the scientific community is still split and thus research is actively ongoing. In such cases I have tried to present objectively both sides of the arguments and I apologize for any bias that may or may not have crept in. In other cases, such as formation models, there are a wide range of results and simulations, and thus it should be remembered that there is considerable variance about the mean view that I have tried to present.

I hope my reader finds this book useful and while I cannot offer the exhilaration of viewing a planet for the first time on a cold, windy hillside, I hope he or she will share my continued enthusiasm for this fascinating area of astronomy.

Patrick G. J. Irwin
Oxford, England, September 2008

Acknowledgements

I would like first of all to thank both my subdepartment of Atmospheric, Oceanic and Planetary Physics and St. Anne's College for granting me sabbatical leave to write the second edition of this book from October 2007 to September 2008. I would also like to thank the U.K. Science and Technology Facilities Council (STFC) for supporting my group's research into the atmospheres of the giant planets, some results of which are presented in this book. I would then like to thank all of my colleagues who have assisted me in collecting and collating the material that I have presented in this book including: David Andrews, Sushil Atreya, Kevin Baines, Bruno Bézard, Simon Calcutt, Gary Davis, Imke de Pater, Remco de Kok, Therese Encrenaz, Uwe Fink, Leigh Fletcher, Tristan Guillot, Franck Hersant, Bill Hubbard, Andy Ingersoll, D. Jewitt, Erich Karkoschka, Emmanuel Lellouch, Steve Lewis, Mark McCaughrean, Julianne Moses, Glenn Orton, Peter Read, Maarten Roos-Serote, Augustin Sánchez-Lavega, Jacques Sauval, Larry Sromovsky, Nick Teanby, Ashwin Vasavada, and Don Yeomans. I am also very grateful to Clive Horwood and John Mason of Springer-Praxis who reviewed the first edition of this book and have continued to support me in writing this second edition. Finally, I would like to thank my wife and young family for putting up with Dad often having to work late and more than occasionally being a little grumpy!

To my wife Dunja,
to my children, Benjamin, Samuel, Emily and James,
and in loving memory of my father,
Patrick Joseph Irwin,
13th May 1916–10th May 1966
and most especially of my wonderful mother,
Joyce Margaret Irwin (née Kingsbury),
1st April 1926–4th November 2006,
who saw the first edition of this book, but sadly will never see the second

Illustrations

Tables

Abbreviations and acronyms

AAO	Anglo-Australian Observatory
ACA	Atacama Compact Array
ACS	Advanced Camera for Surveys
ACURA	Association of Canadian Universities for Research in Astronomy
ALMA	Atacama Large Millimeter Array
AO	Adaptive optics (system)
APEX	Atacama Pathfinder Experiment
AT	Auxiliary telescope
BDRF	Bidirectional reflectivity function
BIMA	Berkeley Illinois Maryland Association
Caltech	California Institute of Technology
CAPE	Convective available potential energy
CAPS	Cassini Plasma Spectrometer
CARMA	Combined Array for Research in Millimeter-wave Astronomy
CDA	Cosmic Dust Analyzer
CELT	California Extremely Large Telescope
CFHT	Canada–France–Hawaii Telescope
CIA	Collision-induced absorption
CIRS	Composite Infrared Spectrometer
CNRS	Centre National de la Recherche Scientifique
COBE	Cosmic Background Explorer
COROT	Convection, rotation and transits
CVF	Continuous Variable Filter
DALR	Dry adiabatic lapse rate
DLR	German Space Agency
DS2	Second dark spot

E-ELT	European Extremely Large Telescope
ECCM	Equilibrium Cloud Condensation Model
ELT	Extremely Large Telescope
EPIC	Explicit Planetary Isentropic-Coordinate Model
ESO	European Southern Observatory
EUV	Extreme ultraviolet
EZ	Equatorial Zone
FIRST	Far-infrared Space Telescope
FOV	Field of view
FP	Fabry–Pérot
FTS	Fourier Transform Spectrometer
FUV	Far-ultraviolet
GCM	General circulation model
GCR	Galactic cosmic ray
GDS	Great Dark Spot
GEISA	Gestion des Etudes des Informations Spectroscopiques Atmosphériques
GMT	Giant Magellan Telescope
GPMS	*Galileo* Probe Mass Spectrometer
GRS	Grcat Red Spot
GWS	Great White Spot
HEB	Hot Electron Bolometer
HGA	High Gain Antenna
HIFI	Heterodyne Instrument for the Far Infrared
HITRAN	HIgh resolution TRANsmission molecular absorption database
HRC	High Resolution Camera
HST	Hubble Space Telescope
IGN	Instituto Geográfico Nacional
ING	Isaac Newton Group of Telescopes
INMS	Ion and Neutral Mass Spectrometer
IPD	Interplanetary dust
IR	Infrared
IRAC	InfraRed Array Camera
IRAM	Institut de RadioAstronomie Millimétrique
IRIS	Infrared Spectrometer and Radiometer
IRS	Infrared Spectrograph
IRTF	Infrared Telescope Facility
ISIM	Integrated Science Instrument Module
ISM	Interstellar medium
ISO	Infrared Space Observatory
ISOCAM	ISO Camera
ISOPHOT	ISO Imaging Photo-polarimeter
ISS	Imaging Science Subsystem
ITCZ	Inter Tropical Convergence Zone

JAC	Joint Astronomy Centre
JAXA	Japanese Aerospace Exploration Agency
JCM	JunoCam
JCMT	James Clark Maxwell submillimetre Telescope
JIRAM	Jupiter Infrared Auroral Mapper
JKR	Jupiter's kilometric radiation
JPL	Jet Propulsion Laboratory
JWST	James Webb Space Telescope
KAO	Kuiper Airborne Observatory
KBO	Kuiper Belt Object
LBT	Large Binocular Telescope
LBTI	LBT Interferometer
LORRI	LOng Range Reconnaissance Imager
LRS	Little Red Spot
LTE	Local thermodynamic equilibrium
LW	Longwavelength
LWS	Long Wavelength Spectrometer
MAG	Dual Technique Magnetometer
MAMA	Multi-Anode Microchannel Array
MHD	Magneto-hydrodynamic
MIMI	Magnetospheric Imaging Instrument
MIPS	Multiband Imaging Photometer
MIRI	Mid Infrared Instrument
MPG	Max Planck Gesellschaft
MWR	Microwave Radiometer
NA	Narrow angle
NAC	Narrow Angle Camera
NAOJ	National Astronomical Observatory of Japan
NEB	North Equatorial Belt
NEBn	North Equatorial Belt-North
NEMESIS	Nonlinear optimal Estimator for MultivariatE Spectral AnalySIS
NEP	Noise equivalent power
NER	Noise equivalent radiance
NGDS	New Great Dark Spot
NICMOS	Near IR Camera and Multi Object Spectrometer
NIMS	Near Infrared Mapping Spectrometer
NIRCam	Near Infrared Camera
NIRSpec	Near Infrared Spectrograph
NMA	Nobeyama Millimeter Array
NPH	North Polar Hexagon
NPS	North Polar Spot
NPZ	North Polar Zone
NTB	North Temperate Belt
NTropZ	North Tropical Zone

NTrZ	North Tropical Zone
NTT	New Technology Telescope
OCIW	Observatories of the Carnegie Institution of Washington
OVRO	Owens Valley Radio Observatory
OWL	Overwhelmingly Large Telescope
PACS	Photodetector Array Camera and Spectrometer
PAH	Polycyclic aromatic hydrocarbon
PC	Planetary Camera
PPR	Photopolarimeter–Radiometer
PSF	Point spread function
QBO	Quasi-biennial oscillation
QQO	Quasi-quadrennial oscillation
RADAR	*Cassini* radar
RF	Radio Frequency
RL	Richardson–Lucy
RPWS	Radio and Plasma Wave Science
RSS	Radio Science Instrument
RTG	Radioisotope Thermal Generator; Radioisotope Thermoelectric Generator
s.v.p.	Saturated vapor pressure
SALR	Saturated adiabatic lapse rate
SBC	Solar Blind Camera
SCIP	Solar composition icy planetesimal
SEB	South Equatorial Belt
SED	Saturn electrostatic discharge
SIAC	Spectrally identifiable ammonia cloud
SIM	Space Interferometry Mission
SIRTF	Space Infrared Telescope Facility
SIS	Superconductor–insulator–superconductor
SKR	Saturn's kilometric radiation
SL	Shortwavelength, Low-resolution module
SL-9	Comet Shoemaker–Levy 9
SOFIA	Stratospheric Observatory for Infrared Astronomy
SPF	South Polar Feature
SPIRE	Spectral and Photometric Imaging Receiver
SPV	South Polar Vortex
SPW	South Polar Wave
SSI	Solid State Imaging
STrBs	South Tropical Belt-South
STFC	Science and Technology Facilities Council
STIS	Space Telescope Imaging Spectrograph
STP	Standard temperature and pressure
STrZ	South Tropical Zone
STZ	South Temperate Zone
SW	Shortwavelength

SWAS	Submillimeter Wave Astronomy Satellite
SWS	Short Wavelength Spectrometer
TMT	Thirty Meter Telescope
TPF	Terrestrial Planet Finder
UKIRT	United Kingdom Infrared Telescope
UT	Unit Telescope
UV	Ultraviolet
UVIS	Ultraviolet Imaging Spectrograph
UVS	Ultraviolet Spectrometer
v.m.r.	Volume mixing ratio
VEEGA	Venus–Earth–Earth Gravity Assist
VIMS	Visible and Infrared Mapping Spectrometer
VISTA	Visible and Infrared Telescope for Astronomy
VLA	Very Large Array
VLBA	Very Large Baseline Array
VLST	VLT Survey Telescope
VLT	Very Large Telescope
VLTI	VLT Interferometer
VVEJGA	Venus–Venus–Earth–Jupiter Gravity Assist
WA	Wide angle
WFC	Wide Field Camera
WFPC	Wide Field/Planetary Camera
YSO	Young stellar object

1

Introduction

1.1 THE GIANT OUTER PLANETS

The giant outer planets—Jupiter, Saturn, Uranus, and Neptune (Figure 1.1)—are by far the largest planetary bodies in the solar system and together comprise 99.56% of the planetary mass. Although very far from the Earth, the enormous physical size of Jupiter and Saturn meant that these planets were easily visible to the ancients. However, the other two "giants", Uranus and Neptune, are significantly smaller and so much farther from the Earth that they were unknown before the advent of telescopes, although Uranus is in fact just visible to the naked eye. Uranus was discovered by accident in 1781 by William Herschel (1738–1822) (later Sir William Herschel). Perturbations in the observed orbit of Uranus led John Couch Adams (1819–1892) and Urbain Jean Joseph Le Verrier (1811–1877) to independently predict the presence of a further planet, and Neptune was subsequently discovered close to its predicted position by Johann Gottfried Galle (1812–1910) in 1846. The mean observable properties of the outer planets are listed in Table 1.1.

All the giant planets are observed to rotate very rapidly and the shape of the planets distort under the centrifugal forces that arise. Hence, all the giant planets are noticeably oblate, especially Jupiter and Saturn, with the pole-to-pole diameter being significantly less than the equatorial diameter. Another key difference between the inner terrestrial planets, such as the Earth, and the giant planets is that the latter have surprisingly low densities, roughly equivalent to water and similar to that of the Sun, which has a density of 1.41 $g\,cm^{-3}$. Hence, while we know that the Earth is a rocky body, the outer planets must be composed predominantly of much lighter materials. In fact the giant planets are now known not to have a solid rocky "surface" at all, but instead are gaseous in nature throughout.

Considering the mass, radius, and density of the giant planets they can be seen to divide naturally into two pairs: Jupiter and Saturn, composed primarily of hydrogen and helium and sometimes known as the "gas giants", and Uranus and Neptune,

Figure 1.1. The giant planets as observed by the *Voyager* spacecraft together with the Earth for comparison. Courtesy of NASA.

composed primarily of ices such as water and methane and sometimes known as the "ice giants". Jupiter and Saturn both have a radius in the range 60,000 km to 72,000 km, while Uranus and Neptune are somewhat smaller with a radius of approximately 25,000 km. These have respectively ten times, and four times the radius of the Earth. By comparison, the radius of the Sun is ten times larger than

Table 1.1. Observed properties of the giant planets and Earth.

Property	*Earth*	*Jupiter*	*Saturn*	*Uranus*	*Neptune*
Solar distance (AU)[a]	1.0	5.2	9.5	19.2	30.1
Sidereal orbital period (yr)	1.0	11.9	29.5	84	165
Orbital eccentricity	0.017	0.048	0.056	0.046	0.009
Equatorial radius R_e (km)[b]	6,378.0	71,492	60,268	25,559	24,764
Equatorial radius relative to Earth	1.0	11.2	9.4	4.0	3.9
Oblateness $(R_e - R_p)/R_e$	0.00034	0.065	0.098	0.023	0.017
Mass (10^{24} kg)	5.97	1,898.8	568.5	86.625	102.78
Mass (relative to Earth)	1.0	318	95	14.5	17.1
Mean density (g/cm^3)	5.515	1.33	0.70	1.27	1.76
Sidereal day	23h 56min	9h 55min (System III)	10h 39min (System III)[c]	17h 14min	16h 6min
Equatorial surface gravity at 1 bar (m/s^2)	9.81	23.1	9.1	8.7	11.0
Escape velocity (km/s)	11.2	58.6	33.1	21.1	23.3
Obliquity	23.5°	3°	27°	98°	29°
Equilibrium radiating temperature T_E (K)	255	110	82	58	47
Mean temperature at 1 bar (K)	288	167	138	79	70
Bolometric temperature[d] T_B (K)	255	124	95	59	59
Emitted/absorbed flux ratio	1.0	1.67	1.78	1.06	2.52

1 AU is one astronomical unit, the mean distance of the Earth from the Sun equal to 1.496×10^8 km.
The equatorial radii quoted refer to the 1 bar pressure level. References are given in Table 2.4.
Recently revised figure from *Cassini* observations for bulk rotation rate is 10 h 32 min (Anderson and Schubert, 2007).
The bolometric temperature is the temperature of a black body sphere which would radiate heat to space at the same rate as that observed for the planet.

Jupiter at 696,265 km. The mass of Uranus and Neptune are similar to each other at around 15 $M_\oplus$ (where $M_\oplus$ is the mass of the Earth), while the masses of Jupiter (318 $M_\oplus$) and Saturn (95 $M_\oplus$) are substantially different indicating that Jupiter is much more compressed than Saturn, given their similar size. Although the giant planets are truly massive, their combined mass is still only 0.1% of the solar mass, which is estimated to be 1.9891×10^{30} kg. However, while most of the solar system mass is accounted for by the Sun, its combined spin angular momentum and orbital momentum about the solar system barycenter account for only 1% of that of the total solar system. Instead, most of the solar system angular momentum is accounted for by the giant planets, with the orbital angular momentum of Jupiter and Saturn contributing 85%. This counter-intuitive observation provides major constraints on models of solar system formation as we shall see in Chapter 2.

The atmospheres of the giant planets are found to be meteorologically active and highly convective with the notable exception of Uranus, which seems normally to have a more sluggish atmospheric circulation, except during equinox periods. A clear indication of convective activity can be seen from observing the mean temperatures of the giant planets. Solar irradiance drops as the inverse square of distance from the Sun, and thus the calculated *effective radiating temperature* (the temperature at which absorbed solar radiation is balanced by the emitted thermal radiation of the planet, discussed in Chapter 3) also decreases with distance from the Sun. While the observed mean *bolometric temperatures* (the mean temperatures at which the planets actually radiate) are indeed found to decrease with distance, the bolometric temperature exceeds the effective radiation temperature for all the giant planets except Uranus. Hence, all the giant planets except Uranus radiate significantly more radiation than they receive from the Sun indicating a substantial source of internal heat. The absorbed solar fluxes vary from a maximum at the subsolar point to zero at the limb and on the night side while the emitted thermal fluxes are found to a first order to be independent of latitude and longitude and are shown as a function of latitude in Figure 1.2. The ratios of emitted to absorbed fluxes are listed in Table 1.1. As can be seen, both Jupiter and Saturn have significant internal heat sources. However, Uranus and Neptune, which are otherwise so similar, are found to be very different in this regard with Uranus having almost no internal heat source and Neptune a very large supply giving it the highest emitted/absorbed flux ratio of all the giant planets. As we shall see in Chapter 2, this internal heat is thought, for all the giant planets except Saturn, to be residual heat left over from the formation of the planet, which is slowly radiating away to space as the planets contract via the *Kelvin–Helmholtz* mechanism (e.g., the radius of Jupiter is estimated to be currently shrinking by approximately 1 mm/yr). The emitted flux of Saturn is thought to be too high for the source to be just residual formation heat since this is estimated to have radiated away almost 2 billion years ago. Instead, the source is thought to arise due to internal differentiation of helium. We will return to this topic in Chapter 3.

1.2 OBSERVED ATMOSPHERES OF THE GIANT PLANETS

The observable atmospheres of the giant planets are dominated by molecular hydrogen and helium, in proportions roughly similar to that found in the Sun. The abundance of "heavy" elements (which in this context refers to elements heavier than helium) is found, or estimated, to be approximately 3–5 times the solar value for Jupiter, ~10 times the solar value for Saturn, increasing to 30–50 times the solar value

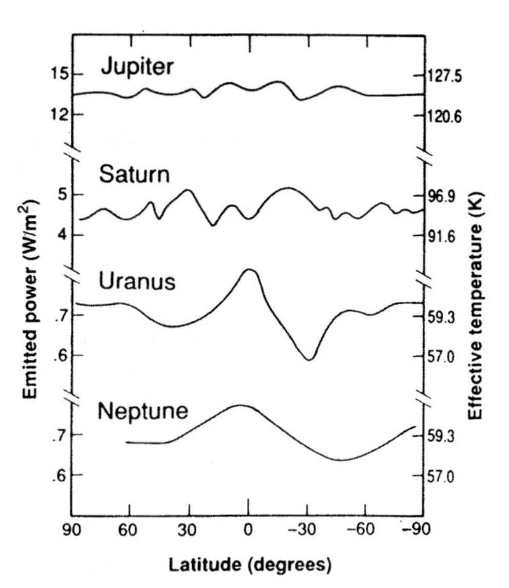

Figure 1.2. Total thermal-infrared radiation flux ($W\,m^{-2}$) emitted by the giant planets as a function of latitude (Ingersoll, 1990). While some belt/zone variations are visible, the emitted flux is to a first approximation independent of latitude. The radiation is emitted predominantly from the 0.3 bar to 0.5 bar pressure levels. On the right-hand axis the radiative flux has been converted to brightness temperature (the temperature of a black body that would emit the same flux). Reprinted with permission from Ingersoll (1990). Copyright 1990 American Association for the Advancement of Science.

for Uranus and Neptune. As we shall see in Chapter 2, the generally favored interpretation of this and the mean size and density measurements is that the outer planets accreted originally from icy planetesimals and became so massive that they were able to attract gravitationally hydrogen and helium from the solar nebula. It would appear that Jupiter and Saturn grew large enough and rapidly enough to capture a huge mass of hydrogen and helium, while Uranus and Neptune were not able to attract so much. Hence, the abundance of icy materials is higher in Uranus and Neptune than in Jupiter and Saturn. In the upper, cooler parts of the giant planet atmospheres that are actually observable, these heavy elements are mainly present in their fully reduced form and thus after H_2 and He the next most abundant molecules inferred or measured (prior to any condensation) are, in decreasing order, H_2O, CH_4, NH_3, and H_2S. In fact, the upper atmospheres of the giant planets are so cold that H_2O, H_2S, and NH_3 condense at various levels forming the cloud decks observed on these giant planets. The upper atmospheres of Uranus and Neptune are so cold that even CH_4 condenses.

The observed atmospheres of the giant planets reveal many very interesting properties which will be briefly described here, and expanded upon in Chapters 2 to 5.

1.2.1 Jupiter

Through a telescope, Jupiter appears as a dusky ochre-colored oblate planet with dark horizontal stripes aligned parallel to the equator. Two of these dark "belts" are especially noticeable on either side of the brighter equatorial "zone", with other thinner "belts" seen closer to the poles. In fact, the atmosphere of Jupiter has the

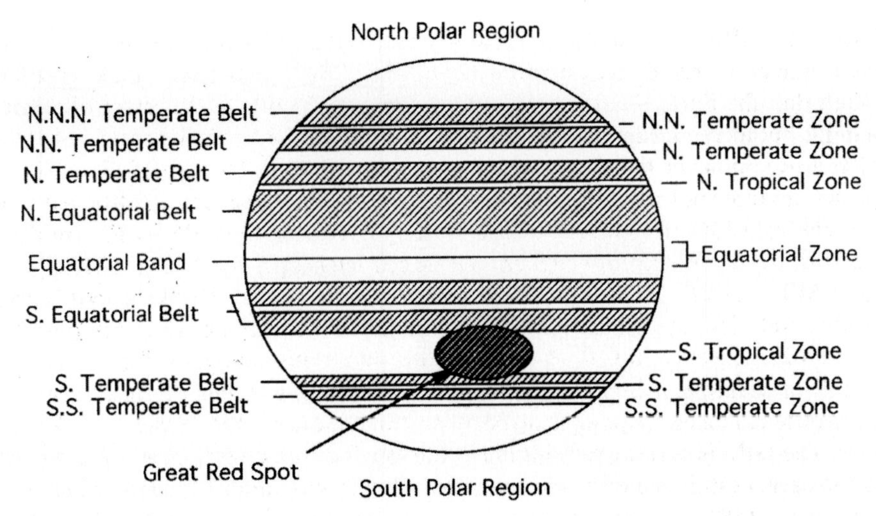

Figure 1.4. Jovian zonal nomenclature (Irwin, 1999). Reprinted with kind permission of Kluwer Academic.

most color contrast of any atmosphere in the solar system, including that of the Earth's (Dowling, 1997); an image of Jupiter recorded by the *Cassini* spacecraft in 2000 is shown in Figure 1.3 (see color section). The general belt/zone structure appears to be very stable and a universally accepted naming scheme is shown in Figure 1.4. Although the general structure is long-lived, the contrast of the different features varies with time. These changes are usually gradual, although the South Equatorial Belt (SEB) often displays dramatic outbursts of cloud activity (Dowling, 1997; Rogers, 1995), the most recent upheaval occurring in 2007. The belt/zone structure is generally thought to be formed by a global circulation system that upwells moist air in the "zones", forming bright cloudy regions and subsides in the belts, forming relatively cloud-free regions that appear dark in the visible, although it has been suggested that this circulation reverses at deeper levels (as we shall see in Chapter 5). The upper observable cloud deck is almost certainly predominantly composed of ammonia crystals, although we shall see in Chapter 4 that these appear to be modified in some way such that their pure spectral absorption features are usually masked. Above the main cloud decks, various processes such as photochemistry act to create hydrocarbon haze particles, of uncertain composition, which gradually settle down through the atmosphere and are eventually pyrolyzed and destroyed at deeper levels.

In addition to the general zonal structure, Jupiter is found to have a number of large oval structures or vortices. Unlike three-dimensional turbulence where one expects a large eddy to split up into smaller ones (a good example is a smoke ring, which can be observed to rapidly dissipate), weather systems are governed predominantly by two-dimensional turbulence, which has the counter-intuitive property that smaller eddies merge into larger ones by a process known as the *backwards energy cascade*. The most famous of these ovals is the Great Red Spot

(GRS). The GRS is a vast anticyclonic weather system that is currently ~20,000 km wide in the east–west direction and ~12,000 km in the north–south, making it large enough that the Earth would easily fit in the middle! Winds in the center of the spot are light, but increase rapidly towards the edge of the spot, reaching speeds of 100 m/s. The GRS appears to be extremely long-lived. Robert Hooke (1635–1703) first reported a large spot in 1665 and "Hooke's spot" was subsequently observed intermittently from 1664 to 1701. Although this may have been the GRS itself, continuous observations of the current GRS can be traced back only to 1831 (Dowling, 1997; Simon-Miller *et al.*, 2002). Indeed it has been argued that Hooke's spot became unstable and dissipated, only for the current GRS to form later (e.g., Simon-Miller *et al.*, 2002). The current GRS has undergone numerous changes since observations began. For example, it became nearly invisible in the 1860s, but within 10 years was very prominent again. It was particularly prominent during the *Pioneer* and *Voyager* flybys. The GRS is currently shrinking in the longitudinal direction. A hundred years ago the east–west diameter of the spot was ~46,000 km, almost twice as wide as it is today. If the GRS continues to shrink in the east–west direction at the current rate then by 2040 the spot will be circular. We shall see in Chapter 5 that such a configuration is thought unlikely to be stable and thus the GRS may actually break up and disappear in our lifetimes, perhaps to spawn the generation of a new great oval!

Other well-known ovals include the South Tropical Belt-South (STrBs) white ovals and the North Equatorial Belt-North (NEBn) brown barges. The current STBs white ovals initiated as a disturbance in the South Temperate Zone (STZ) in 1939 and coalesced into three white-colored ovals. Two of these ovals merged together in 1998, and in March 2000 the two remaining ovals merged to form a single white oval, which took on the same red color as the GRS in 2005. While 90% of the Jovian vortices are anticyclonic, only 10% are cyclonic and the most well-known of these are the NEBn brown barges, which appear at the boundary between the NEB and the North Tropical Zone (NTrZ) and were particularly prominent during the *Voyager* encounters.

We shall see in Chapter 5 that the winds on the giant planets blow almost entirely in the zonal direction (i.e., east to west, or west to east), and the winds alternate in direction in association with the belts and zones. The zonal wind speed on Jupiter varies particularly rapidly with latitude and is puzzlingly strong at the equator, reaching speeds of 100 m/s in the eastward direction. The equatorial region of Jupiter is thus super-rotating (i.e., rotates faster than the bulk of the planet), a state that is difficult to simulate with numerical models pointing to considerable underlying complexity (as we shall see in Chapter 5). Observations of ovals and other atmospheric features by early astronomers such as Jean-Dominique (a.k.a. Gian-Domenico) Cassini (1625–1712) had to be referred to a longitude system and as the equator rotates at a noticeably faster rate than the rest of the planet due to the high wind speeds there, two conventions arose. The System I frame referred to features at equatorial latitudes within 10° of the equator, while the System II frame referred to all other latitudes. Both systems have since been superseded by System III, which is referenced to the bulk rotation of the interior as inferred from radio observations of the rotation of the magnetosphere.

Table 1.2a. Major satellites of Jupiter.

Satellite	*Mass* (kg)	*Radius* (km)	*Density* ($g\,cm^{-3}$)	*P* (days)	A_G	*R* (10^3 km)	*e*	*i* (deg)
Io	8.9×10^{22}	1,815	3.55	1.77	0.61	422	0.004	0.04
Europa	4.8×10^{22}	1,569	3.01	3.55	0.64	671	0.009	0.47
Ganymede	14.8×10^{22}	2,631	1.95	7.15	0.42	1,070	0.002	0.21
Callisto	10.7×10^{22}	2,400	1.86	16.69	0.20	1,883	0.007	0.51
Amalthea	7.2×10^{18}	$\sim$100	1.72	0.50	0.05	181	0.003	0.40

P is the orbital period, A_G is the geometric albedo, *R* is the semi-major axis orbital radius, *e* is the orbital eccentricity, and *i* is the inclination of orbit to the planet's equator. Only satellites with a mean radius greater than 100 km have been included.

Jupiter is accompanied by four large moons, known as the Galilean satellites, after their discoverer Galileo Galilei (1564–1642), which are closely aligned with the planet's equatorial plane. Their alignment and the observed compositional differences (Section 1.3) suggest that these were formed at the same time as Jupiter from the proto-Jupiter accretion disk. Jupiter also has a number of much smaller satellites, which inhabit various eccentric and inclined orbits. These are probably captured planetesimals. The larger satellites of Jupiter, with a radius greater than 100 km, are listed in Table 1.2a. Finally, Jupiter has a small ring that was first observed by *Voyager 1* in 1979.

1.2.2 Saturn

Through a telescope, and ignoring Saturn's magnificent ring system, the observable "surface" of Saturn appears much blander than that of Jupiter with less banding and fewer ovals, although the predominant pale ochre color is similar to that of Jupiter. Like Jupiter, Saturn also has a significant internal heat source. However, although this is a greater fraction of the total heat emitted, the combined total is much smaller than that of Jupiter. Hence, the atmosphere is thought to be less meteorologically active than Jupiter's since the heating rate driving it is less. In addition, any convective structures that are present are more masked by overlying haze due to the colder temperatures found in the upper observable part of Saturn's atmosphere. These cooler temperatures mean that the expected upper cloud deck of ammonia ice (Chapter 4) occurs deeper in the atmosphere than in Jupiter's. In addition, the lower gravitational acceleration of Saturn means that the atmosphere is more vertically extended, and thus the absorption by upper atmospheric haze is enhanced.

Although generally more quiescent than Jupiter, major convective-type events are occasionally observed in Saturn's brighter zonal regions (outlined in Chapter 5). These are known as "brightenings" or "Great White Spots" and have been observed intermittently since 1793. Recent "brightenings" occurred in the Equatorial Zone in

1990 and 1994 and were observed with the Hubble Space Telescope (Figure 1.5, see color section). What appears to be happening in these events is that some kind of disturbance deep below the visible cloud/haze top of Saturn triggers rapid, possibly thunderstorm-style, deep vertical convection, and the resultant formation of thick, very high ammonia clouds. These clouds are subsequently torn apart by the zonal wind flow shear and dissipate over a timescale of a few months.

The zonal wind system of Saturn is similar to that of Jupiter's with the wind direction alternating in association with the belts/zones. However, the widths of Saturn's visible belts and zones (but not those of the deeper clouds, as we shall see in Chapter 5) appear greater than those of Jupiter and the equatorial jet is found to be very much stronger and during the *Voyager* flybys blew at a very rapid 500 m/s in the prograde direction relative to the interior System III rotation rate (although this has since slowed to 400 m/s during the *Cassini* epoch). Thus, the equatorial zone of Saturn is even more super-rotating than that of Jupiter. However, it has been realized recently that the System III rotation rate, as inferred from radio emission at kilometric wavelengths, actually varies with time and is not solely constrained by the bulk rotation rate. Rather, it is determined by the slippage of the magnetosphere relative to the interior (Anderson and Schubert, 2007) and is thus affected by instabilities in Saturn's plasma disk. The latest estimate of Saturn's bulk rotation rate from *Cassini* observations is 10 h 32 min, compared with the 10 h 39 min System III value (Anderson and Schubert, 2007), and so Saturn's equatorial winds may not be as super-rotating as was earlier thought.

The satellite system of Saturn is rather different from that of Jupiter and Table 1.2b lists those satellites with a radius greater than 100 km. Most of the satellites are somewhat smaller than the Galilean satellites with the exception of Rhea, Iapetus, and most notably Titan. Titan is the second largest satellite in the solar system (after Ganymede) with a radius of 2,575 km and a substantial atmosphere composed mainly of N_2 with a surface pressure of $\sim$1.5 bar. However, while its atmospheric composition and surface pressure are similar to the Earth's, its surface temperature of $\sim$90 K is extremely cold. The surface of Titan at visible wavelengths is obscured by atmospheric hazes, although ground-based and orbital telescopes had detected surface features at near-infrared wavelengths prior to the arrival in the Saturnian system of the *Cassini/Huygens* mission, allowing its rotation rate to be shown to be tidally locked. Instruments on the *Cassini* orbiter have now mapped the surface of the world and following its aerobraking capture on January 14, 2005, *Huygens* spent about 2 hours gently descending though Titan's atmosphere, measuring the conditions in its atmosphere and returning breathtaking images of Titan's surface during its descent.

The ring system of Saturn is the most stunning of all the giant planets (Figures 1.6–1.8, see color section for Figures 1.7 and 1.8). It is now known that all the giant planets have ring systems, which may form either from the tidal disruption of captured satellites, which subsequently spread and are dissipated or absorbed by the planet over time, or by the sputtering of particles from the surfaces of satellites. While the very thin ring first observed by the *Voyager* spacecraft around Jupiter is thought to be due to sputtering, the massive ring system of Saturn can only have

Table 1.2b. Major satellites of Saturn.

Satellite	*Mass* (kg)	*Radius* (km)	*Density* ($g\,cm^{-3}$)	*P* (days)	A_G	*R* (10^3 km)	*e*	*i* (deg)
Mimas	4.6×10^{19}	196	1.44	0.94	0.5	186	0.020	1.53
Enceladus	7.4×10^{19}	250	1.13	1.37	1.0	238	0.005	0.00
Tethys	7.4×10^{20}	530	1.19	1.89	0.9	295	0.000	1.86
Dione	1.1×10^{21}	560	1.43	2.74	0.7	377	0.002	0.02
Rhea	2.5×10^{21}	765	1.33	4.52	0.7	527	0.001	0.35
Titan	1.4×10^{23}	2,575	1.89	15.95	0.21	1,222	0.029	0.33
Hyperion	1.7×10^{19}	~130	1.85	21.28	0.3	1,481	0.104	0.43
Iapetus	1.9×10^{21}	730	1.15	79.33	0.2	3,561	0.028	14.72
Phoebe	4.0×10^{17}	110	0.07	550.48R	0.06	12,952	0.163	177
Janus	—	~100	—	0.69	0.8	151	0.007	0.14

Only satellites with a mean radius greater than 100 km have been included. The letter "R" in the period column indicates a retrograde orbit.

formed from the tidal disruption of a satellite and will thus eventually dissipate. Hence, it is purely serendipitous that Saturn's ring system should be so spectacular at this particular moment in the solar system's history, when it can be observed by the people of Earth.

1.2.3 Uranus

Uranus and Neptune are a good deal smaller than their larger siblings Jupiter and Saturn, and a good deal denser, being composed mostly of icy materials, with a much less massive envelope of molecular hydrogen/helium. The greeny-blue color of Uranus (Figure 1.9, see color section), and the blue color of Neptune arises from the greater abundance of red-absorbing methane in the observable atmospheres of these planets and also in the nature of the particles comprising the main observable cloud deck of these planets, which preferentially absorb wavelengths longer than 0.6 μm.

The observable atmospheres of both Uranus and Neptune are much colder than Jupiter and Saturn, and so ammonia is expected to condense very deep in the atmosphere, but at levels accessible to microwave remote sensing. However, the deep abundance of ammonia inferred from ground-based microwave observations is so low that it would seem that most of the ammonia combines at even deeper levels with other molecules to form perhaps aqueous ammonia or ammonium hydrosulphide (NH_4SH) clouds, or perhaps even a water–ammonia ionic ocean well below the main

Figure 1.6. Saturn's rings behind Saturn. The rings appear to bend as they approach Saturn due to Saturn's atmosphere refracting the light.

observed upper cloud deck. The composition of the main cloud deck is unknown, but is most probably hydrogen sulfide (H_2S). Above this main cloud deck, a second very much thinner cloud deck of methane (CH_4) ice is observed at restricted locations on these planets.

The atmospheric circulation of all the giant planets is driven by both solar and internal heating. However, for Uranus the mean global internal heat flux is at most 6% of the solar flux and thus the dominant circulation must be that forced by the uneven distribution of sunlight over the planet, by which the atmosphere attempts to revert to a barotropic state, where temperature is constant on constant pressure surfaces. This thermal forcing is complicated by the fact that Uranus' large obliquity of 98° means that during the course of its orbit Uranus receives direct sunlight over both poles, as well as the equator. In fact, even though both poles experience a night half a Uranian year long, on average they receive 50% more sunlight per unit area than the equator. The *Voyager 2* flyby in 1986 occurred soon after the northern winter solstice when the South Pole was facing almost directly towards the Sun and the North Pole was in complete darkness. If there were no meridional (i.e., latitudinal) circulation, we would have expected there to have been a significant pole-to-pole temperature gradient of the order of 10 K. However, the infrared spectrometer on *Voyager 2* found there to be almost no temperature difference and hence the atmosphere appears to efficiently redistribute absorbed solar heat.

Although only a very low-contrast belt/zone structure was seen on Uranus by *Voyager 2* and the convective overturning of Uranus' atmosphere thus appeared sluggish, occasional small white clouds were observed intermittently, probably resulting from deep convection cells transporting methane-rich air high into the atmosphere, where it condenses, and these discrete clouds could be tracked to determine the zonal wind speeds at the cloud condensation level. Similar cloud events are seen in the atmosphere of Neptune, although these have much higher contrast and appear to be much more vigorous than those seen on Uranus. The zonal wind structure of Uranus shows none of the rapid latitudinal structure associated with belts and zones seen on Jupiter and Saturn. Instead the structure appears fairly symmetric with midlatitude winds blowing at 200 m/s in the prograde direction and equatorial winds blowing at 100 m/s in the retrograde direction, opposite to that of Jupiter and Saturn.

The occasional storm clouds seen by *Voyager 2* became noticeably more vigorous as Uranus approached its northern spring equinox in 2007; in addition, a distinct bright zone appeared at 45°S, which appears to have faded following the equinox, with a new bright zone appearing at 45°N. We shall return to this in Chapter 5.

The large obliquity of Uranus may be evidence of an off-center impact by a single planet-sized body into Uranus towards the end of its accretion phase. The fact that Uranus' compact and regular satellite and ring system closely shares this obliquity suggests that the unusual spin vector was imparted early, some 4.6 Gyr ago. It has even been speculated that this cataclysmic event may have extinguished the internal heat source by effectively turning the planet inside-out causing the planet to release most of its internal energy soon after formation rather than gradually like the other giant planets! The larger satellites of Uranus ($R > 100$ km) are listed in Table 1.2c.

1.2.4 Neptune

Although Neptune is farther from the Sun than Uranus, and thus receives less sunlight, its bolometic temperature is very similar to that of Uranus, indicating a strong source of internal heat. In fact, its ratio of emitted thermal/absorbed solar flux is the highest of any of the giant planets. Neptune appears bluer than Uranus and has some of the most active meteorology and global variability of any of the giant planets (Figure 1.10, see color section).

Table 1.2c. Major satellites of Uranus.

Satellite	*Mass* (kg)	*Radius* (km)	*Density* ($g\,cm^{-3}$)	*P* (days)	A_G	*R* (10^3 km)	*e*	*i* (deg)
Ariel	1.35×10^{21}	579	1.66	2.52	0.34	191	0.003	0.3
Umbriel	1.17×10^{21}	586	1.39	4.14	0.18	266	0.005	0.36
Titania	3.52×10^{21}	790	1.70	8.71	0.27	436	0.002	0.14
Oberon	3.01×10^{21}	762	1.62	13.46	0.24	584	0.001	0.10
Miranda	6.93×10^{19}	240	1.20	1.41	0.27	129	0.003	4.2

Only satellites with a mean radius greater than 100 km have been included.

The vertical cloud structure appears to be similar to that of Uranus. Again ground-based microwave spectra detect very little ammonia at the expected ammonia condensation level suggesting that it combines with either water or H_2S well below the observable cloud decks. Instead, the main cloud deck is again probably composed of H_2S. A thinner methane haze is found at higher altitudes. Like Uranus, no clear belt/zone structure is evident, although the opacity of the methane cloud deck between 30° and 60°, north and south, has increased significantly in the last decade. Unlike Uranus, however, a number of storm systems were observed by *Voyager 2* on Neptune in 1989, including the "Great Dark Spot" at southern midlatitudes. The GDS may have had a similar structure to Jupiter's GRS, but it was short-lived and had disappeared by the time of new Hubble Space Telescope observations in 1994. However, by 1995 a new dark spot had appeared at northern midlatitudes, which itself disappeared shortly after. These features are dark probably because of either a darkening or deepening of the main H_2S cloud top at 3.8 bar. In addition, several smaller white clouds are intermittently seen all over the planet, but mainly at mid-latitudes, and allow estimation of zonal wind speed. The general structure of this seems similar to that of Uranus in that rather than a series of alternating easterlies and westerlies as is found on Jupiter and Saturn, the winds are strongly retrograde at the equator and then reverse direction slowly to become strongly prograde at latitudes of approximately 70° before returning to zero at the poles. However, the strength of the zonal winds on Neptune greatly exceeds those found on Uranus with the retrograde equatorial jet reaching speeds of 400 m/s, while the prograde subpolar jets are estimated to reach 200 m/s. Hence, Neptune has the largest range of atmospheric rotation periods of any of the giant planets. Why the equatorial jets of Jupiter and Saturn should be prograde, and those of Uranus and Neptune are retrograde is a mystery that will be returned to in Chapter 5.

Neptune's satellite system is less compact and organized than Uranus' and the largest satellite, Triton, is in highly inclined orbit, as can be seen in Table 1.2d, which lists the larger Neptunian satellites with a radius greater than 100 km. *Voyager 2* observed Triton to have an extremely thin N_2 atmosphere.

Table 1.2d. Major satellites of Neptune.

Satellite	*Mass* (kg)	*Radius* (km)	*Density* ($g\,cm^{-3}$)	*P* (days)	A_G	*R* (10^3 km)	*e*	*i* (deg)
Triton	2.15×10^{22}	1,353	2.07	5.88R	0.7	355	0.000	157.3
Nereid	2.06×10^{19}	170	1.00	360.14	0.4	5513	0.751	27.6
Larissa	—	~100	—	0.55	0.06	74	0.001	0.2
Proteus	—	~200	—	1.12	0.06	118	<0.001	0.55

Only satellites with a mean radius greater than 100 km have been included. The letter "R" in the period column indicates a retrograde orbit.

1.3 SATELLITES OF THE OUTER PLANETS

The satellites of the giant planets are believed to have formed in two ways: (1) in a circumplanetary accretion disk (accretion disks are discussed in Chapter 2) at the same time as the planet; and (2) later capture of remaining planetesimals in the solar system. Satellites formed by the first mechanism are expected to lie in the equatorial plane of the planet and to have near-circular orbits. Satellites formed by the second mechanism may have any inclination, and are likely to have eccentric orbits.

The satellites of the giant planets show just this dichotomy (as can be seen in Tables 1.2a–d). We will see in Chapter 2 that those satellites that formed directly from the circumplanetary disk should have compositional differences that reflect the temperature distribution of the disk during the period in which the satellites were forming. This is particularly clear for the "Galilean" satellites of Jupiter (Io, Europa, Ganymede, and Callisto), where distinct density and thus compositional differences are observed with the inner satellites containing less water and other volatiles than the outer satellites. To capture interplanetary planetesimals into the eccentric orbits seen requires some kind of friction, which suggests either aerocapture or that the planetesimals were braked by passing through the dense circumplanetary disk and were thus captured early in the planets' formation. Presumably the composition of these captured satellites has not varied much since the beginning of the solar system and so they are very interesting bodies to examine to understand the composition of early planetesimals. The clearest candidate for a captured satellite is Triton, whose physical characteristics are thought to be similar to Pluto and Charon (listed in Table 1.2e) and to the Kuiper Belt objects discussed in Chapter 2.

1.4 EXPLORATION OF THE OUTER PLANETS

How do we know what we do know about the giant planets? The basic mass, density, and size parameters given in Table 1.1 were established from ground-based visible-wavelength observations of the planets themselves, of their orbits around the Sun,

Table 1.2e. Properties of Pluto and Charon.

Body	*Mass* (kg)	*Radius* (km)	*Density* ($g\,cm^{-3}$)	*P* (days)	A_G	*R* (10^3 km)	*e*	*i* (deg)
Pluto	1.5×10^{22}	1151	1.1	—	0.3?	—	—	—
Charon	3.3×10^{21}	593	3.8	6.39	0.5	19.6	<0.001	99.0

and also the orbits of their satellites around them. We shall see in Chapter 2 that even this very simple data led to profound implications for the internal structure of the giant planets.

Information on the temperature, composition, and cloud structure of the giant planets may be determined from ground-based observations at UV/visible through to infrared and microwave wavelengths in a number of spectral windows where the absorption of the Earth's atmosphere is low (Chapter 7). The absorption features of many molecules are observed in these spectra, which may be used to constrain composition. At certain wavelengths dominated by absorption features of well-mixed gases such as hydrogen, helium, and methane, thermal-infrared and microwave observations of the "brightness temperature" may be used to infer atmospheric temperatures over a wide range of pressure levels. Another technique that is sometimes used is stellar occultation. Occasionally the planets move in front of a star as seen from certain points on the Earth, and observations of the star's light curve during one of these occultations provides unique information on the upper atmospheres of these planets.

Ground-based observations have been improved by the advent of adaptive optics, together with data-processing techniques such as speckling and deconvolution (discussed in Chapter 7), which have greatly increased their spatial resolution. In addition, Earth-orbiting telescopes such as the Hubble Space Telescope (HST) and the Infrared Space Observatory (ISO) are capable of not only greater spatial resolution (owing to the lack of an intervening, turbulent terrestrial atmosphere), but are also unencumbered by terrestrial absorptions and so may measure the entire visible and infrared spectrum and also the ultraviolet spectrum, which contains additional information on composition and clouds.

Together with continually improving telescopic observations from the ground or Earth orbit, observations of the giant planets entered the space age on December 3, 1973 when *Pioneer 10* became the first spacecraft to fly by a giant planet, Jupiter. Spacecraft remote observations of planetary atmospheres from ultraviolet to far-infrared wavelengths offer excellent spatial resolution and unrivaled phase angle coverage (the angle between the direction of the Sun and the direction of observation) of these atmospheres. Remote observations by subsequent missions—*Pioneer 11*, *Voyagers 1* and *2*, *Galileo* and *Cassini/Huygens*—have greatly increased our knowledge of the atmospheres of the giant planets. In addition to remote observations, it is possible to record the strength of the radio signal broadcast by these spacecraft as they go behind or come out from behind the planets in their orbital trajectory and such radio occultations provide highly precise measurements of the vertical density profile, from which the thermal structure can be determined. Furthermore, the *Galileo* mission included an entry probe, which parachuted through the atmosphere of Jupiter on December 7, 1995 providing the first ever *in situ* measurements of the atmosphere of a giant planet.

1.5 ORGANIZATION OF THE BOOK

This book is aimed at final-year physics/astronomy undergraduates and first-year postgraduate students of planetary physics. Knowledge of basic physics is assumed, but no previous atmospheric physics knowledge is needed. Formulas are derived where possible or referred if not.

In Chapter 2 we will look at theories of formation of the giant planets, which may be used to interpret their physical and compositional differences and in Chapter 3 we will review how the atmospheres of these planets may have evolved with time. In Chapter 4 we will review what is known about the vertical temperature, composition, and cloud structure of the planets and in Chapter 5 we will look in detail at the meteorology and dynamical processes taking place. Since the only giant planet where *in situ* measurements have been made is Jupiter, most of what we know about the giant planets comes from remote sensing via measurements of the ultraviolet, visible, infrared, and microwave spectra as we mentioned in the previous section. Hence, in Chapter 6 we will examine the observed spectra of the planets and review the physics of the observed spectral features and radiative transfer processes. In Chapter 7 we will review the sources of information that have been used to construct our current understanding of the atmospheres of these planets and outline how these remotely sensed spectra may be inverted via retrieval theory in order to estimate the physical conditions in these atmospheres. Finally, in Chapter 8 we will look to the future and describe further planned measurements of the giant planets and also missions to find extrasolar planets, including giant planets, about other stars.

1.6 BIBLIOGRAPHY

Bagenal, F., T. Dowling, and W. McKinnon (Eds.) (2004) *Jupiter: The Planet, Satellites and Magnetosphere*. Cambridge University Press, Cambridge, U.K.

Beatty, J.K., C. Collins Petersen, and A. Chaikin (Eds.) (1999) *The New Solar System* (Fourth Edition). Cambridge University Press, Cambridge, MA.

de Pater, I. and J.J. Lissauer (2001) *Planetary Sciences*. Cambridge University Press, Cambridge, U.K.

Encrenaz, T., J.-P. Bibring, and M. Blanc (1995) *The Solar System* (Second Edition). Springer-Verlag, Berlin.

Jones, B.W. (2007) *Discovering the Solar System* (Second Edition). John Wiley & Sons, Chichester, U.K.

Miner, E.D. (2000) *Uranus: The Planet, Rings and Satellites* (Second Edition). Springer/Praxis, Chichester, U.K.

Miner, E.D. and R.R. Wessen (2002) *Neptune: The Planet, Rings and Satellites.* Springer/Praxis, Chichester, U.K.

Rogers, J.H. (1995) *The Giant Planet Jupiter*. Cambridge University Press, Cambridge, U.K.

Shirley, J.H. and R.W. Fairbridge (Eds.) (1997) *Encyclopaedia of the Planetary Sciences.* Chapman & Hall, London.

Weissman, P.R., L.-A. McFadden, and T.V. Johnson (Eds.) (1998) *Encyclopaedia of the Solar System.* Academic Press, San Diego, CA.

2

Formation of the giant planets

2.1 FORMATION OF THE UNIVERSE AND PRIMORDIAL CONSTITUENTS

According to current cosmological theories, approximately 14 billion years ago the universe was created at a single point in space-time in the "Big Bang". As the universe expanded and cooled, numerous particle physics processes and particle–antiparticle annihilations occurred (e.g., Krane, 1996). After about $t = 6$ s, the universe consisted of some number N protons, N electrons, $0.16N$ neutrons, $10^9 N$ photons, and $10^9 N$ neutrinos, all at a temperature of approximately 10^{10} K.

As protons and neutrons collided with each other it was possible to form deuterium nuclei via the reaction

$$\mathrm{n} + \mathrm{p} \rightarrow {}^2\mathrm{H} + \gamma. \tag{2.1}$$

However, interactions with photons with energy greater than 2.22 MeV can break up these deuterium nuclei via the reverse reaction. Hence, the universe had to cool to about 9×10^8 K before significant numbers of neutrons could stably combine with protons in this way. At about the same time, the deuterons formed could engage in further reactions such as

$$^2\mathrm{H} + \mathrm{p} \rightarrow {}^3\mathrm{He} + \gamma \tag{2.2}$$

with an energy of formation of 5.49 MeV, and

$$^2\mathrm{H} + \mathrm{n} \rightarrow {}^3\mathrm{H} + \gamma \tag{2.3}$$

with an energy of formation of 6.26 MeV. Both these reactions have energies well above the threshold of deuteron formation and thus photons not energetic enough to break up deuterons would certainly not have been energetic enough to destroy these nuclei.

The final steps in initial nuclei formation were

$$^{3}\text{He} + \text{n} \rightarrow {}^{4}\text{He} + \gamma \tag{2.4}$$

and

$$^{3}\text{H} + \text{p} \rightarrow {}^{4}\text{He} + \gamma. \tag{2.5}$$

There are no stable nuclei with molecular weight 5 and thus no further reactions involving single nucleons were possible. Further reactions involving other nuclei did occur, but their products made up only a very small fraction of the final number of nuclei produced, which was dominated by ^{1}H and ^{4}He. By $t = 250$ s, the original $0.16N$ neutrons had decayed to about $0.12N$, and these combined with $0.12N$ protons to form very nearly $0.06N$ ^{4}He nuclei via the reactions listed above. In other words, the amount of ^{2}H, ^{3}H (which rapidly decays to ^{3}He), ^{3}He, and heavier nuclei left over was very small. Hence, after this time the universe contained $0.82N$ nuclei, of which 7.3% were ^{4}He and 92.7% were protons. This translates to a helium mass fraction of about 24%. While ^{3}He and ^{3}H can be produced by fusion in stars, deuterium (^{2}H or D) is not produced in significant quantities by any cosmic process that has occurred since the Big Bang, although it is destroyed via stellar fusion. Hence, the deuterium nuclei present in the universe now are truly *primordial*, and the primordial value of D/H is estimated (from observing absorption lines in the spectra of very distant, first-generation stars; Burles and Tytler, 1988) to be $(3.4 \pm 0.25) \times 10^{-5}$.

As time progressed, the universe expanded and cooled until at about $t = 700{,}000$ yr the energy of photons had reduced to such a level that the nuclei could combine with electrons to form neutral atoms. At this point the universe became transparent to electromagnetic radiation and astronomy could begin! The residual photons at this time typically had ultraviolet (UV) wavelengths, but as the universe has since expanded these have become considerably redshifted to the microwave part of the electromagnetic spectrum. This residual radiation from the Big Bang is called "cosmic microwave background radiation" and was first observed in 1964 by Arno Penzias and Robert Wilson (Penzias and Wilson, 1965). Cosmic microwave background radiation is found to be almost entirely isotropic (i.e., has almost the same intensity in all directions) and is consistent with black body (or cavity radiation) with a temperature of 2.7 K. Although it was not identified until 1964, the microwave "hiss" of cosmic microwave background radiation is actually responsible for a small fraction (approximately 1%) of the familiar white noise seen on television screens between channels in the days before digital television.

2.2 FORMATION OF THE STARS AND EVOLUTION OF THE INTERSTELLAR MEDIUM

As the universe further cooled and expanded, fluctuations in density initiated the condensation of galaxies and the formation of first-generation stars. Such density variations appear to have been present from the earliest stages of formation and were first detected as ripples in cosmic microwave background radiation by the Cosmic Background Explorer (COBE) spacecraft, launched in 1989. The early, massive first-generation stars had very short lifetimes and ended in Type II supernovae, spreading their fused and remaining unfused molecules into the material occupying the gaps between the stars known as the interstellar medium (ISM). Hence, the composition of this medium changed over time, with the D/H ratio slowly reducing as deuterium was "burnt" in stars and the He/H ratio increasing for the same reason. In addition, the abundance of heavy elements (where the term "heavy" denotes atoms heavier than He) also increased as more and more hydrogen and then helium was fused in stars. Subsequent stars typically had lower mass and longer lives and shed their atmospheres more gently into the ISM when they died. The $^{15}N/^{14}N$ ratio of the material in the ISM is a good indicator of how much the material has evolved since ^{15}N comes mainly from early primary production in stars ending in Type II supernovae, while ^{14}N comes mainly from second-generation lower mass, longer lived stars. Hence, we expect the abundance of ^{14}N to build up over time and thus that the $^{15}N/^{14}N$ ratio slowly reduces.

Hence, all the heavy atoms of our world, and indeed our bodies, such as carbon, oxygen, nitrogen, etc., were all produced in the cores of ancient stars that have long since perished and given up fractions of their atmospheres to the ISM. A review of the chemistry taking place in the ISM and particularly molecular clouds is given by Fraser *et al.* (2002). The estimated abundance, relative to hydrogen, of different elements in the solar system has been the subject of many studies over the years such as Cameron (1982) and the widely used abundance tables of Anders and Grevesse (1989). These abundances have been significantly updated in light of new 3D hydrodynamical models of the solar photosphere, coupled with improved solar line data, non–local thermodynamic equilibrium modeling and observations (e.g., Grevesse and Sauval, 1998; Grevesse *et al.*, 2007). Most recently, Grevesse *et al.* (2007) have revised their estimates of solar metallicity to significantly lower values. As an example, the solar C/H ratio of Anders and Grevesse (1989), referenced by numerous authors discussing the composition of the outer planets, has been revised to 68% of its previous value by Grevesse *et al.* (2007). Hence, when the abundances of elements in the giant planet atmospheres are referred to in various papers in terms of their ratio to solar abundance, the source of the solar data used needs to be clarified to avoid any confusion. In this book, we shall assume the solar abundances of Grevesse *et al.* (2007). It should be noted, however, that the current observed solar photospheric abundances are thought to be less than the original proto-solar abundances by up to 12% due to effects such as gravitational settling (Turcotte and Wimmer-Schweingruber, 2002; Turcotte *et al.*, 1998). However, we shall refer here to the Grevesse *et al.* (2007) observed solar photospheric values.

The molecular form of the elements in the current ISM may be inferred from spectroscopic measurements and modeling. Oxygen is thought to be mostly found within molecules of water ice and carbon, mainly within molecules of CO and some CH_4. Nitrogen is assumed to be mainly in the form of N_2, although this molecule cannot be spectroscopically detected (Chapter 6), and sulfur is thought to exist mainly within H_2S molecules. Ion–molecule reactions forming water molecules directly from the atoms in the ISM would have enriched the deuterium abundance, such that the D/H ratio in water molecules of the pre-solar cloud was increased to something like 7.3×10^{-4}. All recent comets coming from the Oort Cloud (see Section 2.4.1) have been observed to have a D/H ratio of 3×10^{-4} in their water molecules, which has implications for how they formed as we shall see later in Section 2.6.1. For reference the D/H ratio in the Earth's oceans is $\sim 1.5 \times 10^{-4}$. This is greater than would be expected if all the Earth's water came from the local solar nebula at the time of formation, but less than if it all came from comets as has sometimes been suggested. A combination of the two sources thus seems most likely.

2.3 FORMATION OF THE PROTO-SOLAR NEBULA

2.3.1 Collapse of the interstellar cloud

Our solar system formed at the edge of our galaxy in one of its spiral arms about 4.6 billion years ago (as determined by radioisotope dating analysis). During this time the solar system has completed approximately 20 orbits about the galactic center. Since the formation of the galaxy, the composition of the local ISM had been evolving and at the time the solar system formed was composed of approximately 71% by mass of hydrogen, 27% helium, and 2% heavy elements (i.e., elements with molecular weight greater than helium). About 1% of the heavy elements are thought to have existed in a condensed "dust" phase.

The density and temperature of the ISM varies considerably with position. The denser parts have a temperature of the order of 10 K, and a density of 10^{-14} kg m^{-3}, from which the pressure can be calculated to be 3.5×10^{-10} Pa. The pressure in these "dense molecular clouds" is thus considerably less than the best modern laboratory vacuum! These dense clouds typically have a size of a few light-years across and thus contain enough mass to form many hundreds of stars. Figure 2.1 shows a typical dense molecular cloud, Barnard 68, observed in 1999 by the European Southern Observatory Very Large Telescope (VLT). At visible wavelengths, the opacity of dust in these clouds is sufficient to obscure the light of stars behind the cloud. However, since the dust grains are small, the opacity decreases rapidly with wavelength (as we shall see in Section 6.4) such that at near-infrared wavelengths the cloud becomes almost transparent (as can be seen in Figure 2.1).

Jeans' theory of collapse

Under certain circumstances these dense molecular clouds may become unstable to gravitational collapse, leading to the formation of stars. The conditions for gravitational collapse of dense molecular clouds were first considered by Sir James Jeans in 1917. Ignoring all other forces, an isothermal cloud of mass M and temperature T will undergo gravitational collapse if its gravitational potential energy is greater than its internal thermal energy. The thermal energy of a cloud is given roughly by

$$E_T = Nk_BT = \frac{M}{\mu m_H} k_BT \tag{2.6}$$

where μ is the mean molecular weight of the material in the cloud; m_H is the mass of a

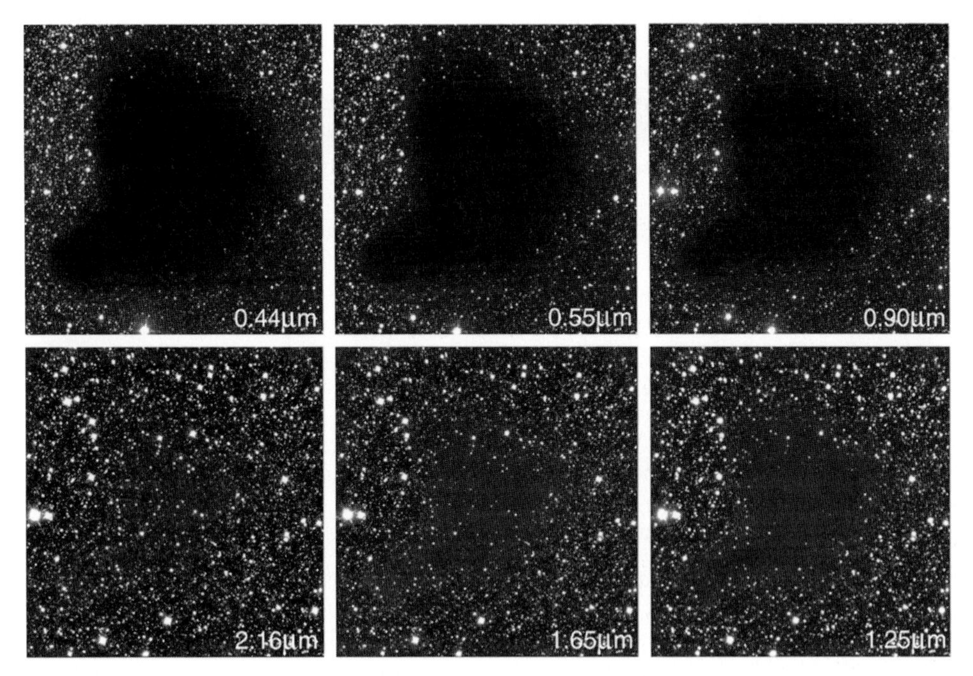

Figure 2.1. Molecular cloud Barnard 68 observed by the ESO Very Large Telescope at a range of visible and near-infrared wavelengths between 0.44 μm and 2.16 μm. Credit European Southern Observatory.

hydrogen atom; N is the total number of molecules in the cloud; and k_B is the Boltzmann constant.

The gravitational binding energy of a cloud of radius R is given by

$$E_G \approx \frac{GM^2}{R} \tag{2.7}$$

and hence an interstellar cloud should collapse if $E_G > E_T$, or

$$\frac{GM^2}{R} > \frac{Mk_BT}{\mu m_H}. \tag{2.8}$$

Assuming that the cloud has a uniform density ρ, its mean radius will be

$$R = \left(\frac{3M}{4\pi\rho}\right)^{3/2}. \tag{2.9}$$

Substituting for R in Equation (2.8), and rearranging, we obtain Jeans' expression for the minimum mass of cloud of temperature T and density ρ that will collapse

$$M_J = \frac{1}{\rho^{1/2}}\left(\frac{k_BT}{G\mu m_H}\right)^{3/2}. \tag{2.10}$$

Substituting a temperature of 10 K, and a density of 10^{-14} kg m^{-3}, we find that $M_J \approx 10^{29}$ kg, or approximately 0.1 $M_\odot$ (where $M_\odot$ is the mass of the Sun), and from Equation (2.9), $R = 1.4 \times 10^{14}$ m (or 0.015 light-years, or ~1,000 AU). Hence, according to Jeans' theory, the denser parts of the ISM should be unstable to Jean's collapse. However, this theory ignores effects such as magnetic fields and gas flow in the clouds that oppose the collapse. In reality it thus appears that in most cases some sort of external compression is also required to initiate collapse, such as collision between two clouds, impact of a shock wave from a nearby exploding star, or the action of the spiral density wave that periodically sweeps through the galaxy (Jones, 2007).

Once the whole cloud starts to contract, the denser parts of the cloud contract more quickly and thus the cloud quickly fragments, with each cloud fragment condensing to form its own star system. Thus new stars seem to form in clusters, as is observed.

2.3.2 Formation and evolution of circumstellar disks

As each cloud fragment collapses, gravitational binding energy is released as thermal energy. In the initial stages the temperature rise is small since the opacity of the nebula is very thin and thus this energy efficiently radiates away. However, the center of the nebula is predicted to collapse more quickly than the edge, and once its density increases to the point where it starts to become opaque, the temperature rise is rapid. Although this rise in temperature tends to slow the rate of collapse, the temperature of the central "proto-star" still continues to rise until the temperature of the core is sufficient to initiate fusion.

The whole cloud fragment will have some net rotation. Hence, as the cloud collapses the inner parts will begin to rotate more rapidly by conservation of angular momentum. Material on or near the net rotation axis will fall freely towards the center whereas the infall elsewhere is moderated by centrifugal force. Hence, a circumstellar disk forms in the plane perpendicular to the rotation axis. Approximately 50% of young stellar objects (YSOs) that have been observed to date are surrounded by disk-like or ring-like structures (André and Montmerle, 1994; Drouart *et al.*, 1999). There are four classes of these:

(1) Class 0. The youngest class of YSO with a large mass of circumstellar material (0.5 $M_\odot$ or more) and a lifetime of approximately 10,000 yr ($M_\odot$ is the current mass of the Sun).
(2) Class I. Mass ~0.1 $M_\odot$, lifetime around 100,000 yr, extending up to few 1,000 AU across.
(3) Class II. Mass ~0.01 $M_\odot$, lifetime around 1 Myr and optically thick at 10 μm with excess thermal emission.
(4) Class III. Optically thin at 10 μm, and radially more compact at around 100 AU.

These observations are consistent with the evolutionary scenario where a massive envelope (Class 0) rapidly collapses to form a proto-star in a timescale of only

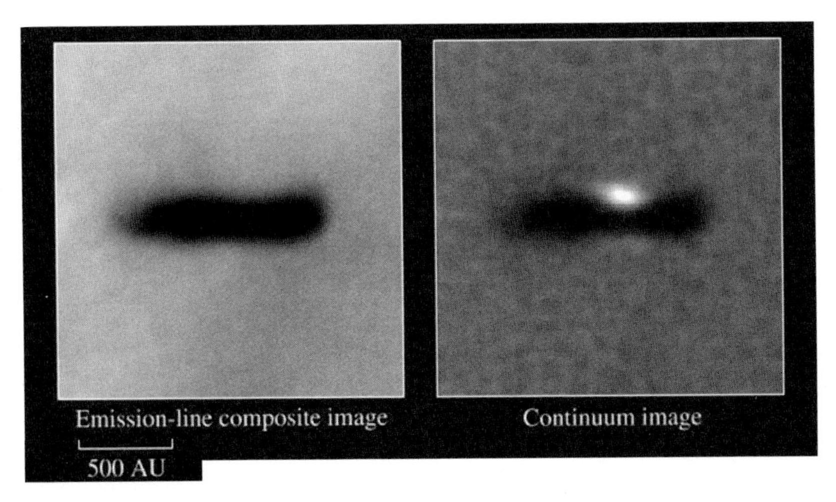

Figure 2.2. Hubble Space Telescope image of a young circumstellar disk (Orion 114–426) in the Orion Nebula (after McCaughrean and O'Dell, 1996). The left-hand panel shows an emission-line composite, made by combining data from three narrow-band filters centered on bright emission lines from the nebula (namely, [O III], H_α and [N II]). The strong emission lines provide a bright background which reveals the circumstellar disks as silhouettes around their young stars. The right-hand panel shows the corresponding continuum image taken through the medium bandwidth F547M filter. The central star shows up most clearly in the right-hand panel, and a faint reflection nebula is also seen above and below the plane of the silhouette disk. Courtesy of Mark McCaughrean and the *Astronomical Journal.*

100,000 years, together with an extended disk of mass between 0.01 $M_\odot$ and 0.1 $M_\odot$ (Class I and Class II), which evolves over a timescale of 1 Myr to 10 Myr into a less massive disk with lower density (Class III). The upper bound of estimated disk masses is consistent with the theoretical upper mass of $\sim$0.3 $M_\odot$ derived from stability arguments of Shu *et al.* (1990). Particularly clear examples of these circumstellar disks are seen in the Orion Nebula. Here, several such nebulae are seen in front of a glowing background interstellar cloud, which is lit by the light of stars already created (Figure 2.2) (McCaughrean and O'Dell, 1996; McCaughrean *et al.*, 1998). As can be seen the circumstellar disk is of the order of 1,000 AU across. This formation scenario requires a mechanism for disk dispersal on timescales of 1 Myr to 10 Myr. Solar mass stars cannot easily blow away dense circumstellar disks and thus accretion onto the star seems more likely. However, in order for this accretion to proceed some way is needed of losing or redistributing angular momentum.

By conservation of angular momentum, one would expect the nebula to rotate fastest towards the center and thus the proto-Sun should initially have been rotating very rapidly. However, the Sun now contains only 1% of the solar system angular momentum (0.5% in spin, and 0.5% in orbital rotation of Sun about the solar system barycenter, which is just outside the Sun) with the bulk (85%) now contained in the orbital angular momenta of Jupiter and Saturn. Hence, to explain the current state of the solar system a means is needed whereby the circumstellar disk was accreted or dissipated and where most of the proto-Sun's angular momentum was lost. Two key processes have been identified.

(1) *Turbulence*. Conditions in the early circumstellar disk are likely to have been very turbulent. This turbulence would have led to a net transfer of mass outward in the outer part of the disk, and inwards towards the proto-Sun nearer the center. Associated with this mass flow would have been a net transfer of angular momentum from the proto-Sun to the disk. The transfer of angular momentum would have caused the disk to spread farther and farther out into space, reducing its density. In the inner disk, the mass loss would have been towards the proto-Sun and also to space via outflow along the rotation axis, in collimated jets which are observed in proto-star nebulae, as shown in Figure 2.3 (McCaughrean *et al.* 1994). These polar jets are probably collimated both by the dense circumstellar disk itself and by the magnetic field of the protostar. They are seen to be episodic in nature, which is consistent with them being fed by turbulent infall of the inner circumstellar disk. However, while this bipolar outflow would have carried off as much as 10% of the nebula mass, it would not have accounted for much of the angular momentum.

(2) *T-Tauri phase*. The Sun is currently about half-way though its *main sequence* (Lewis, 1995). Just prior to the main sequence phase, a period called the *T-Tauri phase* (named after a star currently observed in the constellation of Taurus) occurs, which is marked by a considerable outflow of gas in the solar wind, and also high UV radiation. For a star with the mass of the Sun, this period would have lasted about 10 Myr and would have led to the Sun losing about 10% of its mass, and a significant proportion of its angular momentum. As we shall see later, the high solar wind associated with this phase would have swept away any remaining fragments of the solar nebula, and thus the timing has profound implications on the formation of giant planet atmospheres.

Observations of YSOs suggest that the Sun had accreted most of its mass very rapidly within about 100,000 yr of the start of collapse. At this stage the Sun was surrounded by a circumstellar disk. The inner part of the disk would have been very hot due to a combination of opacity, turbulent frictional heating, and solar luminosity. Temperatures out to approximately 1 AU probably exceeded 2,000 K, evaporating almost every solid constituent. Farther out, temperatures reduced with distance from the Sun and thus the "condensation line" (i.e., the distance from the Sun that different minerals and ices would have condensed from the nebula) occurred at varying distances from the Sun, with more volatile materials condensing farther from the Sun than less volatile materials. As time progressed and the disk cooled and spread out, these condensation lines would have moved inwards towards the Sun at a rate governed by both the radiative heat loss to space, and the frictional heating generated by turbulence. Not all parts of the disk may have been dominated by turbulence, however. In the outer disk, beyond perhaps 50 AU, the nebula density is predicted to have been so low that turbulence would not have been able to play such an important role. In this low-turbulence region, material would rotate about

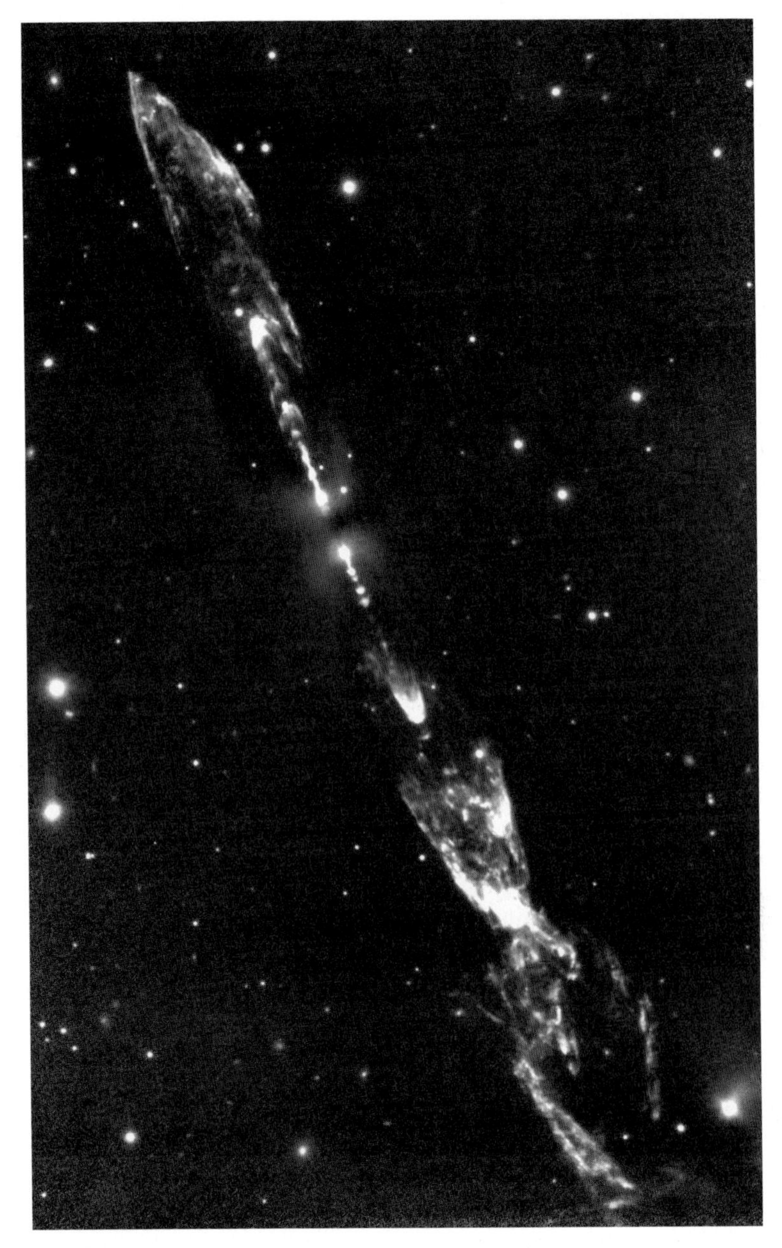

Figure 2.3. Wide-field medium-resolution near-IR (molecular hydrogen line at 2.122 μm) image of HH212. Data taken in December 1996 using the Calar Alto 3.5 m telescope in Spain. The image shows the bi-polar jet from the formation of a star near the center of the frame and hence the circumstellar disk appears to be almost edge-on. Material can be seen to be ejected in pulses and there is a clear symmetry between the pulses in the upper and lower jets. The episodic nature of the outflow points to the turbulent nature with which material falls from the circumstellar disk onto the central star. Courtesy of Mark McCaughrean.

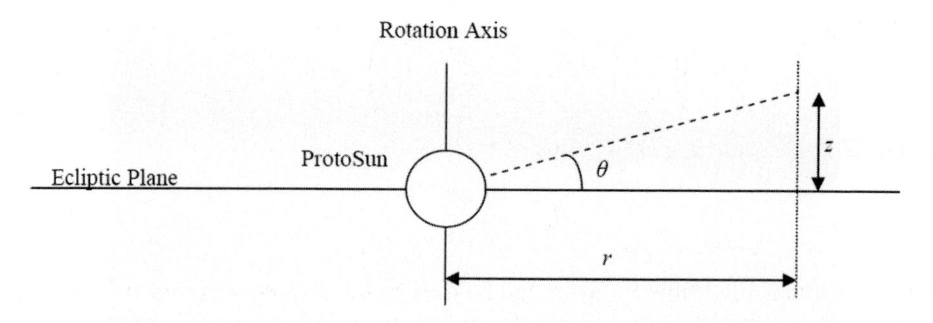

Figure 2.4. Variation of nebula pressure in the circumstellar disk above the ecliptic plane.

the protostar with a period governed by Kepler's orbital laws and thus this part of the nebula is called the Keplerian disk. Such two-component circumstellar disks have been observed (Drouart *et al.*, 1999; Guilloteau *et al.*, 1997). Hence, while we expect the material in the inner turbulent disk to have undergone significant mixing and thermal reprocessing, material in the outer Keplerian disk will have been largely unmixed and unprocessed from its pre-solar form.

Modeling of the evolution of circumstellar disks is usually done with the alpha disk model of Shakura and Sunyaev (1973). In such models the surface density of the disk, $\Sigma(r, t)$ (where r is the distance from the star and t is the time) is solved, where it is assumed that turbulence alone redistributes mass and angular momentum. The level of turbulence is prescribed by a single, dimensionless scaling parameter α and values of between 10^{-2} and 10^{-3} yield disk lifetimes consistent with the inferred lifetimes of circumstellar disks (Hartmann *et al.*, 1998; Hueso and Guillot, 2003). More advanced models (e.g., Huré *et al.*, 2001) use the beta parameterization of turbulence of Richard and Zahn (1999). While the structure of the gaseous part of the nebula is determined via hydrostatic equilibrium and centrifugal forces in the radial direction, the structure in the vertical direction (i.e., in the direction parallel to the rotation axis; Figure 2.4) is determined by hydrostatic forces only. Collapse in this direction stops when the pressure gradient force equals the gravitational force: that is,

$$g_z = g \sin\theta \approx \frac{GM_S z}{r^3} \tag{2.11}$$

where r is again the distance from the proto-star; z is the height of a parcel of gas above the disk plane; M_S is the mass of the proto-star; and θ is the angular elevation of the parcel above the disk plane as seen from the protostar. Assuming hydrostatic equilibrium, the change in pressure between heights z and $z + dz$ above the disk plane is given by

$$dp = -\rho g_z \, dz = \left(-\frac{p\mu}{RT}\right)\left(\frac{GM_S z}{r^3}\right) dz \tag{2.12}$$

where μ is the molecular weight of the gas; and R is the gas constant, and hence in equilibrium

$$p_z = p_c \exp\left(-\frac{z^2}{H^2}\right) \tag{2.13}$$

where H is the *scale height* given by

$$H = \left(\frac{2RTr^3}{\mu GM_S}\right)^{1/2}. \tag{2.14}$$

This structure is very different from that formed by the dust grains in the nebula. The condensed dust/ice grains were initially only 1 μm to 30 μm across and the grains tended to settle on the disk (or ecliptic) plane as a result of the net gravitational field and gas drag. Settling increased as disk turbulence diminished and dust grains grew. The dust sheet was probably of the order of 10,000 km thick near the Sun, increasing slightly with distance (Jones, 2007).

2.4 FORMATION OF THE JOVIAN PLANETS AND COMETS

In Section 2.3 we saw how a circumstellar disk is believed to have formed about the Sun and saw how such disks have been observed forming around other stars. We further saw that these disks are estimated to exist for approximately 10 Myr before dispersing. There is overwhelming evidence that the planets of the solar system formed from such a circumstellar cloud and a number of theories have been put forward to explain the process of planet formation.

Until the mid-1990s, the only known planetary system was the solar system and thus such models had only the observed characteristics of the solar system and circumstellar clouds to constrain them. However, in 1995 the first extrasolar planet was discovered orbiting the star 51-Pegasus by Mayor and Queloz (1995) and since then the number of confirmed "exoplanets" has risen to 306 (as of August 2008) with over 30 stars having more than one planet. What has been most remarkable about these newly discovered planetary systems is how little resemblance most of them have to the solar system! Most exoplanets discovered are of Jupiter mass and orbit within 1 AU of their star. This is not just an observational bias of the detection techniques (Irwin, 2008), but a real difference between these planetary systems and our own. Hence, formation models of the solar system can no longer be considered in isolation, but must be consistent with exoplanetary systems also.

The most commonly accepted planetary formation model is the core accretion theory, which is described below. However, this still has considerable problems, not least of which is the difficulty in forming Jupiter-mass planets within the lifetime of that deduced for circumstellar disks, and also the fact that planets do not appear to remain where they formed in the circumstellar disk, but instead appear to migrate (Alibert *et al.*, 2005c). Hence, alternative theories have been advanced, which will also now be discussed.

2.4.1 Core accretion model

The core accretion theory is the most commonly accepted theory of how all planets, including the giant planets, came to form in our solar system (Mizuno, 1980; Pollack *et al.*, 1996). In this model, planet growth starts with the concentration of solid material in a proto-solar nebula into a sheet in the disk plane about the proto-star. Concentration of the solid material into a sheet increases the chances of collision for small grains in similar orbits, which when they collide with low collision speeds can stick together via small forces such as Van der Waal's forces, in a process called coagulation. While the relative velocities between grains in nearby orbits reduce with distance from the proto-star, so does the collision probability since nebula density also decreases. Hence, the coagulation time increases with distance except at the "ice line", where water first condenses and where there is a small step-like increase superimposed on the general decreasing trend. Hence, it is estimated that by the time bodies grow to 10 mm at 30 AU, bodies at 5 AU can have grown to 0.1 km–10 km, large enough to be called *planetesimals*.

Dust and small particles in a circumstellar disk will orbit the star according to Kepler's laws, but the gas in the nebula orbits at a slightly lower rate due to the radial pressure gradient (Hueso and Guillot, 2003; Weidenschilling, 1977). Hence, small particles experience a drag force that causes them to spiral in towards the proto-star unless they are massive enough (i.e., larger than ~1 km) for the drag force to be negligible. This drag implies that the growth of particles to the size of ~1 km must have been rapid, or that most of the solid material forming the cores of the planets was originally condensed at much greater distances from the proto-Sun.

Conditions in these early nebulas are likely to be highly turbulent and evidence for this comes from our own solar system, with the detection of crystalline silicates in comets coming from the Oort Cloud (Bockelée-Morvan *et al.*, 1998). As discussed below, these comets are thought to have originally condensed in the cooler outer nebula near the orbits of Uranus and Neptune. They were thus expected to have a very low abundance of crystalline silicates since these minerals have high condensation temperatures and should have condensed close to the Sun, not in the outer solar system. Detection of such minerals in these comets thus suggests considerable mixing between material that originally condensed close to the Sun and that which condensed farther away.

The formation of the giant planets from planetesimals and nebula gas is generally thought to occur in three phases (Mizuno, 1980; Pollack *et al.*, 1996) as outlined below.

Phase 1

Once planetesimals start to reach a size of the order of 10 km, their gravitational attractive forces start to become significant. This increases the collision rate between planetesimals and leads to the growth of larger planetesimals at the expense of smaller ones. This is a runaway process and leads to the formation of a number of *embryos* in a timeframe of perhaps 500,000 years after the formation of the star. The embryos account for ~90% of the original mass in the local *feeding zone*, which forms an annular strip covering a small range of distances from the star, centered on the embryo. The remaining 10% of the solid material in each feeding zone is modeled to be composed of a swarm of very much smaller planetesimals. Solar nebula models predict that both the embryo masses and the widths of the feeding zones increase with distance, with a sharp increase at the water ice condensation line (Jones, 2007). Typical modeled embryo masses at 1 AU are of the order of 0.1 $M_{\oplus}$, while at 5 AU they are of the order of 10 $M_{\oplus}$.While it is tempting to think that this is due to the presence of ice as well as refractory elements in the outer parts of the dust sheet, and hence a greater density of solid material, this is in fact not quite correct. The embryo formation process initially requires low-speed collisions between planetesimals in near-identical orbits and these conditions are more easily met in the outer, more slowly rotating part of the nebula, than the inner more rapidly rotating, turbulent part. An additional effect is that at greater distances from the Sun, tidal disruption forces are less, which also allows embryos to form more easily. Hence, the calculated widths of the feeding zones are modeled to increase with distance and thus the predicted embryo masses, and their separations, are larger.

Phase 2

Once the embryos reach a mass of the order of 10 $M_{\oplus}$, they start being able to trap the nebula gas itself, as well as the remaining planetesimals. This second, possibly very slow stage of accretion is predicted to last between 1 Myr and 10 Myr and leads to the accumulation of a considerable envelope of gas and ice about the initial primary ice core. Eventually the mass of some of the planets reaches a critical mass, which is so high that the remaining nebula gas becomes unstable to hydrodynamical collapse, leading to the final phase of formation. Estimates of the critical mass required range from 5 $M_{\oplus}$ (Hubickyj *et al.*, 2005) to 20 $M_{\oplus}$ or more (Pollack *et al.*, 1996).

Phase 3

Once the critical mass is reached, any remaining nebula gas in the region of the planet hydrodynamically collapses onto it. This is believed to be the most rapid phase of formation and is modeled to have lasted 30,000 yr for Jupiter and 20,000 yr for Saturn. Since the time for accretion of the critical mass is predicted to increase with distance from the star, due to the decrease in nebula density, it would appear that Uranus and Neptune in our own solar system never accreted enough mass to leave Phase 2. The energy released by the accretion of giant planets would raise the internal temperatures of the planets, including their new gaseous envelopes, significantly. Hence, most of the icy planetesimals would dissolve into the envelope, with only the more rocky materials accreting onto the core itself. The heat released by this accretion would also initiate substantial convection in the envelope, further inhibiting accretion on to the core.

Evidence in favor of the core accretion model

The timing of these phases is critical in explaining the nature of the giant planets of our solar system. Jupiter is predicted to have reached Phase 3 about 1.5 Myr after the formation of the Sun, and Saturn after something like 11 Myr (Hersant *et al.*, 2001). These times are very much model-dependent, but what is generally accepted with these models is that before Uranus and Neptune could reach their critical mass, the remnants of the circumsolar disk were finally dissipated when the Sun entered its T-Tauri phase, about 16 Myr after the formation of the proto-Sun (Drouart *et al.*, 1999). The high solar wind associated with this phase effectively swept all remaining gas out of the solar nebula and shut off the gas capture process.

The bulk differences between the giant outer planets, and indeed the inner terrestrial planets, are very elegantly explained with this model and an observation that supports this timescale of giant planet formation is that of the proto-planetary disk around a nearby pre–main sequence star (Brittain and Rettig, 2002). The abundance of CO (and presumably other gases) in the inner part of this disk, which is estimated to be 5 Myr to 10 Myr old, appears to be very low out to a distance of approximately 17 AU, but is substantial at larger distances. This suggests that the inner stellar system has already been substantially cleared of gas. However, H_3^+ emission is also detected from the star system. While it is theoretically possible that such emission comes from the inner edge of the circumstellar disk near 17 AU this seems unlikely since the inferred abundance of CO at this distance should very efficiently destroy all H_3^+ molecules there. Instead, it is suggested that the H_3^+ emission comes from the auroral regions of one or more giant planets that have already formed at distances less than 17 AU from the star. This suggestion is consistent with the observation that the only H_3^+ emission observed anywhere in our own solar system comes from the auroral regions of our own giant planets.

Although some of the planetesimals remaining in the nebula would have continued to be captured (as indeed they are today with Jupiter capturing Comet Shoemaker–Levy 9 in 1994), most remaining planetesimals would have been ejected from the solar system by the gravitational perturbations to their orbits exerted by the giant planets. All planetesimals in the Jupiter–Saturn region are predicted to have been accumulated or ejected completely. Planetesimals in the Uranus–Neptune region, however, would have been less violently ejected and are thought to have formed the Oort Cloud, which is believed to exist as a spherical shell of small, irregularly shaped bodies orbiting the Sun well beyond the main planets. The existence of this cloud was postulated in 1950 by the Dutch astronomer Jan Hendrick Oort (1900–1992) who noticed that no comet had ever been observed with an orbit indicating that it came from interstellar space, and that the orbits of most long-period comets had aphelia (greatest distance from the Sun) of about 50,000 AU, and no preferred direction. Hence, Oort proposed that long-period comets come from objects uniformly spread around the Sun at about this aphelia distance, which are perturbed by tiny effects such as gravitational tides exerted by stars in the galactic disk and in the galactic core. Estimates for the mass of material in the Oort Cloud vary from about 40 $M_\oplus$ to perhaps the mass of Jupiter, and most Oort Cloud objects

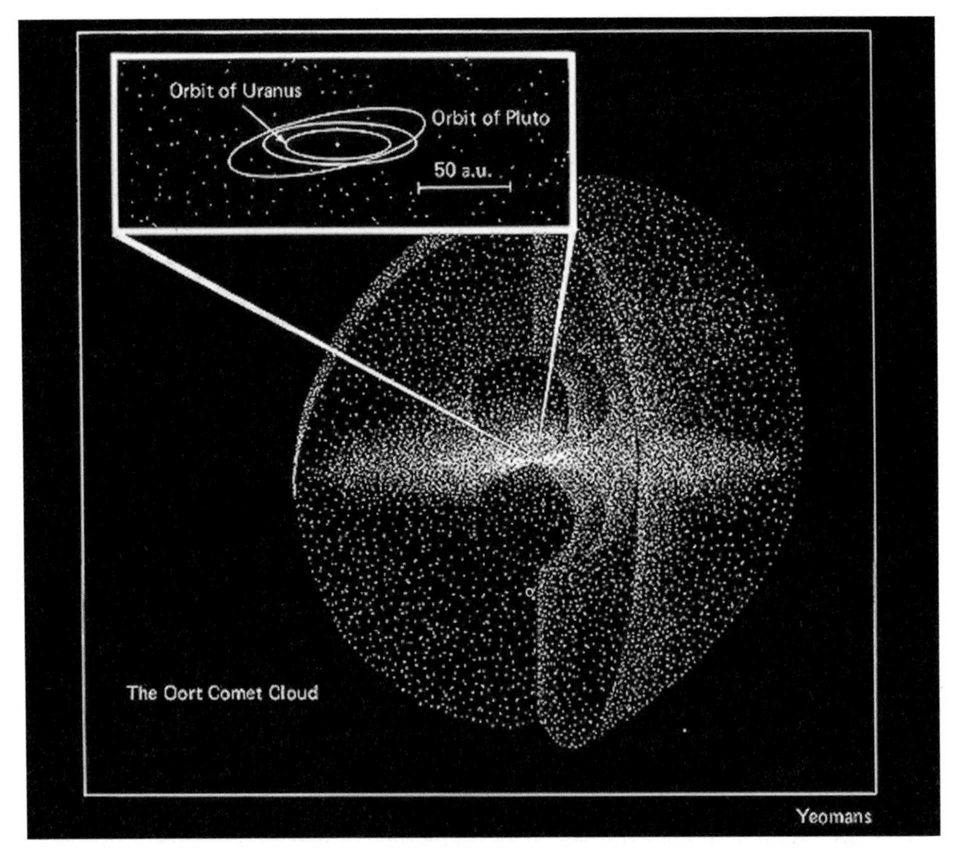

Figure 2.5. The Oort Comet Cloud. Courtesy of Donald Yeomans, Jet Propulsion Laboratory.

are thought to orbit at a distance of between 10,000 AU and 20,000 AU from the Sun, although the cloud extends outwards as far as perhaps 50,000 AU to 70,000 AU (Figure 2.5).

Beyond the orbit of Neptune, the probability of embryo formation appears to have been too small to form a giant planet, perhaps due to the lack of turbulent mixing at this distance. Instead, the remaining planetesimals in this region, which form the Kuiper–Edgeworth Belt, are probably relatively unevolved examples of the planetesimals that originally existed in this region. The Kuiper–Edgeworth Belt is named after Gerard Peter Kuiper (1905–1973) and Kenneth Essex Edgeworth (1880–1972) who independently suggested the presence of small planetary bodies beyond the orbit of Pluto and Neptune, respectively. The Kuiper–Edgeworth Belt (often just called the Kuiper Belt) extends from the orbit of Neptune at 30 AU out to approximately 50 AU and is thought to contain at least 70,000 "trans-Neptunians" or Kuiper Belt Objects (KBOs) with diameters exceeding 100 km, concentrated near the ecliptic plane (Figure 2.6, see color section). New KBOs are continually being discovered and the total now stands at over 1,000. It is thought that the Kuiper Belt may extend farther and merge with the Oort Cloud at a distance of roughly 1,000 AU. The high eccentricity of Pluto's orbit, and the high inclination of Triton's orbit about Neptune (indicating that it is probably a captured satellite, rather than forming from the circumplanetary disk as discussed in Section 2.5) suggest that both these bodies may in fact themselves be KBOs. This conclusion is supported by the recent discovery of several KBOs with diameters of the order of 1,000 km.

Finally, the core accretion model of formation is consistent with almost all of the planets spinning in the same direction as their orbital motion (prograde). The obliquities of the planets probably arose from off-center collisions between planetesimals and perhaps embryos towards the end of formation although they may also have arisen due to spin–orbit resonances (Ward and Hamilton, 2002). One particularly extreme case is Uranus which has an obliquity of 98° and thus spins almost on its side. Such a large obliquity could have arisen from several cumulative off-center impacts, or conceivably a single massive impact towards the end of formation.

2.4.2 Gravitational instability model

Although the core accretion model fits the observed characteristics of our solar system very well, there are two main problems with it. First of all, the model as outlined above appears to be too slow when compared with the lifetimes of circumstellar disks observed around other stars. Haisch *et al.* (2001) have estimated the ages of circumstellar disks about stars in nearby clusters by analyzing the mean color of the stars in these clusters. They find that circumstellar disks appear to evaporate long before the T-Tauri phase is reached and that half of the stars in these clusters lose their disks within 3 Myr, with a mean overall lifetime of 6 Myr. Similarly, Briceño *et al.* (2001) find that stars older than 10 Myr do not have massive, optically thick disks. A second problem with the core accretion model is that it takes no account of the fact that once planets start to form, they appear to migrate, as can be seen by the "Hot Jupiters" of other stars, where very large gas giants are seen orbiting within 1 AU of their star.

An alternate view of giant planet formation that tackles the time of formation problem is that if protostellar disks are dense enough they may become gravitationally unstable. In such an unstable disk, giant planets may collapse directly from the disk in the early period of circumstellar disk evolution (Boss, 1997, 2002, 2004; Mayer *et al.*, 2002). The most widely used criterion for assessing the gravitational stability of a circumstellar disk is the Toomre stability parameter Q (Pickett and Lim, 2004; Toomre, 1964), which depends on a number of factors such as disk surface density. If the Toomre Q parameter is less than approximately unity, a disk can become unstable to the rapid growth of a spiral structure, which can then clump together to form giant planets. However, although this process is fast and can also account for the "Hot Jupiters", the models have considerable difficulty in actually condensing a planet because the disk does not appear to cool rapidly enough (Pickett and Lim, 2004; Rice *et al.*, 2003), although Boss (2004) suggests that convective overturning may assist this. Another drawback of the model is that condensed planets would all have a composition similar to that of the central star, which does not fit at all well with observations of the solar system giant planets, where as we shall see the abundance of heavy elements increases with distance from the Sun. The final drawback to these models is that they are not sensitive to the abundance of dust in the nebula, and thus the process should be equally likely around metal-poor stars as metal-rich ones. This goes against the observation that the exoplanets found so far are preferentially around metal-rich stars (Irwin, 2008).

2.4.3 Migration

The second problem with the core accretion model outlined in the previous section is that it takes no account of the fact that planets appear to migrate during formation. The giant planets of our solar system and of other stars are unlikely to be at the same distance from the central star as when they initially formed due to gravitational interactions with other growing planetesimals (leading to their ejection from the solar system) and frictional interactions with the accretion disk itself. A surprising property of the known KBOs is that almost a quarter of those discovered so far are in a 3:2 orbital resonance with Neptune, meaning that they complete three orbits for every two orbits of Neptune. Pluto also has a 3:2 resonance with Neptune, and for this reason KBOs in this orbit have become known as *Plutinos*. That such a large proportion of KBOs occupy this resonance is strong evidence that Neptune (and probably Uranus) has migrated outwards during the evolution of the solar system, by as much as 7 AU to 8 AU, and swept up a number of KBOs into the 3:2 resonance as it did so.

The phenomenon was first conjectured by Fernandez and Ip (1984), who modeled the scattering of comets and planetesimals in the early solar system and pointed out that the orbital motions of the giant planets are interdependent since comets scattered inwards by one planet may be scattered again by other planets. In this way Fernandez and Ip (1984) found that while Jupiter was modeled to migrate towards the Sun through the ejection of comets, the other giant planets were modeled to migrate outwards. Conditions in the early circumstellar disk were probably very turbulent, perhaps including magneto-hydrodynamic (MHD) effects (Papaloizou *et al.*, 2004) and this turbulence would have led to a net transfer of mass outward in the outer part of the disk, and inwards toward the Sun nearer the center, with a dividing radius at around 10 AU (Ida *et al.*, 2000).

Extrasolar planetary systems with close Jovian planets or "Hot Jupiters" offer clear evidence that in many cases giant planets forming within the critical radius of the disk migrate inwards. Such planets would gravitationally eject or capture all terrestrial planets in the so-called "habitable zone" of the inner planetary system, where surface temperatures would allow the presence of liquid water and thus, perhaps, the evolution of life. Keeping giant planets at distances greater than ~5 AU not only allows the development of an inner planetary system like that in our solar system, but also efficiently expels small planetesimals that would otherwise swarm in this region, continually bombarding the inner planets and impeding the evolution of life. Hence, the "smooth" evolution of life on Earth may actually have required the presence of Jupiter, but not too close! It is thus of great interest to determine whether or not there are other planetary systems with characteristics more like ours, and multiple planetary systems with some resemblance to the solar system are now being discovered (Mason, 2008).

Until recently, the migration of planets in these circumstellar disks was thought to be restricted to two main types. Type I migration is expected when the planet mass is small, such that its Hill radius (i.e., the distance from the planet within which the gravitational field is dominated by the planet itself) is much smaller than the disk thickness. In such cases a proto-planet will tidally interact with the disk via Lindblad resonances (Ward, 1986, 1997) and is modeled to migrate rapidly. Larger planets, however, appear to be controlled by Type II migration, where the planetary mass is large enough to open up a gap in the disk, splitting it up into an inner and outer part. The planet is then locked in with the long-term evolution of the disk (Lin and Papalaizou, 1986a, b) and slowly migrates inwards toward the star (assuming it is within the critical orbital distance described above). In this model, if the planet does not migrate all the way inwards before the disk dissipates, it is likely to have an orbital distance greater than 0.05 AU to 0.2 AU, which is consistent with the observed lack of "Hot Jupiters" with mass $>4\ M_J$ orbiting within 0.3 AU of their stars (Irwin, 2008). However, objects that have undergone this type of migration would be expected to be heavier for smaller orbital radii, since they would have had more time to accrete the disk material, which is not observed amongst known extra-solar giant planets. Hence "Hot Jupiters" do not appear to be explained by either Type I or Type II migration. Instead, "Hot Jupiters" may be explained by an intermediate, or "runaway" mode of migration (Masset and Papalaizou, 2003) where, for certain combinations of planet and disk mass, a planet very rapidly moves either toward or away from the star, changing its orbital radius by a factor of 2 or more. The runaway process does not carry on indefinitely since eventually the planet becomes so large that it clears out its "feeding zone" and is then governed by slow Type II migration.

Finally, one solution to the apparent slowness of the core accretion model is proposed by Rice and Armitage (2003) who suggest that turbulent fluctuations in a proto-planetary disk cause migrating giant planets to perform more of a random walk, with an amplitude of a few tenths of an astronomical unit, than a steady drift towards the star, resulting in an acceleration of the accretion rate by almost an order of magnitude. This scenario is further explored by Alibert *et al.* (2005b) who find that the migration of the early giant planets meant that their feeding zones were never depleted during Phase 2 of their formation, which was consequently much faster than predicted by the classical core accretion model. In their model an embryo was started at 11.5 AU and allowed to migrate inwards, forming a planet with $30\ M_\oplus$ in the envelope and $5\ M_\oplus$ in the core, which is close to the current estimates of internal structure models of Jupiter (Guillot *et al.*, 2004; Saumon and Guillot, 2004).

2.5 FORMATION OF JOVIAN SATELLITES

Material captured by the giant planets during Phases 1 and 2 would have been captured predominantly in the equatorial plane and, prior to accretion, would have formed a circumplanetary disk in exactly the same way as a circumstellar disk of material formed about the Sun. These disks would also have had significant variation of temperature with distance, both due to heating from the central hot planets (generated by the release of gravitational energy by accretions), and by turbulent mixing. Planetesimals captured early in these disks would probably have been disrupted by turbulence and collisions with other planetesimals, and hence the current satellites probably reformed from material that had been disrupted by accretion into the circumplanetary disk and had been thermally reprocessed by the radial temperature gradient. Such a model explains why most of the satellites about the giant planets are in the equatorial plane of the planet. It also explains why the compositions of Jupiter's Galilean satellites (Io, Europa, Ganymede, and Callisto) vary greatly with distance from Jupiter. Satellites of the giant planets that are not in the equatorial plane are thought to be objects captured after most of the proto-planet disk had been dissipated.

2.6 BULK COMPOSITION OF THE OUTER PLANETS AND ISOTOPE RATIOS

Using the estimated solar system elemental abundances given in Table 2.1, we can calculate the expected composition of an atmosphere of solar abundance assuming a cold dense atmosphere where all elements appear in their fully hydrogenated form. The expected solar composition abundances of the most abundant molecules (relative to H_2) are compared with those actually estimated in the giant planet atmospheres in Table 2.2a. The sources of these data will be reviewed in Chapter 4. The same abundances, expressed as volume mixing ratios (or mole fractions), are listed in Table 2.2b. As can be seen the atmospheres of the giant planets have compositions rather different from a pure solar mixture, and it can also be seen that the proportion of heavy elements increases as we move outwards through the solar system.

2.6.1 Constraints on formation: bulk composition X/H

In addition to accumulating hydrogen and helium, the outer planets are also observed to have accumulated significant quantities of other "heavy" elements such as carbon, nitrogen, and sulfur. The predominant pre-solar nebula form of these elements was in uncondensed, gaseous molecules (C as CO or CH_4, N mainly as N_2 with perhaps 10% as NH_3/HCN, S as H_2S, O as H_2O, etc.) and thus a solar composition abundance of these constituents might be expected. However, the ratio of these elements with respect to hydrogen for the giant planets is found to be significantly supersolar with values of approximately 3–5× the solar value for Jupiter for C/H, N/H, S/H, Ar/H, Kr/H, Xe/H. The C/H ratio is estimated to increase to 5–11× the solar value for Saturn and of the order of 40× the solar value for Uranus and Neptune. Although there remain gaps in the data, at first glance it appears that the other heavy molecules are enriched to approximately the same degree on Jupiter (Table 2.2a) and it is suspected that this is the case for all the other giant planets (Owen and Encrenaz,

Table 2.1. Solar system abundances of the elements.

Atomic number	*Name*		*Atomic weight*	*Mass fraction*	*Number of atoms X/H*	*Number of atoms X/H_2*	*X/H_2 (G&S 1998)*	*X/H_2 (A&G 1989)*	*Ratio (New/ G&S)*	*Ratio (New/ A&G)*
1	Hydrogen	H	1.008	0.738296	1.000	1.000	1.000	1.000	1.000	1.000
2	Helium	He	4.0026	0.249524	0.08511	0.1702	0.1954	0.1950	0.871	0.873
8	Oxygen	O	15.9994	0.005355	4.571×10^{-4}	9.142×10^{-4}	0.001352	0.001706	0.676	0.536
6	Carbon	C	12.0111	0.002159	2.455×10^{-4}	4.909×10^{-4}	6.623×10^{-4}	7.240×10^{-4}	0.741	0.678
10	Neon	Ne	20.179	0.001023	6.918×10^{-5}	1.384×10^{-4}	2.405×10^{-4}	2.466×10^{-4}	0.575	0.561
26	Iron	Fe	55.847	0.001153	2.818×10^{-5}	5.637×10^{-5}	6.325×10^{-5}	6.452×10^{-5}	0.891	0.874
7	Nitrogen	N	14.0067	0.000618	6.026×10^{-5}	$1.205 \times 10-4$	1.664×10^{-4}	2.244×10^{-4}	0.724	0.537
14	Silicon	Si	28.086	0.000666	3.236×10^{-5}	6.472×10^{-5}	7.262×10^{-5}	7.168×10^{-5}	0.891	0.903
12	Magnesium	Mg	24.305	0.000603	3.388×10^{-5}	6.777×10^{-5}	7.604×10^{-5}	7.699×10^{-5}	0.891	0.880
16	Sulfur	S	32.06	0.000339	1.380×10^{-5}	2.761×10^{-5}	3.170×10^{-5}	3.692×10^{-5}	0.871	0.748
28	Nickel	Ni	58.71	4.97×10^{-5}	1.698×10^{-6}	3.396×10^{-6}	3.557×10^{-6}	3.534×10^{-6}	0.871	0.877
18	Argon	Ar	39.948	6.66×10^{-5}	1.549×10^{-6}	3.098×10^{-6}	5.024×10^{-6}	7.240×10^{-6}	0.676	0.469
20	Calcium	Ca	40.08	5.32×10^{-5}	2.692×10^{-6}	5.383×10^{-6}	4.477×10^{-6}	4.380×10^{-6}	0.871	0.884
13	Aluminium	Al	26.9815	5.72×10^{-5}	1.950×10^{-6}	3.900×10^{-6}	6.181×10^{-6}	6.086×10^{-6}	0.871	0.890
11	Sodium	Na	22.9898	3.14×10^{-5}	1.862×10^{-6}	3.724×10^{-6}	4.179×10^{-6}	4.115×10^{-6}	0.891	0.905
15	Phosphorus	P	31.00	5.7×10^{-6}	2.512×10^{-7}	5.024×10^{-7}	7.262×10^{-7}	7.455×10^{-7}	0.692	0.674
32	Germanium	Ge	72.71	2.07×10^{-7}	3.890×10^{-9}	7.781×10^{-9}	8.532×10^{-9}	8.530×10^{-9}	0.912	0.912
36	Krypton	Kr	83.903	3.3×10^{-13}	1.905×10^{-9}	3.811×10^{-9}	4.083×10^{-9}	3.226×10^{-9}	0.933	1.181
54	Xenon	Xe	131.28	1.03×10^{-12}	1.862×10^{-10}	3.724×10^{-10}	2.958×10^{-10}	3.369×10^{-10}	1.259	1.105
33	Arsenic	As	75.00	1.07×10^{-8}	1.950×10^{-10}	3.900×10^{-10}	4.688×10^{-10}	4.703×10^{-10}	0.832	0.829

Table 2.2a. Bulk composition of the Jovian planets (relative to H_2).

	Solar	*Jupiter*		*Saturn*		*Uranus*		*Neptune*	
	fraction f	*fraction f*	*f/solar*	*fraction f*	*f/solar*	*fraction f*	*f/solar*	*fraction f*	*f/solar*
H_2	1.0	1.0	1	1.0	1.0	1.0	1.0	1	1
He	0.1702	0.157	0.92	0.135	0.79	0.18	1.06	0.18	1.06
H_2O	9.142×10^{-4}	$\sim 4.9 \times 10^{-4}$	~ 0.54	2×10^{-7}	0.00	?	?	?	?
CH_4	4.909×10^{-4}	2.37×10^{-3}	4.83	5.33×10^{-3}	10.86	0.019	39	0.027	54
Ne	1.384×10^{-4}	2.3×10^{-5}	0.17	?	?	?	?	?	?
NH_3	1.205×10^{-4}	6.64×10^{-4}	5.51	5.68×10^{-4}	4.71	?	?	?	?
H_2S	2.760×10^{-5}	8.9×10^{-5}	3.22	3.86×10^{-4}	14.00	7.38×10^{-4}	27	7.38×10^{-4}	27
Ar	3.098×10^{-6}	1.82×10^{-5}	5.87	?	?	?	?	?	?
PH_3	5.024×10^{-7}	1.2×10^{-6}	2.39	6.7×10^{-6}	13.3	?	?	?	?
GeH	7.781×10^{-9}	7.0×10^{-10}	0.1	4.0×10^{-10}	0.05	?	?	?	?
Kr	3.811×10^{-9}	9.3×10^{-9}	2.44	?	?	?	?	?	?
Xe	3.724×10^{-10}	8.9×10^{-10}	2.39	?	?	?	?	?	?
AsH_3	3.900×10^{-10}	2.2×10^{-10}	0.56	3.0×10^{-9}	7.69	?	?	?	?

lar composition is assumed to be as estimated by Grevesse *et al.* (2007) and listed in Table 2.1. Sources of planetary undances are listed in Chapter 4.

2006). Hence, these molecules cannot have been accreted from their gaseous phase, but instead must have been concentrated in some other way. Current theories suggest that significant quantities of these molecules were incorporated into water ice, which then accreted in Phase 1 and 2 to produce the observed enhancement. However, there is considerable disagreement as to the form of the ice and its temperature of formation.

A number of laboratory experiments have been conducted to assess how well pre-solar nebula gases might be trapped by water ice. At very low temperatures ($T < 100$ K), water condenses as *amorphous* ice; laboratory experiments have shown that to trap the quantities of heavy elements observed in Jupiter's atmosphere the ice formation temperature must have been less than 40 K and may have been as low as 30 K (Bar-Nun *et al.*, 1988; Owen and Encrenaz, 2003, 2006; Owen *et al.*, 1999). This is much lower than the predicted nebula temperatures at 5 AU to 6 AU during the time that Jupiter formed, assuming it did indeed form at this distance, and suggests that either (i) the pre-solar ice grains (which are almost certainly amorphous) never vaporized in the Jupiter region (i.e., the nebula was much colder than is currently

Table 2.2b. Bulk composition of the Jovian planets (as mole fractions).

	Solar mole fraction	*Jupiter*	*Saturn*	*Uranus*	*Neptune*
H_2	0.853	0.862	0.877	0.833	0.829
He	0.145	0.135	0.118	0.150	0.15
H_2O	7.8×10^{-4}	$>4.2 \times 10^{-4}$	$>1.7 \times 10^{-7}$	?	?
CH_4	4.2×10^{-4}	2.0×10^{-3}	4.7×10^{-3}	0.016	0.022
Ne	1.2×10^{-4}	2.0 × 10-5	?	?	?
NH_3	1.0×10^{-4}	5.7×10^{-4}	$>5.0 \times 10^{-4}$	?	?
H_2S	2.4×10^{-5}	7.7×10^{-5}	3.4×10^{-4}	6.2×10^{-4}	6.2×10^{-4}
Ar	2.9×10^{-6}	1.6×10^{-5}	?	?	?
PH_3	4.3×10^{-7}	1.0×10^{-6}	5.9×10^{-6}	?	?
GeH	6.6×10^{-9}	6.0×10^{-10}	3.5×10^{-10}	?	?
Kr	3.2×10^{-9}	8.0×10^{-9}	?	?	?
Xe	3.2×10^{-10}	7.7×10^{-10}	?	?	?
AsH_3	3.3×10^{-10}	1.9×10^{-10}	2.6×10^{-9}	?	?

Solar abundances from Grevesse *et al.* (2007).

modeled), or (ii) that Jupiter originally formed as far out as 30 AU and later migrated in. The former scenario does not seem to tie in with the good evidence for high inner solar nebula temperatures and evidence of turbulent mixing, indicated amongst other things by the presence of crystalline silicates in comets. The latter possibility seems at first sight unlikely owing to the excessively long formation time of a Jupiter-sized planet at 30 AU and the difficulty in moving the planet in by almost 25 AU, but we know from Section 2.4.3 from the characteristics of exoplanetary systems and from our own solar system that forming planets do probably migrate. An alternative explanation has been put forward that utilizes crystalline water ice. Here the water ice is assumed to have vaporized on entry into the nebula and then subsequently re-condensed at the "ice line". The partial pressure of water vapor in the nebula was low and thus water initially condensed at about 150 K. Further cooling would have led to more and more water vapor molecules condensing onto the ice grains as the temperature dropped. Water condensing at 150 K is necessarily *crystalline* and as such may trap other gas molecules, either as hydrates for molecules such as ammonia (NH_3–H_2O would have formed at approximately 85 K at pressures of 10^{-8} bar), or clathrate–hydrates for other molecules, where gas molecules are trapped in "cages" of water molecules (Alibert *et al.*, 2005a, b; Gautier *et al.*, 2001a, b; Hersant *et al.*, 2004). There are two classes of clathrate–hydrate depending on the ratio of the number of water molecules to the number of trapped gas molecules in each cage. This ratio is 5.75 for Class I and 5.66 for Class II. As the temperature dropped, more and more ice condensed onto the grains until the temperature was so low that amorphous ice started to form. However, since the saturated vapor pressure decreases exponentially as the temperature reduces it can be seen that the bulk of the condensed ice grains must have been crystalline. Laboratory studies have shown that the heavy molecules in the solar nebula could be trapped in clathrate–hydrates in quantities needed to account for the observed enrichments of the outer planets at far higher temperatures than would be needed to trap equivalent amounts in amorphous ice (Gautier *et al.*, 2001a, b). The clathrate–hydrate formation scenario is consistent with the current turbulent solar system models and is also consistent with observations of a circumstellar disk where 90% of the ice is observed to be crystalline in a region whose temperature is 30 K to 60 K (Malfait *et al.*, 1999). In addition, crystalline ice has also been detected in Comet Hale–Bopp (Lellouch *et al.*, 1998) and it also provides a good explanation of the low Jovian Ne/H ratio since neon is poorly trapped in clathrates, unlike the other noble gases.

Owen and Encrenaz (2006) provide a detailed comparison of the predictions of the clathrate–hydrate model of Gautier *et al.* (2001a, b) with their own model based on amorphous ice condensed at ~40 K, using the principal of Ockham's Razor (also known as Occam's Razor) that the most simple solution that matches the observed data is most likely to be the correct one. Owen and Encrenaz (2006) note that most elements are enriched on Jupiter to approximately the same degree of ~4× the solar value. We will see later in Section 2.6.3 that the observed $^{15}N/^{14}N$ ratio strongly suggests that nitrogen was incorporated into Jupiter in the form of N_2 and not NH_3 and the fact that the observed N/H and Ar/H ratios are also ~4× the solar value argues against the clathrate–hydrate model since temperatures as low as 40 K would be needed to trap these gases in clathrates. Although Owen and Bar-Nun (1995) had suggested that N_2 was the most likely form of nitrogen in nebula, they did not anticipate the parity of the N/H and C/H ratios on Jupiter since this had not been observed in any comets. As a result, Owen and Encrenaz (2003) suggested that the icy planetesimals forming Jupiter did not have the same composition as that estimated in current comets, but instead were solar composition icy planetesimals (SCIPs), which contained solar relative abundances of O, N, and other heavy elements, but not solar abundances of H and He, with all the hydrogen atoms present in the form of H_2O.

Starting with the assumption that all the giant planets were formed from first accreting SCIPs to form their embryos before collapsing the solar nebula gas, Owen and Encrenaz (2006) use this simple model to calculate the mass of SCIPs that would need to be incorporated into the giant planets to explain the measured enrichments of C/H for Jupiter, Saturn, Uranus, and Neptune. They find that the required mass is similar for all four giant planets and is approximately equal to 10 $M_\oplus$, which matches well the expected embryo mass needed in the core accretion theory to begin Phases 2 and 3. The simple model of Owen and Encrenaz (2006) also predicts well, with the possible exception of Saturn, the D/H ratios of the giant planets (Section 2.6.2) and is also consistent with the measured Jovian $^{15}N/^{14}N$ ratio (Section 2.6.3).

Owen and Encrenaz (2006) then went on to compare their model with the clathrate–hydrate approach. An implicit assumption of the clathrate–hydrate approach is that the giant planets must have considerably higher O/H ratios than other elements through the need to trap other molecules in the clathrate "cages". While some formation models predict very high enrichments of O/H near the "ice line", where water first condenses at about 5 AU, the clathrate model requires at least $10\times$ the solar water abundance (whereas the SCIP model would have equal O/H enrichments to the other heavy elements). However, interior models such as those of Guillot (1999b) rule out O/H $>10\times$ the solar value. Unfortunately, water condenses in the observable part of the giant planet atmospheres and thus the deep abundance cannot be directly measured. Its abundance can be inferred indirectly, however, from the abundance of other molecules. Estimates of the abundance of CO at Jupiter's tropopause (Bézard *et al.*, 2002) suggests that O/H $<9\times$ solar value, which is again inconsistent with the clathrate–hydrate model. While the clathrate–hydrate and SCIP models give similar results for Jupiter, with similar enrichments of all heavy elements, the two models diverge in their predictions for the other giant planets, where the clathrate–hydrate model predicts varying relative enrichments depending on the local thermal conditions in the pre solar nebula at the distance from the Sun where the planets formed. There are two problems with this. First of all, the clathrate–hydrate model implicitly neglects the migration of planets during formation, which we suspect to be an important effect. Second, although measurements are not yet precise enough from the enrichments of elements on the other giant planets to discriminate between these models for most elements, the clathrate–hydrate model predicts a Saturnian C/H enrichment of $\sim 2.5\times$ the solar value, which is very much less than that observed by *Cassini* CIRS of $\sim 11\times$ solar value. Finally, the clathrate model predicts an S/H enrichment for Jupiter that is twice too large. Attempts to explain this by suggesting that H_2S corrodes Fe alloy grains in the inner nebula and that this depleted sulfur gas is turbulently mixed out to Jupiter are disputed by Atreya *et al.* (2003) and Owen and Encrenaz (2003).

Hence, on balance it currently appears that the SCIP model of Owen and Encrenaz (2006) is more consistent with available measurements than the clathrate–hydrate model (Alibert *et al.*, 2005a, b; Gautier *et al.*, 2001a, b; Hersant *et al.*, 2004). A key discriminator between these models is the O/H ratio, which is not well known for any of the giant planets. However, the arrival of the *Juno* mission at Jupiter in a low perijove orbit (Section 8.6.1) will be able to constrain the internal structure and finally determine whether the O/H ratio for that planet is $>10\times$ the solar value or not. Another discriminator between the SCIP model and others is that the SCIP model predicts equal enhancements for all heavy elements. While the C/N ratio for Jupiter is approximately solar, the estimated C/N ratio for Saturn is considerably supersolar. Fletcher *et al.* (2008b) suggest that a supersolar C/N ratio on Saturn is consistent with the known deficiencies of N_2 in cometary material (Mizuno, 1980), and with the suggestion that N_2 could be clathrated at 5 AU, but not at 10 AU, so that Saturn's nitrogen content came only from ammonia hydrate

(Hersant *et al.*, 2004). In addition, Fletcher *et al.* (2008b) find that the C/S and C/P ratios also vary between Jupiter and Saturn and show the opposite trend to the C/N ratio, with the Jovian values being supersolar, and the Saturnian value being subsolar. A caveat to this is that the C/P estimates are poorly known, since the P/H values are determined from the abundance of phosphine, which is a disequilibrium species in the upper tropospheres of the giant planets (as we will see in Chapter 4). The C/S trend, however, suggests that H_2S may have been more readily available in the outer nebula, as suggested by Hersant *et al.* (2004), than closer to the Sun. Given the substantially different C/N, C/S, and C/P ratios on Jupiter and Saturn, Fletcher *et al.* (2008b) suggest that these planets incorporated icy planetesimals from the immediate vicinity of a nebula whose composition varied with position rather than from a single common reservoir as the SCIP model maintains.

Finally, a key drawback of the SCIP model is the question of where the SCIPs are now. Owen and Encrenaz (2006) suggest that they may perhaps be hiding in the Oort Cloud or in the Kuiper Belt. Alternatively, they suggest that they might have been reprocessed in the same way as the building blocks of inner planets were reprocessed into meteorites and asteroids. As ever, with planetary formation models, the real processes involved in forming planetary systems may in fact incorporate aspects of both ice condensation models. Both models have some advantages and some key drawbacks and until better measurements exist of the planets' bulk composition it will be impossible to discriminate between them. As mentioned earlier, a key discriminator will be the measurement of the deep O/H ratio on Jupiter by the *Juno* mission, which is planned to arrive in orbit about Jupiter in 2016.

2.6.2 Constraints on formation: D/H ratio

As the presolar cloud collapsed into a circumstellar disk incoming material arriving from the presolar cloud would have been subjected to appreciable reprocessing both through the collapse process itself, and also through shocking as the material entered the dense nebula. Both these effects are likely to have been more substantial near the Sun and less important at greater distances. Some models suggest that the water molecules in the ices were completely dissociated by this process and then recombined. This theory may be tested by examining the D/H ratio of the various bodies currently in the solar system. As we mentioned earlier, deuterium is a truly primordial isotope, which was created just after the Big Bang and has not been produced since. After H_2 and He, the most abundant molecule in the ISM, and thus the presolar cloud, is H_2O, either in the gaseous or condensed phases. In dense, neutral media such as the solar nebula, fractionation of D and H in water occurs though reactions between the gas phases

$$\mathrm{HD} + \mathrm{H_2O} \rightleftharpoons \mathrm{HDO} + \mathrm{H_2} \tag{2.15}$$

At higher temperatures the D/H ratio in molecular hydrogen and water is the same, but at lower temperatures ($T < 500$ K), deuterium becomes concentrated in water (Drouart *et al.*, 1999). While the equilibrium D/H ratio in water molecules and other molecules increases rapidly as the temperature drops, the rate of

Table 2.3. Solar system D/H ratios.

Molecule	*Species*	*Value* ($\times 10^{-5}$)	*Reference*
Proto-solar	H_2	2.1 ± 0.4	Geiss and Gloeckler (1998)
Proto-solar	H_2O	~ 70	Deloule *et al.* (1998), Mousis *et al.* (2000)
Comets	H_2O	~ 30	Balsiger *et al.* (1995), Eberhardt *et al.* (1995), Bockelée-Morvan *et al.* (1998), Meier *et al.* (1998)
	H_2O	32 ± 3	Meier and Owen (1999), Bockelée-Morvan *et al.* (2005)
Earth	H_2O	14.9 ± 0.3	Lécuyer *et al.* (1998)
Jupiter	H_2	14.9×0.3	Lellouch *et al.* (2001)
	CH_3D	2.2×0.7	Lellouch *et al.* (2001)
Saturn	H_2	$1.85^{+0.85}_{-0.60}$	Lellouch *et al.* (2001)
	CH_3D	$2.0^{+1.4}_{-0.7}$	Lellouch *et al.* (2001)
	CH_3D	1.6 ± 0.2	Fletcher *et al.* (2008b)
Uranus	H_2	$5.5^{+3.5}_{-1.5}$	Feuchtgruber *et al.* (1999)
Neptune	H_2	$6.5^{+2.5}_{-1.5}$	Feuchtgruber *et al.* (1999)

isotopic exchange between neutral molecules in the nebula falls rapidly, and effectively disappears for $T < 200\,\mathrm{K}$. This means that the enrichment factor $f_{H_2O} = (D/H)_{H_2O}/(D/H)_{H_2}$ in water molecules equilibrating with molecular hydrogen in the solar nebula is unlikely to be greater than 3. For the model where water molecules reformed from the dissociation products of formation, the ratio would be close to 1.0. However, the $(D/H)_{H_2O}$ ratio of some solar system objects is found to be greatly in excess of the theoretical limit of three times the proto-solar $(D/H)_{H_2}$ ratio (as can be seen in Table 2.3 and Figure 2.7). This suggests that some other interaction must have taken place to increase the f_{H_2O} ratio.

While the enrichment factor for $(D/H)_{H_2O}$, f_{H_2O}, may not exceed ~ 3 for neutral molecule–molecule interactions, other interactions may occur in the ISM which lead to different enrichments. If the ISM is partially ionized (which is often the case) then ion–molecule reactions can occur where the additional ionization energy serves to get over the activation energy of the fractionation reactions and hence leads to much greater levels of enrichment. In some formation models (Drouart *et al.*, 1999; Mousis *et al.*, 2000) it is assumed that while water ice is probably vaporized when it enters the inner part of the circumstellar disk (out to distances of 30 AU to 50 AU) it is not actually dissociated and thus retains its pristine, high f_{H_2O} ratio. Subsequent neutral gas interactions with molecular hydrogen in the hot inner nebula quickly reduce f_{H_2O} to 1, but farther from the Sun the predicted ratio tends to 1 more slowly, due to the

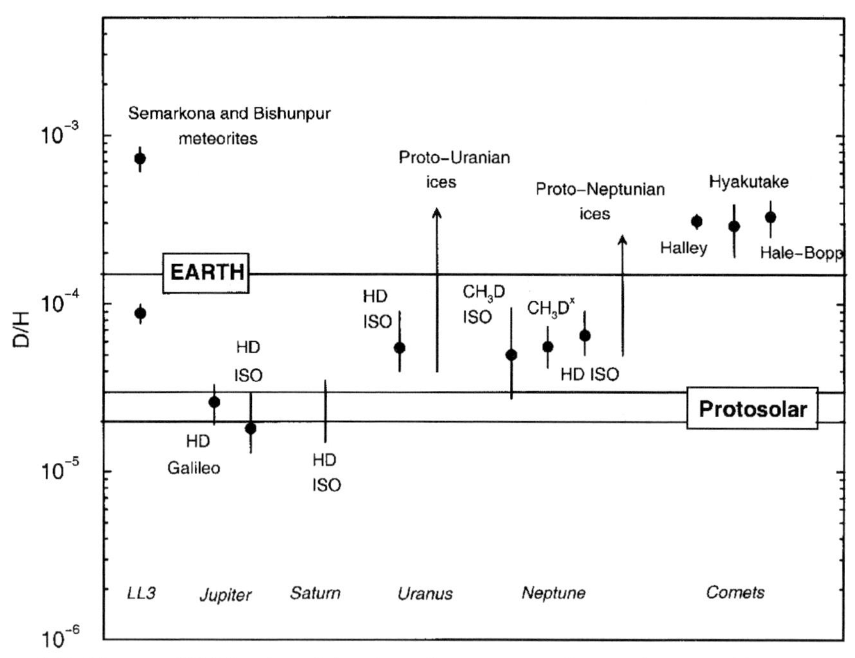

Figure 2.7. Measured D/H ratios of the giant planets, meteorites, and comets. From Hersant *et al.*, 2001. Reprinted with permission from the *Astrophysical Journal*.

lower temperatures and through turbulent mixing in the nebula between low-f ice water near the Sun and high-f ice farther out. The initial degree of D enrichment of presolar ice grains may be indicated by measurements of the composition of Semarkona and Bishunpur LL3 meteorites which are thought to have originally condensed at around 3 AU from the Sun. Water is incorporated as clays in these meteorites and these are found to be composed of two components: roughly 15% has an f_{H_2O} equal to approximately 25, while the remainder has an f_{H_2O} equal to approximately 3. This observation is consistent with the idea of mixing between pristine unvaporized presolar ice grains with $f_{H_2O} = 25$, together with water that has equilibrated with the local solar nebula.

Using this initial value of f_{H_2O}, and making reasonable assumptions about the nebula cooling rate and degree of turbulent mixing, the formation models of Drouart *et al.* (1999) and Mousis *et al.* (2000) predict that by the time that Uranus, Neptune, and the comets were forming in the outer solar nebula, the mean D enrichment of water molecules could have fallen to approximately 10 which is what is observed in the Halley, Hyakutake, and Hale–Bopp comets. The D/H ratio observed in the molecular hydrogen outer envelopes of the giant planets comes not only from the D/H of the molecular hydrogen in the presolar cloud but also through mixing and equilibrium with the D-enriched water ices making up the planet. For Jupiter and Saturn, the mass of hydrogen far outweighs that of ice and thus the D/H ratio should be representative of the presolar ratio in molecular hydrogen. For Uranus and

Neptune, however, the observed ratio is much higher and if we assume that the atmospheres have thoroughly mixed throughout their whole depth and that the D/H exchanged between water and molecular hydrogen, then, providing we know the relative masses of hydrogen and ice, we can calculate the D/H ratio of pre-Neptune and pre-Uranus ices. The relative masses may be calculated from interior models and recent estimates of the mass of hydrogen in these planets is 4.2 $M_\oplus$ for Uranus and 3.2 $M_\oplus$ for Neptune (Mousis *et al.*, 2000; Podolak *et al.*, 2000), where $M_\oplus$ is the mass of the Earth. Using these values, the f_{H_2O} enrichment of pre-Neptune, pre-Uranus ices is consistent with the cometary value of 10. Similar arguments may be used to estimate the presolar D enrichment in HCN and subsequent evolution.

An alternative to the approach of attempting to model the variable composition of the solar nebula with distance is the SCIP model of Owen and Encrenaz (2006), discussed in Section 2.5. The hydrogen in SCIPs is assumed to be purely in the form of water ice and Owen and Encrenaz (2006) assume that the $(D/H)_{H_2O}$ in SCIPs was the same as that currently determined in comets coming from the Oort Cloud and estimated to be $(3.2 \pm 0.3) \times 10^{-4}$ (Bockelée-Morvan *et al.*, 2005; Meier and Owen, 1999). With this assumption and matching the mass of SCIPs needed to match the observed C/H ratios of the giant planets, Owen and Encrenaz (2006) estimate the global D/H ratio for the giant planets to increase as 2.1, 2.2, 4.6, and 6.8×10^{-5} as we move from Jupiter to Neptune, which match quite well the estimated values shown in Table 2.3. Owen and Encrenaz (2006) also note that models of Uranus and Neptune with more than 100× the solar value (Hersant *et al.*, 2004; Lodders and Fegley, 1994) would give $D/H > 9 \times 10^{-5}$, which is the upper bound of present observations. [NB: A SCIP model would have O/H = C/H (i.e., 40–50).]

The only planet for which the predicted D/H ratio of the SCIP model does not appear to agree with measurements is Saturn, for which the most recent estimates from *Cassini* CIRS (Fletcher *et al.*, 2008b) put the D/H ratio in methane to be $(D/H)_{CH_4} = 1.6 \times 10^{-5}$. The equilibration of deuterium in methane is governed by a similar equilibration reaction to that for molecular hydrogen, shown in Equation (2.15); that is

$$HD + CH_4 \rightleftharpoons CH_3D + H_2. \tag{2.16}$$

To determine the associated $(D/H)_{H_2}$ ratio requires knowledge of the methane fractionation factor $f_{CH_4} = (D/H)_{CH_4}/(D/H)_{H_2}$, which for Jupiter is estimated to be $f_{CH_4} = 1.25 \pm 0.12$, and for Saturn is estimated to be $f_{CH_4} = 1.34 \pm 0.19$ (Lellouch *et al.*, 2001). Using this estimate of the f_{CH_4} factor for Saturn, Fletcher *et al.* (2008b) estimate the $(D/H)_{H_2}$ ratio to be 1.2×10^{-5}, which is considerably less than that estimated for Jupiter and goes against the expected trend of the SCIP model. Fletcher *et al.* (2008b) note that this again suggests different origins for the trapped volatiles that formed each planet and argues against the simple SCIP approach.

2.6.3 Constraints on formation: nitrogen $^{15}N/^{14}N$ ratio

Since the Jovian atmosphere is composed almost entirely of captured solar nebula gas, it is expected that the $^{15}N/^{14}N$ ratio of the Sun and Jupiter are the same. The upper value of this ratio has been measured in the solar wind to be $\leq 2.8 \times 10^{-3}$ (Hashizume *et al.* 2000). The $^{15}N/^{14}N$ ratio in ammonia on Jupiter was estimated from Infrared Space Observatory Short Wavelength Spectrometer (ISO/SWS) observations (Fouchet *et al.*, 2000) to be $(1.9^{+0.9}_{-1.0}) \times 10^{-3}$ and an analysis of the *Galileo* Galileo Probe Mass Spectrometer (GPMS) measurements (Owen *et al.*, 2001) yielded an estimate of $(2.3 \pm 0.3) \times 10^{-3}$. More recently, analysis of the *Cassini* CIRS observations of Jupiter have yielded an estimate of the $^{15}N/^{14}N$ ratio of $(2.22 \pm 0.52) \times 10^{-3}$ (Fouchet *et al.*, 2004b) and $(2.23 \pm 0.31) \times 10^{-3}$ (Abbas *et al.*, 2004). All these measurements are consistent with the solar wind estimate. In addition, both Abbas *et al.* (2004) and Fouchet *et al.* (2004b) found negligible variation of the $^{15}N/^{14}N$ ratio with latitude, suggesting that this ratio is not affected by the fractionation effects caused by the condensation of ammonia and thus that this ratio represents the true bulk ratio of Jupiter.

The $^{15}N/^{14}N$ ratio in HCN in the current local ISM is estimated to be $(2.2 \pm 0.5) \times 10^{-3}$ (Dahmen *et al.*, 1995). As mentioned earlier, the $^{15}N/^{14}N$ ratio is expected to reduce over time as the ^{15}N produced by early Type II supernovae is diluted with ^{14}N coming from the death of longer lived, intermediate-mass stars. Hence, we would expect that the $^{15}N/^{14}N$ ratio should have been higher in the presolar cloud than in the current ISM. The $^{15}N/^{14}N$ ratio in HCN for Comet Hale–Bopp was estimated to be roughly 3×10^{-3} (Jewitt *et al.*, 1997), which, assuming the ratio for this comet is representative of the proto-solar value, is consistent with this expectation. We also expect that the $^{15}N/^{14}N$ ratio in molecules such as HCN and NH_3 in the ISM was higher than in N_2 due to the same ion–molecule reactions that increase the D/H ratio. Hence, the lower $^{15}N/^{14}N$ ratio found in ammonia in Jupiter's atmosphere and in the solar wind suggest that nitrogen was captured by the Sun and by Jupiter mainly in the form of N_2 and not as NH_3 or HCN (Owen *et al.*, 2001). The fact that the terrestrial value of the $^{15}N/^{14}N$ ratio of 3.7×10^{-3} is substantially higher than that of Jupiter suggests that much of the Earth's nitrogen probably arrived later in the form of HCN and NH_3 in comets.

2.6.4 Constraints on formation: carbon $^{12}C/^{13}C$ ratio

Estimates of the $^{12}C/^{13}C$ ratio exist for all of the giant planets, with the exception of Uranus, and are summarized in Table 2.4. It can be seen that the estimates agree to within error for all the giant planets and are also consistent with the terrestrial $^{12}C/^{13}C$ ratio. Hence, it seems likely that all carbon accreted by both terrestrial and giant planets has the same source.

2.7 INTERIORS OF THE GIANT PLANETS

Much of what has been outlined previously regarding the formation of the planets is based not only upon their current atmospheric composition, but also on what we know about the interiors of the giant planets. But how can we tell what the conditions are in the interior of the giant planets, well below the visible cloud decks? There are a

Table 2.4. Solar system $^{12}C/^{13}C$ ratios.

Molecule	*Species*	*Value* $(\times 10^{-5})$	*Reference*
Earth	CH_4	$89.9^{+2.6}_{-2.4}$	International Union for Pure and Applied Chemistry (IUPAC, 1991)
Jupiter	CH_4	$92.6^{+4.5}_{-4.1}$	Niemann *et al.* (1996)
Saturn	CH_4	71^{+25}_{-18}	Combes *et al.* (1975), revised by Courtin *et al.* (1983)
	C_2H_6	99^{+43}_{-23}	Sada *et al.* (1999)
	CH_4	$91.8^{+5.5}_{-5.3}$	Fletcher *et al.* (2008b)
Uranus			Unknown
Neptune	CH_4	87 ± 26	Orton *et al.* (1992), corrected by Sada *et al.* (1996)

number of techniques, most notably the precise measurement of the planets' gravitational fields and estimation of the planets' moment of inertia, as we shall now see.

2.7.1 Gravitational data

The shape of the Earth and the other terrestrial planets is well approximated by spheres and thus the gravitational acceleration all over the surface points almost exactly towards the center of the planet. However, as we have seen, the outer planets spin very rapidly and, since the interiors of these planets are essentially fluid, the planets bulge out at the equator in response to the centrifugal force. This non-sphericity means that the shape of the gravitational field is modified from the spherical form with which we are more accustomed.

In general the gravitational acceleration at some point in a gravitational field is given by the gradient of the gravitational potential energy function U,

$$\mathbf{g} = -\nabla U. \tag{2.17}$$

and outside a planet U satisfies Laplace's equation,

$$\nabla^2 U = 0. \tag{2.18}$$

In spherical polar coordinates, Laplace's equation has a number of well-known solutions. If we limit ourselves to solutions which are axially symmetric (i.e., those that do not depend on the azimuth angle) then we find

$$U = \sum_{n=0}^{\infty} (A_n r^n + B_n r^{-(n+1)}) P_n(\cos\theta) \tag{2.19}$$

where θ is the zenith angle; and P_n are Legendre polynomials. If we further limit ourselves to solutions that possess north/south symmetry (which to a very good approximation all the giant planets have) then we only need consider even powers of n. Additionally we may put $A_n = 0$ for all n, since the potential must tend to zero as r tends to infinity. Adding in centripetal effects we then obtain (Lindal *et al.*, 1985)

$$U(r, \phi) = -\frac{GM}{r}\left[1 - \sum_{n=1}^{\infty} J_{2n}\left(\frac{R}{r}\right)^{2n} P_{2n}(\sin\phi)\right] - \tfrac{1}{2}\Omega^2 r^2 \cos^2\phi \qquad (2.20)$$

where ϕ is the *planetocentric* latitude, which is $(90 - \theta)$. Taking the gradient of Equation (2.20) we obtain an expression for the gravitational acceleration as a function of radial distance r and planetocentric latitude ϕ in the radial direction

$$g_r(r, \phi) = -\frac{GM}{r^2}\left(1 - \sum_{n=1}^{\infty}(2n+1)J_{2n}\left(\frac{R}{r}\right)^{2n} P_{2n}(\sin\phi)\right) + \tfrac{2}{3}\Omega^2 r[1 - P_2(\sin\phi)] \qquad (2.21)$$

and in the latitudinal direction

$$g_\phi(r, \phi) = -\frac{GM}{r^2}\left(\sum_{n=1}^{\infty} J_{2n}\left(\frac{R}{r}\right)^{2n} \frac{dP_{2n}(\sin\phi)}{d\phi}\right) - \tfrac{1}{3}\Omega^2 r \frac{dP_2(\sin\phi)}{d\phi} \qquad (2.22)$$

and the total gravitational acceleration is $g = \sqrt{(g_r^2 + g_\phi^2)}$. For the approximately spherical terrestrial planets the so-called "J-coefficients" are negligible and thus $g_r \gg g_\phi$. However, the oblateness of the giant planets means that J-coefficients are substantial and thus the "surface" gravitational field does not point directly towards the center of the planet, but is instead displaced by a small angle ψ where $\psi = \arctan(g_\phi/g_r)$, which varies with ϕ. This offset leads to a second definition of latitude and longitude, the *planetographic* system. The planetographic latitude is defined as the inclination of the local normal to the equatorial plane. This is equal to $\phi_g = \phi + \psi$ and is shown in Figure 2.8. The curve described by the limb of a planet is, to a first approximation, an ellipse and thus the planetocentric and planetographic latitudes are more simply related by

$$\tan\phi_g = \left(\frac{R_e}{R_p}\right)^2 \tan\phi \qquad (2.23)$$

where R_e and R_p are the equatorial and polar radii, respectively. The other difference between the planetocentric and planetographic systems is that while planetocentric longitudes run eastwards, planetographic longitudes run in the direction opposite to the rotation. Hence, for planets which have prograde spins, the longitudes run westwards, while for planets with retrograde spins, such as Venus and Uranus, they run eastwards.

Measurements of the J-coefficients of the giant planets, by observing the gravitational perturbations acting on satellites, rings, and passing spacecraft, can be used to determine the distribution of mass in the interior. In addition to the J-coefficients themselves, we can tell even more about the interior of the planets if we can measure, or estimate, the polar moment of inertia C. To directly measure C requires that we observe the precession of the rotation axis, which is not possible for

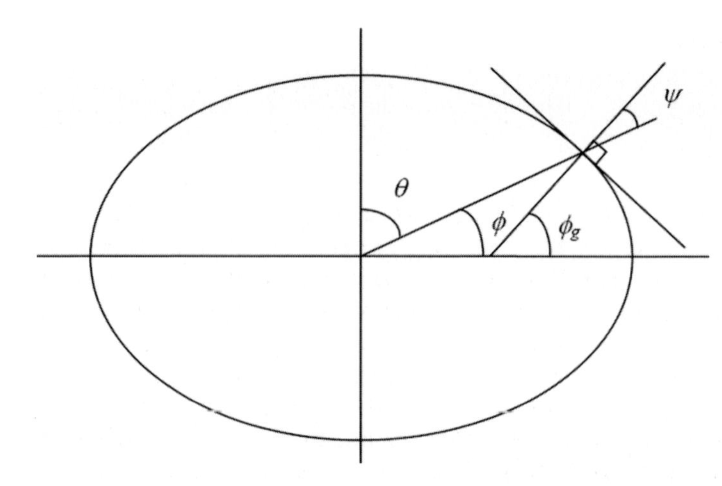

Figure 2.8. Definition of the planetographic and planetocentric latitude systems. Here ϕ is the planetocentric latitude, and ϕ_g is the planetographic latitude.

the giant planets. However, C may be calculated if we can measure the mass M of the planet, its equatorial radius R_e, its sidereal rotation period T, and the J_2-coefficient. This calculation assumes that the interior of the planet is in hydrostatic equilibrium (i.e., that the interior has no shear strength and responds to tidal forces essentially as a liquid). The moment of inertia ratio C/MR_e^2 is particularly useful since it indicates the degree of mass concentration towards the center. For a hollow sphere (Jones, 2007) the ratio is 2/3, while for a sphere with uniform density it is 0.4, and for a sphere where all the mass is at the center it is 0. The gravitational constants of the giant planets are listed in Table 2.5.

Table 2.5. Gravitational and magnetic properties of Earth and giant planets.

Planet	*Equatorial radius R_e* (km)	J_2 ($\times 10^{-2}$)	J_4 ($\times 10^{-4}$)	J_6 ($\times 10^{-4}$)	C/MR^2	*Magnetic dipole moment* (A m^2)
Earth	6,378	1.901	0.00161	?	0.3308	7.9×10^{22}
Jupiter	71,492	1.4697(1)	−5.84(5)	0.31(20)	0.264	1.54×10^{27}
Saturn	60,268	1.6291	−9.35	0.85	0.21	4.6×10^{25}
Uranus	25,559	0.3513(3)	−0.318(5)	?	0.23	3.8×10^{24}
Neptune	24,764	0.3538(9)	−0.380(1)	?	0.29	2.0×10^{24}

Source of gravity data: Jupiter, Campbell and Synnott (1985); Saturn, Anderson and Schubert (2007); Uranus, French *et al.* (1988); Neptune, Owen *et al.* (1991). Moment of inertia and magnetic data after table 4.2 of Jones (1999).

2.7.2 Magnetic field data

Magnetic fields are caused by electrical currents in the centers of the planets and thus much can be learned about the planetary interiors by looking at the B-fields. The mechanism by which planets create magnetic fields is not well understood. The centers of the planets are thought to be electrically conducting and convective. Any stray magnetic fields that might be present will modify the motion of moving electrical charges and will thus induce electrical currents, which will themselves generate their own magnetic field. What appears to happen is that some kind of positive feedback mechanism then acts to increase the strength of the magnetic field in a particular direction, which encourages the currents to flow in one particular sense. This magneto-hydrodynamic effect, called the *self-exciting dynamo*, is thought to be responsible for generating the magnetic fields of all the planets (Jones, 2007). Since viscous dissipation tends to reduce the field over time, internal rotational energy must be continually "tapped" by the magnetic field in order to sustain it. To do this, most models of the process require that the axis of the magnetic field is not coincident with the rotational axis (Cowling theorem) and this is observed for all the planets with the one exception of Saturn. Even when a field is "stable", as it is for the Earth, it is seen that in fact it evolves and varies over time and even undergoes complete reversals. Hence, the generation of the magnetic field can be seen to be a dynamic and ongoing process. The mean dipole moment of the giant planets' magnetic fields are listed in Table 2.5.

The internal currents generating the magnetic field are distributed, and hence the field is not a simple dipole, but has higher terms (just like the J-coefficients for gravity fields). Hence, observation of the shape of the magnetic field can be used to map the current flows in the interior.

2.7.3 Internal structure of Jupiter and Saturn

The mean internal structures of Jupiter and Saturn are shown in Figure 2.9 after Guillot (1999b). Jupiter is significantly oblate (0.065), and its estimated moment of inertia ratio (C/MR_e^2) of 0.264 indicates a significant concentration of mass towards the center of the planet. Saturn is even more oblate (0.098) and in fact is the most oblate body in the solar system. Its moment of inertia ratio of 0.21 is the lowest of all the planets suggesting even greater mass concentration. These figures, together with the measured J-coefficients of the gravitational field, may be fitted with internal models that model how the density, pressure, temperature, and composition vary with depth. Hydrostatic equilibrium is assumed, and from the fitted pressure–density profile, conclusions may be drawn on the internal structure (e.g., Guillot 1999a, b; Guillot *et al.*, 2004; Saumon and Guillot, 2004). Internal modeling studies suggest that Jupiter is significantly heavier than would be expected were the elements to be present in solar abundance. The mass of heavy elements (i.e., those with mass greater than helium) is estimated to be in the range 8 $M_\oplus$ to 39 $M_\oplus$ (Saumon and Guillot, 2004), which is 1–6× more than would be expected if Jupiter had a solar composition. Saturn is estimated to be even more enriched in heavy elements containing as much as

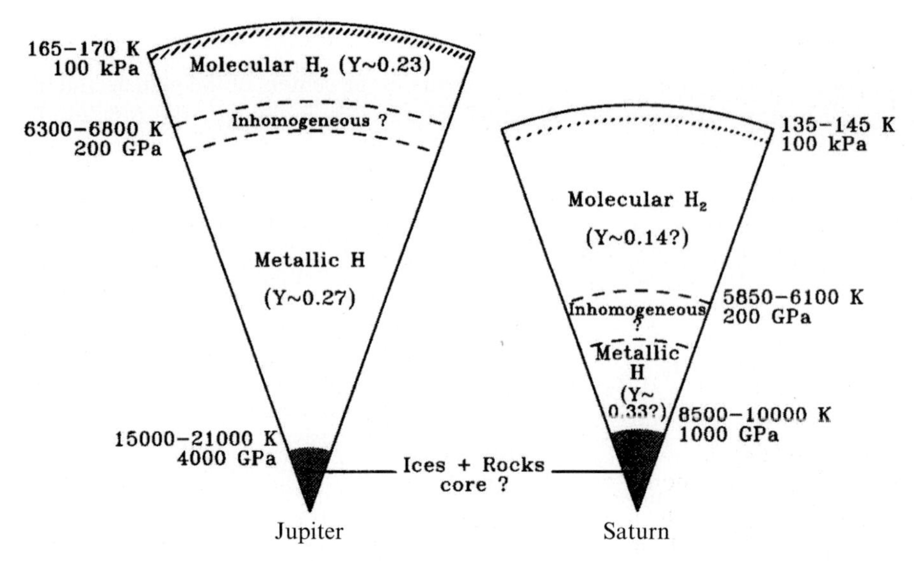

Figure 2.9. Interior models of Jupiter and Saturn. From Guillot (1999b). Reprinted with permission of the American Association for the Advancement of Science. Copyright 1999.

$13\,M_\oplus$ to $28\,M_\oplus$ (Saumon and Guillot, 2004), which is 6–14× more than would be expected if Saturn had a solar composition.

If Saturn is so much less massive than Jupiter, why is its volume approximately the same and thus its density so low? Hydrogen is a light and compressible substance and its compressibility is found to be little affected by temperature (provided that the temperature is not too high). Hence, a low-temperature sphere of hydrogen with a mass similar to Jupiter's has a *characteristic radius* that is nearly independent of total mass and interior temperature (Hubbard, 1997a). In other words, hydrogen is so compressible that a large increase in mass produces almost no change in radius. For a pure-hydrogen planet in the giant planet mass range the characteristic radius is 80,000 km, for a pure-helium planet it is 35,000 km, and for a "heavy" element planet it is typically 25,000 km. For an approximately solar composition, the characteristic radius would be 70,000 km, close to the observed radii of both Jupiter and Saturn.

Pressures and temperatures rise quickly towards the center of both planets and at temperatures greater than around 3,000 K and pressures greater than around 1.4 Mbar, hydrogen is thought to change to an electron-degenerate state of pressure-ionized protons and electrons called "metallic hydrogen". Experiments have been performed in the laboratory, using shock compression of hydrogen samples with gas-guns and lasers, which have confirmed this transition (Nellis, 2000; Cauble *et al.*, 2000; Weir *et al.*, 1996; Nellis *et al.*, 1995). The phase transition appears to be continuous rather than first-order and thus there is probably not a sharp boundary between the two phases in the interiors of these planets (Nellis, 2000). Rather, molecular hydrogen begins to dissociate at ~0.4 Mbar and is completely dissociated at ~3 Mbar. The mid-point pressure of 1.4 Mbar corresponds to a fractional radius of

0.9 for Jupiter and 0.5 for the much less compressed Saturn (Nellis, 2000). The transition region between molecular and metallic hydrogen is thus predicted to occur at a depth of ~7,000 km for Jupiter and ~30,000 km for Saturn. Thus, the overwhelming bulk of Jupiter's hydrogen is thought to exist in the metallic phase.

Saturn's low moment of inertia requires a high degree of differentiation (i.e., concentration of the heavy elements towards the center). Approximately 1 $M_\oplus$ to 8 $M_\oplus$ of heavy elements are thought to reside in the hydrogen/helium envelope (Saumon and Guillot, 2004), with a considerable mass (9–22 $M_\oplus$) thought to reside in a rocky/icy core. A large part of the uncertainty in Saturn's core mass arises from uncertainties in Saturn's J_4 gravitational coefficient. Guillot (1999a) noted that the error in the available value of J_4 prior to the arrival of the *Cassini* spacecraft gave rise to a possible variation of 10 $M_\oplus$ in the core mass (Baraffe, 2005). Through analyzing the trajectory of *Cassini* in the Saturnian system, Anderson and Schubert (2007) have revised the gravitational coefficients, improving their accuracy by a factor of 10–20 in the case of J_4, which should provide much better constraints to interior models.

Jupiter's higher moment of inertia implies less differentiation and it is estimated that perhaps 0 $M_\oplus$ to 11 $M_\oplus$ of heavy elements reside in a dense core, with the remaining 1 $M_\oplus$ to 39 $M_\oplus$ of heavy elements evenly distributed throughout the hydrogen/helium envelope. It is interesting to note that models with no core are still consistent with the estimated errors of current gravitational coefficients. One of the main aims of the future *Juno* mission (Section 8.6.1) will be to better constrain these coefficients through tracking the spacecraft's motion in its low perijove polar orbit, to determine once and for all if Jupiter has a core.

These estimates of the core and heavy element masses are consistent with the core accretion formation model outlined earlier, where an icy embryo forms first which then attracts further ice and gas before entering the final runaway gas collapse phase. Clearly Saturn was able to acquire less hydrogen and helium during this final stage, which accounts for its lighter mass, and greater heavy-element enrichment compared with Jupiter.

The estimated interior convective velocities and the calculated conductivity of metallic hydrogen are more than adequate to sustain a magneto-hydrodynamic dynamo of the size needed to account for Jupiter's very powerful magnetic field, especially since the conducting metallic region extends over 90% of the planetary radius. The convective timescale is estimated to be of the order of 100 yr and thus changes in the field are likely to occur on this timescale also. Long-term monitoring of the magnetic field may eventually provide clues on the deep currents. Charged particles from the solar wind and other sources become trapped and accelerated in this field leading to powerful synchrotron radio wave emission at decametric wavelengths, which are detected at Earth and led to the first measurements of the internal bulk rotation rate as inferred from observations of Jupiter's kilometric radiation (JKR) (System III). This was observable since Jupiter's magnetic field is sufficiently misaligned with respect to the rotation axis that diurnal changes in radio wave emission are easily detectable.

In Saturn the observed heat flux must similarly drive convection in most of the liquid interior. However, it is possible that composition differences across the metallic/molecular hydrogen phase boundary inhibit convection and thus transport of heat across this stably stratified boundary might be via conduction. This might explain why Saturn's magnetic field is so closely aligned to within 1° of the rotation axis since the non-axially symmetric parts of the magnetic field generated via the magneto-hydrodynamic dynamo in the convective interior of the metallic-hydrogen zone (which only accounts for 50% of the radius) may be screened out by a stably stratified conducting layer at the top of this region (Stevenson, 1980). The smaller size of the metallic-hydrogen conducting core may also explain why the magnetic field of Saturn is generally weaker than Jupiter's, and why the higher order field components are more greatly reduced (Hubbard, 1997b; Nellis, 2000). The close alignment between the magnetic and rotation axes makes it difficult to determine Saturn's magnetic field rotation period from the Earth and thus this was not determined until the *Pioneer 11* flyby in 1979 (Russell and Luhmann, 1997). The current estimate of the System III rotation rate of 10 h 39 min was derived from observations of Saturn's kilometric radiation (SKR) during the *Voyager* encounters. More recently, during the *Cassini* approach, the SKR rotation rate was observed to be 10 h 45 min (Gurnett *et al.*, 2005). Detailed observations since then of the magnetic field have yielded a magnetospheric rotation rate of 10 h 47 min (Giampieri *et al.*, 2006). Clearly, System III, derived from SKR observations, does not seem to be a reliable proxy for the bulk rotation rate of Saturn since the magnetosphere seems to slip relative to the interior, perhaps due to centrifugally driven instabilities in Saturn's plasma disk. Most recently, the gravitational data acquired during the Cassini mission have been analysed by Anderson and Schubert (2007) to yield a bulk internal rotation rate of 10 h 32 min, which is now thought to be the most reliable estimate.

The magnetosphere of Jupiter, where Jupiter's magnetic field dominates over the interplanetary magnetic field, is vast and if visible to the naked eye would have a diameter roughly twice that of the Moon as seen from the Earth. This great size is due not only to the high strength of Jupiter's magnetic field, but also to the low density of the solar wind at 5 AU and the additional source of charged particles from Io, which resides within the magnetosphere. The magnetosphere of Saturn is roughly five times smaller than that of Jupiter.

2.7.4 Internal structure of Uranus and Neptune

While Jupiter and Saturn have many similarities to each other, they are both very different from Uranus and Neptune, which form a separate pair of planets. The mean internal structures of Uranus and Neptune are shown in Figure 2.10 (after Guillot, 1999b). The small radii of Uranus and Neptune tell us immediately that, although hydrogen and helium account for the bulk of the molecules in the observable atmosphere, the planets cannot be predominantly composed of hydrogen and helium since the density would then be far too low. Instead, their radius is close to the characteristic radius of icy materials and thus these planets are thought to be composed predominantly of ice, with only a thin outer envelope of hydrogen and helium. It would be possible to match the planets' mean density by having a hydrogen atmosphere over a small rocky core, but this would have a moment of inertia which is very

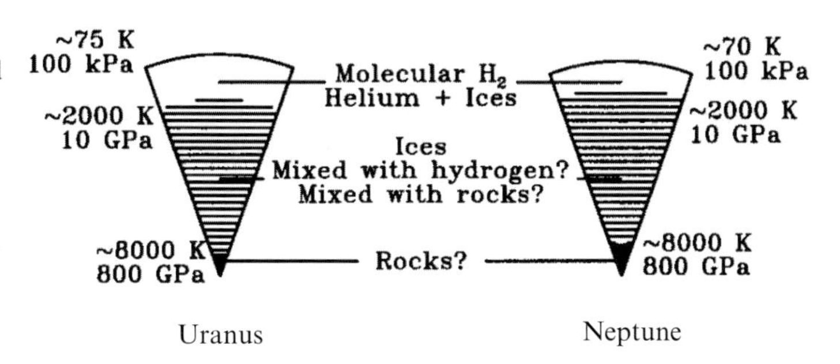

Figure 2.10. Interior models of Uranus and Neptune. From Guillot (1999b). Reprinted with permission of the American Association for the Advancement of Science. Copyright 1999.

much smaller than that observed. The "ices" occurring in the high-pressure, high-temperature interior of Uranus and Neptune would actually exist as a hot molecular fluid, not the solid cold ice that we are more familiar with. Interior models of these planets suggest that their interiors are rather homogeneous and there is no hard evidence for a dense rocky core. This might suggest that the accumulation of these planets was slow, allowing internal heat to radiate away and thus that the ice–rock planetesimals have not been subsequently differentiated. Alternatively, the interiors may have been thoroughly mixed through convective overturning. Either way, the fact that Uranus and Neptune appear to have such similar interiors when their subsequent evolution and current reservoirs of internal heat are so very different is curious.

The high pressure–temperature environments of the internal regions of Uranus and Neptune have been experimentally modeled by "synthetic Uranus" models where a liquid solution of water, ammonia, and isopropanol (with "solar" molar abundances of H, O, C, and N) is subjected to single and double shockwave experiments conducted up to 2.2 Mbar and $T > 4{,}000$ K (Hubbard, 1997c, d; Radousky *et al.*, 1990). The compression curves derived from these experiments (i.e., the relationship between pressure and fluid density) closely match those required by interior models of these planets to explain the observed density and J-coefficients, again suggesting that the bulk of these planets are made of icy materials. For the outer part of these atmospheres at pressures less than 100 kbar, the modeled compression curve matches that of nebular hydrogen and helium. The material in the "mantle" of Uranus and Neptune, if it has a solar composition of heavy elements, should be composed of about 40% water ice, 25% methane ice, approximately 25% iron and rocks, with ammonia making up part of the remaining 10%. The abundance of rocky material would make this material denser than that modeled to exist in the mantle and assumed by "synthetic Uranus" models. Hence, to reduce the density of the mantle material to that required by interior models requires either that the rocks and iron differentiate from the icy material in the mantle and form a rocky core, of which there is currently no evidence, or that sufficient low-density material such as hydrogen and helium is also mixed throughout the mantle to offset the effect of the denser rocks. Roughly 1 $M_\oplus$ to 2 $M_\oplus$ of hydrogen/helium would be required to do this (Hubbard, 1997d).

The hydrogen-rich outer layer appears to be rather thin and comprises no more than 15% (3,500 km) of the radius for Neptune (amounting to 0.5–1.0 $M_{\oplus}$) and 20% (5,000 km) of the radius for Uranus (amounting to less than 3 $M_{\oplus}$). Uranus and Neptune thus appear to be only slightly evolved from the original embryos that gravitationally swept up the gas in the outer parts of the proto-solar nebula during Phase 2 of formation according to core accretion theory (Section 2.4.1). As mentioned earlier, the longer time required to accrete planets at greater distances from the Sun meant that Uranus and Neptune appear never to have reached Phase 3 of formation before the solar nebula dissipated, either on its own, or when the T-Tauri phase of the Sun swept the solar system clean of any remaining nebula gas.

A small fraction of the hydrogen in the envelopes of these planets may have resulted from the decomposition of hydrogen-bearing ices in the interior. However, the He/H_2 ratio in Uranus' atmosphere is found to be very close to the solar value and hence this fraction is probably very small. The high abundance of methane (40–50 times greater than would be expected for a solar C/H fraction) in the observable atmospheres of these planets, however, is good evidence of the presence of an enormous ice reservoir beneath the hydrogen-rich envelope, and considerable mixing between the two which also explains the high D/H ratio in H_2 found for these planets. Given the other similarities between Uranus and Neptune, we might then expect Neptune's He/H_2 ratio to be similar to that of Uranus. However, initial estimates of the Neptune's He/H_2 ratio were supersolar, which was somewhat puzzling. These ratios were determined from the infrared spectra of the planets recorded by *Voyager 2* and the analysis made the assumption that there was no N_2 in the atmosphere (Conrath *et al.*, 1991). However, the detection at microwave wavelengths of HCN in the stratosphere of Neptune (and not in the stratosphere of Uranus) suggested that there may be N_2 in the Neptunian atmosphere. A later study suggested that a mole fraction of only 0.3% of N_2 would be sufficient to reduce the He/H_2 ratio inferred from the far-infrared spectrum of Neptune to that of Uranus (Conrath *et al.*, 1993). Burgdorf *et al.* (2003) find from ISO measurements that they can fit the observed spectra with an upper limit on the abundance of nitrogen of 0.7%, which would make the N/H enrichment intriguingly similar to that of C/H. One possibility is that the stratospheric N_2 of Neptune comes from nitrogen exospherically lost from the atmosphere of Neptune's moon Triton. However, it would appear that this process has difficulty in matching the amounts of N_2 required. An alternative, probably more plausible explanation is that N_2 is dredged up from the interior of Neptune via the vigorous convection known to exist on that planet. If the C/N ratio really is approximately solar for Neptune, then it would provide support for the SCIP formation model described in Section 2.6.1.

2.8 BIBLIOGRAPHY

Bagenal, F., T. Dowling, and W. McKinnon (Eds.) (2004) *Jupiter: The Planet, Satellites and Magnetosphere.* Cambridge University Press, Cambridge, U.K.

Beatty, J.K., C. Collins Petersen, and A. Chaikin (Eds.) (1999) *The New Solar System* (Fourth Edition). Cambridge University Press, Cambridge, MA.

de Pater, I. and J.J. Lissauer (2001) *Planetary Sciences*. Cambridge University Press, Cambridge, U.K.

Encrenaz, T., J.-P. Bibring, and M. Blanc (1995) *The Solar System* (Second Edition). Springer-Verlag, Berlin.

Jones, B.W. (2007) *Discovering the Solar System* (Second Edition). John Wiley & Sons, Chichester, U.K.

Lewis, J.S. (1997) *Physics and Chemistry of the Solar System*. Academic Press. San Diego, CA.

Mason, J.W. (Ed.) (2008) *Exoplanets: Detection, Formation, Properties, Habitability*. Springer/Praxis, Heidelberg, Germany/Chichester, U.K.

Shirley, J.H. and R.W. Fairbridge (Eds.) (1997) *Encyclopaedia of the Planetary Sciences*. Chapman & Hall, London.

Weissman, P.R., L.-A. McFadden and T.V. Johnson (Eds.) (1998) *Encyclopaedia of the Solar System*. Academic Press, San Diego, CA.

The following websites are also useful:

Kuiper Belt Objects: *http://www.ifa.hawaii.edu/faculty/jewitt/kb.html*
http://www.nineplanets.org/kboc.html

Oort Cloud: *http://www.nineplanets.org/kboc.html*
http://www.solarviews.com/eng/oort.htm

Extrasolar planets: *http://www.obspm.fr/encycl/encycl.html*

3

Evolution processes in outer planet atmospheres

3.1 INTRODUCTION

Subsequent to their formation from the collapse of the solar nebula, the atmospheres of all the planets, including the giant planets, have (to a greater or lesser extent) evolved over time through the action of a number of possible processes including thermal escape, cometary bombardment, and internal differentiation. In this chapter we will outline the principal evolution mechanisms and estimate their effects on the present-day composition of the giant planet atmospheres.

3.2 THERMAL ESCAPE

3.2.1 Jeans' formula

In the upper atmospheres of planets, density becomes very low and temperature becomes very high due to absorption of solar ultraviolet light and, for the giant planets, other sources such as the viscous damping of vertically propagating gravity waves (Matcheva and Strobel, 1999; Young *et al.*, 1997). Hot, fast-moving molecules may escape should their kinetic energy (in the vertical direction) exceed their gravitational potential energy: that is,

$$\tfrac{1}{2}mv^2 > \frac{GMm}{r} \tag{3.1}$$

where M is the mass of the planet; G is the gravitational constant; m is the mass of the molecule; v is its velocity in the upward direction; and r is the planetary radius at the altitude of the thermosphere. Since the gravitational acceleration at radius r is given by $g = GM/r^2$, this condition may be rewritten as

$$\tfrac{1}{2}v^2 > gr \tag{3.2}$$

or

$$v > \sqrt{2gr}. \tag{3.3}$$

Molecules moving upwards with speeds in excess of this calculated escape velocity will, however, only escape if they do not collide with any other molecules on the way. Hence, substantial escape of molecules only occurs when the vertically integrated density of air molecules above a certain *critical level*, z_c, accounts for one mean free path: that is,

$$\int_{z_c}^{\infty} \sigma n_a(z)\, dz = 1 \tag{3.4}$$

where $n_a(z)$ is the number density profile of *all* molecules at an altitude z above a reference level in the planet's atmosphere; and σ is the mean collisional cross-section. Assuming the atmosphere to be in hydrostatic equilibrium (Section 4.1.1) such that $n_a(z) = n_a(z_c)\exp(-z/H^*_{ca})$, where H^*_{ca} is the mean atmospheric number density scale height discussed in Section 3.2.2, Equation (3.4) may be integrated to give

$$\int_{z_c}^{\infty} \sigma n_a(z)\, dz = \sigma n_a(z_c)\int_{z_c}^{\infty} e^{-z/H^*_{ca}}\, dz = \sigma n_a(z_c)H^*_{ca} = 1. \tag{3.5}$$

For the giant planet atmospheres, the most abundant molecule in the upper atmosphere is atomic hydrogen, and thus it is this molecule that largely determines the altitude of the critical level, otherwise known as the *exobase*. Assuming a Maxwell–Boltzmann distribution of speeds, the probability that a molecule of molecular weight m has a speed in the range c to $c + dc$ is given by

$$P(c)\, dc = 4\left(\frac{\alpha^3}{\pi}\right)^{1/2} c^2 \exp(-\alpha c^2)\, dc \tag{3.6}$$

where $\alpha = m/2RT$; R is the gas constant; and T is the temperature. For such a gas, the number of molecules passing upwards through unit area per second at the exobase with speeds in the range c to $c + dc$ is then given by the well-known kinetic theory (e.g., Flowers and Mendoza, 1970) expression

$$dF = \tfrac{1}{4}n(z_c)cP(c)\, dc. \tag{3.7}$$

Assuming that all such molecules with speed greater than the escape velocity $v_e = \sqrt{2gr}$ will escape the atmosphere, we may integrate Equation (3.7) to calculate the flux of escaping molecules, known as the *Jeans' flux*

$$\begin{aligned} F_{\text{Jea}} &= n(z_c)\left(\frac{\alpha^3}{\pi}\right)^{1/2}\int_{v_e}^{\infty} c^3 \exp(-\alpha c^2)\, dc \\ &= \tfrac{1}{2}n(z_c)\left(\frac{\alpha}{\pi}\right)^{1/2}\left(v_e^2 + \frac{1}{\alpha}\right)\exp(-\alpha v_e^2) \end{aligned} \tag{3.8}$$

or

$$F_{\text{Jea}} = \frac{n(z_c)U}{2\sqrt{\pi}}(1 + \lambda)\exp(-\lambda) \tag{3.9}$$

where the most probable speed $U = \sqrt{2RT/m}$; and the escape parameter λ is defined as $\lambda = v_e^2/U^2$.

The rate at which the concentration of molecules is reduced may then be calculated by considering that the total number of molecules per unit area above the exobase (and thus which may escape) is given by $N = n(z_c)H_c^*$, where H_c^* is the number density scale height of the escaping molecule or atom at the critical level. Thus, expressing the Jeans' flux as $F_{\text{Jea}} = \beta n(z_c)$, we obtain

$$F_{\text{Jea}} = \left(\frac{\partial N}{\partial t}\right)_{\text{Jea}} = -\frac{\beta}{H_c^*}N \tag{3.10}$$

which may be integrated to give

$$N = N_0 \exp\left(-\frac{\beta}{H_c^*}t\right). \tag{3.11}$$

From this expression we can see that there is a *characteristic escape time* for thermal escape given by $\tau_e = H_c^*/\beta$. Alternatively, we can define a mean velocity at which molecules or atoms escape upwards from the exobase, known as the *expansion velocity* given by $v_{\text{ex}} = H_c^*/\tau_e = \beta$. The calculated characteristic escape times for various gases in the atmospheres of the giant planets, the Earth, and also Titan and Triton are listed in Table 3.1. As can be seen, compared with the smaller planets, the masses of the giant planets are so large, and their exospheric temperatures so cool, that negligible exospheric escape is calculated and thus these planets have effectively lost none of their atmospheres. This is not the case for the Earth, Titan, and Triton, where significant loss of the lighter atoms is calculated and whose atmospheres have thus significantly evolved over time. The relatively unevolved atmospheres of the giant planets thus offer a unique picture of the composition of the solar nebula at the time of the planets' formation, provided that no other processes have acted to modify the composition. We shall return to this in Sections 3.3 and 3.4.

3.2.2 Diffusion and limiting flux

For planets such as the Earth, and giant planet satellites such as Titan and Triton, where exospheric escape is of importance, the rate of exospheric escape is actually limited by the rate at which molecules may be transported to the exobase from below by processes such as diffusion, as described by Chamberlain and Hunten (1987). The approach of Chamberlain and Hunten (1987) is reproduced here for the reader's convenience.

Suppose a minor constituent has a density distribution $n_i(z)$ in an atmosphere where its density distribution in diffusive equilibrium would be $n_{iE}(z)$, then by the

Table 3.1. Calculated exospheric escape times for the giant planets, Earth, Titan, and Triton.

	Earth	*Jupiter*	*Saturn*	*Uranus*	*Neptune*	*Titan*	*Triton*
R_0 (km)	6,378	71,492	60,268	25,559	24,764	2,575	1,353
z_c (km)	500	2,000	2,000	2,000	2,000	1,000	300
T (K)	1,480	700	420	700	700	300	300
g_0 (m s^{-2})	9.81	23.1	9.1	8.7	11	1.41	0.78
v_e (km s^{-1})	10.77	56.6	32.5	20.3	22.4	2.28	1.31
τ(H) (yr)	5.6×10^{-4}	2.6×10^{112}	1.2×10^{59}	1.5×10^{9}	2.6×10^{12}	1.2×10^{-4}	$1.7 \times 10^{-}$
τ(He) (yr)	1.1×10^{-2}	2.2×10^{471}	2.2×10^{256}	3.0×10^{54}	9.3×10^{67}	5.7×10^{-4}	$1.3 \times 10^{-}$
τ(C) (yr)	5.5×10^{17}	1.4×10^{1430}	1.1×10^{784}	8.6×10^{176}	6.30×10^{217}	5.5×10^{-1}	$5.7 \times 10^{-}$
τ(N) (yr)	5.5×10^{21}	8.9×10^{1669}	1.1×10^{916}	4.2×10^{207}	2.2×10^{255}	3.62	$9.4 \times 10^{-}$
τ(O) (yr)	5.6×10^{25}	5.6×10^{1909}	1.1×10^{1048}	2.2×10^{238}	7.6×10^{292}	24.3	$1.6 \times 10^{-}$
τ(Ar) (yr)	2.1×10^{74}	5.8×10^{4787}	5.0×10^{2632}	1.8×10^{607}	8.2×10^{743}	5.5×10^{11}	1.8
τ(Kr) (yr)	9.8×10^{163}	5.4×10^{10064}	3.1×10^{5538}	1.7×10^{1284}	2.6×10^{1571}	2.0×10^{31}	2.5×10

NB. R_0 is the equatorial radius at the 1 bar pressure level, and z_c is the altitude of the critical level above R_0.

process of molecular diffusion there will be an upward flux equal to

$$\phi_i = n_i w_i = -D_i n_{iE} \frac{\partial(n_i/n_{iE})}{\partial z} \tag{3.12}$$

where D_i is the molecular diffusion coefficient for the ith gas; and w_i is the mean effective vertical speed. From kinetic theory considerations, the coefficient of diffusion for a single gas may be shown to be

$$D_i = \tfrac{1}{3}\lambda_i \bar{c}_i \tag{3.13}$$

where $\bar{c}_i = \left(\dfrac{8RT}{\pi m_i}\right)^{1/2}$ is the mean speed of the molecules of molecular weight m_i; $\lambda_i = 1/\sqrt{2}n\sigma_i$ is the mean free path; and σ_i is the collision cross-section. Hence, the coefficient may be written as $D_i = \bar{c}_i/3\sqrt{2}n\sigma_i$ or more generally in the semi-empirical form

$$D_i = \frac{b_i}{n} = \frac{A_i T^{S_i}}{n} \tag{3.14}$$

where b_i is the *binary collision parameter* expressed in terms of the coefficients A_i and S_i, which are fitted to experimental data.

For an atmosphere in hydrostatic equilibrium (Section 4.1.1), the pressure distribution $p(z)$ is given by $p = p_0 e^{-z/H}$, where the scale height is to a first approximation $H = RT/\bar{m}g$ (assuming negligible variation of temperature with height) and $\bar{m}$ is the mean molecular weight. Similarly, it is straightforward to show that the number density $n_{iE}(z) = p(z)/kT$ has a similar distribution given by

$$n_{iE} = n_0 e^{-z/H^*} \tag{3.15}$$

where H^* is the number density scale height, given by

$$\frac{1}{H^*} = \frac{1}{H_i} + \frac{1}{T}\frac{dT}{dz}. \tag{3.16}$$

Substituting for $n_{iE}(z)$ into Equation (3.12) we obtain

$$\phi_i = -n_i D_i \left(\frac{1}{n_i}\frac{dn_i}{dz} + \frac{1}{H_i} + \frac{1}{T}\frac{dT}{dz} \right) \tag{3.17}$$

or

$$\phi_i = n_i w_i = n_i D_i \left(\frac{1}{H_i^*} - \frac{1}{H_{iE}^*} \right) \tag{3.18}$$

where we have also equated the flux to the product of the number density times the mean vertical flux velocity w_i. A mean estimate for the diffusion time is given by the expression $\tau_i \sim H_i^*/w_i$ or, since w_i is of the order of D/H, $\tau_i \sim H^2/D_i$.

Defining the *volume mixing ratio*, or *mole fraction*, of the ith element as $f_i = n_i/n_a$, and differentiating we obtain

$$\frac{1}{f_i}\frac{df_i}{dz} = \frac{1}{n_i}\frac{dn_i}{dz} - \frac{1}{n_a}\frac{dn_a}{dz}. \tag{3.19}$$

Substituting this into Equation (3.17) we get

$$\phi_i = \phi_L - D_i n_a \frac{df_i}{dz} \tag{3.20}$$

where the *limiting flux* ϕ_L is given by

$$\phi_L = -n_i D_i \left(\frac{1}{n_a}\frac{dn_a}{dz} + \frac{1}{H_i} + \frac{1}{T}\frac{dT}{dz} \right). \tag{3.21}$$

Hence, exospheric loss is in fact moderated by the rate at which molecules may diffuse up to the exobase. This consideration is of particular importance for estimating the rate of escape of H and H_2 from the atmospheres of Titan and Triton. Note that in Equations (3.20) and (3.21) the eddy diffusion coefficient should also be added to the molecular diffusion coefficient to give total "diffusion". Eddy diffusion is discussed in Chapter 4.

3.2.3 Hydrodynamic escape

We saw earlier that only light atoms can escape in significant amounts by Jeans' escape mechanism from the atmospheres of Titan and Triton. However, if the flux of light atoms is large then heavier atoms may also be driven off by "blowoff" (Chamberlain and Hunten, 1987). Substituting the empirical expression for the diffusion coefficient (Equation 3.14) into the diffusion equation (Equation 3.17) we have

$$\phi_i = -n_i \frac{b}{n}\left(\frac{1}{n_i}\frac{dn_i}{dz} + \frac{1}{H_i} + \frac{1}{T}\frac{dT}{dz}\right) \tag{3.22}$$

where b is the empirical binary collision parameter. This equation may be rearranged to give

$$\frac{dn_i}{dz} = -\left(\frac{\phi_i n}{b} + \frac{n_i}{H_i} + \frac{n_i}{T}\frac{dT}{dz}\right) \tag{3.23}$$

This equation assumes that the individual gases may be treated separately. However if two gases are considered, moving with different fluxes of, respectively, ϕ_1 and ϕ_2, then their combined diffusion equations, ignoring the dT/dz terms which are found to be negligible (Chamberlain and Hunten, 1987), are

$$\left.\begin{aligned} \frac{dn_1}{dz} &= -\frac{n_1}{H_1} + \frac{1}{b}(n_1\phi_2 - n_2\phi_1) \\ \frac{dn_2}{dz} &= -\frac{n_2}{H_2} + \frac{1}{b}(n_2\phi_1 - n_1\phi_2). \end{aligned}\right\} \tag{3.24}$$

With a bit of manipulation the flux ϕ_2 of the heavier gas of mass M_2, blown by a flux ϕ_1 of lighter gas molecules of mass M_1 may be shown to be

$$\phi_2 = \frac{f_2}{f_1}\phi_1\left(\frac{M_c - M_1}{M_c - M_2}\right) \tag{3.25}$$

where f_n is the mole fraction of component n and

$$M_c = M_1 + \frac{kT\phi_1}{bgf_1}$$

is called the *crossover mass* and represents the heaviest species that can be removed in this way. This effect is clearly negligible in the giant planet atmospheres since the Jeans' flux of even the lightest atom hydrogen is so low.

3.3 IMPACTS WITH COMETS AND PLANETESIMALS

Planets have undergone significant bombardment by comets and planetesimals in the past and in fact, as the collision of Comet Shoemaker–Levy 9 with Jupiter in 1994 showed, these impacts are an ongoing process in the evolution of planetary atmospheres. During such an impact, in addition to introducing new material to the planet from the planetesimal itself, such as "soot" and trace species (discussed in Chapters 4 and 5), a small fraction of the planet's atmosphere may be driven off if it acquires sufficient kinetic energy to escape the planet's gravity.

We can make a simple calculation of the rate at which the atmosphere is ejected in this way. Suppose the impactor has a radius R and speed v_s and strikes an atmosphere of density σ_a per unit area above the level where the impactor's energy is spent. The kinetic energy acquired by the air through which the comet passes may be estimated to be roughly

$$E = \frac{1}{2}\pi R^2 \sigma_a v_s^2. \tag{3.26}$$

Assuming all this energy is converted to eject a mass M_e of atmosphere at the escape velocity v_e, we can obtain an upper estimate of atmospheric mass loss

$$M_e = \frac{\pi R^2 \sigma_a v_s^2}{v_e^2}. \tag{3.27}$$

This process, unlike thermal escape and hydrodynamic escape, does not discriminate with respect to mass and hence does not cause fractionation of species. However, its effect on the evolution of giant planet atmospheres is likely to be very small due to the large escape velocities of these planets.

3.4 INTERNAL DIFFERENTIATION PROCESSES

The final way in which the composition of the outer planet atmospheres may evolve is though differentiation, where heavier materials preferentially fall towards the center of the planet decreasing their abundance in the exterior. This is a separate effect to the initial differentiation of the giant planets arising from the different phases of their growth described in Chapter 2. This secondary differentiation is inhibited by convection which tends to make the interior of the planets well-mixed and homogeneous. If differentiation does occur, however, it should be observable both via observed depletions in the outer parts of the atmosphere and through the additional source of internal heat that such gravitational settling would release. To evaluate the degree of internal heating currently active in the interiors of the giant planets we need to compare their current observed bolometric temperatures with those expected were the planets to be in thermal equilibrium with incident solar radiation.

3.4.1 Effective radiating temperature of planets

The Sun radiates essentially as a black body, with an effective surface temperature $T_S = 5{,}750$ K and radius $R_S = 700{,}000$ km. Applying Stefan–Boltzmann's law we find that the total power radiated by the Sun in all directions is

$$E_S = \sigma(4\pi R_S^2)T_S^4 = 3.8 \times 10^{26}\ \text{W}. \tag{3.28}$$

where σ is the Stefan–Boltzmann constant. The flux of sunlight (W m^{-2}) arriving at a planet at distance D from the Sun is simply this power divided by the surface area of a sphere of radius equal to the distance to the Sun: that is,

$$F_S = \sigma\left(\frac{R_S}{D}\right)^2 T_S^4. \tag{3.29}$$

At the Earth's distance from the Sun the solar constant F_S is equal to $1.37\,\mathrm{kW\,m^{-2}}$. The total amount of sunlight absorbed by the planet is then equal to

$$P_{\mathrm{abs}} = (1 - A_B)\pi R_P^2 F_S \tag{3.30}$$

where R_P is the planetary radius; πR_P^2 is the projected disk area; and A_B is the *Bond albedo*, which is the fraction of sunlight reflected by the planet in all directions. Assuming no other sources of heat, in order for the temperature of the planet to be in equilibrium, this absorbed power must be balanced by the thermal radiation emitted to space. Since the observable temperatures of the giant planets are so much smaller than the Sun's, the Planck functions of incident solar and emitted thermal radiation show negligible overlap and thus the two fluxes may be considered separately. The total power emitted by the planet (of *effective radiating temperature* T_E) to space is equal to its total surface area multiplied by the infrared flux $F_{\mathrm{IR}} = \sigma T_E^4$; that is,

$$P_{\mathrm{emm}} = 4\pi R_P^2 F_{\mathrm{IR}}. \tag{3.31}$$

Equating the total absorbed power to the emitted power we find

$$\sigma(4\pi R_P^2)T_E^4 = (1 - A_B)\pi R_P^2 F_S \tag{3.32}$$

and rearranging for T_E, the effective radiating temperature, we obtain

$$T_E = \left(\frac{(1 - A_B)}{4\sigma} F_S\right)^{1/4} \tag{3.33}$$

or by substituting F_S from Equation (3.29)

$$T_E = T_S\left(\frac{(1 - A_B)R_S^2}{4D^2}\right)^{1/4}. \tag{3.34}$$

For the Earth T_E is estimated to be approximately 255 K. The mean *surface* temperature on the Earth is fortunately significantly higher than this due to the greenhouse effect provided by the IR absorptivity of carbon dioxide and water vapor.

The albedo used here for effective radiating temperature calculations is the *Bond albedo* (or *planetary bolometric albedo*) and is defined as the fraction of incident solar radiation scattered in all directions. When measuring the reflectivity of planets, in particular the giant planets, all we see is the radiation reflected in the direction back to Earth. If the Earth, Sun, and planet are all in the same line, the albedo measured is called the *geometric albedo*. Should the surface of the planet reflect light equally well in all directions (and thus be a perfect *Lambert* reflector) then the Bond and geometric albedos are in fact identical. Otherwise, they differ depending upon the scattering properties of the planet's surface or cloud layers.

Table 3.2. Thermal balance of the Earth and giant planets.

	Earth	*Jupiter*	*Saturn*	*Uranus*	*Neptune*
D (AU)	1	5.20	9.56	19.22	30.11
Solar flux F_S ($\mathrm{W\,m^{-2}}$)	1,370.0	50.66	14.99	3.71	1.51
Geometric albedo A_G	0.3?	0.274	0.242	0.215	0.215
Bond albedo A_B	0.3	0.343	0.342	0.300	0.290
Absorbed flux $F_{\mathrm{ab}}((1 - A_B) \times F_S/4)$	205.5	8.32	2.47	0.65	0.27
Observed F_{IR}	205.5	13.89	4.40	0.69	0.72
Energy balance ($F_{\mathrm{IR}}/F_{\mathrm{ab}}$)	1.0	1.67	1.78	1.06	2.61
T_E (K)	255	109.5	82.4	58.2	46.6
T_{IR} (K)	255	124.4	95.0	59.1	59.3
$F_{\mathrm{int}}(F_{\mathrm{IR}} - F_{\mathrm{ab}})$	0.0	5.57	1.93	0.04	0.45

Adapted from table 7 of Pearl and Conrath (1991).

The measured albedos and thermal infrared fluxes of the giant planets are listed in Table 3.2. The data in this table show that the calculated effective radiating temperatures of all the Jovian planets except Uranus are significantly smaller than observed bolometric temperatures. This implies, therefore, that these planets have significant internal sources of heat, which may come from three possible sources:

(1) Residual heat of formation arising from continued cooling, and shrinking of the planets via the Kelvin–Helmholtz mechanism.
(2) Fractionation in the interior, with heavy elements settling towards the center and converting gravitational energy into thermal energy.
(3) Radioisotope heating.

The contribution of all these sources to the observed bolometric temperatures of the giant planets will now be discussed.

3.5 EVOLUTION OF THE GIANT PLANET ATMOSPHERES

3.5.1 Jupiter

The bolometric temperature of Jupiter (124 K) is consistent with that expected by the planet radiating its primordial heat to space continually since formation via the *Kelvin–Helmholtz* mechanism. The interior heat flux remains sufficiently high to keep the liquid metallic interior highly convective with the result that most of planet is expected to mix thoroughly on a timescale of the order of 100 years. As we saw in Chapter 2, this vigorous convection of the metallic-hydrogen interior easily accounts for Jupiter's very high magnetic field.

The phase transition between molecular hydrogen and metallic hydrogen may, however, have an effect upon the atmospheric abundances in the observable atmosphere. Some calculations suggest that solar abundances of helium cannot be dissolved in metallic hydrogen under current Jovian temperature conditions at the lowest pressure where metallic hydrogen exists. Droplets of liquid helium would form in these regions and drop towards the center of the planet converting their gravitational energy into thermal energy and reducing the abundance of helium in the observable atmosphere. However, these solubility calculations are highly model-dependent and thus it is not clear whether or not there has been significant depletion of helium. The current helium mass fraction of 0.238 (von Zahn *et al.*, 1998) is almost the same as the current observable solar value, although the primordial helium fraction is believed to have been slightly larger, perhaps 0.25 (Grevesse *et al.*, 2007). Hence, Jupiter's atmosphere could be depleted if the Sun underwent a similar process of helium separation over time with the Sun fusing helium in its core. The atmospheric depletion of neon observed by the *Galileo* entry probe would be consistent with helium differentiation since neon is highly soluble in liquid helium. However, the low abundance of neon may also be explained by the clathrate–hydrate theory of formation outlined in Chapter 2, since neon is not easily trapped in clathrates.

The deep circulation of the interior is not well understood. Most models assume that the molecular and metallic regions are homogeneous (i.e., well-mixed) since Jupiter (and Saturn) are still emitting more energy than they receive from the Sun, which implies active convection at great depth. This is a good assumption provided that the hydrogen–helium mix is sufficiently opaque such that convective heat transfer is more efficient than radiative heat transfer. This is almost certainly true in the metallic-hydrogen region, which is thought to be highly opaque to thermal photons, but not particularly conductive for a metal. However, in the molecular-hydrogen region some models suggest that opacity may be sufficiently small at kilobar pressure levels to allow the presence of a thin radiative zone, which would serve as a barrier to convection. This may also occur in Saturn. Likewise the molecular-metallic interface (Figure 2.9), if it exists as a discrete phase boundary, may also act as a barrier (Hubbard, 1997a), although recent laboratory experiments (Nellis, 2000) suggest that the molecular-metallic phase boundary is in fact somewhat smooth as was noted earlier in Chapter 2. In these "barrier" regions, stable to convection, the fluid interior would be stably stratified. If there were negligible turbulent eddy mixing (see Chapter 4) then gravitational settling of the heavier molecules towards the base of the layer may occur in these regions. Hence, a composition gradient would be set up leading to a net transfer of molecules between the well-mixed convective layers above and below the stable region and thus lead to long-term evolution of the composition of the observed atmosphere. In reality, interior models suggest that it is very unlikely that eddy mixing is so low that gravitational settling occurs. However, these "barrier" regions may have an effect on the observed composition in another way. Eddy mixing is not as efficient at transporting material as convective mixing and thus if the material above and below the convective "barrier" initially has a different composition then the barrier would greatly inhibit the transfer of material across it, thus maintaining compositional differences over long periods of time. During the formation of the giant planets deeper material is thought, from the core accretion model, to have been more icy than material accreted later, which contained proportionally more and more nebula gas. If the whole interior was convectively unstable then these differences would be rapidly eliminated. However, the presence of convective barriers may mean that the material above and below the barriers has still not come to equilibrium today and thus that the composition of the observable atmosphere may still be evolving.

3.5.2 Saturn

The current Saturn thermal flux is 1.9 W m^{-2}. Although this is smaller than the corresponding figure for Jupiter, this is due to Saturn's much lower mass and in fact, unlike Jupiter, most models suggest that the figure is too large to be explained by the Kelvin–Helmholtz mechanism alone. Cooling via the Kelvin–Helmholtz mechanism decreases exponentially over time and for Saturn is predicted by most models to have mostly disappeared after approximately 2.5 Gyr. Hence, an additional source of heat seems to be required, which is released later during the planet's evolution as it cools to lower temperatures. Although Saturn has more rocky material than Jupiter, radioactive heating from the 13 $M_\oplus$ to 28 $M_\oplus$ of heavy material (Section 2.7.3) in Saturn would account for no more than 1% of this and may be discounted. One possibility is that, like Jupiter, at the lower temperatures found in the metallic-hydrogen region of Saturn, helium becomes slightly immiscible and droplets of helium form, which "rain out" towards the center. This gradual precipitation, with resultant release of gravitational energy would account for the additional internal heat source, and is also consistent with estimates of the He/H_2 ratio of 0.135 which is significantly smaller than the ratio of 0.157 measured in Jupiter's atmosphere.

Saturn may also have a radiative zone in its molecular-hydrogen region and, together with the possible convective barrier at the metallic/molecular-hydrogen boundary (Figure 2.9), these may act to inhibit the transfer of material between the interior and exterior of the planet. The extent of this is unknown.

3.5.3 Uranus and Neptune

Uranus radiates at most only 6% more energy than it receives from the Sun and its internal heat flux is estimated to be less than 0.04 W m^{-2}. Hence, to a first approximation it has almost no internal heat source at all, although by the Kelvin–Helmholtz mechanism as much as 1 W m^{-2} would currently be expected. This low value of Uranus' internal flux is extremely puzzling, especially when it is considered that Neptune, which is otherwise so similar to Uranus, has a very strong internal heat source. In fact Neptune's total thermal emission to space is almost equal to that of Uranus' even though Neptune is much farther from the Sun. Uranus is estimated to include as much as 4 $M_\oplus$ of rocky materials, and assuming that this rock is heated by radioactive decay at the same rate as terrestrial rocks, then a flux of 0.02 W m^{-2} is calculated (Hubbard, 1997b), which is a substantial fraction of that estimated.

There is no consensus on why Uranus' flux is so low. One suggestion is that chemical composition gradients, such as that at 5,000 km depth at the ice/hydrogen–helium interface (Figure 2.10) may act as a convection barrier and thus inhibit the transport of heat from the planet's hot interior. However, why this might happen in Uranus' atmosphere and not in Neptune's is unclear. Another possibility is that towards the end of its formation, Uranus suffered a cataclysmic off-center impact with another planet-sized body. Such an impact would account for Uranus' abnormally high obliquity and compact, equatorially aligned satellite system formed out of collision debris. It might also have had the effect of greatly accelerating the release of internal heat by effectively turning the planet inside out! Neptune, on the other hand, may have suffered much more centered impacts and thus rather than turn the planet on its side, the collision energy was converted into additional internal heat which would explain Neptune's high internal heat source. It should be remembered that these "giant collision" theories are highly speculative and indeed Uranus' high obliquity could have been imparted by several off-center collisions with much smaller bodies during formation and not necessarily a single giant collision.

Although methane and water in the deep interior should dissociate at the high temperatures and pressures found there, it does not appear that the hydrogen this releases permanently escapes the interior and mixes with the nebular hydrogen in the outer hydrogen-helium envelope since the observed He/H_2 ratio is close to solar. There is, however, good evidence of considerable mixing of nebula hydrogen with this dissociated hydrogen since the $(D/H)_{H_2}$ ratio is observed to be high (as was discussed in Chapter 2). The case for Neptune is similar although initial estimates of the He/H_2 ratio were supersolar, which was very puzzling. It may be possible that the vigorous convection in Neptune's atmosphere, consistent with its high internal heat flux, brings substantial quantities of N_2 to the observable atmosphere, which has a substantial effect on the He/H_2 ratio detection technique employed by *Voyager 2*. As little as 0.3% abundance of N_2 reduces the estimated He/H_2 ratio to that of Uranus (as we saw in Chapter 2). However, an alternative explanation is that the supersolar estimate of the He/H_2 ratio was based on far-IR absorption data that have recently been revised for the low temperatures found in Uranus' atmosphere (Orton *et al.*, 2007b). Using these revised data, the far-IR spectrum of Uranus is consistent with the expected solar He/H_2 ratio.

Although Uranus' internal heat source appears very low, and thus vertical convection is presumably sluggish, the atmosphere is still very dynamic and the circulation efficiently redistributes relatively warm polar air over the planet. During the *Voyager 2* flyby, Uranus was close to northern winter solstice with the South Pole permanently in sunlight and the North Pole permanently in darkness. If there were no atmospheric motion then we would have expected the South Pole to become very much warmer than the North by as much as 7 K. However, *Voyager 2* found almost no temperature difference at the 0.5 bar to 1 bar pressure level indicating a very dynamic circulation. The dynamics of Neptune's atmosphere were observed by *Voyager 2* to be even more vigorous with winds approaching speeds of 400 m/s, as we shall see in Chapter 5. Subsequent observations of these planets from the ground and from Earth-orbiting telescopes have confirmed the dynamic nature of these planetary atmospheres.

3.6 BIBLIOGRAPHY

Andrews, D.G. (2000) *An Introduction to Atmospheric Physics*. Cambridge University Press, Cambridge, U.K.

Atreya, S.K. (1986) *Atmospheres and Ionospheres of the Outer Planets and Their Satellites*. Springer-Verlag, Berlin.

Chamberlain. J.W. and D.M. Hunten (1987) *Theory of Planetary Atmospheres: An Introduction to Their Physics and Chemistry* (Second Edition). Academic Press, San Diego, CA.

Houghton, J.T. (1986) *The Physics of Atmospheres* (Second Edition). Cambridge University Press, Cambridge, U.K.

Shirley, J.H. and R.W. Fairbridge (Eds.) (1997) *Encyclopaedia of the Planetary Sciences*. Chapman & Hall, London.

Weissman, P.R., L.-A. McFadden, and T.V. Johnson (Eds.) (1998) *Encyclopaedia of the Solar System*. Academic Press, San Diego, CA.

4

Vertical structure of temperature, composition, and clouds

4.1 PRESSURE AND TEMPERATURE PROFILES

Much of what we know about the giant planets comes from observations of the conditions in the upper parts of their molecular-hydrogen/helium envelopes. Remote-sounding and *in situ* observations of the vertical profiles of pressure, temperature, and composition provide constraints on vertical motion, mixing, heat sources, deep composition, and photochemistry. These observations are complemented by observations of vertical cloud structure.

4.1.1 Pressure

In planetary atmospheres, where vertical wind velocities are generally very much less than horizontal wind velocities, the assumption of hydrostatic equilibrium is a very good one. Thus, the vertical pressure difference dp across a slab of air at altitude z of density ρ and thickness dz subject to a gravitational acceleration g is

$$dp = -\rho g \, dz. \tag{4.1}$$

Assuming the air behaves as an ideal gas, the density may be determined from temperature T and pressure p as

$$\rho = \frac{\bar{m}p}{RT} \tag{4.2}$$

where R is the molar gas constant; and $\bar{m}$ is the mean molecular weight of the atmosphere. Substituting for ρ in Equation (4.1) and integrating (assuming T is constant with height) we obtain

$$p = p_0 e^{-z/H} \tag{4.3}$$

Table 4.1. Mean pressure/temperature properties of the giant planet atmospheres.

Planet	*Location*	T (K)	c_p ($\mathrm{J\,kg^{-1}\,K^{-1}}$)	g ($\mathrm{m\,s^{-2}}$)	Γ_d ($\mathrm{K\,km^{-1}}$)	H (km)
Jupiter	Equator (1 bar)	163	10,998	23.1	2.10	25.5
	Pole (1 bar)	163	10,998	26.9	2.45	21.9
Saturn	Equator (1 bar)	135	10,658	9.1	0.85	54.5
	Pole (1 bar)	135	10,658	12.1	1.13	41.0
Uranus	Equator (1.5 bar)	87.5	8,643	8.7	1.01	32.8
	Pole (1.5 bar)	87.5	8,643	8.9	1.03	32.3
Neptune	Equator (1.5 bar)	107.5	8,187	11.0	1.34	29.1
	Pole (1.5 bar)	107.5	8,187	11.4	1.39	28.0

where p_0 is the pressure at altitude $z = 0$; and $H = RT/\bar{m}g$ is known as the *scale height*. As we saw in Chapter 2, the giant planets are significantly oblate and rapidly rotating. Thus net gravitational acceleration (which includes centripetal acceleration) varies significantly with latitude as does the scale height. This can be seen in Table 4.1.

4.1.2 Temperature

Lower troposphere

In the lower troposphere of the outer planets, air is heated from the interior. If the infrared (IR) optical depth to space is high, then heat may not escape radiatively and instead the air rises convectively in order to transfer the heat upwards. This region of the atmosphere is called the troposphere (after the Greek word *tropo* for "turning"). As the air parcels rise they expand and, assuming negligible thermal contact with neighboring air masses, they cool adiabatically to give a temperature profile that may be calculated as follows. Consider a parcel of air moving vertically containing one mole of gas. If we assume that there is no net exchange of heat between the parcel and its surroundings, then the expansion may be considered to be adiabatic and reversible, and hence isentropic: that is,

$$dS = \left(\frac{\partial S}{\partial T}\right)_p dT + \left(\frac{\partial S}{\partial p}\right)_T dp = 0 \tag{4.4}$$

or

$$dS = C_p \frac{dT}{T} - \left(\frac{\partial V}{\partial T}\right)_p dp = 0 \tag{4.5}$$

where C_p is the molar heat capacity at constant pressure of the parcel; S is the entropy of the parcel; and where the second partial derivative in Equation (4.4) has been replaced using one of Maxwell's relations (e.g., Finn, 1993). Assuming the gas is ideal, we may calculate $\partial V/\partial T$ from the ideal gas equation $pV = RT$ and derive

$$C_p \frac{dT}{T} = \frac{R}{p} dp \tag{4.6}$$

where R is again the molar gas constant. Substituting for dp from the hydrostatic equation (Equation 4.1) we find

$$C_p \frac{dT}{dz} = -\bar{m}g \tag{4.7}$$

where $\bar{m}$ is the mean molecular weight of the atmosphere and thus

$$\frac{dT}{dz} = -\frac{\bar{m}}{C_p} g = -\frac{g}{c_p} = -\Gamma_d \tag{4.8}$$

where c_p is the *specific* heat capacity at constant pressure ($\mathrm{J\,kg^{-1}\,K^{-1}}$); and Γ_d is called the *dry adiabatic lapse rate* (DALR).

In parts of the atmosphere where the air contains volatiles, cooling caused by upward motion of air parcels causes the condensation of cloud particles and the subsequent release of latent heat, which causes the temperature to drop more slowly with height than expected for a dry atmosphere. For air at temperature T, where several gas components are saturated, the saturated adiabatic lapse rate (SALR) may be shown to be (Andrews, 2000; Atreya, 1986)

$$\Gamma_s = -\frac{g}{c_p} \frac{\left[1 + \frac{1}{RT} \sum L_i x_i\right]}{\left[1 + \frac{1}{\bar{m} c_p T^2} \sum \frac{L_i^2 x_i}{R}\right]} \tag{4.9}$$

where L_i is the *molar* latent heat of vaporization (or sublimation) of the ith condensing component; R is the molar gas constant; $\bar{m}$ is the mean molecular weight of the air (excluding condensates); and x_i is the saturated volume mixing ratio (or *mole fraction*) of the ith component defined as $x_i = p_i/p$, where p is the total pressure and p_i is the saturation vapor pressure of the constituent at the local temperature. In the terrestrial atmosphere, Γ_s is usually significantly less than Γ_d (e.g., for the Earth, $\Gamma_s = 6$ K/km to 9 K/km, while $\Gamma_d \sim 9.8$ K/km). For Jupiter and Saturn, the difference is usually small since the volume mixing ratios of condensing species is small. However, for Uranus and Neptune the abundance of condensing species is much higher and thus significant differences between Γ_d and Γ_s exist at levels of major cloud formation. A major difference between giant planet atmospheres and the terrestrial atmosphere is that condensing species in the giant planet atmospheres (H_2O, NH_3, CH_4, etc.) are all *heavier* than the bulk of the hydrogen–helium air. The opposite is true for the Earth's atmosphere where water vapor is lighter than the bulk of the nitrogen–oxygen atmosphere. As a result "moist" air is naturally buoyant in the terrestrial atmosphere and tends to rise, whereas "moist" air is naturally dense in giant planet atmospheres and tends to sink. Hence, while clouds may form in the Earth's atmosphere simply by moist, buoyant air rising to its condensation level, the formation of clouds in the giant planet atmospheres is more complicated. Another difference is that c_p, and thus Γ_d, can vary significantly with height, especially at cold temperatures where, as we will see in Section 4.1.3, the heat capacity is a strong function of temperature and of the ortho:para ratio of molecular hydrogen. This effect is difficult to observe in the warmer tropospheres of Jupiter and Saturn, but is significant in the cooler atmospheres of Uranus and Neptune. These latter planets have the additional complication that methane, which has a significant abundance, freezes out at roughly the 1 bar level. This leads to significant changes in both c_p and the mean molecular weight of the atmosphere at this level.

In addition to single condensate clouds, two-component clouds may also form in the atmospheres of the giant planets, of which the most important is probably solid ammonium hydrosulfide NH_4SH, which may form by the reaction

$$NH_3 + H_2S \rightleftharpoons NH_4SH(s). \tag{4.10}$$

The formation of the cloud releases additional heat that affects the lapse rate, and it may be shown that the equation for the saturated adiabatic lapse rate must be modified to (Atreya, 1986)

$$\Gamma_s = -\frac{g}{c_p}\frac{\left[1+\dfrac{1}{RT}\left(\sum L_i x_i + \dfrac{2(x_{NH_3}\cdot x_{H_2S})}{(x_{NH_3}+x_{H_2S})}L_{NH_4SH}\right)\right]}{\left[1+\dfrac{1}{\bar{m}c_p T^2}\left(\sum\dfrac{L_i^2 x_i}{R}+\dfrac{2(x_{NH_3}\cdot x_{H_2S})}{(x_{NH_3}+x_{H_2S})}L_{NH_4SH}\cdot B_{NH_4SH}\right)\right]} \tag{4.11}$$

where x_{NH_3} is the saturated partial pressure of ammonia above solid NH_4SH particles; x_{H_2S} is the saturated partial pressure of hydrogen sulfide above the same particles; the molar heat of formation $L_{NH_4SH} = 46{,}025.0$ J/mol; and B_{NH_4SH} is a constant of the reaction rate equation (see Section 4.3.3) whose value is 10,833.6 K^{-1}.

Upper troposphere and stratosphere

At higher levels in the atmosphere, the opacity of the overlying air becomes progressively smaller and thus radiation becomes more efficient than convection at transporting heat. Hence, at these levels the temperature profile is determined by radiative equilibrium. Consider a very thin layer of opacity ε high in the atmosphere with negligible atmospheric opacity above it. The atmosphere below this layer effectively emits as a black body of temperature equal to the bolometric temperature T_B. Hence, the heating rate of the thin layer per unit area is simply εT_B^4. However, heat from this layer may be emitted both upwards and downwards and thus in equilibrium; assuming Kirchoff's law, we find

$$\varepsilon\sigma T_B^4 = 2\varepsilon\sigma T_S^4. \tag{4.12}$$

Thus the limiting temperature of the thin slab, known as the *stratospheric temperature*, is given by

$$T_S = \frac{T_B}{2^{1/4}} \approx 0.841 T_B. \tag{4.13}$$

With more detailed analysis it can be shown that the temperature in the upper atmosphere should tend gradually to this stratospheric temperature via the Milne–Eddington equation (Atreya, 1986)

$$T_S^4 = \frac{T_B^4}{4}(2 + 3\tau) \tag{4.14}$$

where τ is the IR optical thickness between the layer and space, known as the IR *optical depth*. Calculating $T(\tau)$ from Equation (4.14) we find that the rate of increase of temperature with depth as we go down through the atmosphere is initially small and much less than the DALR. Hence, the upper atmosphere is convectively stable since parcels displaced vertically will tend to return to their original altitudes. However, since τ increases quickly with depth in the atmosphere, the radiative equilibrium vertical temperature gradient increases rapidly until at some point it exceeds the dry adiabatic lapse rate. Such a temperature gradient is highly unstable and convection in the atmosphere quickly reduces the temperature gradient to the DALR. The boundary between the convective and radiative regions is known as the *radiative–convective boundary*.

According to Equation (4.14), the temperature in the upper atmosphere should tend to a constant value in the absence of other sources of heat. However, when we look at the temperature profiles of the planets it is found that the temperatures decrease with height to roughly the stratospheric temperature at a certain level known as the *tropopause* (after the Greek words for "turning" and "stop") and then increase again. The region above the tropopause, where temperature increases with height, is very stable to convection and is known as the *stratosphere*, since the air forms stably stratified layers. The region between the tropopause and the radiative–convective boundary is known as the *upper troposphere*. This is actually a slight misnomer since the lapse rate in this region is less than the DALR and the region is thus also stable to convective overturning, although turbulent overturning is important. [NB: Eddy mixing, discussed in Section 4.2 is found to be a minimum at the tropopause so this is perhaps not such a misleading word after all!] Unforced convective overturning only occurs in the atmosphere below the radiative–convective boundary, in the *lower troposphere*.

The increase of temperature with height in the stratosphere implies the presence of additional energy sources. These sources include

(1) absorption of ultraviolet radiation from the Sun via gaseous photodissociation reactions;
(2) absorption of sunlight by stratospheric aerosols; and
(3) absorption of near-IR sunlight by methane gas absorption bands.

To achieve thermal balance these sources of energy must be transported or radiated away. The stratospheres of all the giant planets appear to be close to radiative equilibrium and the cooling of the lower stratosphere appears to be due mainly to the thermal emission from ethane and acetylene molecules, with the upper stratosphere cooled by methane emission in the case of Jupiter and Saturn (Yelle *et al.*, 2001). Acetylene and ethane are photochemical products derived from the photolysis of methane and are observed in the stratospheres of all the giant planets. Thus, the temperature structure in the stratospheres of the giant planets depends critically on the vertical distribution of photochemical products in the same way that the stratospheric temperature profile of the Earth depends critically on the abundance of another photochemical product, ozone. The stratospheres of Mars and Venus, which do not contain significant quantities of photochemical products such as ozone, do not have nearly so well a defined tropopause. Hence, in a peculiar way the terrestrial stratosphere has, in this sense, more in common with the stratospheres of the giant planets than it does with the stratospheres of the other terrestrial planets (Yelle *et al.*, 2001)! From Figure 4.1 it can be seen that the stratospheric temperatures of all the giant planets are rather similar, which is puzzling considering that the solar flux at Neptune is 33 times smaller than at Jupiter (Table 3.1). Chamberlain and Hunten (1987) note that this may be due to the positioning of the main ethane and acetylene absorption bands with respect to the Planck function at the stratospheric temperature. Although these gases are not very abundant in the stratosphere, their absorption bands overlap with the Planck function much better than the main ν_4 methane band at 7.7 μm and this is why stratospheric temperatures depend so critically on their abundances. At 150 K, the Planck function peaks at 19 μm, and only slightly overlaps with the main acetylene band at 13.7 μm and even less with

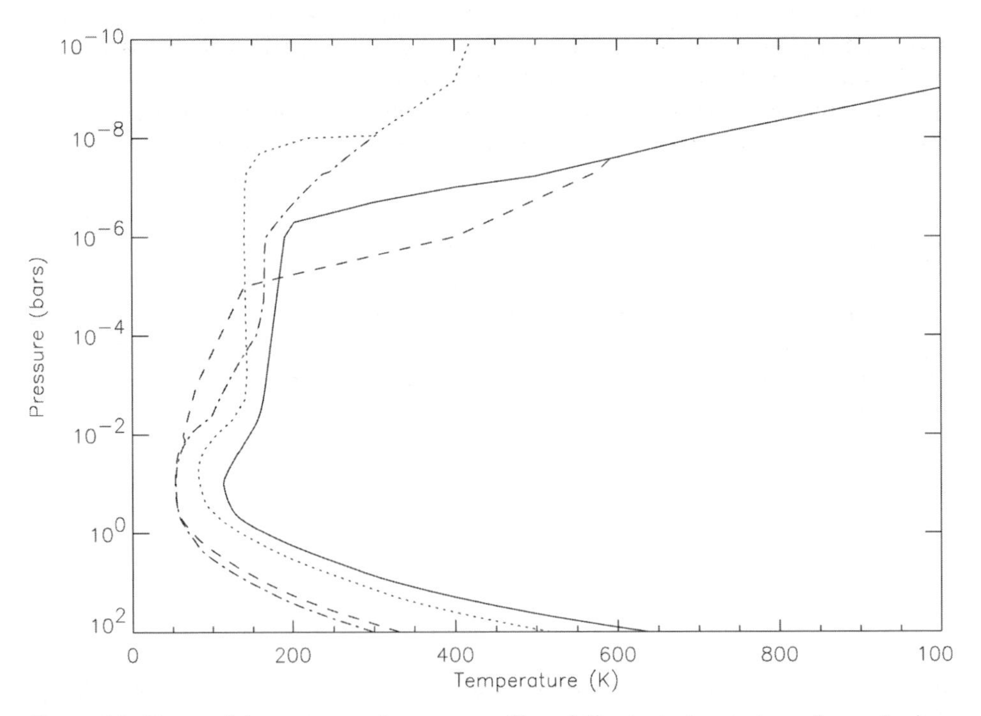

Figure 4.1. Equatorial temperature/pressure profiles of the giant planet atmospheres: Jupiter, solid line; Saturn, dotted line; Uranus, dashed line; Neptune. dot-dashed line.

the main ethane band at 12.2 μm. If the stratosphere cools, this overlap rapidly disappears, effectively shutting off the cooling to space by ethane and acetylene. Hence the stratospheric temperatures of the giant planets would appear to be almost "thermostatted" by the presence of these molecules.

At even higher altitudes, significant ionization (and thus heating) of the atmosphere is allowed via solar extreme-ultraviolet (EUV) photons and bombardment with charged particles from the solar wind and cosmic rays. The breaking of gravity waves propagating vertically from the troposphere may also be important. A description of the ionospheres of the outer planets is beyond the scope of this book, but is well described by Atreya (1986).

4.1.3 Secondary effects on temperature/pressure profiles

Heat capacity and ortho-hydrogen/para-hydrogen

We saw earlier that the lapse rate in the troposphere depends on c_p and g. The gravitational acceleration g is a function of latitude on all the giant planets, while c_p is dominated by the heat capacity of both helium (which is monatomic and thus has only the three translational degrees of freedom leading to $C_p = 2.5R\,\mathrm{J\,mol^{-1}\,K^{-1}}$, where $C_p = C_v + R$) and molecular hydrogen, which is a diatomic molecule and thus also has rotational degrees of freedom. For Jupiter and Saturn the contribution of other molecules to mean heat capacity is negligible. However, for Uranus and Neptune the heat capacity of methane and water vapor are also important considerations below their respective condensation levels.

The rotation of linear molecules may be modeled by the motion of a rigid rotator. From quantum mechanics (e.g., Rae, 1985) the allowed energy levels of such a rotator are

$$E_l = \frac{\hbar^2}{2I} l(l+1) \tag{4.15}$$

where l is an integer; I is the moment of inertia; and $\hbar$ is Planck's constant divided by 2π. Equation (4.15) may conveniently be re-expressed as $E_l = l(l+1)k_B\Phi_R$, where k_B is the Boltzmann constant, and $\Phi_R = \hbar^2/2k_B I$ is known as the *rotational temperature*. These rotational energy levels are degenerate with a degeneracy factor $g_l = (2l+1)$, and thus the rotational partition function is

$$Z_{\text{rot}} = \sum_{l=0,1,\ldots} (2l+1)\exp[-l(l+1)\Phi_R/T] \tag{4.16}$$

and the contribution of rotational energy to the molar internal energy U_{rot} is

$$U_{\text{rot}} = RT^2 \frac{d(\ln Z_{\text{rot}})}{dT}. \tag{4.17}$$

Assuming that the rotational energy of diatomic molecules is independent of their translational energy, the contribution to heat capacity at constant volume, $C_v = dU/dT$, may be simply added to the translational heat capacity at constant volume $C_v = 3R/2$. Using this expression for the partition function we can calculate

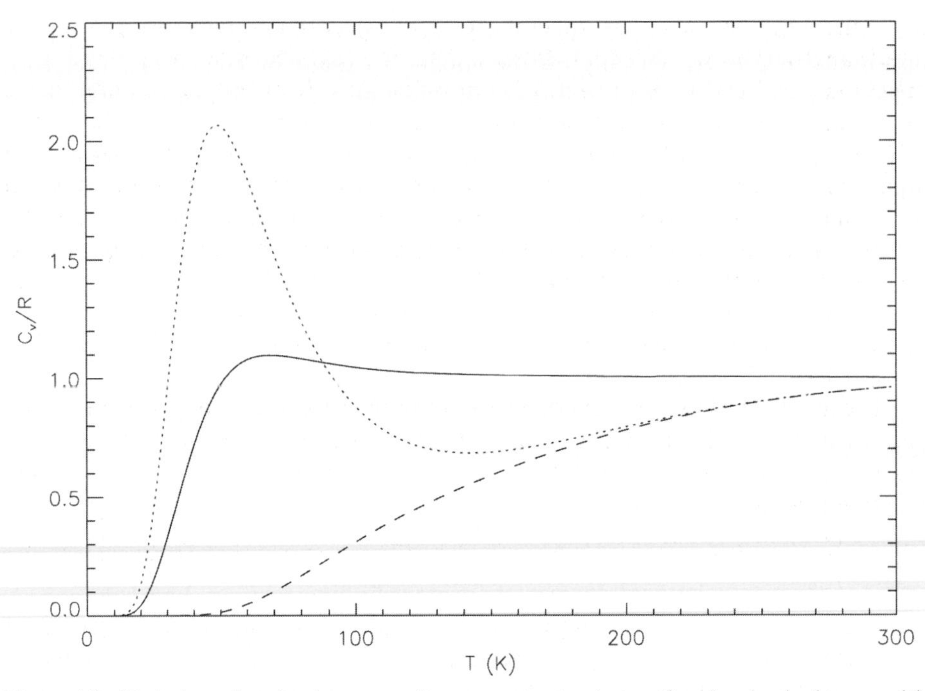

Figure 4.2. Variation of molar heat capacity at constant volume of molecular hydrogen with temperature. Solid line is the expected curve from using the expression for the partition function of Equation (4.16). The dotted line is the heat capacity calculated from the revised partition function of Equation (4.18). This is the curve measured in the presence of a catalyst. For cases where ortho:para equilibration is so slow that the ortho:para hydrogen ratio may be considered to be frozen at its high-temperature 3:1 ratio, the dashed line is obtained from Equation (4.19), in good agreement with measurement.

the rotational heat capacity as a function of temperature for H_2 (where $\Phi_R = 85\,\mathrm{K}$) shown in Figure 4.2 (solid line). As can be seen the rotational contribution to molar heat capacity tends to R ($\mathrm{J\,mol^{-1}\,K^{-1}}$) for temperatures well above the rotational temperature and tends to zero for $T \ll \Phi_R$. Although the predicted heat capacity using this formula agrees well with the measured rotational heat capacity for linear molecules such as CO and NO (adjusting the rotation temperature appropriately), something strange is found to happen for molecular hydrogen. If hydrogen at room temperature is cooled fairly rapidly and its heat capacity measured as a function of temperature, the dashed curve of Figure 4.2 is obtained, which is clearly rather different from that expected. Even more puzzling is that if this experiment is instead done very slowly, or in the presence of a catalyst such as activated charcoal, the dotted curve is obtained. What can be going on?

Although the theory outlined above is satisfactory for heteronuclear diatomic molecules such as CO and NO, it is not applicable for homonuclear diatomic molecules, such as H_2, where the two nuclei are identical. The nuclei of the H_2 molecule are protons, which are fermions and must thus be described by antisymmetric wavefunctions. Thus, the wavefunction of the molecule must change if the two protons are interchanged. The wavefunction may in fact be separated into the product of a "rotation part" and a "spin part". The rotation part describes the rotation of the two nuclei round each other, and it is found that the states $l = 0, 2, 4, \ldots$ have even exchange parity, while the states $l = 1, 3, \ldots$ have odd exchange parity. Hence, the even, symmetric rotation states must have odd, antisymmetric spins with total spin $S = 0$ in order for the total wavefunction to be antisymmetric, and likewise the odd, asymmetric rotational states must have even, parallel spins with total spin $S = 1$. Hydrogen molecules with spins antiparallel ($S = 0, l = 0, 2, 4, \ldots$) are known as para-hydrogen, while hydrogen molecules with spins parallel ($S = 1, l = 1, 3, \ldots$) are known as ortho-hydrogen. The $S = 0$ state can be shown to be a singlet state, while the $S = 1$ state is a triplet state.

Because there is such a fundamental difference between ortho-hydrogen and para-hydrogen, a simple summing over the rotational states in Equation (4.16) is incorrect. Instead, assuming that the hydrogen is in thermal equilibrium, the partition function is actually given by

$$Z_{\mathrm{rot}} = \sum_{l_{\mathrm{even}}} (2l+1) \exp[-l(l+1)\Phi_R/T] + 3 \sum_{l_{\mathrm{odd}}} (2l+1) \exp[-l(l+1)\Phi_R/T]. \quad (4.18)$$

The heat capacity derived from this partition function accurately fits the dotted curve of Figure 4.2, where hydrogen is slowly cooled in the presence of a catalyst to ensure equilibrium. What can be said about the remaining curve, however? It turns out that conversion between ortho-hydrogen and para-hydrogen via collision processes is actually quite difficult since to change l by ± 1 requires that the total spin is also changed, which is not easy unless there is a third body, such as a surface, to take away the spin. Hence, ortho-hydrogen and para-hydrogen behave almost as different gases and may be treated quasi-separately. At high temperatures, the sums in Equation (4.18) are both equal, and thus high-temperature ($T \gg \Phi$), thermally equilibrated hydrogen has an ortho:para ratio of 3:1, or equivalently a para-H_2 fraction $f_p = 0.25$. Room temperature is "high" in this case. The heat capacity of hydrogen that is cooled rapidly enough such that negligible ortho–para conversion occurs will have a heat capacity given by

$$C_v = 0.75 C_{v,\mathrm{ortho}} + 0.25 C_{v,\mathrm{para}} \quad (4.19)$$

where the ortho part is obtained from the rigid rotor partition function (Equation 4.16) summed only over odd l, and the para part is obtained also from Equation (4.16) by summing only over even l, and this then accurately models the dashed curve in Figure 4.2. In the giant planet atmospheres, air upwelling from the deep interior will have the deep ortho:para ratio of 3:1 and this ratio will slowly change as ortho–para conversion proceeds at a rate governed mainly by the availability of aerosol surfaces to exchange spin angular momentum with (Fouchet *et al.*, 2003; Massie and Hunten, 1982). Hence, measurement of the ortho:para ratio in the giant planet atmospheres provides information on the rate of vertical upwelling and on the presence of catalytic aerosol surfaces. Observations of this fraction will be discussed later. An additional effect of so-called "lagged" ortho–para conversion is that the latent heat release can act to stabilize the vertical profile of temperature and thus inhibit convection.

Molecular weight

The scale height of an atmosphere depends on temperature, gravity, and also mean molecular weight. If no condensation occurs and the atmosphere is well-mixed then the molecular weight remains constant. However, the condensation of non-negligible constituents such as water vapor or, for Uranus and Neptune, methane, leads to significant vertical variation of molecular weight and thus density and scale height.

4.1.4 Temperature/pressure profiles of the outer planets

The temperature/pressure profiles observed and inferred in the upper parts of the outer planet atmospheres in the equatorial regions were shown previously in Figure 4.1. The stratospheric temperatures calculated from Equation (4.13) are 104.6 K, 79.9 K, 49.7 K, and 49.9 K, respectively, for Jupiter, Saturn, Uranus, and Neptune, and it can be seen that the observed tropopause temperatures are indeed close to these values. Below the radiative–convective boundary, the temperature profiles are observed to follow an adiabatic lapse rate, and in this figure the SALRs have been assumed. For Jupiter and Saturn it is not possible to determine the hydrogen ortho:para ratio from the measured lapse rate since the heat capacities (and hence lapse rates) of both equilibrium and frozen (3:1) hydrogen are indistinguishable from the measured lapse rate at the observable tropospheric temperatures of these planets (Figure 4.2). However, Uranus and Neptune are much colder and so there is a big difference between the two calculated lapse rates. The atmospheric circulation of Uranus is apparently very sluggish and, as we shall see in Section 4.4.3, the ortho:para ratio is found to be roughly in equilibrium. However, the observed lapse rate is more consistent with the frozen 3:1 ortho:para ratio. It has been suggested that the two observations may be compatible if in general the ortho:para ratio is equal to the equilibrium value at a particular level, but vertical displacements are small and rapid enough such that negligible ortho:para redistribution occurs during the motion of individual parcels. Thus

$$C_v = (1 - f_{\text{eqm}})C_{v,\text{ortho}} + f_{\text{eqm}} C_{v,\text{para}} \tag{4.20}$$

where

$$f_{\text{eqm}} = \frac{\sum_{\text{even}} (2l+1)\exp[-l(l+1)\Phi_R/T]}{\sum_{\text{even}} (2l+1)\exp[-l(l+1)\Phi_R/T] + 3\sum_{\text{odd}} (2l+1)\exp[-l(l+1)\Phi_R/T]}. \tag{4.21}$$

This "intermediate" hydrogen (de Pater and Massie, 1985) has a very similar heat capacity to "frozen" 3:1 ortho:para ratio hydrogen and may thus explain why the observed lapse rate on Uranus is more consistent with "frozen" ortho–para hydrogen than with equilibrium ortho–para hydrogen.

All the planets have clearly defined upper tropospheres, which start at approximately 500 mbar to 600 mbar for Jupiter, 400 mbar to 500 mbar for Saturn, and 1 bar to 2 bar for both Uranus and Neptune. The tropopauses occur at 100 mbar, 60 mbar, 100 mbar, and 50 mbar, respectively. The stratospheric temperatures in Saturn's atmosphere are generally lower than those found in Jupiter's, which might be expected from Saturn's increased distance from the Sun. However, the stratospheric temperatures of Uranus and Neptune are noticeably and puzzlingly different. Although the temperature profiles of these planets is similar in the lower troposphere, the stratospheric temperatures in the 10 mbar to 0.01 mbar pressure region is of the order of 40 K warmer in Neptune's atmosphere than in Uranus. The difference may be due to the abundance of stratospheric hazes (see Section 4.4.4), which perhaps absorb more sunlight in Neptune's atmosphere than in Uranus'. Some absorption of solar radiation by stratospheric aerosols in Uranus' atmosphere does occur, however, leading to the small temperature maxima observed by *Voyager 2* radio occultation in Uranus' atmosphere near 10 mbar. Neptune's stratosphere may also be warmer than Uranus' due to the higher abundance of stratospheric methane in Neptune's atmosphere, which absorbs sunlight in the visible and near-IR.

4.2 VERTICAL MIXING–EDDY MIXING COEFFICIENTS

In Chapter 2 we outlined the bulk composition of the outer gaseous envelopes of the giant planets. These compositions refer to the "deep" atmosphere, although for Jupiter and Saturn this refers to pressures only up to 10 bar to 20 bar, and up to approximately 100 bar for Uranus and Neptune. In the upper, observable parts of the atmosphere, the composition of certain gas species vary as a function of height due to processes such as photochemistry and condensation. As we have seen in Section 4.1.3 for ortho-hydrogen/para-hydrogen the composition is also a function of the rate at which air is uplifted from the warm interior since if the upwelling is rapid enough, non-equilibrium "quenched" molecules may be present. Hence, the rate of vertical motion has a major effect on the vertical profiles of composition, and also cloud structure.

Air parcels may be transported vertically in atmospheres by three main mechanisms: (1) convection; (2) atmospheric waves; and (3) turbulence. To understand how these processes affect the measured abundance profiles, consider the continuity equation in the vertical direction (Atreya, 1986; Yung and DeMore, 1999) for a certain gas species which has a number density n_i

$$\frac{\partial n_i}{\partial t} + \frac{\partial \phi_i}{\partial z} = P_i - L_i \tag{4.22}$$

where ϕ_i is the vertical flux of molecules; P_i is the chemical production rate; and L_i is the chemical loss rate. If the mean vertical wind speed w is known at a certain location, then the vertical flux of molecules is simply $\phi_i = n_i w$. For the organized belt/zone circulation of Jupiter, where zones are interpreted as regions of general upwelling and belts are regions of general subsidence, the mean vertical wind may sometimes be calculated from departures of the temperature from the radiative equilibrium temperature (Gierasch *et al.*, 1986; Conrath *et al.*, 1998). However, even with these general flows, there are superimposed smaller convective events, such as isolated convective plumes observed in Jupiter's North Equatorial Belt, and a good deal of turbulence that tends to mix the air vertically, even in regions which from dynamical models (Chapter 5) appear to be regions of general uplift and subsidence only. Hence, to understand the effects of these processes on the vertical abundance profiles, we need a more general way to parameterize the vertical flux of molecules.

According to *Prandtl's mixing length theory* (Holton, 1992), a parcel of air displaced vertically will carry the mean abundances of its original level for a characteristic distance l' analogous to the mean free path in molecular diffusion. This displacement will create a turbulent fluctuation in the composition of the new level, whose magnitude will depend on l' and on the vertical gradient of the mean composition. Thus, the process is very similar to molecular diffusion and may be modeled in the same way if we define an eddy-mixing coefficient K, analogous to the molecular diffusion coefficient D. Using such a model the vertical flux of species i may be calculated as being due both to molecular and eddy diffusion via an equation very similar to Equation (3.17), presented on p. 63

$$\phi_i = n_i\left[-D_i\left(\frac{1}{n_i}\frac{\partial n_i}{\partial z}+\frac{1}{H_i}+\frac{1}{T}\frac{\partial T}{\partial z}\right)-K\left(\frac{1}{n_i}\frac{\partial n_i}{\partial z}+\frac{1}{H_a}+\frac{1}{T}\frac{\partial T}{\partial z}\right)\right] \tag{4.23}$$

where D_i is the molecular diffusion coefficient of the ith component; T is temperature; H_i and H_a are the pressure scale heights of the ith species and bulk atmosphere, respectively ($H_i = RT/m_i g, H_a = RT/\bar{m}g$); and K is the eddy diffusion coefficient determined via observation. When molecular diffusion dominates, each species tends to its own profile with its own scale height. When eddy diffusion dominates, the gas is well-mixed and has a single bulk scale height. The level where molecular diffusion becomes dominant is called the *homopause*.

In the stratospheres of the outer planets it is thought that the principal mechanism for eddy mixing is the dissipation and break-up of vertically propagating gravity waves. The kinetic energy of non-dissipative waves should remain constant and this is, by simple analysis, proportional to ρa_v^2, where ρ is the density and a_v is the amplitude of the velocity. Thus the amplitude of these waves should be roughly proportional to $\rho^{-0.5}$ or equivalently $n^{-0.5}$, where n is the number density (Andrews *et al.*, 1987). Hence, the amplitude of these waves grows rapidly with height and eventually becomes so high that the wave is unstable to either convective instability or shear instability. This leads to wave "breaking" and turbulent mixing of the air. Such waves are generated in the troposphere due to convective turbulence, or for the terrestrial planets from deflection of air over surface features. Small-amplitude waves clearly travel higher into the stratosphere before breaking than high-amplitude waves and thus waves break at a range of altitudes in the stratosphere, providing a source of eddy mixing throughout the region. Gravity waves will be discussed in more detail in Chapter 5. By mixing length theory (Holton, 1992) the eddy-mixing coefficient should be proportional to the mean mixing length and thus we expect that K is proportional to the amplitude of breaking gravity waves and thus $K \propto n^{-0.5}$. In the troposphere,

Table 4.2. Measured estimates of the eddy-mixing coefficient in the giant planet atmospheres.

lanet	*K* ($cm^2 s^{-1}$)	*Density* (cm^{-3})	*Pressure* (bar)	*Altitude* (km)
upiter	$1.4^{+0.8}_{-0.7} \times 10^6$	1.4×10^{13}	10^{-6}	385 (homopause)
	4.6×10^2	9.3×10^{18}	0.14	40 (troposphere)
	1.5×10^4	2.4×10^{19}	0.42	20 (troposphere)
aturn	$1.7^{+4.3}_{-1.0} \times 10^8$	1.2×10^{11}	5×10^{-9}	1,140 (homopause)
	$\leq (3.5-8.7) \times 10^3$	1.1×10^{19}	0.13	80 (troposphere) (40°N, 12°N)
Jranus	$(5–10) \times 10^3$	$2–1 \times 10^{15}$	$4–2 \times 10^{-5}$	354–390 (homopause)
Jeptune		$\sim 10^{13}$	2×10^{-7}	585–610 (homopause)

apted from Atreya *et al.* (1999).

convective processes dominate and thus it is assumed that K is either roughly a constant, or increases linearly with density. Hence, the functional dependence of $K(z)$ in the atmosphere is usually parameterized as

$$K(z) = K_H \left[\frac{n_H}{n(z)}\right]^{\gamma}, \qquad P \leq P_T, \tag{4.24}$$

$$K(z) = K_T(n/n_T) \text{ or } K(z) = K_T, \quad P > P_T \tag{4.25}$$

where P_T is the pressure at the tropopause; $n(z)$ is the total number density; n_H is the number density at the homopause; K_H is the associated eddy-mixing coefficient; n_T is the number density at the tropopause; and γ is a coefficient close to 0.5 (Atreya, 1986). Note that this formulation allows for a discontinuity in $K(z)$ at the tropopause. Typical estimated values of $K(z)$ at different levels for the outer giant planets are listed in Table 4.2.

From Chapter 3 we saw that the molecular diffusion coefficient may be calculated as

$$D_i = \frac{1}{3\sqrt{2}n\sigma_i}\bar{c}_i \tag{4.26}$$

and thus D_i varies with density as n^{-1}. However, since K varies typically as $n^{-0.5}$ it can be seen that, although K dominates at higher pressures in the *homosphere*, it increases more slowly with height than D_i and hence at some altitude for a given species, D_i becomes greater and dominates at all higher altitudes. It should thus be noted that the homopause level, where $D_i = K$, is actually dependent on the molecule under consideration. For the giant planets, it is presumed that, unless stated otherwise, the word homopause refers to the *methane* homopause. The time constants for reaching diffusive equilibrium by the two processes are $\tau_D = H_a^2/D_i$ and $\tau_K = H_a^2/K$, respectively, via reasoning introduced in Section 3.2.2.

The effect of eddy mixing on vertical profiles may be conveniently explained with the example of ammonia on Jupiter and Saturn. Ammonia should condense at approximately 700 mbar for Jupiter and 1.8 bar for Saturn and above that, in the absence of eddy mixing, the partial pressure initially follows the saturated vapor pressure curve. At higher altitudes, solar ultraviolet (UV) radiation photolyzes ammonia and leads to an even greater rate of decrease of abundance with height. If the eddy-mixing coefficient is high, then more ammonia would be expected at the photolysis altitudes because fresh ammonia is transported there faster than it can be photolyzed. Conversely if the eddy-mixing coefficient is low, then very little ammonia is expected at the photolysis altitudes since the new ammonia will be photolyzed as fast as it can arrive there. Hence, by measuring the ammonia mixing ratio profile, we can estimate the value of $K(z)$ in the upper troposphere. Another good indicator of the upper tropospheric eddy-mixing coefficient for Jupiter and Saturn is the vertical profile of phosphine, which is also photolyzed in the region of the tropopause. The profiles of ammonia and phosphine may be estimated from thermal-IR and near-IR remotely sensed measurements for Jupiter and Saturn, but not for Uranus and Neptune, where abundances are too low. At higher altitudes, methane photochemistry becomes important, which produces hydrocarbons such as ethane and acetylene. Measurements of the vertical distribution of these gases may be derived from thermal-IR measurements for *all* of the giant planets and used to infer the value of $K(z)$ at these altitudes. Finally, estimates of the value of $K(z)$ at the methane homopause may be made from observations at two UV wavelengths (Atreya, 1986): Lyman-α (1,216 Å), and helium 584 Å. The Jovian planets are relatively bright at Lyman-α wavelengths, primarily through the resonance scattering of incident sunlight by hydrogen atoms. Methane is a strong UV absorber and hence only those hydrogen atoms that are above the methane homopause level may contribute to this radiation. Hence, for atmospheres with large eddy diffusion coefficient (high homopause) fewer hydrogen atoms may contribute to the resonance scattering and thus the measured intensity is less. Similarly helium atoms in the Jovian planets resonantly scatter photons at 584 Å. However, in this case as $K(z)$ increases, the measured intensity increases (rather than decreases as it does for Lyman-α) since the abundance of helium at the top of the homosphere is greater. However, to interpret the He 584 Å observations independent measurements of the temperature in the scattering region are also required, which are difficult to estimate.

A combination of many of these measurements has been used to estimate the variation of $K(z)$ with pressure for all of the giant planets, which is shown in Figure 4.3 (after Fouchet *et al.*, 2003), where the estimates in table 4.2 of Atreya *et al.* (1999) are also shown as cross symbols. Also plotted as dotted lines are the profiles of the molecular diffusion coefficient for methane calculated from the semi-empirical formula (after Moses *et al.*, 2000):

$$D_{\mathrm{CH_4}} = \frac{2.3 \times 10^{17} T^{0.765}}{n} \tag{4.27}$$

where the temperature profiles shown earlier in Figure 4.1 have been used. The greater rate of increase with height of D than K is clearly seen and the level where

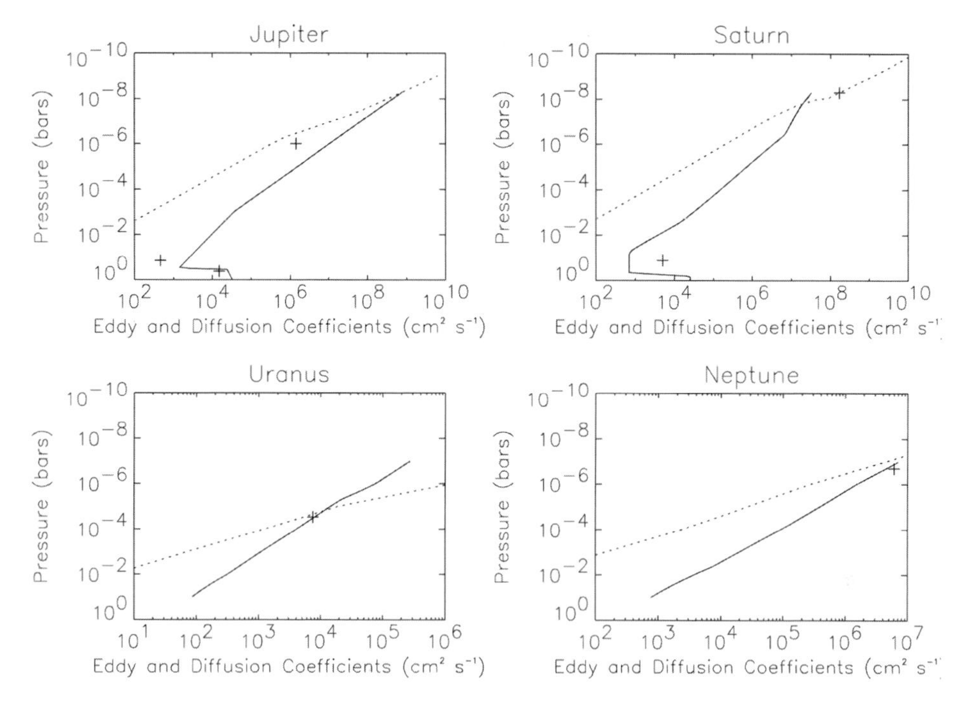

Figure 4.3. Variation of eddy mixing and molecular diffusion coefficients with height in the giant planet atmospheres. Solid lines are eddy diffusion coefficient profiles (from Fouchet *et al.*, 2003). Dotted lines are the molecular diffusion coefficients of methane calculated from Equation (4.27). Cross symbols are estimates of eddy diffusion coefficients from Table 4.2.

the curves meet is the methane homopause. It can be seen that the methane homopause occurs at a pressure of approximately 10^{-8} bar for both Jupiter and Saturn, indicating active vertical eddy mixing, but that turbulence in Neptune's stratosphere is less active with the methane homopause at 10^{-6} bar. The vertical mixing in Uranus' atmosphere can be seen to be particularly sluggish with the homopause at $\sim 10^{-5}$ bar.

4.3 COMPOSITION PROFILES: GENERAL CONSIDERATIONS

Now that we have discussed the vertical structure of planetary atmospheres in terms of temperature, pressure, and vertical mixing, we are in a position to discuss and interpret the vertical composition profiles of the outer planets.

4.3.1 Disequilibrium species

The most stable equilibrium chemical form that different elements exist in depends upon the temperature and on the abundance of other molecules and atoms with which reactions may occur. Similarly, the equilibrium ortho:para ratio of hydrogen is a function of temperature. The rate at which equilibrium is reached depends upon the temperature itself, density, and sometimes the presence of catalyzing aerosols. Hence, in rapidly overturning atmospheres the composition of air which has been rapidly uplifted from warmer, denser levels may be partially "quenched" at its deeper equilibrium composition. Measurement of the abundances of so-called "disequilibrium" species thus provides information on the upwelling rates, and hence on vertical circulation, and is a very important diagnostic tool in understanding the circulation of the giant planet atmospheres.

Nitrogen and carbon

The molecular forms of carbon and nitrogen are good examples of disequilibrium species in giant planet atmospheres. The chemical form of nitrogen and carbon that is observed depends upon the following equilibrium reactions:

$$CH_4 + H_2O \rightleftharpoons CO + 3H_2 \tag{4.28}$$

$$2NH_3 \rightleftharpoons N_2 + 3H_2. \tag{4.29}$$

Above about 1,000 K, the right-hand side of these reactions dominates, whereas at lower temperatures the left-hand side dominates. Hence, in the observable, cool parts of the giant planet atmospheres we do not expect to see much CO or N_2 unless vertical transport is particularly vigorous. This conclusion seems to be true for all of the giant planets with the notable exception of Neptune, where there is some evidence that small quantities of both molecules are present in the observable atmosphere, suggesting rapid upwelling, which is consistent with Neptune's observed high internal heat source.

Carbon dioxide may also appear in disequilibrium in the upper troposphere and stratosphere through the reaction

$$CO + H_2O \rightleftharpoons CO_2 + H_2 \tag{4.30}$$

although carbon dioxide has not been detected in the lower tropospheres of any of the giant planets.

Germane, arsine, and phosphine

The abundances of phosphine, arsine, and germane in the atmospheres of Jupiter and Saturn are higher than would be expected if the atmospheres were in chemical equilibrium. These gases are produced by equilibrium chemistry at $p \sim 1$ kbar, $T \sim 1{,}000$ K, and convert slowly at the cold temperatures found in the upper parts of these atmospheres to the chemical forms given below, with a timescale of the order of 100 days. The chemistry here is highly complicated and there are lots of intermediate steps. The reader is referred to, for example, Atreya (1986) for more details.

$$GeH_4 + H_2S \rightleftharpoons GeS(s) + 3H_2 \quad (4.31)$$

$$4PH_3 + 6H_2O \rightleftharpoons P_4O_6 + 12H_2 \quad (4.32)$$

$$4AsH_3 \rightleftharpoons As_4(s) + 6H_2 \quad (4.33)$$

Measurements of the vertical profiles of these species provide constraints in the eddy-mixing coefficient as mentioned earlier. Mapping the spatial abundance of these materials provides information on the general upwelling/downwelling motions.

Ortho-hydrogen/para-hydrogen

We have seen already that molecular hydrogen may exist in two distinct ortho and para states. At high temperatures (T greater than roughly 300 K) the equilibrium para-H_2 fraction f_p is 0.25, and this value increases as the temperature decreases, as may be determined from Equation (4.21). The vertical profile of the ortho:para ratio in the atmospheres of the giant planets may be estimated from collision-induced and quadrupole absorptions in the far-IR (the nature of dipole and quadrupole absorption lines are discussed in Chapter 6) and also from visible wavelength hydrogen quadrupole absorption lines. Any departure of the measured para-H_2 fraction from the equilibrium value calculated from Equation (4.21) using the local temperature provides information on the vigorousness of vertical convection, with more convectively active atmospheres having an f_p closer to the deep value of 0.25 than to the equilibrium value. However, the conversion of ortho-hydrogen to para-hydrogen is catalyzed by the presence of aerosols, as mentioned in Section 4.1.3, and thus knowledge of the aerosol distribution is essential to fully understand the observed f_p profile.

The ortho/para fraction in the upper tropospheres of around 300 mbar of all the giant planets was estimated as a function of latitude from *Voyager* IRIS far-IR data by Conrath *et al.* (1998), while Fouchet *et al.* (2003) used observations of far-IR hydrogen quadrupole lines with the Infrared Space Observatory Short Wavelength Spectrometer (ISO/SWS) to determine disk-averaged values of f_p in the stratosphere at pressure levels between 1 mbar and 10 mbar. These results and others are discussed in Section 4.4.

4.3.2 Photolysis

Photodissociation of molecules by solar UV radiation is an extremely important driver on the composition profiles measured in the upper parts of the giant planet atmospheres. It is beyond the scope of this book to provide a detailed exposition on the finer points of this topic and the reader is referred to books such as Atreya (1986) for a more complete discussion. Here we shall limit ourselves to the main photochemical reactions governing the upper atmospheric composition profiles of important giant planet gases. However, before we can discuss photodissociation, we must briefly introduce Rayleigh scattering, and discuss how it affects the flux of UV photons reaching the upper atmospheres of the giant planets.

Rayleigh scattering

Photolysis of different molecules requires UV photons of different frequencies. For example, photolysis of methane requires photons with wavelengths less than 160 nm, whereas photolysis of ammonia, phosphine, and hydrazine requires photons in the wavelength range 160 nm to 230 nm. Solar photons with wavelengths less than 160 nm are dominated by Lyman-α emission at 121.6 nm. The penetration of UV photons into planetary atmospheres is strongly regulated by the Rayleigh scattering of air molecules, which is strongly wavelength-dependent. In general, the Rayleigh scattering cross-sectional area per dipole (Goody and Yung, 1989) is given by

$$\sigma_R = \frac{8\pi}{3}\left(\frac{2\pi}{\lambda}\right)^4 |\alpha|^2 \tag{4.34}$$

where α is the polarizibility, which relates the electric dipole induced on a molecule or atom to the local electric field strength by $\mathbf{p} = \alpha\mathbf{E}_{\text{local}}$; and λ is the wavelength. It can immediately be seen that shorter wavelength photons are much more efficiently scattered than longer wavelength photons and indeed this is why the Earth's sky appears blue from the ground. For atoms and molecules where the polarizibility is independent of the molecule's orientation with respect to the incident electric field, the polarizibility is related to the refractive index m via the equation

$$\alpha = \frac{m-1}{2\pi N} \tag{4.35}$$

where N is the number of molecules per unit volume. Hence, substituting into Equation (4.34) we find

$$\sigma_R = \frac{32\pi^3}{3\lambda^4}\frac{(m-1)^2}{N^2} \simeq \frac{8\pi^3}{3\lambda^4}\frac{(m^2-1)^2}{N^2} \tag{4.36}$$

where the last approximation assumes m is close to unity, which for a gas it is. For all but spherical top molecules (Chapter 6), the polarizibility is not actually independent of the molecule's orientation with respect to the incident electric field and thus Equation (4.36) must be modified to

$$\sigma_R = \frac{8\pi^3}{3\lambda^4}\frac{(m^2-1)^2}{N^2} f_{\text{anisotropic}} \tag{4.37}$$

where $f_{\text{anisotropic}}$ is a parameter describing the non-isotropy of the atom or molecule (Goody and Yung, 1989). Assuming that a number of atoms/molecules are randomly orientated with respect to the incident electric field, the correction factor $f_{\text{anisotropic}}$ may be shown to be equal to

$$f_{\text{anisotropic}} = \frac{3(2+\Delta)}{(6-7\Delta)} \tag{4.38}$$

where Δ is known as the depolarization factor, which may be measured in the laboratory and is listed in Table 4.3 for relevant Jovian gases. The refractive indices

Table 4.3. Refractive index parameters and depolarization factors for giant planet gases.

Molecule	A	B (m^2)	Δ
He	3.48×10^{-5}	2.3×10^{-15}	0
H_2	13.58×10^{-5}	7.52×10^{-15}	0.02
CH_4	42.7×10^{-5}	10.0×10^{-5}	0.02
N_2	29.06×10^{-5}	7.7×10^{-5}	0.03

Data from Allen (1976).

of hydrogen, helium, and methane at standard temperature and pressure (STP) may be calculated from the semi-empirical formula

$$m - 1 = A(1 + B/\lambda^2) \tag{4.39}$$

and values of the coefficients A, B are also listed in Table 4.3. Since m is close to 1.0 and since from the *Clausius–Mossotti* relation (e.g., Bleaney and Bleaney, 1976) we expect $m^2 - 1$ to vary linearly with N, we may use m evaluated at STP in Equation (4.37) provided N is also calculated at STP. In practice for the giant planet atmospheres, where composition is dominated by near-solar hydrogen–helium, it is found that Equation (4.37) may be reasonably accurately approximated by (Atreya, 1986)

$$\sigma_R = \frac{7.5 \times 10^{-17}}{\lambda^4} \tag{4.40}$$

where the wavelength λ is assumed to be in units of angstroms (Å) and where the calculated cross-section is in units of m^2. The pressure level of unit optical depths at different UV wavelengths for the giant planets is listed in Table 4.4, where it can be seen that longer UV wavelengths penetrate to much deeper levels than shorter wavelengths.

Table 4.4. Pressure level of unit optical depth for Rayleigh scattering in the giant planet atmospheres.

Planet	1,000 Å	2,000 Å	3,000 Å
Jupiter	10 mbar	200 mbar	1,000 bar
Saturn	6 mbar	80 mbar	400 mbar
Uranus	5 mbar	70 mbar	300 mbar
Neptune	6 mbar	90 mbar	400 mbar

Photodissociation of important giant planet gases

Now that we have seen how Rayleigh scattering affects the incident fluxes of UV photons we will now discuss the photodissociation of the three most important photo-active gases in the upper tropospheres and stratospheres of the giant planets: namely: ammonia, phosphine, and methane.

Ammonia

If ammonia is transported to sufficiently high altitudes in an atmosphere (where pressures are of the order of 100 mbar) then it may be photodissociated to form hydrazine (N_2H_4) via the reactions:

$$NH_3 + h\nu \rightleftharpoons NH_2 + H \tag{4.41}$$

$$NH_2 + NH_2 + M \rightleftharpoons N_2H_4 + M \tag{4.42}$$

where M is any other molecule. Hence, we expect the abundance of ammonia to decrease above the 100 mbar level. In fact, the ammonia decreases at lower altitudes also due to vertical eddy mixing and, as we saw in Section 4.2, the rate of decrease with height is thus determined by the strength of sunlight and the degree of eddy mixing.

Hydrazine itself may also be photolyzed into other products and the photo-absorption cross-sections of ammonia, hydrazine, and phosphine are shown in Figure 4.4. Hydrazine should condense at the temperatures found in Jupiter and Saturn's upper troposphere and the resultant ice particles may thus be a constituent of the hazes found in the upper tropospheres of these planets.

Phosphine

Photolysis of phosphine may lead to the formation of diphosphine P_2H_4 via the reactions:

$$PH_3 + h\nu \rightleftharpoons PH_2 + H \tag{4.43}$$

$$PH_2 + PH_2 + M \rightleftharpoons P_2H_4 + M. \tag{4.44}$$

The main pressure level of photolysis is again around 100 mbar and thus the abundance of phosphine is expected to decrease throughout the upper troposphere and lower stratosphere of all the giant planets with a scale height governed by the vertical eddy-mixing coefficient and solar insolation.

If temperatures are low enough for the diphosphine to condense, then this may provide an extra source of haze materials in the upper tropospheres of Jupiter and Saturn. However, if the diphosphine does not condense then further reactions are possible which, provided that the levels of scavenging molecules such as acetylene (C_2H_2) and ethylene (C_2H_4) are not too high, give rise to the production of the solid phosphorus allotrope P_4(s). The P_4(s) allotrope is bright red and it has been suggested that this product might be able to explain the red coloration of Jupiter's GRS. However, this identification is highly speculative and unproven.

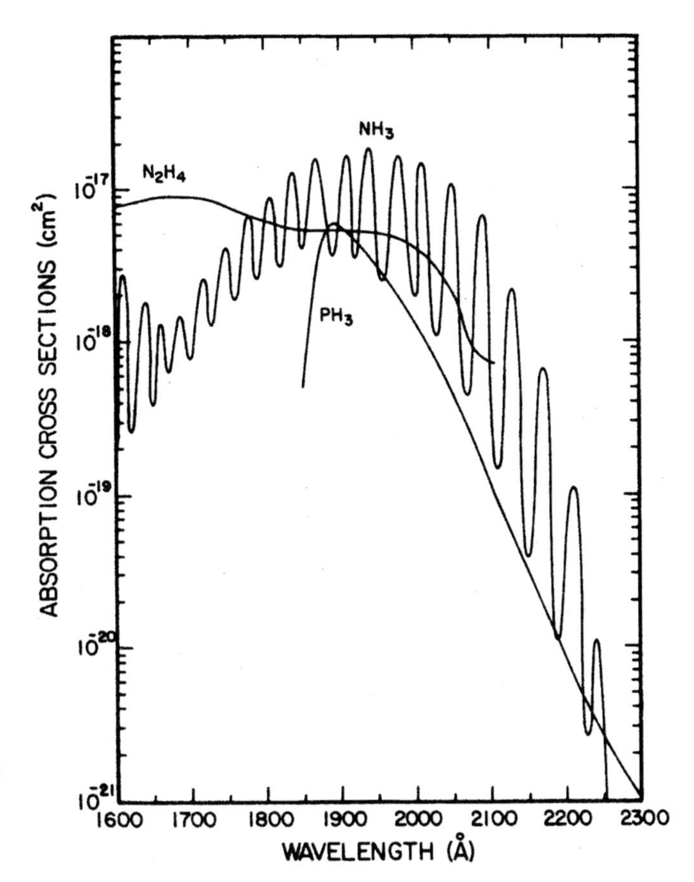

Figure 4.4. UV cross-sections of different gases relevant to giant planet atmospheres. From Atreya (1986).

Unfortunately, insufficient thermodynamic data exists for diphosphine to predict whether or not it does actually condense in the upper tropospheres of the giant planets.

Methane

Methane photochemistry is highly complicated with a number of possible branch reactions shown in Figure 4.5.

From Table 4.4, we can see that the UV photons capable of initiating methane photolysis (mainly Lyman-α at 121.6 nm) may penetrate to a depth of only 10 mbar in the atmospheres of all the giant planets due to Rayleigh scattering. However, the photoabsorption cross-section of methane is very high and thus the peak level of methane photodissociation is actually at somewhat higher levels (depending on the methane abundance) and for Jupiter extends between 0.1 μbar and 0.1 mbar. While this is the main region of methane photodissociation, to combine the resulting products into hydrocarbons requires a number of two-body and three-body reactions, which are only efficient at higher pressures (0.1 mbar and higher). The

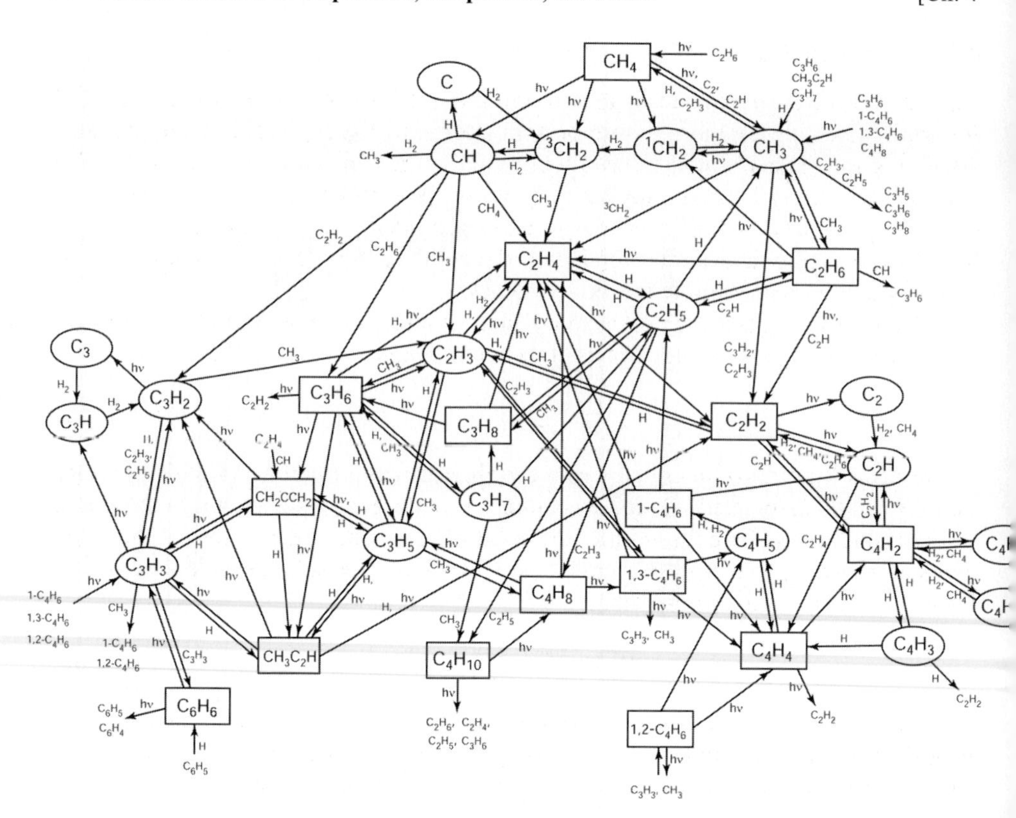

Figure 4.5. Methane photochemistry paths. From Moses *et al.* (2000). Reprinted with kind permissio Elsevier.

main resultant products of methane photodissociation are acetylene (C_2H_2), ethane (C_2H_6), and polyacetylenes ($C_{2n}H_2$). The intermediate molecule ethylene (C_2H_4) is more unstable, but has been detected in the atmosphere of Jupiter. Further reactions are possible leading to more complicated molecules such as methyl acetylene (CH_3C_2H) and benzene (C_6H_6), which have been detected in some of the giant planet atmospheres. The intermediate product CH_3 has now also been detected in all the giant planet atmospheres except Uranus.

Haze particles, similar to terrestrial "smog", are probably produced at these altitudes by further complex chemical reactions that produce long-chain polymers that ultimately form aerosol particles. In addition, the lower stratospheres of Uranus and Neptune are so cold that some of the photochemically produced hydrocarbons freeze out directly. Both the hydrocarbons and the hazes spread vertically to other pressure levels through eddy mixing. At deeper pressures, the haze particles may also start to coagulate and settle gravitationally through the atmosphere. Eventually the photochemical products will reach the deep warm atmosphere where they are pyrolyzed back into methane again.

4.3.3 Condensation

As material is transported upwards through the troposphere, temperatures drop and for some gases the partial pressure becomes equal to the saturated vapor pressure. Assuming the presence of cloud condensation nuclei, which appear abundant in the giant planet atmospheres, cloud layers may form and the gaseous abundance of the condensing molecules falls according to the saturated vapor pressure. For simple liquid/vapor and solid/vapor transitions, the condensation levels of clouds may be easily estimated from simple thermodynamics.

The Clausius–Clapeyron equation, which governs how pressure p and temperature T vary along a first-order phase transition, is given by (e.g., Finn, 1993)

$$\frac{dp}{dT} = \frac{\Delta S}{\Delta V} \tag{4.45}$$

where ΔS is the change in entropy of a certain quantity of material changing phase; and ΔV is the associated change in volume. For the liquid–vapor and solid–vapor phase boundaries, where the volume of the gas phase greatly exceeds that of the liquid or solid phases, ΔV is approximately the volume of the gas phase alone and thus Equation (4.45) is well approximated by

$$\frac{dp}{dT} = \frac{Lp}{RT^2} \tag{4.46}$$

where the ideal gas equation has been assumed and where L is the latent heat of vaporization *per mole* and R is the molar gas constant. If L is assumed not to vary with temperature then this equation may be simply integrated to give

$$\ln p = A - \frac{L}{RT} = A - \frac{B}{T} \tag{4.47}$$

where A and B are constants that may be fitted to the measured saturated vapor pressure curves. More generally, the latent heat L varies with temperature also, which introduces extra terms into Equation (4.47). However, these are secondary effects and for the case of giant planet atmospheres, where many constituents may condense whose phase boundaries have not been measured with very great precision, they are usually neglected. The fitted coefficients A and B for a number of first-order phase transitions relevant to the giant planet atmospheres are listed in Table 4.5.

In addition to the simple condensation of vapor into liquid or solid aerosols, more complex two-component reactions may occur such as the formation of solid ammonium hydrosulfide (Equation 4.10). The thermodynamics of this are more complicated, but the variation in vapor pressure of both reactants with temperature may be adequately approximated by a similar equilibrium constant equation to Equation (4.47)

$$\ln(p_{NH_3}) + \ln(p_{H_2S}) = 34.15 - \frac{10{,}833.6}{T} \tag{4.48}$$

where p_{NH_3} and p_{H_2S} are the partial pressures (in bar) of NH_3 and H_2S, respectively. Since this equation has two unknowns, we need additional information, which comes from the fact that we know that one molecule of NH_3 reacts with one molecule of H_2S

Table 4.5. Coefficients A, B for various sublimation and vaporization curves relevant to giant planet atmospheres.

Gas	A_{ice}	B_{ice}	T_{triple}	A_{liq}	B_{liq}
NH_3	17.347	3,930.6	195.45	11.901	2,850.4
H_2O	17.477	6,164.65	273.16	14.149	5,257.2
H_2S	12.884	2,702.4	187.65	11.347	2,411.4
CH_4	10.682	1,163.8	90.65	9.382	1,045.5
C_2H_2	14.636	2,757.7	191.65	—	—
C_2H_6	—	—	90.0	10.685	1,959.6
C_2H_4	—	—	104.14	10.831	1,822.4
C_4H_2	16.284	4,366.7	238.25	11.137	3,144.6
CH_3C_2H	-	-	170.45	11.884	2,803.6
PH_3	10.057	1,865.9	140.65	—	—

NB: The form of C_3H_4 found in the atmospheres of the giant planets is as above, CH_3C_2H, known as methyl acetylene (or alternatively propyne). The alternative isomer of C_3H_4, called allene (or alternatively propadiene) is not found.

to form NH_4SH. Hence, if by substituting the uncondensed values of p_{NH_3} and p_{H_2S} into Equation (4.48) we find that the left-hand side exceeds the right-hand side we solve for the amount Δ by which we need to reduce both p_{NH_3} and p_{H_2S}; that is,

$$\ln(p_{NH_3} - \Delta) + \ln(p_{H_2S} - \Delta) = 34.15 - \frac{10{,}833.6}{T}. \tag{4.49}$$

To a first-order approximation, the general shape of the abundance profile of a condensable species may be determined from an equilibrium cloud condensation model (ECCM) by considering a parcel of deep air that is lifted right up through the atmosphere without mixing with surrounding air. At first the mixing ratio remains fixed at its deep level, but at a certain level it becomes equal to the saturated volume mixing ratio (v.m.r.) and thus the gas starts to condense. Moving the parcel to higher altitudes—and thus lower temperatures—more and more of the gas condenses to form aerosols and thus the mixing ratio profile follows the saturated v.m.r. curve. At the tropopause, the temperature stops decreasing with height and instead starts to rise again. However, if we assume that cloud particles condensed at lower altitudes are not carried with the parcel, but instead fall through the atmosphere, then the v.m.r. cannot rise again above the tropopause by re-evaporation of the aerosols and instead remains fixed at the tropopause value. Hence, the tropopause acts as a "cold trap" to molecules that condense in the troposphere and limits stratospheric abundances. The altitude where the deep fixed v.m.r. meets the saturated v.m.r. curve determines the base level of the condensed cloud. Clearly if the deep v.m.r. is higher then the cloud base pressure is higher and *vice versa.*

In reality, the mixing of rising air parcels with descending dry air, both vertically and horizontally, means that the v.m.r. profile derived from spatially averaged remotely sensed data is usually substantially subsaturated even in areas of localized rapid convection. Hence, while this model is useful for estimating the approximate level of the cloud bases, and thus where the volatile species start to condense, it does not model well the rate of decrease of gas abundance with height. In addition, the technique gives no indication on the vertical extent or optical thickness of the cloud. These depend on two things: (1) the rate of uplift, or vertical mixing; and (2) the rate of formation of cloud aerosols and their size, which governs how quickly they fall back down through the atmosphere towards warmer regions where they may again evaporate. Thus, accurately estimating the vertical distribution of cloud particle density (sometimes expressed in terms of a *cloud scale height*, analogous to the pressure scale height mentioned earlier) and cloud particle size distribution, both of which are vital parameters to know if remote-sensing observations are to be used to interpret cloud structure, is an extremely difficult microphysical problem.

As air is uplifted from the deep atmosphere, in addition to the observable clouds of material such as water and ammonia, a number of rather exotic layers may form at deeper pressures. For example, in Jupiter's atmosphere, clouds of magnesium silicates are predicted to condense near 2,000 K, and clouds of silver and gold are predicted to condense in thin layers near 1,000 K!

4.3.4 Extraplanetary sources

The ISO mission discovered the presence of gaseous H_2O in the stratospheres of all four giant planets (Feuchtgruber *et al.*, 1997), and the abundances are listed in Tables 4.6–4.9. Since the tropopauses of these planets are very cold, there is effectively a "cold trap", which in the absence of any other sources should keep the stratospheric water abundance very low, as discussed in Section 4.3.3. The fact that significant levels of water, and in addition oxygenated species such as CO_2, exist at these levels suggests that there is an external source of oxygen. The source of this stratospheric H_2O and CO_2 was initially thought to be due to the arrival of interplanetary icy micrometeoroids, and a flux of approximately 10^6 mol cm^{-2} s^{-1} was modeled to be required for all four giant planets. Species such as CO and CO_2 may be contributed directly from these micrometeoroids or through subsequent reactions between water vapor and stratospheric hydrocarbons. Note that the presence of CO does not on its own suggest external sources since rapid convection from the interior may inject this disequilibrium species into the stratosphere. The presence of stratospheric CO_2, however, does imply the presence of stratospheric water vapor.

Although an approximately similar flux rate of interplanetary dust (IPD) was required for all four giant planets, Feuchtgruber *et al.* (1997) noted that Jupiter and Saturn actually required a flux approximately 10 times greater than Uranus and

Neptune. It was initially suggested that this might be due to variations in IPD density as a function of distance from the Sun. However, it is now thought that the differences are due to additional sources of water. For Saturn, the extra source would appear to be the erosion of ring material, which becomes ionized and then spirals into the atmosphere along connecting magnetic field lines causing increased abundances of water at midlatitudes (Prangé *et al.*, 2006). For Jupiter, the extra source of water, and CO_2, would appear to have been the collision of Comet Shoemaker–Levy 9 (SL-9) with Jupiter in 1994, which injected large abundances of oxygen-rich molecules into the stratosphere (Lellouch *et al.*, 2002). While the latitude dependence of Jovian stratospheric water remains unclear, the abundance of stratospheric CO_2 ten years after the impact of SL9 was observed (Kunde *et al.*, 2004) to decrease by a factor of 7 from southern midlatitudes, where SL9 struck, to northern midlatitudes (Section 5.4.2). Further exploration of the latitudinal variation of such stratospheric gases will require more observations, either from ground-based microwave observatories, or more likely from future projects such as SOFIA and *Herschel*, which are described in Chapter 8.

4.4 COMPOSITION AND CLOUD PROFILES OF THE GIANT PLANETS

Now that we have described the various processes affecting the vertical composition and cloud profiles of the giant planet atmospheres, we will now review the most recent measurements of these profiles from remote and *in situ* observations.

4.4.1 Jupiter

Composition profiles

The bulk composition of Jupiter (Figure 4.6) was discussed in Chapter 2. Of all the giant planets it is the one that most closely approximates a proto-solar composition, but even here the abundance of heavy elements such as carbon, sulfur, and nitrogen is found to be three to five times greater than would be expected in a purely solar composition atmosphere. This is consistent with the core accretion theory (Mizuno, 1980; Pollack *et al.*, 1996) that Jupiter initially formed from icy planetesimals, which then formed an embryo big enough to gravitationally attract a very large quantity of the surrounding nebula gas.

When we talk about "deep" compositions, which are listed in Tables 2.2a and 2.2b, it is worth clarifying how "deep" we really mean. Jupiter is the only giant planet from which *in situ* measurements of the atmospheric composition have been made by the *Galileo* entry probe in 1995. These measurements extend down to pressures of approximately 20 bar. While this is very high compared with remotely sensed infrared and microwave observations, which extend down to 10 bar at most, it is still only scraping the surface of the enormous Jovian atmosphere. Hence, all we can really measure is the composition of the top of the molecular-hydrogen region. Any composition gradients that may occur in radiative zones, or at the metallic-

Figure 4.6. *Voyager 1* image of Jupiter. The Great Red Spot (GRS) is clearly visible near the center of the image. Courtesy of NASA.

hydrogen/molecular-hydrogen phase boundary are very difficult to detect and may only be inferred from interior models matching the observed oblateness, rotation rate, and gravitational *J*-coefficients.

Figure 4.7 shows the results of a calculation using a simple equilibrium cloud condensation model (ECCM) of Jupiter. In these models (e.g., Atreya, 1986; Lewis, 1995), as described in Section 4.3.3, a parcel of air is raised upwards and if the partial pressure of a gas exceeds the saturated vapor pressure (s.v.p.), the excess is assumed to condense as cloud droplets and be lost from the parcel. Here, the temperature profile observed in the upper troposphere has been extended downwards towards the interior along the SALR, consistent with the condensation of water, NH_4SH, and

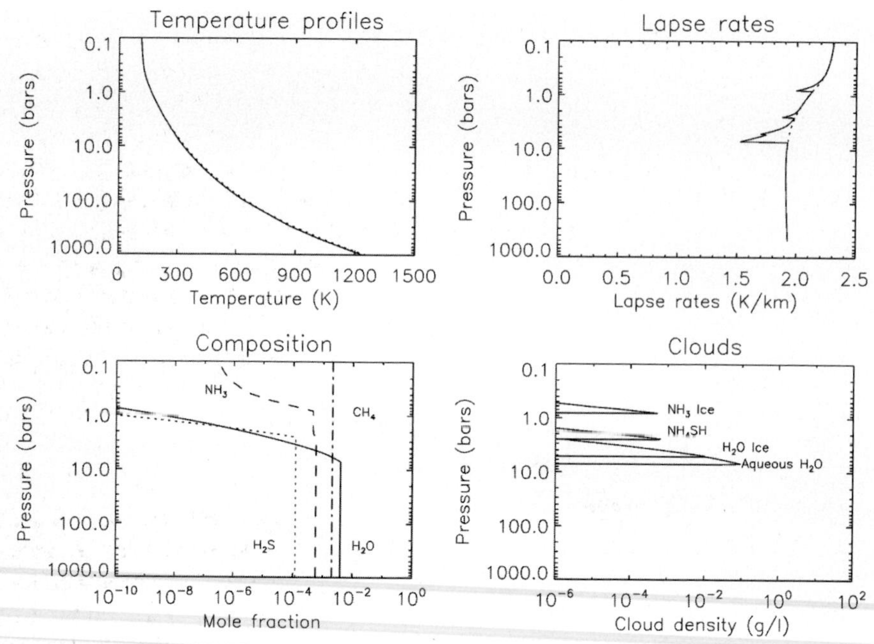

Figure 4.7. Equilibrium cloud condensation model of Jupiter's atmosphere. (1, top left) Temperature profile follows SALR below radiative–convective boundary. Dotted line follows DALR below 1 bar level. (2, top right) Variation of calculated lapse rates with height. Dotted line is DALR, solid line is SALR. (3, bottom left) Composition profiles: H_2O (solid), H_2S (dotted), NH_3 (dashed), CH_4 (dot-dashed). (4, bottom right) Cloud densities: H_2O cloud (water, then ice) at ~ 7 bar, NH_4SH at ~2 bar, and NH_3 ice at ~0.7 bar. Assumed composition: O/H, S/H = 5× the solar value, N/H = 5.5× the solar value, C/H = 4.8× the solar value.

NH_3 clouds at the different levels shown in the figure. The temperature profile that is calculated assuming a DALR is shown as the dotted line for comparison. The deep abundances of O, N, S, and C (relative to H) are assumed to be 5×, 5.5×, 5×, and 4.8× the solar value, respectively. The lapse rates calculated at different altitudes in the troposphere are also shown in Figure 4.7. The DALR can be seen to increase with height, due almost entirely to the decrease of molecular-hydrogen heat capacity with height as the temperature falls. The slight reduction in gravitational acceleration with height also tends to increase the lapse rate, but this effect is small. When condensation is included, the SALR can be seen to be smaller than the DALR near the bases of the main clouds due to the release of latent heat, although the resultant differences in the calculated dry and saturated temperature profiles is small since the main condensates have low deep abundances. The abundance profiles of NH_3, H_2S, H_2O, and CH_4 are calculated by limiting the partial pressures to the s.v.p. when condensation occurs (and for the NH_4SH cloud via Equation 4.49), and the associated cloud densities are also shown. We will discuss the calculated cloud profiles in the next section ("Clouds and hazes").

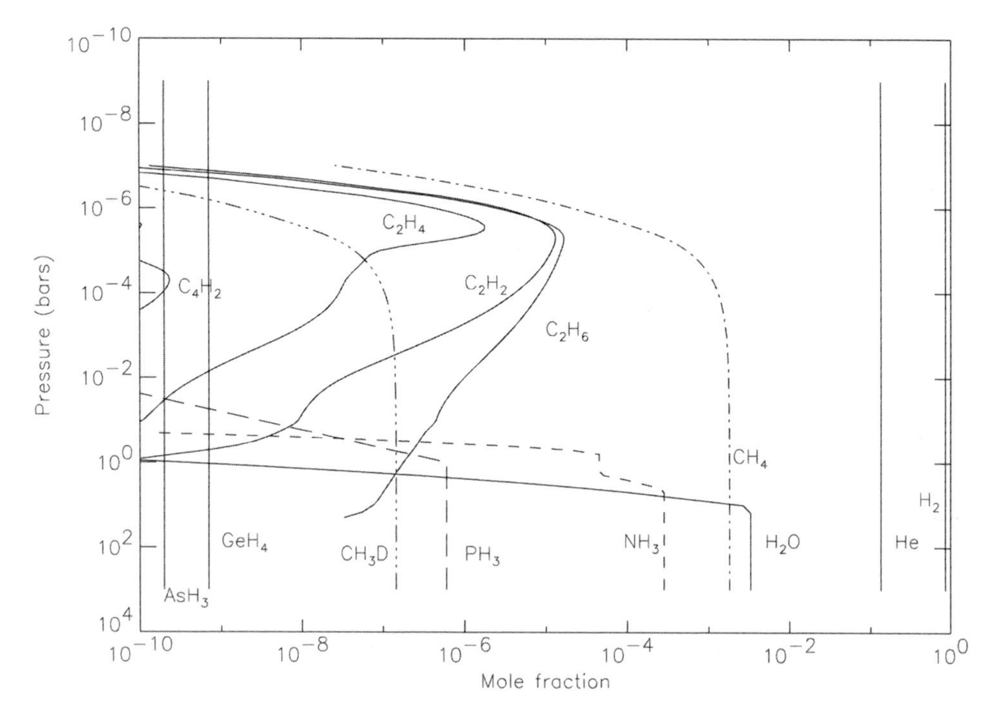

Figure 4.8. Observed and modeled abundance profiles in the atmosphere of Jupiter.

While such ECCM models are useful for making an initial estimate of the basic tropospheric abundance profiles of condensing molecules, observations of the composition profiles are rather different as can be seen in Figure 4.8 where the observed and best-modeled composition profiles of condensable gases and other species are shown. Sources of estimated Jovian composition data and references are listed in Table 4.6. In particular, analysis of the 5 μm part of the infrared spectrum measured by *Voyager*, and more recently *Galileo*, has found water to be significantly less abundant than expected. This is probably due to this spectral region being most sensitive to cloud-free areas, which appear to be volatile-depleted regions either due to subsidence or column-stretching (as we shall see in Chapter 5). Typical levels of saturation in the 5 bar region are found to be of the order of 10%. It was hoped that the question of the deep abundance of water would have been answered by the *Galileo* entry probe. However, the probe unfortunately descended through just such a cloud-free 5 μm bright region and so its estimates of volatile abundances are similarly depleted. The probe Mass Spectrometer found that the abundance of water increased with depth and had reached a value of ~0.5× the solar value at 19 bar. A similar depleted profile was inferred from the probe's Net Flux Radiometer. Presumably the water vapor profile in upwelling, cloudy regions is similar to the calculated ECCM case, but until estimates of water vapor can be made in such regions we cannot be sure.

Table 4.6. Composition of Jupiter.

Gas	*Mole fraction*	*Measurement technique*	*Reference*
He	0.135	*Galileo* probe GPMS and HAD	Von Zahn *et al.* (1998) Niemann *et al.* (1998)
NH_3	4.4×10^{-4} at deep levels decreasing to 1.3×10^{-4} near 1 bar level	Ground-based microwave	Marten *et al.* (1980)
	2.2×10^{-4} at $p > 2$ bar	Radio occultation	Lindal *et al.* (1981)
	$(2.5–3.4)\times10^{-4}$ at $p > 2$ bar decreasing to $(3–3.8)\times10^{-5}$ at $p < 1.5$ bar	Ground-based microwave	de Pater and Massie (1985)
	Sub-saturated in upper troposphere	*Voyager* mid-IR	Carlson *et al.* (1993)
	$(7\pm1)\times10^{-4}$ at $p > 7$ bar	*Galileo* probe signal	Folkner *et al.* (1998)
	2.5×10^{-4} at $p > 5$ bar, falling to 1.5×10^{-4} at 2.5 bar and then declining rapidly with height	*Galileo* probe NFR	Sromovsky *et al.* (1998)
	~10% relative humidity above condensation level	*Galileo* NIMS 5 μm	Irwin *et al.* (1998) Roos-Serote *et al.* (1998)
	$(6.1\pm2.8)\times10^{-4}$ in 8.6–12.0 bar region	*Galileo* probe GPMS	Atreya *et al.* (1999)
	$\sim3.5\times10^{-4}$ at $p > 5$ bar, reducing to $\sim2.5\times10^{-5}$ between 2 and 0.5 bar and falling with f.s.h of 0.15 above	ISO/SWS 10 μm	Fouchet *et al.* (2000)
	$\sim2.8\times10^{-4}$ at $p > 5$ bar, reducing to $\sim4.5\times10^{-5}$ between 2 and 0.5 bar and falling with f.s.h. of 0.15 above	ISO/SWS 5 μm	Fouchet *et al.* (2000)
	3.1×10^{-4} (5–8 bar)	Ground-based 5 μm	Bézard *et al.* (2002)
	$(5.74\pm0.22)\times10^{-4}$ at deep levels from 8.9–11.7 bar	*Galileo* probe GPMS	Wong *et al.* (2004b)
H_2S	Never detected	Ground-based microwave	de Pater *et al.* (1991)
	6.7×10^{-5} at $p > 16$ bar	*Galileo* probe GPMS	Niemann *et al.* (1998)
	$(7.69\pm1.81)\times10^{-5}$	*Galileo* probe GPMS	Wong *et al.* (2004b)
H_2O	5.2×10^{-4} at 19 bar and increasing	*Galileo* probe GPMS	Niemann *et al.* (1998) Atreya *et al.* (1999)
	$(4.23\pm1.38)\times10^{-4}$ at deep levels and increasing	*Galileo* probe GPMS	Wong *et al.* (2004b)
	1.7×10^{-9} at $p < 10$ mbar	ISO/SWS	Lellouch *et al.* (1997)

Gas	*Mole fraction*	*Measurement technique*	*Reference*
CH_4	1.81×10^{-3} $(2.04 \pm 0.49) \times 10^{-3}$	Galileo probe GPMS *Galileo* probe GPMS	Niemann *et al.* (1998) Wong *et al.* (2004b)
CH_3D	3.5×10^{-7} $(1.2–2) \times 10^{-7}$ 1.8×10^{-7}	*Voyager* 5 μm ISO/SWS Ground-based 5 μm	Kunde *et al.* (1982) Lellouch *et al.* (2001) Bézard *et al.* (2002)
PH_3	6×10^{-7} (5–8 bar) f.s.h. above 1 bar level = 0.3 f.s.h. above 1 bar level = 0.27 Not detected $(1.04 \pm 0.10) \times 10^{-6}$	*Voyager* 5 μm *Voyager* mid-IR *Galileo* near-IR (NIMS) Ground-based millimeter *Cassini* mid-IR	Kunde *et al.* (1982) Carlson *et al.* (1993) Irwin *et al.* (1998) Weisstein and Serabyn (1994) Irwin *et al.* (2004)
Ne	1.99×10^{-5}	*Galileo* probe GPMS	Niemann *et al.* (1998)
Ar	$(1.57 \pm 0.35) \times 10^{-5}$	*Galileo* probe GPMS	Mahaffy *et al.* (2000)
Kr	$(8.04 \pm 3.46) \times 10^{-9}$	*Galileo* probe GPMS	Mahaffy *et al.* (2000)
Xe	$(7.69 \pm 2.16) \times 10^{-10}$	*Galileo* probe GPMS	Mahaffy *et al.* (2000)
AsH_3	$(1.90 \pm 0.95) \times 10^{-10}$ (5–8 bar) 2.4×10^{-10} (5–8 bar) f.s.h. above 1 bar level ~0.7	Ground-based 5 μm Ground-based 5 μm	Noll *et al.* (1990) Bézard *et al.* (2002)
GeH_4	$(6.05 \pm 1.73) \times 10^{-10}$ (5–8 bar) 4.5×10^{-10} (5–8 bar) f.s.h. above 1 bar level ~1.6	*Voyager* 5 μm Ground-based 5 μm	Kunde *et al.* (1982) Bézard *et al.* (2002)
CO	1.6×10^{-9} (5–8 bar) $(1.0 \pm 0.2) \times 10^{-9}$ at 6 bar rising to perhaps 4×10^{-9} at $0.2 < p < 100$ mbar	Ground-based 5 μm Ground-based 5 μm	Noll *et al.* (1988) Bézard *et al.* (2002)
HCN	$(0.9–3.6) \times 10^{-9}$ (upper troposphere) $<8 \times 10^{-10}$ (above 55 mbar) Not detected $>8 \times 10^{-9}$ at peak at 45°S	Ground-based mid-IR Ground-based mid-IR Ground-based submillimeter Ground-based mid-IR	Tokunaga *et al.* (1981) Bézard *et al.* (1995) (prior to SL-9) Davis *et al.* (1997) Griffith *et al.* (2004) (post SL-9)

(continued)

Table 4.6 (*cont.*)

Gas	*Mole fraction*	*Measurement technique*	*Reference*
N_2	—		
C_2H_6	3.5×10^{-6} (~10 mbar) ~3×10^{-7} (200 mbar) increasing by factor of ~2 towards poles ~4×10^{-6} (5 mbar) increasing by factor of ~2 towards poles	Mid-IR *Cassini* mid-IR *Cassini* mid-IR	Moses *et al.* (2004) Nixon *et al.* (2007) Nixon *et al.* (2007)
C_2H_2	3.5×10^{-8} (~10 mbar) ~3×10^{-9} (200 mbar) and roughly constant with latitude ~4×10^{-8} (5 mbar) at 20°N, decreasing to ~1×10^{-8} at poles	Mid-IR *Cassini* mid-IR *Cassini* mid-IR	Moses *et al.* (2004) Nixon *et al.* (2007) Nixon *et al.* (2007)
C_4H_2	~10^{-12} (~10 mbar) Detected	Modeled *Cassini* mid-IR	Moses *et al.* (2004) Kunde *et al.* (2004)
C_2H_4	3×10^{-10} (~10 mbar)	Mid-IR and modeled	Moses *et al.* (2004)
CH_3C_2H	2×10^{-10} (~10 mbar)	Mid-IR and modeled	Moses *et al.* (2004)
C_3H_8	2.6×10^{-8} (5 mbar)	Ground-based IR	Greathouse *et al.* (2006)
CH_3	Detected	*Cassini* mid-IR	Kunde *et al.* (2004)
CO_2	3.5×10^{-10} at $p < 10$ mbar	ISO/SWS	Encrenaz (1999)
HF	$<2.70 \times 10^{-11}$	*Cassini* far-IR	Fouchet *et al.* (2004a)
HCl	$<2.30 \times 10^{-9}$	*Cassini* far-IR	Fouchet *et al.* (2004a)
HBr	$<1.00 \times 10^{-9}$	*Cassini* far-IR	Fouchet *et al.* (2004a)
HI	$<7.60 \times 10^{-9}$	*Cassini* far-IR	Fouchet *et al.* (2004a)

The ammonia profile is expected to have two "knees" since the abundance is first expected to be depleted at the level of the ammonium hydrosulfide cloud layer at around 1 bar to 2 bar and then to remain fixed until the formation of an ammonia ice cloud at 700 mbar to 500 mbar. Such a profile is consistent with ground-based microwave observations of the disk-averaged spectrum, where the deep ammonia abundance was estimated to be roughly 3× the solar value. However, like water, ammonia is observed to be severely depleted in 5 μm hotspot areas, both from measurements of the near-IR and 5 μm spectrum and also from the *in situ* observations of the *Galileo* entry probe. However, one surprising result of the probe analysis was that although the ammonia abundance was found to be severely depleted above about 2 bar, it increased rapidly with depth to a maximum value of over 5× the solar value at 19 bar. de Pater *et al.* (2001) have reanalyzed their ground-based disk-averaged microwave spectra in terms of this new estimated ammonia profile (assuming it to apply globally) and have found that it is also consistent with their spectra. If the estimated 5 μm hotspot ammonia profile is really representative of the globally averaged ammonia profile, then some means of globally depleting the abundance of ammonia in the upper troposphere of the Jupiter is required. de Pater *et al.* (2001) suggest that more ammonia might dissolve in the aqueous ammonia cloud than is currently expected (laboratory data do not exist at Jovian temperatures and so must be extrapolated from room temperature measurements), or alternatively that the ammonia molecules combine with more than one hydrogen sulfide molecule to produce products such as ammonium sulfide $(NH_4)_2S$ (where each H_2S molecule combines with two NH_3 molecules). Another possibility is that more ammonia may adsorb onto the solid NH_4SH particles than is currently estimated. At higher altitudes, analysis of mid-IR spectra show that the ammonia abundance in the upper troposphere decreases much more rapidly with height than would be expected from saturation alone. This is due to the horizontal averaging effect, described in Section 4.3.3, and also because ammonia is photolyzed at roughly 100 mbar and vertical mixing brings this ammonia-depleted air to lower altitudes. At these altitudes (~400 mbar), mid-IR observations by *Cassini* (Achterberg *et al.*, 2006; Irwin *et al.*, 2004) have revealed a clear latitudinal variation in ammonia abundance, which matches well the observed belt/zone structure with moist air rising in the zones and descending in the belts. Achterberg *et al.* (2006) also found increased levels of ammonia over the Great Red Spot, indicating rapid upwelling. This belt/zone difference is also apparent in ground-based microwave images (de Pater, 1986; de Pater *et al.*, 2001), shown in Figure 6.21.

Although phosphine does not condense in the Jovian atmosphere, its abundance is found to decrease rapidly above the 1 bar level due to photodissociation near the tropopause and vertical mixing. Its "deep" abundance is estimated from *Cassini* CIRS observations to be 2–3 times the solar value (Irwin *et al.*, 2004), which is consistent with the observation that the enrichment of almost every other element appears to be ~3–5× the solar value. Of course, since phosphine is a disequilibrium species, the deep bulk P/H ratio may actually be higher. The abundance of phosphine from *Cassini* CIRS observations also appears to vary with latitude in the manner just described for ammonia and there is also, as for ammonia, an indication of increased abundance over the Great Red Spot (Irwin *et al.*, 2004). In addition to phosphine, the other main disequilibrium species observed are germane (GeH_4) and arsine (AsH_3). The detected tropospheric (5–8 bar) mole fractions are approximately 6.1×10^{-10} and 1.9×10^{-10}, respectively, and indicate rapid vertical uplifting. Unfortunately, it has not been possible with existing measurements to determine how the abundances of germane and arsine vary with height or latitude. Similarly, a higher than expected level of carbon monoxide has also been detected in the 5 bar to 8 bar region with a mole fraction of approximately 1×10^{-9}, which requires rapid upwelling from the deep interior. However, the abundance of CO in the stratosphere has been found to be higher than that found in the troposphere, rising to perhaps 4×10^{-9} just above the tropopause (Bézard *et al.*, 2002). This observation requires an external source of CO in addition to the internal source and, like Noll *et al.* (1997), Bézard *et al.* (2002) conclude that this may be formed by shock chemistry from the infall of kilometer-size to subkilometer-size Jupiter family comets. Other gases that appear to be introduced to Jupiter's atmosphere by cometary impacts include HCN and CO_2, which were detected after the impact of Comet Shoemaker–Levy 9 in 1994. Observation of how the abundance of such gases has changed subsequently can reveal much about the circulation of Jupiter's atmosphere, which we will return to in Chapter 5.

Another indicator of vertical and meridional flow in the atmospheres of the giant planets is the observed para-H_2 fraction, f_p. The latitudinal variation of temperature and f_p at roughly the 300 mbar level has been calculated from *Voyager* IRIS measurements by Conrath *et al.* (1998) for all four giant planets. For Jupiter, the upper tropospheric temperatures are found to be cool above zones and warm above belts while f_p is found to be low and high, respectively, supporting the view of uplift in the zones and subsidence in the belts. At higher altitudes, in the stratosphere, Fouchet *et al.* (2003) used disk-averaged Infrared Space Observatory Short Wavelength Spectrometer (ISO/SWS) observations to determine the value of f_p at 1 mbar and 10 mbar and found that it is close to the value at the tropopause and does not vary with pressure. Fouchet *et al.* (2003) propose that since ortho/para conversion is mainly accomplished through the catalytic effects of aerosols, and since the Jovian stratosphere is observed to be relatively clear of aerosols, the stratospheric f_p value is frozen at its tropopause value.

In the stratosphere the photolysis products of methane, ethane, and acetylene, are observed with peak abundances occurring towards the lower part of the main photolysis region between 1 μbar and 0.1 mbar. Smog-like haze particles are probably also produced at these altitudes by further complex chemical reactions. Both the hydrocarbons and the hazes spread vertically to other pressure levels through eddy mixing. The mole fractions of ethane and acetylene are estimated in this lower region (1–10 mbar) to be 4×10^{-6} and 3×10^{-8}, respectively (Nixon *et al.*, 2007). Hydrocarbon abundances decrease with increasing pressure due to vertical mixing and conversion back to methane. Nixon *et al.* (2007) find that at 5 mbar, the abundance of acetylene is highest at 20°N and then decreases towards both poles by a factor of $\sim$4, while at lower altitudes (200 mbar level) its abundance changes little with latitude. In contrast, the abundance of ethane shows no north–south asymmetry, and instead of decreasing towards the poles its abundance is seen to increase by a factor of $\sim$2. Nixon *et al.* (2007) argue that these observations are due to a combination of the facts that the UV production rate of these molecules decreases towards the poles and that the chemical lifetime of acetylene in the Jovian atmosphere is much shorter than that of ethane. It is argued that meridional circulation transports air from low latitudes and high altitudes, where hydrocarbons are produced, towards the poles and down to lower altitudes faster than the hydrocarbons can decay, in the case of ethane, but not fast enough in the case of acetylene.

Water vapor has been observed in Jupiter's stratosphere, as was outlined in Section 4.3.4, and is believed to have an external source.

Clouds and hazes

If we follow a parcel of Jovian air, with the assumed "deep" composition, traveling up through the atmosphere with its temperature decreasing adiabatically, three main cloud decks are calculated to form by an ECCM. Atreya *et al.* (1999) calculate the following cloud layers: (1) an aqueous-ammonia cloud blending into a water ice cloud at higher levels based at approximately 7 bar with a maximum column density of $\sim$1,000 kg m^{-2}; (2) a solid ammonium hydrosulfide cloud based at 2.5 bar with a maximum column density of 22 kg m^{-2}; and (3) an ammonia ice cloud based at 0.8 bar with a maximum column density of 12 kg m^{-2}. While the cloud bases calculated by an ECCM are fairly reliable, the cloud densities are likely to greatly exceed the actual mass density of the condensed clouds since they neglect the precipitation (and thus re-evaporation) of condensed aerosols and also horizontal mixing with nearby dry air.

In reality, the Jovian cloud structure in the troposphere appears to be much more complicated than this, and a post-*Voyager* review was made by West *et al.* (1986) and a post-*Cassini* review made by West *et al.* (2004). One immediately obvious problem is that the clouds of Jupiter appear colored, while pure water, ammonium hydrosulfide, and ammonia condensates produced in the laboratory are all pure white. Hence, the yellow-ochre appearance of Jupiter is intriguing. The origin of the colors, or *chromophores*, observed in the Jovian atmosphere has always been a source of

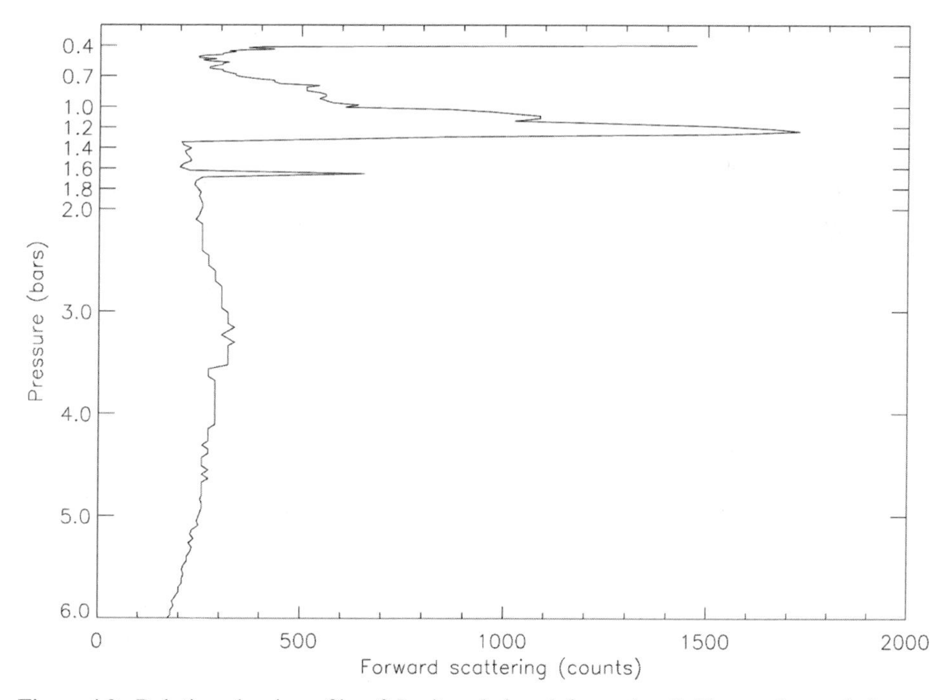

Figure 4.9. Relative cloud profile of Jupiter deduced from the *Galileo* probe nephelometer experiment cloud results. From Ragent *et al.* (1998).

much speculation. The red appearance of the Great Red Spot (GRS) is often ascribed to triclinic phosphorous P_4(s) and candidates for the various yellow, red, and brown colors seen elsewhere include allotropes of sulfur or hydrocarbon "smog" particles produced by photolysis in the stratosphere, such as tholins (Cruikshank *et al.*, 2005). However, Atreya and Wong (2005) point out that since H_2S condenses well below the UV penetration depth, the existence of free sulfur at the cloud tops is unlikely. It should also be remembered that the pictures of Jupiter, and the other giant planets, that we have all become accustomed to are heavily enhanced and colour-stretched. The "true" appearance of the giant planets is much blander, and indeed it has been argued that the apparent redness of the GRS has more to do with how the human eye perceives color in different lighting conditions than with a real intrinsic redness (Young, 1985)!

For a long time, the main cloud deck responsible for observed albedo contrasts in the visible and near-IR was believed to be the ammonia ice cloud with a base at about 0.8 bar. The first spectral indication of this was made by Brooke *et al.* (1998), who found that the 3 μm part of the ISO/SWS disk-averaged spectrum was well-modeled with the inclusion of ammonia ice particles. However, Irwin *et al.* (2001) pointed out that this would introduce features in other parts of the infrared spectrum that are not observed. Observations with *Cassini* CIRS (Wong *et al.*, 2004a) in the mid-IR (10 μm) report a more wide-area detection of ammonia ice absorption in the North Tropical Zone at latitudes between 22°N and 25°N, but only if the particles are assumed to be non-spherical. However, why such an absorption feature would be generally visible at 10 μm and not at any other wavelength is again not clear.

Wide-area visible and near-IR measurements by the Galileo orbiter of the height of the main cloud deck reveal a structure which is significantly different from a simple ECCM calculation. Estimates from *Galileo* SSI observations (Banfield *et al.*, 1998; Simon-Miller *et al.*, 2001) place the lower cloud deck just above the 1 bar level, while *Galileo* NIMS observations (Irwin and Dyudina 2002; Irwin *et al.*, 2001) place the main cloud just beneath the 1 bar level. Irwin *et al.* (2005) showed that part of this discrepancy was due to the SSI analysis assuming a discrete thin lower cloud, while the NIMS analysis assumed an extended cloud. Analysis of mid-IR (7.2 μm) CIRS Jupiter observations by Matcheva *et al.* (2005), which also assumed an extended cloud distribution, also placed the main cloud somewhere in the 1.1 bar to 0.9 bar range. Hence, the consensus does appear to be that the main cloud deck of Jupiter is somewhere close to the 1 bar level, which is inconsistent with it being composed of ammonia ice unless the N/H ratio is significantly supersolar at these altitudes, which as we saw in the last section it does not seem to be. Instead, it has been suggested by a number of authors that this cloud might perhaps be the top of the NH_4SH cloud, or that the ammonia ice particles are contaminated, perhaps by haze material, which increases their sublimation temperature.

Comparison of visible or near-IR images of Jupiter with those made at 5 μm show a clear anti-correlation with regions that are bright in the visible and near-IR appearing dark at 5 μm, while regions that are dark in the visible and near-IR appear bright at 5 μm. This observation is consistent with visibly bright regions being cloudier, which thus reflect sunlight better, but blocks the escape of thermal radiation from deeper levels. Furthermore, the close anti-correlation between visible/near-IR and 5 μm brightness suggests either that a single cloud deck is responsible for the opacity at both visible/near-IR and 5 μm wavelengths, or that the opacity of overlying different cloud layers is closely correlated. However, analysis of the limb darkening of *Galileo* NIMS 5 μm observations (Roos-Serote and Irwin, 2006) concludes that the main cloud deck responsible for the variations in opacity seen at 5 μm must exist at pressures of less than 2 bar, suggesting that the same cloud deck is responsible for the observed opacity variations at both 5 μm and visible/near-IR wavelengths.

The scenario where the clouds are made predominantly of contaminated ammonia ice particles would be consistent with the observed absence of ammonia ice spectral features (discounting the ISO detection at 3 μm of Brooke *et al.*, 1998) throughout the infrared spectrum of Jupiter (Atreya *et al.*, 2005; Irwin *et al.*, 2005; Kalogerakis *et al.*, 2008), except in small, localized regions of rapid uplift (Figure 4.10, see color section), which condense pure spectrally identifiable ammonia clouds (SIACs) at high altitude (Baines *et al.*, 2002), seen by *Galileo* NIMS to cover less than 1% of Jupiter. It is thus suggested that pure ammonia ice particles are only condensed in regions of rapid uplift, but that they are then quickly coated with haze materials, which mask their identity. In the stratosphere, the primary constituent of the haze is expected to be polycyclic aromatic hydrocarbons (PAHs) (Atreya and Wong, 2005), which slowly settle down through the atmosphere and then probably mix with the hydrazine haze expected to form in the upper troposphere through the photolysis of ammonia. Such a material is expected to have a grayish color and would provide an ideal coating material for the ammonia ice particles. As well as being seen in *Galileo* NIMS observations (Baines *et al.*, 2002), the appearance, evolution, and disappearance of an SIAC was observed by the *New Horizons* spacecraft during its flyby of Jupiter in February 2007 (Reuter *et al.*, 2007). A small white cloud with an identifiable ammonia ice signature was seen to appear at 35°S, and then expand, broaden, and disappear within the space of five Jupiter days.

At deeper levels, there is now increasing evidence for the detection of a deep water cloud at roughly 5 bar. This has been detected both by *Galileo* NIMS 5 μm measurements (Nixon *et al.*, 2001), and by visible/near-IR observations of Jovian thunderstorm clouds, which have bases in excess of 4 bar (Figure 4.11, see color section) and have clearly detected lightning activity (Banfield *et al.*, 1998; Dyudina and Ingersoll, 2002; Gierasch *et al.*, 2000; Little *et al.*, 1999) shown in Figure 4.12. In addition, the possible spectral absorption of water ice has now been observed in *Voyager* far-IR spectra (Simon-Miller *et al.*, 2000), which suggests that water ice particles may on occasion be lifted up to pressures less than 1 bar. Figure 4.13 shows a summary of the estimated cloud structure of Jupiter and of the other giant planets.

At higher altitudes, above the radiative–convective boundary but below the tropopause, the "modified ammonia" cloud appears to blend into the haze layers formed possibly by the dissociation products of ammonia and phosphine, and also methane haze products settling from the stratosphere. The upper tropospheric haze layers are seen mostly over the GRS and the northern edge of the Equatorial Zone (EZ) and appear much more zonally spread out than the convective clouds seen in the

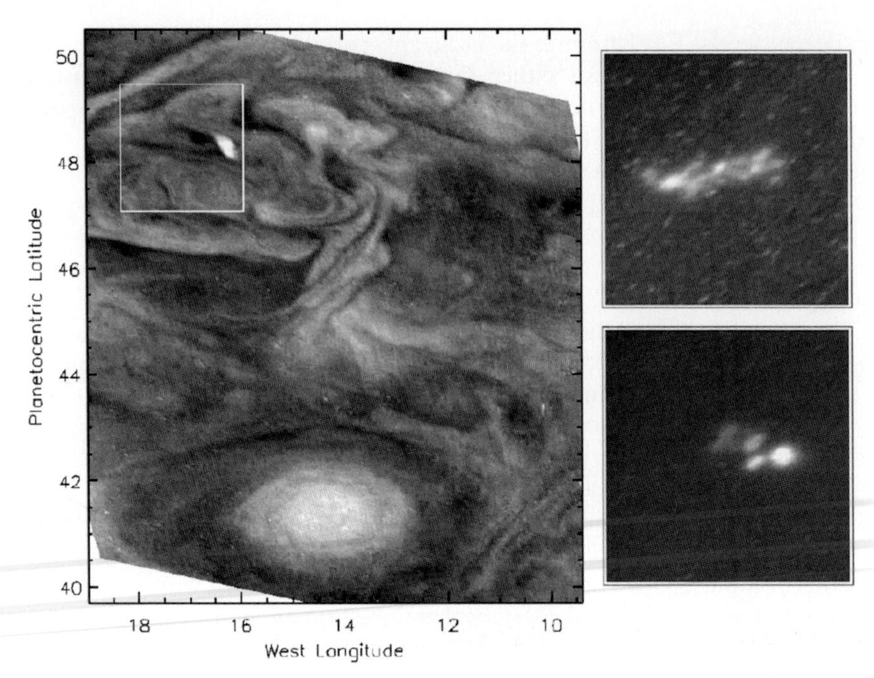

Figure 4.12. *Galileo* SSI images of a convective storm (left panel) and the associated lightning (right panels) in Jupiter's atmosphere. The left-hand image shows the dayside view of a storm cloud while the right-hand images show a close-up of lightning strikes from the storm some 2 hours later when the feature had rotated around to the nightside. The two lightning images were taken about 4 minutes apart. The dayside image was recorded at 727 nm, while the night images were recorded with the "red" filter to improve the throughput. Courtesy of NASA.

troposphere (Figure 4.14, see color section). The main altitude of ammonia photodissociation occurs in the 30 mbar to 300 mbar region, and peaks at the 80 mbar pressure level for phosphine. The main photochemical product of ammonia is likely to be hydrazine, which should form ice particles that slowly settle through the atmosphere and are pyrolyzed at deeper levels. Similarly, diphosphine particles may also be present in the upper tropospheric hazes. In addition, red phosphorus P4(s) may also be produced (as was mentioned in Section 4.3.2).

Observations of stratospheric hazes have been made both from ground-based methane band measurements (West, 1979a, b; West and Tomasko, 1980) and more recently from *Galileo* (Rages *et al.*, 1999). Such observations are described more fully in the context of Saturn's stratospheric hazes (Section 4.4.2). These hazes may be produced from the higher mass photolysis products of methane. However, the stratospheric aerosols of Jupiter (and Saturn) in the polar regions are found to be highly UV-absorbing and very different from those seen at other latitudes. In these regions it would appear that haze production results from a different mechanism possibly by charged particles from the solar wind and magnetosphere which travel down the magnetic field lines to strike air molecules in the upper atmosphere causing ionization.

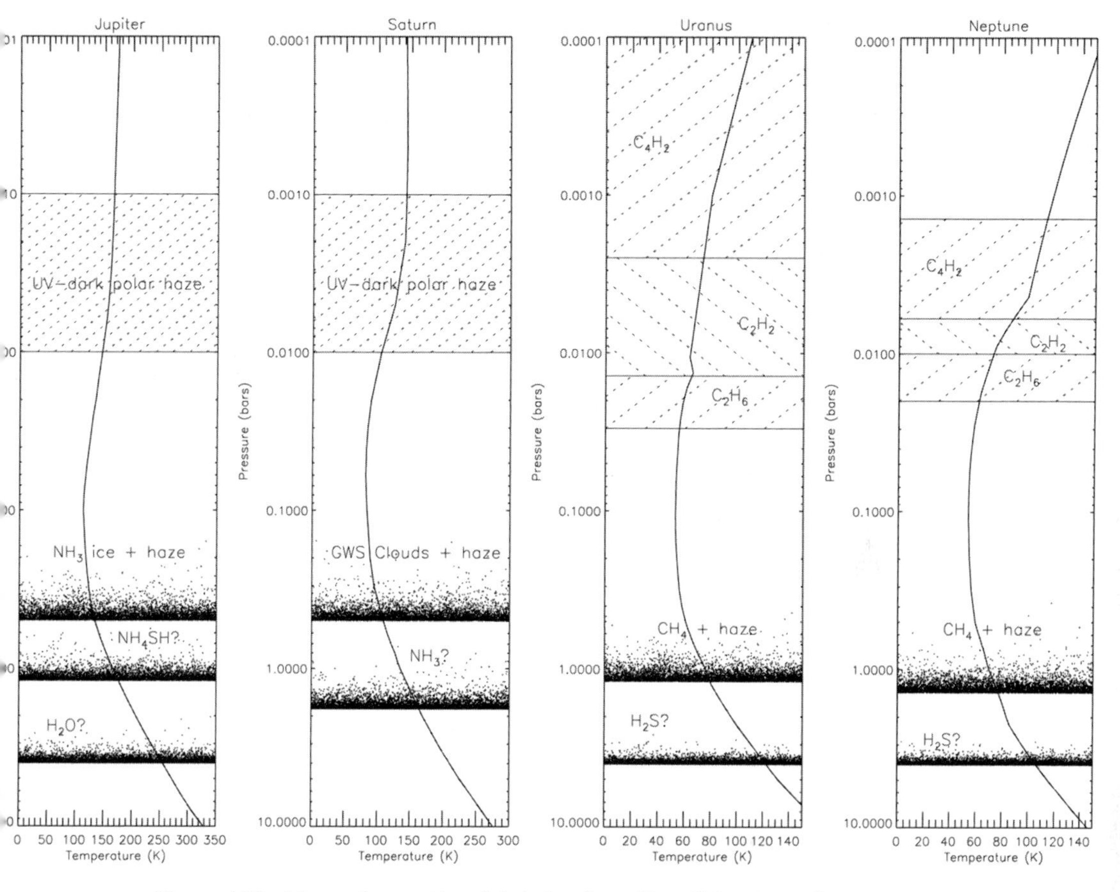

Figure 4.13. Mean observed/modeled cloud profiles of the giant planets.

The only *in situ* measurements that have been made of Jupiter's cloud structure with the *Galileo* entry probe are unfortunately ambiguous since the probe sampled an unrepresentative 5 μm hotspot, which has abnormally low cloud cover and a low abundance of volatiles (we will return to the phenomenon of 5 μm hotspots in Chapter 5). The cloud structure measured by the *Galileo* probe nephelometer experiment was found to be very tenuous with a main cloud layer based at 1.4 bar, a thinner cloud below at 1.6 bar and the suggestion of a cloud base at 0.4 bar (Figure 4.9, see p. 107). The three clouds have tentatively been identified as ammonium hydrosulfide, water ice, and ammonia ice, respectively, but they are at higher altitudes than expected from ECCMs. As we shall see in Chapter 5 these hotspots have been modeled either as regions of extremely rapid downdraft or more recently, and probably more plausibly, as part of an atmospheric wave system, where the air column is vertically stretched and compressed with the hotspots occurring at the stretching phase. Both explanations are consistent with low volatile abundances and low cloud cover.

4.4.2 Saturn

Composition profiles

Tables 2.2a and 2.2b list the "deep" composition of the Saturnian atmosphere. Like Jupiter there is good evidence that Saturn (Figure 4.15) initially formed from icy planetesimals before reaching sufficient mass to collapse and condense the solar nebula in its feeding zone. The fact that Saturn is much less massive than Jupiter suggests that it was able to attract a much smaller mass of H_2 and He from the nebula, and thus the mixing ratios of the heavier elements (X/H) are expected to be correspondingly higher. The observed estimated value of the deep C/H ratio, from *Cassini* CIRS observations, of $\sim 11\times$ the solar value (Fletcher *et al.*, 2008b) is thus entirely consistent with this expectation. Similarly, the deep abundance of ammonia has been estimated from ground-based microwave observations to be approximately 4–5$\times$ the solar value (de Pater and Massie, 1985), although 5 µm observations suggest a lower value. The estimated abundance of phosphine from submillimeter observations suggest that the deep P/H ratio is 7–14$\times$ the solar value (Orton *et al.*, 2000, 2001) and this is consistent with the figure estimated from *Cassini* CIRS observations (Fletcher *et al.*, 2007b, 2008b) which is roughly 13$\times$ the solar value.

The troposphere of Saturn is colder than that of Jupiter and thus the condensation levels of different tropospheric gases is correspondingly lower. Figure 4.16 shows the results of calculation of a Saturn ECCM. The deep abundances of O, N, S, and C (relative to H) are assumed to be 10$\times$, 10$\times$, 14$\times$, and 11$\times$ the solar value, respectively. It can be seen that water vapor should start to condense near 18 bar and thus the composition of this molecule falls rapidly with height and has a v.m.r. of only 2×10^{-7} at a pressure of 3 bar. The ammonia v.m.r. remains fixed until approximately 5 bar, where it is partially depleted by the formation of a putative

Figure 4.15. Image of Saturn recorded by the HST/WFPC-2 instrument in 1990. Courtesy of NASA.

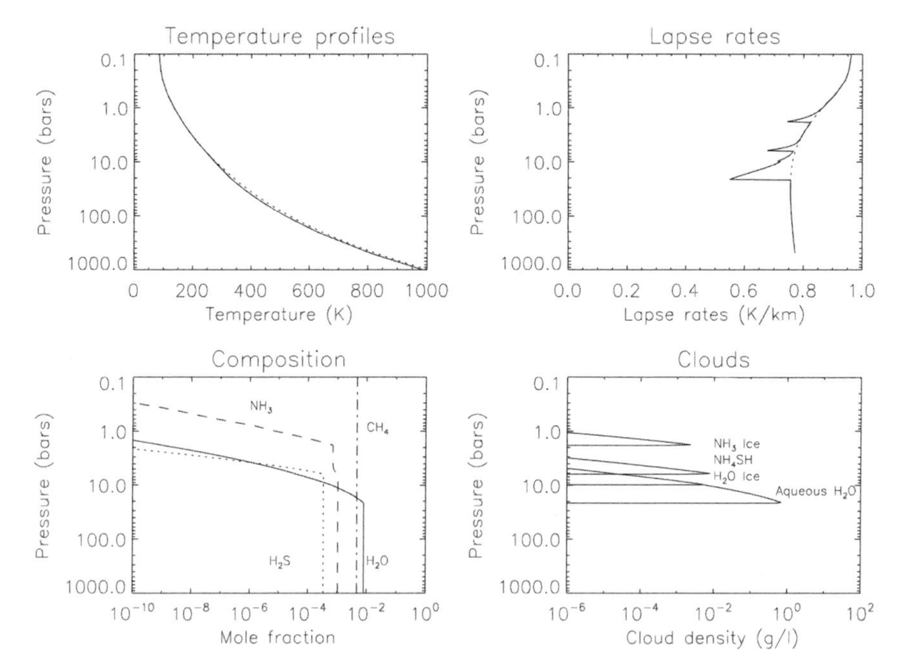

Figure 4.16. Equilibrium cloud condensation model of Saturn's atmosphere (as Figure 4.7). Calculated cloud layers: H_2O cloud (water, then ice) at ~18 bar, NH_4SH at ~5 bar, and NH_3 ice at ~1.8 bar. Assumed composition: O/H, N/H = 5× the solar value, S/H = 14× the solar value, and C/H = 11× the solar value.

NH_4SH cloud. Just as for Jupiter, the ammonia v.m.r. in this model is expected to remain fixed until the ammonia cloud condensation level of approximately 1.8 bar, and to then fall rapidly above this due to the same combination of condensation, photolysis, and mixing which defines Jupiter's ammonia profile.

Estimates of the Saturn composition profiles (and references thereto) are listed in Table 4.7 and best-fit profiles are shown in Figure 4.17. The abundance of water at 3 bar is found to be similar to that calculated from the ECCM. Ammonia, however, is again found to be less abundant than predicted by the ECCM with an estimated deep mole fraction of only ~1× the solar value from ISO 5 μm observations (de Grauuw *et al.*, 1997), although earlier 5 μm observations put the figure at less than 2.5× the solar value, while microwave observations (de Pater and Massie, 1985) suggest a value as high as 5× the solar value. The variation of ammonia above the condensation level is more difficult to detect than for Jupiter since the spectral absorption features in the mid-IR are swamped by those of phosphine, which has a much higher abundance in Saturn's atmosphere than it does in Jupiter's. An ammonia profile with approximately 50% humidity is consistent with ISO measurements (de Graauw *et al.*, 1997). The phosphine profile appears to be fixed up to a pressure level of ~600 mbar and falls rapidly above this due to photodissociation and mixing. Significant levels of CO are detected in the troposphere (of the order of 10^{-9}, if uniformly mixed, according to

Table 4.7. Composition of Saturn.

Gas	*Mole fraction*	*Measurement technique*	*Reference*
He	0.118	*Voyager* far-IR	Conrath and Gautier (2000)
NH_3	3×10^{-4} at $p > 3$ bar	Ground-based microwave	de Pater and Massie (1985)
	$\sim 5 \times 10^{-4}$	Ground-based microwave	de Pater and Massie (1985)
	$< 3 \times 10^{-4}$ (deep)	Ground-based 5 μm	Noll and Larson (1991)
	1.1×10^{-4} (deep)	ISO 5 μm	de Graauw *et al.* (1997)
	50% humidity	ISO mid-IR	de Graauw *et al.* (1997)
	1.1×10^{-4} at $p > 1.0$ bar, decreasing to $(4.5 \pm 1.0) \times 10^{-9}$ at 500 mbar with fractional scale height $f = 0.07$	*Cassini* IR	Fletcher *et al.* (2007a)
H_2S	Not detected	Ground-based microwave	de Pater *et al.* (1991)
	$(3.4 \pm 0.4) \times 10^{-3}$	Ground-based 1.3, 2, 6, 21, and 70 cm	Briggs and Sackett (1989)
H_2O	2.3×10^{-7} (upper troposphere)	ISO 5 μm	de Graauw *et al.* (1997)
	$(2–23) \times 10^{-9}$ at $p < 0.3$ mbar	ISO	Feuchtgruber *et al.* (1997)
CH_4	$\sim 5 \times 10^{-3}$	*Voyager* mid-IR	Courtin *et al.* (1984)
	$(4.5 \pm 0.9) \times 10^{-3}$	CIRS far/mid-IR	Flasar *et al.* (2005)
	$(4.7 \pm 0.2) \times 10^{-3}$	CIRS far/mid-IR	Fletcher *et al.* (2008b)
CH_3D	$(3.3 \pm 1.5) \times 10^{-7}$	Ground-based 5 μm	Noll and Larson (1991)
	$(3.0 \pm 0.2) \times 10^{-7}$	CIRS mid-IR	Fletcher *et al.* (2008b)
PH_3	1×10^{-6} at $p < 400$ mbar, $7(+3,-2) \times 10^{-6}$ at $p > 400$ mbar	Ground-based 5 μm	Noll and Larson (1991)
	5×10^{-6}	ISO 5 μm	de Graauw *et al.* (1997)
	3×10^{-6} (upper troposphere cut off at 100 mbar)	Ground-based 8.9 cm^{-1}	Weisstein and Serabyn (1994)
	7.4×10^{-6} at 645 mbar, 4.3×10^{-7} at 150 mbar	Ground-based 8.9 and 26.7 cm^{-1}	Orton *et al.* (2000/1) Lellouch *et al.* (2001)
	3×10^{-7} at 150 mbar, 4×10^{-6} at 250 mbar, 6×10^{-6} below 600 mbar	ISO 5 μm	
	7×10^{-6} in deep atmosphere	ISO 43–197 μm	Burgdorf *et al.* (2004)
	$(5.9 \pm 0.2) \times 10^{-6}$ at $p > 500$ mbar decreasing with fractional scale height $f = 0.31$ to $(3.2 \pm 0.4) \times 10^{-7}$ at 100 mbar	*Cassini* IR	Fletcher *et al.* (2008b), Fletcher *et al.* (2007a)

Gas	*Mole fraction*	*Measurement technique*	*Reference*
AsH_3	$(2.64 \pm 0.88) \times 10^{-9}$ 2.3×10^{-9} (troposphere)	Ground-based 5 µm ISO 5 µm	Noll and Larson (1991) de Graauw *et al.* (1997)
GeH_4	$(3.5 \pm 3.5) \times 10^{-10}$ (troposphere) 2.3×10^{-9} (troposphere)	Ground-based 5 µm ISO 5 µm	Noll and Larson (1991) de Graauw *et al.* (1997)
CO	1×10^{-9} (if uniformly mixed) 2.5×10^{-8} (if in stratosphere only at $p < 80$ mbar) $<10^{-7}$ (stratosphere) 2.1×10^{-9}	Ground-based 5 µm Ground-based 1.3 µm ISO 5 µm	Noll and Larson (1991) Rosenqvist *et al.* (1992) de Graauw *et al.* (1997)
HCN	—		
N_2	—		
C_2H_6	4.5×10^{-6} at $p < 10$ mbar) 1.3×10^{-5} (0.5 mbar), 8.6×10^{-6} (0.1–3 mbar) 6.5×10^{-6} (1 mbar) 1.5×10^{-5} (0.5 mbar) 1.5×10^{-5} (0.5–1 mbar) increasing towards pole	ISO mid-IR ISO mid-IR and modeled *Voyager* mid-IR ISO mid-IR *Cassini* mid-IR	de Graauw *et al.* (1997) Moses *et al.* (2000) Sada *et al.* (2005) Greathouse *et al.* (2005) Howett *et al.* (2007)
C_2H_2	4×10^{-6} (0.1 mbar), 3×10^{-7} (1 mbar) 1.4×10^{-6} (0.3 mbar), 3.2×10^{-7} (1.4 mbar) 1.6×10^{-7} (1.6 mbar) 2.0×10^{-6} (0.3 mbar), 4.1×10^{-7} (1.4 mbar) 1.9×10^{-7} (2 mbar), 3.1×10^{-7} (1.4 mbar) and decreasing towards pole	ISO mid-IR ISO mid-IR and modeled *Voyager* mid-IR ISO mid-IR *Cassini* mid-IR	de Graauw *et al.* (1997) Moses *et al.* (2000) Sada *et al.* (2005) Greathouse *et al.* (2005) Howett *et al.* (2007)
C_4H_2	1×10^{-10} at $p < 10$ mbar	ISO mid-IR	de Graauw *et al.* (1997)
C_2H_4	—		
CH_3C_2H	7×10^{-10} at $p < 10$ mbar	ISO mid-IR	de Graauw *et al.* (1997)
CH_3	Column density in the range $1.5 \pm 7.5 \times 10^{13}$ mol cm^{-2}	ISO/SWS	Bézard *et al.* (1998)
CO_2	3.4×10^{-10} at $p < 10$ mbar	ISO mid-IR	de Graauw *et al.* (1997)

(*continued*)

Table 4.7 (*cont.*)

Gas	*Mole fraction*	*Measurement technique*	*Reference*
HF	$<8.0 \times 10^{-12}$	*Cassini* far-IR	Teanby *et al.* (2006)
HCl	$<6.7 \times 10^{-11}$	*Cassini* far-IR	Teanby *et al.* (2006)
HBr	$<1.3 \times 10^{-10}$	*Cassini* far-IR	Teanby *et al.* (2006)
HI	$<1.4 \times 10^{-9}$	*Cassini* far-IR	Teanby *et al.* (2006)

Noll and Larson, 1991), which indicates rapid upwelling and vertical mixing. Indeed, the inferred stratospheric eddy-mixing coefficient profile is found by some studies (e.g., Atreya *et al.*, 1999) to be much greater than that of Jupiter's. This mixing may arise from gravity waves generated in the convective troposphere. It is presumed that since CO is found, N_2 may also be present in the stratosphere, although this is impossible to detect directly. However, at the time of writing, photochemical products such as HCN, which are observed in Neptune's atmosphere indicating N_2, have not been observed in Saturn's atmosphere. In addition to the aforementioned disequilibrium species PH_3 and CO, arsine, and germane are also observed in the Saturnian atmosphere, but at higher v.m.r.s (both are estimated to be

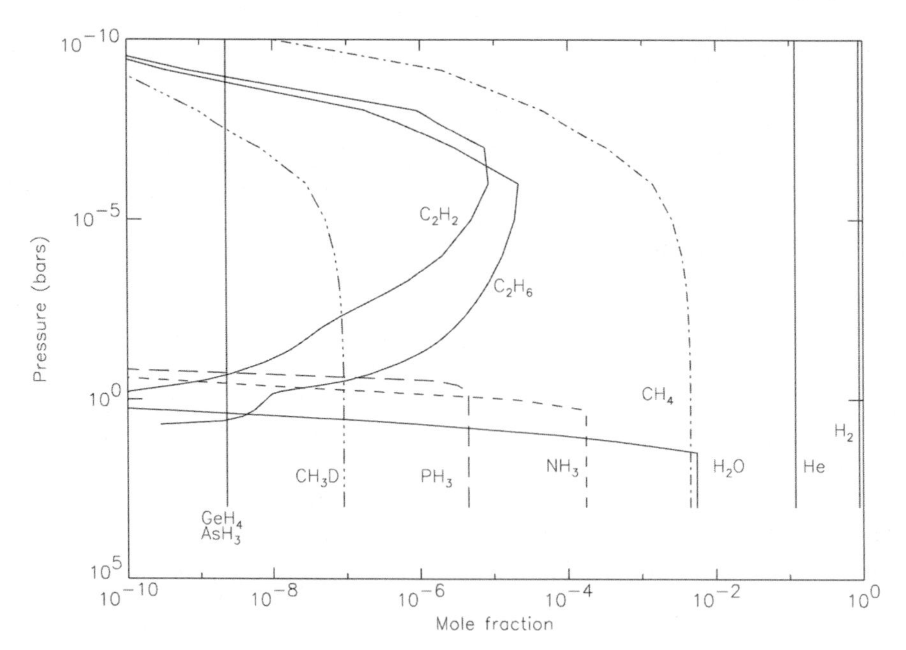

Figure 4.17. Observed and modeled abundance profiles in the atmosphere of Saturn.

2.3×10^{-9}) than seen in Jupiter's atmosphere, consistent with the generally increased abundance of heavy elements in Saturn's atmosphere than in Jupiter's. The latitudinal variation of PH_3 was investigated by Fletcher *et al.* (2007a), who used *Cassini* CIRS data to show that at the 250 mbar level the abundance of phosphine is greater at the equator than at midlatitudes, suggesting upwelling at the equator, as was also seen for Jupiter by Irwin *et al.* (2004). This characteristic has also been seen with *Cassini* VIMS observations (Baines *et al.*, 2005). At midlatitudes, Fletcher *et al.* (2007a) saw an anti-correlation between PH_3 at 250 mbar and temperature, while polewards of 60°S, PH_3 was seen to be depleted at 500 mbar, but reduced more slowly with altitude than at other latitudes, suggestive of subsidence and reduced PH_3 photolysis rates in subpolar regions. Around the poles themselves (within 3°) Fletcher *et al.* (2008a) find the deep phosphine abundance to be significantly depleted implying strong local subsidence in the cores of polar vortices. We will return to the polar vortices of Saturn in Chapter 5.

For their analysis Fletcher *et al.* (2007a) assumed a phosphine profile that was well-mixed up to 600 mbar and then fell off with a defined fractional scale height above that due to UV photolysis (i.e., the scale height with which partial pressure decreases with height is set as a specified fraction of the pressure scale height). This fits well with the estimate of Bézard *et al.* (1984) from *Voyager* IRIS observations that the radiative–convective boundary (which gives a measure of the penetration depth of sunlight) on Saturn is in the 500 mbar to 600 mbar range and is also consistent with *Cassini* CIRS observations, which find the radiative–convective boundary to be in the 350 mbar to 500 mbar range, depending on latitude (Fletcher *et al.*, 2007b). Furthermore, the assumed phosphine profile is consistent with the modeling calculations of Pérez-Hoyos and Sánchez-Lavega (2006b), who find that due to haze absorption, sunlight (integrated over all wavelengths) may only penetrate to the 600 mbar level.

The variation of the para-H_2 fraction, f_p, with latitude in the upper troposphere was determined from *Voyager* IRIS measurements by Conrath *et al.* (1998). Considerable north/south asymmetry was seen, showing Saturn to be significantly affected by seasonal forcing. In addition, some regions with low upper tropospheric temperatures, such as at 60°S, were concluded to have low f_p indicating upwelling, while other regions, having higher than average temperatures, such as 15°S, were seen to have higher f_p, indicating downwelling. More recent estimates of the para-H_2 fraction in the upper troposphere of Saturn have been determined from *Cassini* CIRS observations by Fletcher *et al.* (2007b). The ortho/para ratio is found to be approximately in equilibrium at all latitudes except for the equator, and latitudes polewards of 40°N. It is argued that at the equator this is due to rapid upwelling, which would fit with the observation of high abundances of PH_3 (Fletcher *et al.*, 2007a). Elsewhere, the high abundance of hazes in Saturn's atmosphere acts to effectively catalyze the ortho/para equilibration reaction at all latitudes except polewards of 40°N, where the abundance of larger tropospheric haze particles (which would appear to be more efficient catalysts of ortho:para equilibration) is observed to decrease (as we shall see in the next section, "Clouds and hazes"). In the stratosphere, Fouchet *et al.* (2003) used disk-averaged ISO/LWS observations to show that while f_p had the same value at both the 1 mbar and 10 mbar levels, this value was higher than the tropopause value, but lower than the equilibrium value. They proposed that some catalysis occurs on lower stratospheric equatorial haze, which is observed to be confined below 10 mbar.

In the upper stratosphere, methane photochemistry again produces hydrocarbons, such as ethane and acetylene, whose abundances may be estimated from mid-IR spectroscopy (de Graauw *et al.*, 1997; Greathouse *et al.*, 2005; Moses *et al.*, 2000; Sada *et al.*, 2005). The mole fraction of acetylene has most recently been determined from *Cassini* CIRS observations (Howett *et al.*, 2007) to be 2×10^{-7} at 2 mbar, increasing with altitude, and is seen to decrease towards the south pole by a factor of almost 2, similar to the observed distribution seen in Jupiter's stratosphere (Nixon *et al.*, 2007). Similarly, the abundance of ethane is found to be 1.5×10^{-5} at 1 mbar and increases towards the south pole by a factor of 2.5 (Howett *et al.*, 2007). Like Jupiter, the reason for these latitudunal differences is again believed to be due to the stratospheric circulation bringing hydrocarbon-rich air from equatorial high altitudes towards the poles and descending, with the acetylene abundance seen to drop as it decays faster than it can be enriched by this circulation.

In addition to the general micrometeoroid source of stratospheric oxygen discussed in Section 4.3.4, an additional source of stratospheric water in Saturn's atmosphere appears to be material falling from the rings onto the planet at specific latitudes (magnetically connected to rings). Although it is difficult to measure the latitudinal variation of water vapor, measurements have been made at UV wavelengths (Prangé *et al.*, 2006) and it is also possible to observe a decrease in hydrocarbon abundance at certain latitudes. This could be due to reactions between water and hydrocarbons which lead to other, so far undetected molecules. This may be an indirect signature of water from the rings.

Clouds and hazes

Using an ECCM, and assuming all heavy elements are enriched to the levels similar to those previously described, Atreya *et al.* (1999) predict the same three main cloud decks to occur in Saturn's atmosphere as are expected in the Jovian atmosphere. However, since the Saturnian temperatures are lower than those of Jupiter the clouds are based at correspondingly deeper levels. Hence, Atreya *et al.* (1999) find that an aqueous-ammonia cloud (blending into a water ice cloud at higher levels) is expected to start condensing near 20 bar with a column density of 12,000 kg m^{-2}, followed by an ammonium hydrosulfide cloud at $\sim$6 bar with a column density of 200 kg m^{-2}, and an ammonia ice cloud at 1.8 bar with a column density of $\sim$100 kg m^{-2} (again these cloud densities are again greatly overestimated, but they do provide a guide for the *relative* maximum cloud thicknesses). Hence, part of the reason that the visible cloud features of Saturn have a lower contrast than those of Jupiter is that the main cloud decks of Saturn lie at considerably deeper pressures. A number of estimates of Saturn's cloud structure have been made from ground-based telescopes, HST, *Voyager*, and most recently *Cassini* observations, which are summarized in Figure 4.13. While to a first order the presumed ECCM cloud structure is roughly consistent with the measured abundances of water vapor and ammonia, observations suggest that, like Jupiter, the depletion in abundance of these volatiles occurs at deeper pressures and higher temperatures than predicted by a simple ECCM. For example, ground-based microwave observations by de Pater and Massie (1985) suggest that, globally, ammonia is depleted between 3 bar and 1.5 bar, which would push the NH_3 cloud to higher altitudes and reduce its opacity. Indeed, cloud models of Saturn that are consistent with measured visible and near-infrared reflectance spectra place a thin haze layer at roughly 500 mbar and a deeper optically thick cloud top at 1.5 bar (de Graauw *et al.*, 1997). The other factor contributing to the low-contrast appearance of Saturn's belts and zones is that Saturn's pressure scale height is more than twice that of Jupiter. Assuming that the tropospheric haze has roughly the same number density in both atmospheres, the column abundance of haze above the ammonia condensation level is estimated to be 5× greater in Saturn's atmosphere than in Jupiter's leading to lower contrast of convective cloud features (Smith *et al.*, 1981). The discrete cloud features that are observed appear to be the tops of active convection systems that push their way up into the overlying semitransparent tropospheric haze region.

Just as in Jupiter's atmosphere, molecules such as ammonia and phosphine are photolyzed near the tropopause, probably contributing to the production of upper tropospheric haze. Likewise, in the stratosphere the photolysis of methane leads to detectable levels of ethane, acetylene, and other hydrocarbon products. These products diffuse downwards, and heavier hydrocarbons may possibly condense near the tropopause, where the temperatures are lowest. The highest concentrations of these smog products might thus be expected at the subsolar latitude, which varies during the period of Saturn's orbit from 26.7°N to 26.7°S (Beebe, 1997). During the *Voyager* flybys, Saturn's year was just entering northern spring and thus this simple model would predict that the haze would be thickest over the equator. However, measurements by *Voyager* in the UV (Smith *et al.*, 1982) found the stratospheric haze to be thickest over the North Pole, moderately thick in northern mid-equatorial latitudes and almost absent at southern midlatitudes (the South Pole was in darkness at the time). This north/south asymmetry was also observed in ground-based observations in the near-IR methane absorption bands (West *et al.*, 1982) and would thus seem at odds with the simple photolyzed methane haze model.

Ground-based observations of the reflection spectrum of Saturn have now been made from near-infrared to visible wavelengths for almost 30 years (Karkoschka and Tomasko, 1992; Ortiz *et al.*, 1993, 1995, 1996; West *et al.*, 1982) and have been extended into the ultraviolet by HST (Karkoschka and Tomasko, 1993, 2005; Pérez-Hoyos *et al.*, 2005). The vertical and horizontal haze structure may be inferred from these observations as follows. At visible continuum wavelengths, reflection from clouds and hazes at all levels is seen, while in the near-IR methane absorption bands, only light reflected from the upper haze layers may be seen which appear *bright* against a *dark* background. In the UV, Rayleigh scattering from the air molecules becomes important and thus, as for the near-IR methane absorption bands, only light reflected from the upper atmosphere is seen. However, since Rayleigh scattering from the air molecules is conservative, high-altitude hazes appear *dark* against a *bright* background. Hence, images of giant planets at ultraviolet and methane absorption wavelengths should, to a first approximation, be complementary and thus any differences that are present may be used to infer particle size and absorption properties.

The results of these studies suggest that there are two distinctly different types of hazes in Saturn's atmosphere (West *et al.*, 1983): a stratospheric and a tropospheric haze layer. The thin stratospheric haze layer is comprised of small particles (0.1–0.2 μm), which are highly absorbing in the UV (Karkoschka and Tomasko, 1993, 2005; Muñoz *et al.*, 2004; Ortiz *et al.*, 1996; Pérez-Hoyos *et al.*, 2005), while the tropospheric haze layer is made up of larger particles (1–2 μm), which are optically thicker (Karkoschka and Tomasko, 1993, 2005). In some models, such as Karkoschka and Tomasko (1993), the tropospheric haze spreads between the tropopause and the expected ECCM base of the ammonia cloud at 1.8 bar, while in most recent studies, there is a clear gap between the bottom of the tropospheric haze at the radiative–convective boundary at 400 mbar to 600 mbar and the lower condensation clouds (Pérez-Hoyos and Sánchez-Lavega, 2006a). This region is thought to be cleared by eddy-mixing processes and convection.

The tropospheric haze is probably associated with the ammonia cloud, but if it is, its spectral features are not clear. Kim *et al.* (2006) have tentatively identified a feature at 2.96 μm, which they attribute to ammonia ice particles below 390 mbar to 460 mbar. However, Kerola *et al.* (1997) used lower resolution 3 μm observations to exclude the possibility of an ammonia haze unless it lies beneath the 700 mbar level. The expected ammonia ice absorption features at 9.4 μm and 26 μm have not been observed by CIRS.

The tropospheric haze appears to be thickest over the Equatorial Zone (EZ) and in general seems correlated with the belt/zone structure. Karkoschka and Tomasko (2005) find that there are significant seasonal variations in the size of troposperic haze particles, with the particles being largest in summer and smallest in winter. Hence, aerosol sizes and their scattering phase functions (Section 6.5) inferred at a particular season should not be taken to be representative of Saturn's atmosphere at other seasons. Pérez-Hoyos *et al.* (2005) find that the tropospheric haze extends from 75 mbar to 400 mbar with an optical thickness (at 814 nm) that varies from 20–40 at the equator to only 5 at the pole.

Stratospheric haze is found to be most abundant in polar regions and is highly UV-absorbing (Karkoschka and Tomasko, 2005) with its opacity appearing to be directly correlated with the level of solar irradiation (Pérez-Hoyos *et al.*, 2005). Stratospheric aerosols are thus thought to be produced directly from gas by auroral processes in the polar regions, as is the case for Jupiter. At the equator, stratospheric aerosols may derive directly from the photochemical products of methane, or they may perhaps arise through the "overshooting" of tropospheric particles. This latter interpretation is perhaps supported by the observation of Ortiz *et al.* (1993, 1995, 1996) that the brightness of the EZ in near-IR methane absorption bands increased dramatically from 1991 to 1992 and appeared to be decaying in 1993, suggesting an increase in the optical depth and/or height of the tropospheric haze. This period was just after the "Equatorial Disturbance" (or Great White Spot) of 1990. The 1994

Equatorial Disturbance was shown earlier in this book in Figure 1.5 and such disturbances are discussed more fully in Chapter 5. At continuum wavelengths a north/south asymmetry was observed with the southern hemisphere appearing darker at longer wavelengths, suggesting smaller particle sizes in the tropospheric haze. These latitudes had just emerged from the shadow of the rings, which may have had an effect. The distribution of clouds and hazes observed by HST in 1998 is shown in Figure 4.18 (see color section).

The nature of tropospheric haze particles is puzzling. Simplistically, one would expect these to be composed predominantly of ammonia. However, as is the case on Jupiter (except in small regions of localized vigorous upwelling), there has been no unambiguous spectroscopic identification of pure ammonia ice on Saturn. One suggestion that has been made is that, as for Jupiter, the ammonia crystals perhaps become coated with stratospheric haze material settling down from above. Alternatively, it may be that the thermal history of tropospheric particles hides their identity, or perhaps the photochemical products of ammonia and phosphine produced near the tropopause combine with ammonia ice in some way to produce a hybrid particle. However, it may simply be that the tropospheric haze layer on Saturn is not composed or based on ammonia at all since current haze models have a clear gap between the bottom of the haze layer and the top of the ammonia cloud (Pérez-Hoyos and Sánchez Lavega, 2006a).

The presence of a relatively optically thick tropospheric haze layer increases the absorption of sunlight just below Saturn's tropopause and is probably responsible for the local maximum in temperature, or temperature "knee" observed in the 150 mbar to 300 mbar region by *Voyager* (Hanel *et al.*, 1981, 1982). This is consistent with the scattering model analysis of Pérez-Hoyos and Sánchez-Lavega (2006b), who conclude that the penetration level of sunlight in the 0.25–1.0 μm range is at ~250 mbar. *Cassini* has made more precise observations of the latitudinal and vertical extent of this knee (Fletcher *et al.*, 2007b) and has found it to be higher and less vertically extended over the equator than other latitudes, suggestive of upwelling.

4.4.3 Uranus

Composition profiles

In Chapter 2 we saw that the observable atmosphere of Uranus (Figure 4.19, see color section) has much higher levels of methane than Jupiter and Saturn, and a much greater D/H ratio. The methane v.m.r. was estimated to be 2.3% from *Voyager 2* radio occultation measurements (Lindal *et al.*, 1987). Later ground-based visible hydrogen quadrupole measurements suggested a figure closer to 1.6% (Baines *et al.*, 1995b), and more recent estimates from ground-based near-IR observations put the abundance as most likely between 1% and 1.6% (Sromovsky and Fry, 2008). Hence the C/H ratio appears to be between 40× and 50× the solar value. This, and other indicators, suggest that the hydrogen–helium atmosphere observed is merely a thin shell, accounting for only 20% of the radius and 20% of the mass, and that the bulk of Uranus is composed of "icy" materials such as water and methane, albeit hot and fluid.

The troposphere of Uranus (and Neptune) is much colder than that of Jupiter and Saturn, and together with the higher abundances of condensable species such as water, ECCM calculations predict that clouds such as water and NH_4SH condense at correspondingly deeper levels, as can be seen in Figure 4.20. The deep abundances are here assumed to be: O/H = 100× the solar value, N/H = the solar value, S/H = 11× the solar value, and C/H = 40× the solar value. The reasons for these assumptions are outlined in the following paragraphs. Water is expected to condense at very deep levels, with methane condensing near 1.5 bar. Considering the measured (or assumed) high abundances of these gases, the SALR can be seen to be very different from the DALR and thus the dry and wet temperature profiles differ significantly. The sharp variation in the DALR at 1.5 bar is due to the reduction in atmospheric heat capacity caused by the condensation of methane. It should also be noted that the sudden change in the mean molecular weight at this cloud base (and at the base of the deeper water cloud) causes a substantial change in the pressure scale height. The ortho-hydrogen/para-hydrogen was here assumed to be in the "intermediate" state outlined in Section 4.1.4.

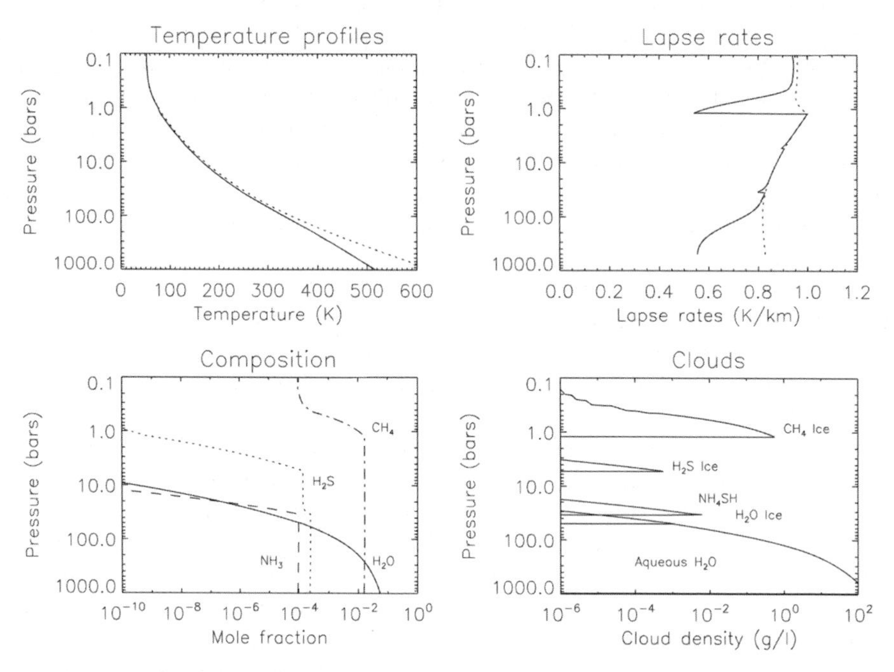

Figure 4.20. Equilibrium cloud condensation model of Uranus' atmosphere (as Figure 4.7). Calculated cloud layers: H_2O cloud (water, then ice) at $p > 1{,}000$ bar, NH_4SH at ~40 bar, H_2S ice at ~5 bar, and CH_4 ice at ~1.2 bar. Assumed composition: O/H = 100× the solar value, N/H = the solar value, S/H = 11× the solar value, C/H = 40× the solar value.

Because the observable atmospheres of Uranus and Neptune are extremely cold, it is difficult to determine composition profiles using thermal-infrared spectroscopy since the emitted spectrum has such low power. Hence, we know a lot less about the composition of the atmospheres of Uranus and Neptune than we do about the warmer atmospheres of Jupiter and Saturn. What has been determined about Uranus' composition is outlined in Table 4.8 and Figure 4.21. Methane is indeed found to condense near the 1.5 bar level and the v.m.r. drops very rapidly with height above this level reaching a minimum of approximately $(0.3–1) \times 10^{-4}$ at the tropopause. In the stratosphere, photodissociation of methane occurs between 0.1 mbar and 1 mbar, giving rise to hydrocarbon products. Acetylene (C_2H_2) was detected by ISO (Encrenaz *et al.*, 1998) with a maximum v.m.r. of 4×10^{-7} peaking at the 0.1 mbar pressure level and ethane (C_2H_6) has been detected by *Spitzer* (Burgdorf *et al.*, 2006) with a v.m.r. at the 0.1 mbar level of $(1.0 \pm 0.1) \times 10^{-8}$, along with methyl acetylene (CH_3C_2H or C_3H_4) and diacetylene (C_4H_2).

Ground-based microwave observations of Uranus between 1 mm and 20 cm indicate that both ammonia and water vapor are substantially subsolar (by a factor of several hundred) down to pressures of approximately 50 bar, but that H_2S is much more abundant than in the atmospheres of Jupiter and Saturn at levels of (10–35)× the solar value. The low abundance of water vapor is not surprising since it is expected to have mostly condensed by 100 bar, but the low abundance of ammonia is very surprising, especially when the abundance of methane is so high, and considering that both Jupiter and Saturn have significant quantities of ammonia. It would appear that either almost all the available ammonia reacts with H_2S to form NH_4SH at levels of approximately 40 bar (which would require the S/N ratio to be greater than 4× the solar value), or that substantial quantities are incorporated into a massive aqueous ammonia cloud at deep levels. It has even been suggested that ammonia might dissolve into water at even greater depths (Atreya *et al.* 2006), forming a water–ammonia ionic ocean, which may also help explain features of the observed magnetic field.

Ground-based microwave observations with the VLA have also revealed that the deep abundance of ammonia (5–50 bar) appears to vary with latitude by almost an order of magnitude with higher levels detected at equatorial latitudes (de Pater *et al.*, 1991; Hofstadter and Butler, 2003). Such a variation may be indicative of a large-scale Hadley cell with air rising at the equator and descending at the poles or alternatively that the atmosphere is convectively overturning at low latitudes, but convectively suppressed closer to the poles, with the transition latitude at ±45°. VLA observations in 2005 also show some banding and structure at midlatitudes and equatorial latitudes (Orton *et al.*, 2007c). Higher in the atmosphere, the *Voyager* IRIS observations suggest that the meridional circulation seems to be somewhat different with upwelling at midlatitudes and subsidence at the poles and equator (Flasar *et al.*, 1987; Orton *et al.*, 2007c), a topic we will return to in Chapter 5. If cloud absorption is neglected, the observed microwave spectra suggest that the atmospheric temperature profile becomes isothermal at depth (de Pater *et al.*, 1989). Such a profile would be consistent with Uranus having a very low internal heat flux since atmospheric dynamics would then be driven primarily by absorbed sunlight. How-

Table 4.8. Composition of Uranus.

Gas	*Mole fraction*	*Measurement technique*	*Reference*
He	0.15 at $p < 1$ bar	*Voyager* far-IR	Conrath *et al.* (1987)
f_{eH_2} [a]	$0.85 < f_{eH_2} \leq 1.0$	Visible hydrogen quadrupole	Baines *et al.* (1995b)
NH_3	Solar/(100–200) at $p < 10$–20 bar	Ground-based microwave	de Pater and Massie (1985)
H_2S	(10–30)× solar	Ground-based microwave	de Pater *et al.* (1991)
S/N	>5× solar	Ground-based microwave	de Pater *et al.* (1991)
H_2O	$(6–14) \times 10^{-9}$ at $p < 0.03$ mbar	ISO/SWS	Feuchtgruber *et al.* (1997)
CH_4	0.023 at $p > 1.5$ bar 0.016 at $p > 1.5$ bar $(0.3–1) \times 10^{-4}$ at tropopause, $<3 \times 10^{-6}$ at 0.1 mbar	Radio occultation Visible hydrogen quadrupole ISO/SWS	Lindal *et al.* (1987) Baines *et al.* (1995b) Encrenaz *et al.* (1998)
CH_3D/CH_4	$3.6^{+3.6}_{-2.4} \times 10^{-4}$	Ground-based 6,100–6,700 cm^{-1}	de Bergh *et al.* (1986)
PH_3	<6× solar (2.2×10^{-6}). No evidence of strong supersaturation $<8.3 \times 10^{-7}$	Ground-based 1–1.5 mm HST 5 μm	Encrenaz *et al.* (1996) Encrenaz *et al.* (2004)
AsH_3	—		
GeH_4	—		
CO	$<4 \times 10^{-8}$ (stratosphere) $<10^{-8}$ (troposphere) $<5 \times 10^{-7}$ (troposphere) $<1.7 \times 10^{-8}$ (at 3.1 bar cloud tops) increasing to $\sim 2.5 \times 10^{-8}$ at tropopause	Ground-based mm Ground-based 2.6 mm Ground-based 1–1.5 mm HST 5 μm	Rosenqvist *et al.* (1992) Marten *et al.* (1993) Encrenaz *et al.* (1996) Encrenaz *et al.* (2004)
HCN	—		
N_2	—		

Gas	*Mole fraction*	*Measurement technique*	*Reference*
C_2H6	3×10^{-6} at 0.1 mbar $<2 \times 10^{-8}$ in stratosphere $\sim 4 \times 10^{-7}$ in stratosphere $(1.0 \pm 0.1) \times 10^{-8}$ at 0.1 mbar	Modeled Ground-based 12 μm Ground-based 12 μm *Spitzer* 10–20 μm	Encrenaz *et al.* (1998) Orton *et al.* (1987) Hammel *et al.* (2006) Burgdorf *et al.* (2006)
C_2H_2	10^{-8} 4×10^{-7} at 0.1 mbar	*Voyager* UV ISO/SWS	Yelle *et al.* (1989) Encrenaz *et al.* (1998)
C_4H_2	$(1.6 \pm 0.2) \times 10^{-10}$ at 0.1 mbar	*Spitzer* 10–20 μm	Burgdorf *et al.* (2006)
C_2H_4	—		
CH_3C_2H	$(2.5 \pm 0.3) \times 10^{-10}$ at 0.1 mbar	*Spitzer* 10–20 μm	Burgdorf *et al.* (2006)
CH3	Not detected	*Spitzer* 10–20 μm	Burgdorf *et al.* (2006)
CO_2	$(4 \pm 0.5) \times 10^{-11}$ at 0.1 mbar	*Spitzer* 10–20 μm	Burgdorf *et al.* (2006)

[a] Fraction of H_2 with eqm ortho/para.

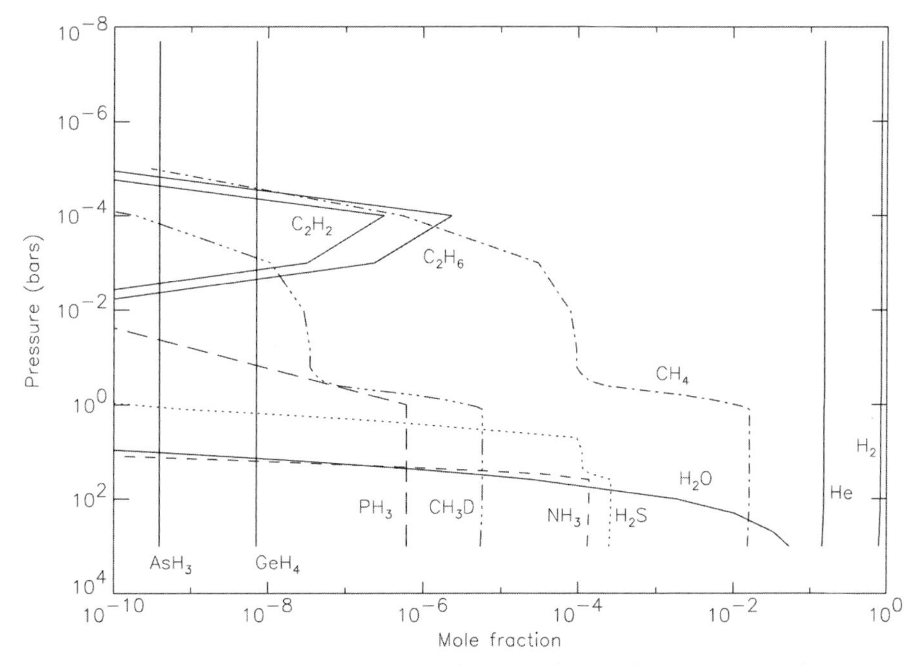

Figure 4.21. Observed and modeled abundance profiles in the atmosphere of Uranus.

ever, the observed spectrum may also be explained by the presence of a deep, very optically thick cloud, such as the expected water cloud. Hence, whether or not the deep atmosphere of Uranus is really isothermal is debatable. In Figure 4.20 we have assumed that the temperature profile follows an SALR at depth.

The low IR flux of Uranus means that spectral features of possible disequilibrium species such as phosphine, germane, and arsine are very difficult to detect. Ground-based observations of phosphine lines in the very far IR between 1 mm and 1.5 mm suggest the P/H to be $<6\times$ the solar value, while HST observations at 5 μm (Encrenaz *et al.*, 2004) revise the v.m.r. (at the 3 bar level) to be less than 8×10^{-7} ($1.3\times$ the solar value). We might expect the deep P/H ratio to be somewhat higher, given the high C/H ratio, and its low abundance suggests sluggish vertical mixing in Uranus' atmosphere. This sluggish motion is supported by the low detected abundance of CO at the cloud tops from the same 5 μm HST observations of less than 1.8×10^{-8} increasing to $\sim 2.5 \times 10^{-8}$ at the tropopause, pointing to an additional external supply of oxygen atoms.

The sluggish motion of Uranus' atmosphere is also indicated by the observation by *Voyager* IRIS (Conrath *et al.*, 1998) that the ortho:para hydrogen ratio appeared to be close to equilibrium at all altitudes, although the ratio was found to be lowest at southern latitudes and highest at northern latitudes, suggesting upwelling in the south and subsidence in the north during the *Voyager 2* flyby. The near-equilibrium of the ortho:para hydrogen ratio in Uranus' stratosphere was also concluded by ISO (Fouchet *et al.*, 2003), who further point out that, in addition to sluggish motion, the ortho–para conversion process is efficiently catalyzed by small aerosols, which are thought to be very abundant in Uranus' stratosphere.

Clouds and hazes

Using the ECCM described earlier a massive water cloud is expected to condense anywhere between 100 bar and 1,000 bar (the exact level depends on the H_2O/H_2 ratio and the deep temperature profile, both of which are uncertain); an ammonia hydrosulfide cloud somewhere around 40 bar; either an ammonia ice cloud, or a hydrogen sulfide ice cloud somewhere around 8 bar (which depends on whether the deep v.m.r. of ammonia is greater than or less than that of hydrogen sulfide, assuming that the minor species is effectively mopped up in the NH_4SH cloud leaving just the more abundant species to condense at lower temperatures); and a methane ice cloud at approximately 1.5 bar (de Pater *et al.*, 1991). As has already been discussed, the apparent depletion of tropospheric ammonia suggests that almost all the ammonia that is not dissolved in an aqueous ammonia cloud or water–ammonia ionic ocean at deeper levels reacts with H_2S to form NH_4SH leaving an H_2S cloud to condense near 2 bar to 8 bar.

Observationally only two convective cloud decks were observed in the Uranian atmosphere during and immediately after the *Voyager 2* flyby in 1986 (Figure 4.13): a thin cloud near the 1.5 bar level, and an optically thick cloud beneath with a cloud top at approximately 2.7 bar to 3.1 bar, detected using observations of the hydrogen 4-0 and 3-0 quadrupole lines, and the 681.8 nm methane line (Baines and Bergstralh,

1986; Baines *et al.*, 1995b). The observed abundance of methane rapidly reduces with height at the 1.5 bar level indicating that this is indeed the methane cloud, but the visible optical depth of the methane cloud was found to be very thin ($0.4 < \tau < 0.7$) at low latitudes, indicating either very weak vertical mixing at this level or that particles rapidly grow and precipitate in this cloud leading to low visible reflectance. The visible opacity of the methane haze was found to increase to approximately 2.4 at 65°S in Uranus' "bright" South Polar zone and the mean particle size in the methane layer was estimated to be of the order of 1 μm at all latitudes observed (Rages *et al.*, 1991). The lower cloud is probably the top of the expected hydrogen sulfide cloud, although no positive spectral identification has been made. All we do know is that the aerosols in this cloud appear bright in the blue–green, but darken significantly at wavelengths longer of 0.6 μm (Baines and Bergstrahl, 1986; Baines and Smith, 1990). This, combined with methane gas becoming increasingly absorbing at longer wavelengths in the visible spectrum leads to Uranus' dominant blue–green color. The identity of this chromophore material is unknown but may arise from "tanning" of aerosols by incident UV sunlight.

Disk-averaged observations of the near-IR spectrum of Uranus from the 1970s (Fink and Larson, 1979) were used by Sromovsky *et al.* (2006) to show that reflectance from the expected methane cloud at 1.5 bar remains small in the 1 μm to 2 μm range, but rather than finding a single cloud at ~3 bar, reflectance was needed from two pressure levels: ~2 bar and 6–8 bar. This conclusion was also reached from near-IR observations recorded in 2006 by UKIRT (Irwin *et al.*, 2007), and the Keck Observatory in 2004 (Sromovsky and Fry, 2007) and in 2006 (Sromovsky and Fry, 2008). Reflection from the upper cloud was determined to increase towards the southern hemisphere, reaching a peak at 45°S, where a bright polar collar has been visible in the last decade.

As mentioned earlier it is believed that methane observed in the stratosphere arrives there almost entirely through eddy mixing from the tropopause. Although small convective clouds are seen (Karkoschka, 1998b) these are not as bright as the similar small clouds seen on Neptune, although they have become increasingly common and brighter in the lead-up to, and during, Uranus' Northern Spring Equinox in 2007 (Hammel *et al.*, 2005a, b; Sromovsky *et al.*, 2007). The pressure at the tops of these discrete clouds is estimated to vary from 0.5 bar to as little as 0.2 bar, with particle sizes of the order of 1 μm, similar to the properties of the methane haze layer. Hence, like Neptune, these clouds are thought to be convective methane cumulus clouds. However, unlike Neptune (as we will see), these clouds are not thought to be vigorous enough to penetrate the tropopause and thus increase the stratospheric methane to levels greater than the "cold trap value" of the saturated v.m.r. at the tropopause. In fact, the stratospheric abundance of methane in Uranus' stratosphere is the lowest of any of the giant planets indicating very weak vertical mixing. Methane photochemistry, which is important between 10 mbar and 0.1 mbar, produces hydrocarbons such as ethane (C_2H_6), acetylene (C_2H_2), ethylene (C_2H_4), and polyacetylenes ($C_{2n}H_2$, $n = 2, 3, 4$). These products diffuse through the atmosphere via eddy mixing, but unlike Jupiter and Saturn where the products diffuse through the tropopause without further processing, the temperature in Uranus' stratosphere is so low that these products condense to form stratospheric hydrocarbon haze layers at lower altitudes. Diacetylene ice (C_4H_2) is expected to start condensing at $p > 0.1$ mbar, acetylene ice at $p > 2.5$ mbar, and ethane ice haze at $p > 15$ mbar. Once haze particles start to condense, they begin to coalesce to form larger particles, which may then gravitationally sediment out of the atmosphere. Hence the haze layers gradually thin out at pressures greater than roughly 30 mbar. The mean haze particle size is estimated to be of the order of 0.1 μm (West *et al.*, 1991) and the visible optical depth of the combined haze layers is estimated to be very low at only 0.01. The main component of the haze is modeled to be acetylene ice, although the haze particles are found not to be the pure white hydrocarbon condensates that are expected, but instead are quite dark (imaginary refractive index of 0.01 in the visible). The cause of this may possibly be UV-induced polymerization or "tanning", which appears consistent with the dark particles that are also found in Neptune's stratosphere. The submicron size of the haze particles means that they precipitate out of the stratosphere only very slowly on timescales of 10 to 100 years. Once they reach the troposphere they are evaporated and eventually pyrolyzed back to methane at sufficiently high temperatures. No very great change in the stratospheric haze optical depth with latitude has been found, which is in stark contrast to the stratospheric hazes of Jupiter and Saturn, which are strongly UV-absorbing near the poles. Presumably auroral processes are not so important in Uranus' atmosphere.

The disk-averaged albedo of Uranus has been monitored at visible wavelengths since the 1950s. The light curve is found to be approximately sinusoidal with peaks at the solstices (Lockwood and Jerzykiewicz, 2006) and is well fitted by a simple model where the albedo of Uranus at latitudes equatorwards of 45°S and 45°N is darker than that polewards of 45°S and 45°N by a factor of f (Hammel and Lockwood, 2007). Different parts of the observed light curve at 472 nm and 551 nm require different f factors of between 0.75 and 0.98, pointing to considerable interseasonal variability, probably caused by variations in convective activity at midlatitudes. Variations in stratospheric temperature have also been monitored by a number of methods (Hammel and Lockwood, 2007) and temperatures were seen to increase up to the solstice in 1986, but have since decreased. Variations in microwave brightness have also been monitored since the early 1970s (Klein and Hofstadter, 2006) and the variations do not appear to be due to geometric effects alone, but are also due to temporal variations in temperature or ammonia abundance down to pressure levels of tens of bars. In particular, there is evidence of a rapid planetary-scale change from 1993 to 1994. These changes coincided with significant changes seen in the cloud and haze layers at visible/near-IR wavelengths by HST (Rages *et al.*, 2004), where the South Pole faded during 1994–2002 and a bright zone appeared at 45°S, together with a less bright zone appearing in the intervening years at 70°S.

4.4.4 Neptune

Composition profiles

Neptune is the most remote and most difficult to observe of the giant planets (Figure

Figure 4.22. Neptune observed by *Voyager 2* in 1989. Courtesy of NASA.

4.22). The mean composition profiles of tropospheric and stratospheric gases in Neptune's atmosphere are shown in Figure 4.23. Just like Uranus, Neptune appears to have a higher abundance of methane in its atmosphere than Jupiter and Saturn, and a much greater D/H ratio. *Voyager* radio occultation profiles and ground-based observations of hydrogen quadrupole lines indicate a deep methane v.m.r. of 2.2% (Baines and Hammel, 1994), indicating a C/H enrichment of $\sim 50\times$ the solar value. Since interior models of Neptune suggest that it contains a greater proportion of heavy elements than Uranus, it is likely that Neptune's methane mixing ratio is indeed greater than that of Uranus. Again the hydrogen–helium atmosphere observed is merely a thin shell, accounting for roughly 15% of the radius and 6% of the mass.

As for Uranus, ECCM calculations predict that clouds such as water and NH_4SH condense at very deep levels (as can be seen in Figure 4.24) with methane condensing near 2 bar. The deep abundances are here assumed to be: $O/H = 100\times$ the

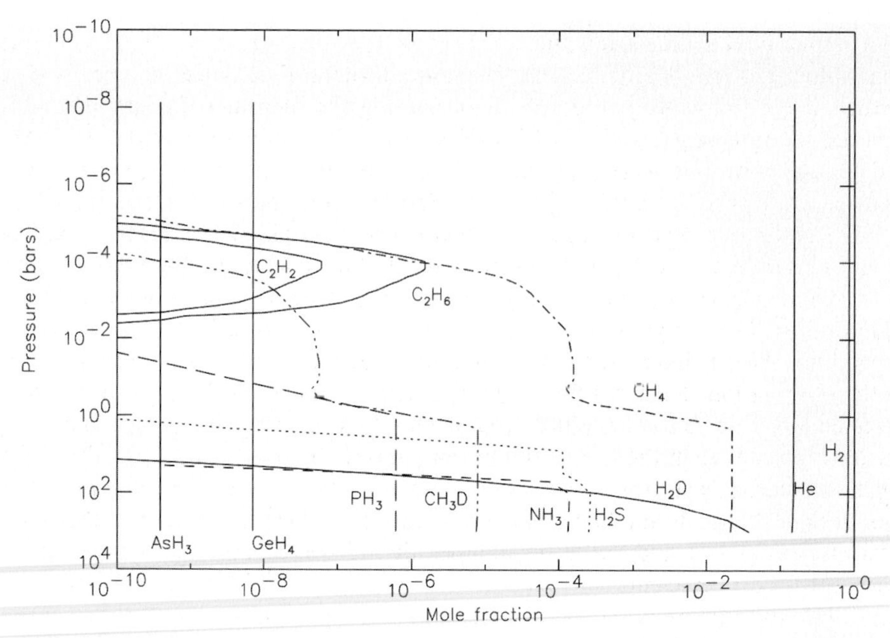

Figure 4.23. Observed and modeled abundance profiles in the atmosphere of Neptune.

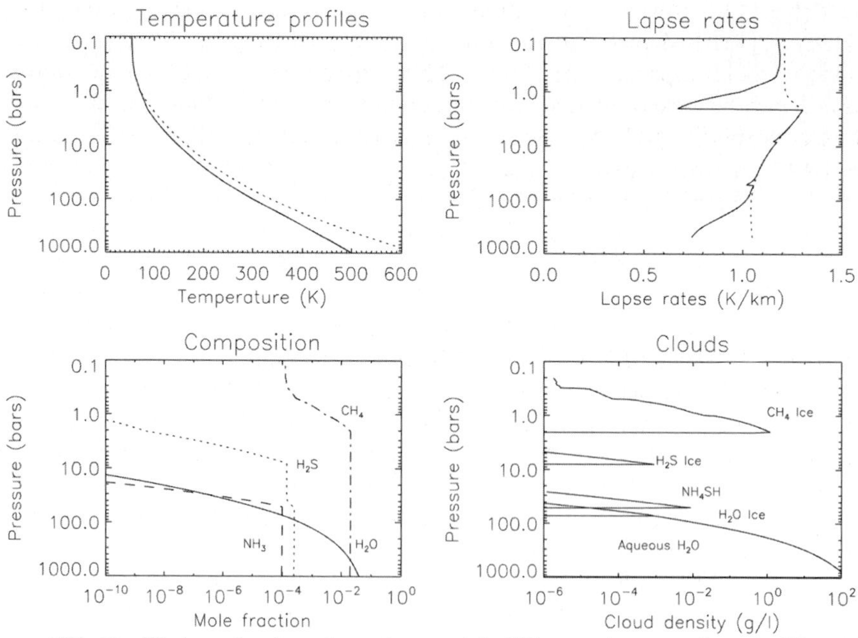

Figure 4.24. Equilibrium cloud condensation model of Neptune's atmosphere (as Figure 4.7). Calculated cloud layers: H_2O cloud (water, then ice) at $p > 1{,}000$ bar, NH_4SH at $\sim$50 bar, H_2S ice at $\sim$8 bar, and CH_4 ice at $\sim$2 bar. Assumed composition: O/H = 100× the solar value, N/H = the solar value, S/H = 11× the solar value, C/H = 50× the solar value.

solar value, N/H = the solar value, S/H = 11× the solar value, and C/H = 50× the solar value. The low N/H value and the choice of other abundances will be discussed further in the following paragraphs. Considering the measured (or assumed) high abundances of these gases, the SALR can be seen to be very different from the DALR and thus the dry and wet temperature profiles differ significantly. The ortho:para hydrogen ratio was assumed to be in the "frozen" state outlined in Section 4.1.3.

The ortho:para hydrogen ratio in Neptune's atmosphere was determined from *Voyager* IRIS observations by Conrath *et al.* (1998), who noted that Neptune has a fairly symmetric temperature structure in the upper troposphere, which is coolest at midlatitudes. These regions were found to correspond with a low f_p, which is consistent with these being regions of rapid uplift, a deduction that is further confirmed by the observation that vigorous convective activity is seen at these latitudes. In the stratosphere, Fouchet *et al.* (2003) found that f_p decreases with height, but is significantly greater than the local equilibrium para-H_2 fractions. Although the optical depth of Neptune's stratospheric hazes is known to be greater than those of Uranus (next section, "Clouds and hazes"), the particle sizes are estimated to be rather larger (Pryor *et al.*, 1992) and thus the number density of aerosols is less, leading to presumably less efficient catalysis. In addition, the eddy-mixing coefficient in Neptune's atmosphere is 10× greater than in Uranus' atmosphere leading to rapid vertical transport of high-f_p air from the tropopause.

Observations of Neptune's composition profiles are listed in Table 4.9. Ground-based microwave observations suggest that the atmosphere of Neptune is, like Uranus, greatly depleted in ammonia by a factor of roughly 100 relative to solar down to levels of approximately 50 bar. This suggests again that large quantities of ammonia may be locked up in an aqueous ammonia cloud, or that the abundance of H_2S exceeds that of NH_3 by a factor of at least 5 and thus that the formation of an NH_4SH cloud at ~40 bar effectively removes all remaining ammonia from the atmosphere. The H_2S abundance is estimated from these ground-based microwave studies to be (10–35)× the solar value. However, alternative explanations exist. For example, it may just be that Neptune (and Uranus) has been deficient in nitrogen since formation since N_2 is not efficiently trapped in amorphous ice unless the ice formation temperature is very cold (see Chapter 2). However, it would appear unlikely that Jupiter and Saturn should be nitrogen-rich and Uranus and Neptune nitrogen-poor if these planets formed in the neighborhood of their current distances from the Sun. Instead, we would have to form Jupiter and Saturn initially at the edge of the solar system (at ~30 AU) and then migrate inwards, and form Uranus and Neptune at 5 AU to 10 AU and then migrate outwards, which seems unlikely. Another explanation for the low abundance of ammonia in Neptune's atmosphere is that nitrogen may instead be mostly in the form of N_2 in the observable atmosphere, which is difficult to detect spectroscopically. This scenario is consistent with the observed levels of stratospheric HCN (v.m.r. $\sim 1 \times 10^{-9}$, Marten *et al.*, 2005), which is most likely formed from the photolysis by-products of CH_4 and nitrogen atoms. High levels of molecular nitrogen, a disequilibrium species in the observable atmosphere of Neptune, suggest rapid convection, which is consistent with Neptune's strong internal heat flux, the non-equilibrium ortho:para hydrogen

Table 4.9. Composition of Neptune.

Gas	*Mole fraction*	*Measurement technique*	*Reference*
He	0.19 at $p < 1$ bar 0.15 if N_2 v.m.r. is 0.3% $0.149^{+0.017}_{-0.022}$ CH_4 v.m.r. is 2% and assuming N_2 v.m.r. <0.7%	*Voyager* far-IR *Voyager* far-IR ISO SWS/LWS	Conrath *et al.* (1991) Conrath *et al.* (1993) Burgdorf *et al.* (2003)
f_{eH_2} [a]	$0.63 < f_{eH_2} \leq 1.0$	Visible hydrogen quadrupole	Baines *et al.* (1995b)
NH_3	Solar/(100–200) at $p <$ 10–20 bar 6×10^{-7} (saturated) at ~130 K, 6 bar May be supersaturated (w.r.t. NH_4SH) at $p <$ 20–25 bar, hence greater than Uranus	Ground-based microwave Radio occultation Ground-based microwave	de Pater and Massie (1985) Lindal *et al.* (1992) de Pater *et al.* (1991)
H_2S	(10–30)× solar, same for Uranus and Neptune	Ground-based microwave	de Pater *et al.* (1991)
S/N	>5× solar	Ground-based microwave	de Pater *et al.* (1991)
H_2O	Up to 440× solar <100–200× solar $(1.7–4.1) \times 10^{-9}$ at $p < 0.6$ mbar	Modeled to allow 1 ppm of CO Interior models ISO	Lodders and Fegley (1994) Podolak and Marley (1991) Feuchtgruber *et al.* (1997)
CH_4	~0.01 at $p > 1.5$ bar 0.02 at $p > 1.5$ bar 0.022 at $p > 1.5$ bar, 3.5×10^{-4} (stratosphere) possibly as high as 1.7×10^{-3}	Radio occultation Radio occultation Visible reflectance	Tyler *et al.* (1989) Lindal *et al.* (1992) Baines and Hammel (1994)
CH_4/ CH_3D	$6^{+6}_{-4} \times 10^{-4}$ $(3.6 \pm 0.5) \times 10^{-4}$	Ground-based 6,100–6,700 cm^{-1} Ground-based mid-IR	de Bergh *et al.* (1990) Orton *et al.* (1992)
PH_3	No evidence of strong supersaturation, deep abundance unmeasurable	Ground-based 1–1.5 mm	Encrenaz *et al.* (1996)
AsH_3	—		

Gas	*Mole fraction*	*Measurement technique*	*Reference*
GeH_4	—		
CO	$(6.5 \pm 3.5) \times 10^{-7}$ (stratosphere) seems present in troposphere, too	Ground-based 1–1.3 mm	Rosenqvist *et al.* (1992)
	$(1.2 \pm 0.4) \times 10^{-6}$	Ground-based 1 mm	Marten *et al.* (1993)
	$(0.6–1.5) \times 10^{-6}$ (0.5–2 bar)	Ground-based 2.6 mm	Guilloteau *et al.* (1993)
	$(0.7–1.3) \times 10^{-6}$ (troposphere)	Ground-based 0.87 mm	Naylor *et al.* (1994)
	2.7×10^{-7} (30–800 mbar)	UV reflectance HST	Courtin *et al.* (1996)
	$<1 \times 10^{-6}$ (6×10^{-7} preferred)	Ground-based 1–1.5 mm	Encrenaz *et al.* (1996)
	$(1.0 \pm 0.2) \times 10^{-6}$	Ground-based 1 mm	Marten *et al.* (2005)
	0.5×10^{-6} at $p > 20$ mbar, 1.0×10^{-6} at $p < 20$ mbar	Ground-based 1 mm	Lellouch *et al.* (2005)
	$(0.6 \pm 0.4) \times 10^{-6}$ in upper troposphere/lower stratosphere, increasing to $(2.2 \pm 0.5) \times 10^{-6}$ in upper stratosphere	Ground-based 1 mm	Hesman *et al.* (2007)
HCN	$(3 \pm 1.5) \times 10^{-10}$ (stratosphere)	Ground-based 1–1.3 mm	Rosenqvist *et al.* (1992)
	$(1.0 \pm 0.3) \times 10^{-9}$ (0.003–30 mbar), confined to stratosphere	Ground-based 1 mm	Marten *et al.* (1993)
	$(3.2 \pm 1.5) \times 10^{-10}$ at 2 mbar, approx. constant with height, condenses at 3 mbar	Ground-based 1.1 mm	Lellouch *et al.* (1994)
	1.5×10^{-9} at $p < 0.3$ mbar, decreasing at lower altitudes	Ground-based 1 mm	Marten *et al.* (2005)
N_2	<0.6%	*Voyager* far-IR	Conrath *et al.* (1993)
	<0.7%	ISO SWS/LWS	Burgdorf *et al.* (2003)
C_2H_6	1.5×10^{-6} (stratosphere)	*Voyager* mid-IR	Bézard and Romani (1991)
	$(0.2–1.2) \times 10^{-6}$ (stratosphere)	Ground-based mid-IR	Orton *et al.* (1992)
C_2H_2	6×10^{-8} (stratosphere)	*Voyager* mid-IR	Bézard and Romani (1991)
	$(0.6–7.1) \times 10^{-8}$ (stratosphere)	Ground-based mid-IR	Orton *et al.* (1992)
C_4H_2	—		
C_2H_4	—		

(*continued*)

Table 4.9 (*cont.*)

Gas	*Mole fraction*	*Measurement technique*	*Reference*
CH_3C_2H	—		
CH_3	1.6×10^{-13} mol cm^{-2} above the 0.2 mbar level	ISO/SWS	Bézard *et al.* (1999)
CO_2	6×10^{-8} at $p < 5$ mbar	ISO/SWS	Feuchtgruber *et al.* (1997)

[a] Fraction of H_2 with eqm ortho/para.

ratio determinations (Conrath *et al.*, 1998), and with the presence of significant levels of tropospheric and stratospheric CO, whose v.m.r. has been estimated from millimeter observations to be 2.2×10^{-6} in the upper stratosphere, decreasing to just under 1×10^{-6} in the lower stratosphere/upper troposphere (Hesman *et al.*, 2007). Such a CO profile requires both an internal and external source of CO with the lower stratosphere/upper troposphere abundance requiring an O/H value of several hundred (Lodders and Fegley, 1994). The source of upper stratospheric CO is unclear. The material is unlikely to come from Neptune's moons and rings, but it might possibly come from water arriving in the form of meteorites and interplanetary grains. However, the flux of water required is significantly higher than has been detected in Neptune's stratosphere from ISO observations (Feuchtgruber *et al.*, 1997), and Lellouch *et al.* (2005) suggest that the CO may instead have arrived from a cometary impact in the last few hundred years. This suggestion is supported by the observation that when Comet Shoemaker–Levy 9 struck Jupiter's atmosphere in 1994, most of the comet's water was converted to CO by shock chemistry.

If a significant fraction of nitrogen in Neptune's atmosphere does exist mostly in the form of N_2 and not ammonia, then this may explain why the He/H_2 ratio derived from *Voyager* far-IR measurements apparently exceeds the solar value, an observation which is almost impossible to explain theoretically. However, an N_2 mole fraction of only 0.3%, which is equivalent to an N/H value of $\sim$55$\times$ the solar value (i.e., similar to the C/H ratio), reduces the derived He/H_2 ratio to solar, which is much more plausible and is also consistent with the Uranian estimate. Once transported to the stratosphere, nitrogen atoms may be produced by the dissociation of N_2 molecules by galactic cosmic ray (GCR) impacts. An alternative source of nitrogen atoms may be from atoms escaping from neighboring Triton, which are then captured by Neptune.

Clearly N_2 is very much a disequilibrium species in the cold reducing atmosphere of Neptune and thus as fast as it may be uplifted, a certain fraction per second will convert to NH_3. Why then do we still not see much ammonia in Neptune's atmosphere? It is likely that ammonia formed from N_2 at pressures less than $\sim$40 bar will react with H_2S (which may be more abundant in the atmospheres of Neptune and

Uranus by a factor of $S/N > 5$) to form NH_4SH. Alternatively, NH_3 formed at pressures less than ~8 bar should freeze out to form ammonia crystals. It is interesting to note that the *Voyager 2* radio occultation experiment estimated the ammonia v.m.r. at ~130 K and 6 bar to be 6×10^{-7}, which is close to the s.v.p. of ammonia under those conditions, suggesting the presence of ammonia crystals.

Of the other possible tropospheric species—H_2O, AsH_3, GeH_4, and PH_3—only PH_3 has any thermal-IR features in regions of the spectrum where the extremely cold Neptune atmosphere emits in any strength. To date no phosphine has been detected, although ground-based millimeter studies suggest that the possibility of supersaturation is very small (Encrenaz *et al.*, 1996). Some formation models of Neptune suggest that the bulk H_2O/H_2 fraction must be several hundred times the solar ratio and some models find that just such a large fraction is necessary to account for the detected high abundance of CO in Neptune's upper troposphere/upper stratosphere (Lodders and Fegley, 1994) outlined earlier, although this does not fit well with the SCIP model of Owen and Encrenaz (2003), described in Chapter 2.

The abundance of methane in the stratosphere is estimated to be 3.5×10^{-4} (Baines and Hammel, 1994), but may be as high as 1.7×10^{-3}. This is much greater than the saturated vapor pressure of methane at the temperature of Neptune's tropopause determined by *Voyager*, which is $(1\text{–}3) \times 10^{-5}$. The implication of this may be that methane is transported to the stratosphere not just as a gas, but also as ice crystals from the troposphere, which subsequently sublimate at the higher stratospheric temperatures. It has been proposed that the high internal heat flux of Neptune drives moist convection that produces convection cells vigorous enough to punch their way up through the tropopause and into the stratosphere. There is good observational evidence of localized, high, thick methane clouds and other transient storms, especially at midlatitudes, which are considered to be sufficiently vigorous to lift methane ice crystals to the stratosphere. The shadows cast by these clouds on the underlying main 3.8 bar cloud deck were occasionally observed during the *Voyager 2* flyby (Figure 4.25) and were used to determine that their cloud tops were near the 1 bar level, consistent with them being composed of methane crystals. An alternative explanation for the high abundance of methane in Neptune's stratosphere comes from ground-based observations of Neptune's South Polar region in 2006, which showed the temperature near the South Pole to be greatly enhanced at the 100 mbar level (Orton *et al.*, 2007a). If this hotspot is due to localized solar heating, as Orton *et al.* (2007a) suggest, and not due to adiabatic heating from downwelling, then the "cold trap" to methane is lifted in the polar regions, allowing methane to diffuse easily up into the stratosphere with a v.m.r. as high as 1%, before being transported latitudinally and mixed with less methane–enriched air.

Once in the stratosphere, methane is photolyzed at high altitudes to form hydrocarbons. The only hydrocarbons that have been detected so far are ethane and acetylene with v.m.r.s of $\sim 1 \times 10^{-6}$ and 6×10^{-8}, respectively (Bézard and Romani, 1991; Orton *et al.*, 1992) and also CH_3 (Bézard *et al.*, 1999). Emission from stratospheric methane and ethane can be seen in ground-based observations, and this is seen to be enhanced polewards of 70°S (Hammel *et al.*, 2007; Orton *et al.*, 2007a). The enhancements are too large to be explained by increased methane and

Figure 4.25. Close-up of Neptune's methane clouds, as observed by *Voyager 2*, showing shadows cast by them on the main cloud deck beneath. Courtesy of NASA.

ethane abundance, and instead are probably due to a temperature increase in the stratosphere of $\sim$3 K. In addition, disk-averaged observations of the emission from stratospheric methane and ethane have been used to monitor a steady rise in emission since 1985 (Hammel and Lockwood, 2007; Hammel *et al.*, 2006; Marten *et al.*, 2005).

Both stratospheric water ($\sim 3 \times 10^{-9}$ at $p < 0.6$ mbar) and CO_2 (6×10^{-8} at $p < 5$ mbar) have been detected by ISO/SWS, indicating an external source of oxygen, probably via the continuing capture of interplanetary dust (Feuchtgruber *et al.*, 1997).

Clouds and hazes

To first order, the composition and pressure levels of the major cloud layers are similar to those of Uranus. Cloud layers of CH_4, H_2S or NH_3, NH_4SH, and H_2O

(ice and aqueous) are expected at about 2 bar, 8 bar, 50 bar, and greater than 50 bar, respectively. The H_2O cloud in particular is likely to be extremely massive and heavily precipitating since the deep O/H ratio may be several hundred times greater than the solar ratio. However, similar to Uranus, only two observable cloud decks have been confirmed (Figure 4.13): a very thin cloud near the 1.5 bar level ($\tau \sim 0.1$), and an optically thick cloud beneath with a cloud top at ~3.8 bar, detected using observations of the hydrogen 4-0 and 3-0 quadrupole lines and 681.8 nm methane line (Baines and Hammel, 1994; Baines *et al.*, 1995a). The observed abundance of methane rapidly reduces with height at the 1.5 bar level indicating that this is indeed the methane cloud. However, the optical depth is again very much less than that predicted by ECCM models. It is thought that particle growth in this cloud may be very rapid and thus that "raindrops" quickly form, which drop back through to below the condensation level thus keeping the visible and near-IR optical depth of this cloud low. The deeper thick cloud is most likely to be the top of the expected H_2S cloud, although it has an unexpectedly strong absorption at red wavelengths, which contributes to Neptune's blue color. However, forming an H_2S cloud at ~3 bar requires ammonia to be severely depleted at this level. While some ammonia can be absorbed in forming aqueous ammonia and NH_4SH cloud decks, the amount that can be trapped by these mechanisms is insufficient to explain a pure H_2S cloud at ~3 bar unless the S/N ratio is significantly supersolar, as was alluded to earlier. Atreya *et al.* (2006) suggest that the ammonia may instead be absorbed in a water–ammonia ionic ocean at very deep levels, first mooted by Hubbard (1984) to explain aspects of Neptune's and Uranus' magnetic fields. Clearly Neptune's main cloud deck is more red-absorbing than Uranus' since Neptune appears significantly bluer than Uranus (Baines and Bergstrahl, 1986; Baines *et al.*, 1995a). The blueness of the deep cloud is puzzling since the coloration is not experimentally observed in H_2S and NH_3 ices. Irradiated frosts do have a dip in reflectance of 0.6 μm, but the albedo recovers at longer wavelengths whereas the observed clouds are dark for $\lambda > 0.7$ μm. An alternative interpretation is that the 3.8 bar cloud is partially transparent (Sromovsky *et al.*, 2001c) and thus that the properties may have more to do with the wavelength dependence of scattering efficiency.

The general background appearance of Neptune is that the midlatitudes are slightly darker than the equator and poles (as can be seen in Figure 1.6). This variation is due to variable reflectance from the deeper cloud and could be due to variations in the scattering efficiencies of the particles in this cloud, variations in the cloud top height, or perhaps due to variations in the tropospheric methane abundance. However, superimposed on this background structure are small convective clouds (presumably composed of methane), which were observed to be common during the *Voyager 2* flyby in 1989 and widely occur at many latitudes. Subsequently, convective activity appears to be increasing, and such clouds are observed to be most common at midlatitudes from 25–60°S and 20–40°N (due to Neptune's obliquity, it is not possible currently to view latitudes polewards of ~40°N from the ground). Sromovsky *et al.* (2001c, d) noted a clear increase in midlatitude cloud activity from HST observations between 1996 and 1998 and Max *et al.* (2003) used ground-based observations to show that cloud activity was continuing to increase and estimated the cloud tops to be just below the tropopause. Gibbard *et al.* (2003) analyzed ground-based spectral observations between 2 μm and 2.3 μm to estimate that northern midlatitude clouds had cloud tops in the lower stratosphere, while those at southern mid-latitudes had cloud tops near the tropopause at between 0.1 bar and 0.14 bar. In addition, clouds were also seen near 70°S with cloud tops between 0.27 bar and 0.17 bar. Gibbard *et al.* (2003) used these observations to suggest that upwelling was occurring at southern midlatitudes, and suggested that northern midlatitude clouds were due to a subsidence of stratospheric haze material. The small clouds near the South Pole were inferred to be isolated convective events in a region of general subsidence.

Higher up in the stratosphere methane is photolyzed into hydrocarbons, which diffuse down through the atmosphere and freeze out as haze layers as the temperature decreases towards its minimum of 50 K at the tropopause. Using photochemical models and estimated eddy-mixing coefficients to calculate the hydrocarbon v.m.r. profiles consistent with measurements, it is predicted that C_4H_2 condenses at $p > 2$ mbar, C_2H_2 condenses at $p > 6$ mbar, and C_2H_6 condenses at $p > 10$ mbar (Baines *et al.*, 1995a). Neptunian hydrocarbon hazes appear to be more abundant and optically thicker than their counterparts in the Uranian atmosphere, and indeed absorption of sunlight by these aerosols may explain why the lower stratosphere of Neptune is some 40 K warmer than Uranus. It is estimated that 6% to 14% of incident solar UV and visible flux is absorbed by stratospheric hazes. However, other sources of heating are possible such as tidal heating by Triton, or the breaking of vertically propagating gravity waves generated in the more vigorous Neptunian troposphere. The mean particle size of stratospheric hazes is estimated to be 0.2 μm, and the average visible optical depth (619 nm) is estimated to be 0.025 (Baines *et al.*, 1995a; Pryor *et al.*, 1992). Particle formation requires the presence of foreign condensation nuclei or ions. At lower stratospheric altitudes these may be provided by hazes "drizzling" down from above, but at the highest altitudes the likely sources of condensation nuclei, such as ions generated by meteoritic impacts or UV and cosmic-ray ionization, are limited. Hence, it is possible that hydrocarbon vapors may become significantly supersaturated before condensation starts. Indeed it has been postulated that haze formation may be episodic with partial pressures slowly rising to levels greatly in excess of the s.v.p., triggering the onset of condensation and the rapid formation of haze particles, which reduce the supersaturation level to 1.0 and then fall down though the atmosphere.

The disk-averaged albedo of Neptune has been monitored since 1950 and distinct trends are seen. Sromovsky *et al.* (2003) considered observations from 1970 onwards and found a good correlation between observed trends and a lagged seasonal model where disk-averaged albedo was greatest near the solstices. However, Lockwood and Jerzykiewicz (2006) found this model to be inconsistent with a longer time series of data and Hammel and Lockwood (2007) note that the brightness is better correlated with subsolar latitude.

A particularly curious feature of Neptune's stratospheric aerosols was that for a long time their reflectance appeared to be correlated with solar activity (Baines, 1997b; Baines *et al.*, 1995a; Lockwood *et al.*, 1991), whereas no such correlation was seen in Uranus' stratosphere, which is otherwise so similar to Neptune's. It was proposed that over time the hydrocarbon aerosol particles of both Uranus and Neptune are "tanned" by radiation from the Sun, which makes them more absorbing, and the rate of "tanning" depended on solar activity. On Neptune, the higher optical depth of stratospheric hazes means they contribute more to disk-integrated reflectivity, and the particles sizes are large enough (0.2 μm) that tanned particles fall reasonably quickly through the atmosphere to be replaced by fresh white particles, which themselves slowly tan. Hence, a fair degree of correlation might be expected between solar activity and disk-averaged reflectivity. On Uranus, however, the haze particle size is estimated to be only of the order of 0.1 μm and thus these particles remain for much longer in the stratosphere before settling. If haze particle residence time is a sufficiently large fraction of the solar cycle period of $\sim$11 years then there might not be expected to be such a strong tracking between mean aerosol brightness and reflectivity. In addition, the smaller optical depth of stratospheric hazes in Uranus' atmosphere would mean variations in stratospheric haze reflectivity would contribute a smaller fraction of total disk reflectance. Although an elegant and appealing theory, Lockwood *et al.* (1991) noted that anti-correlation was becoming harder to observe as the disk-averaged albedo of Neptune continued to increase in the early 1990s. Subsequently, Lockwood and Jerzykiewicz (2006) note that anti-correlation started to break down in 1990 and is no longer visible. Whether the "tanning" theory is fundamentally flawed, or whether the effect is now being masked by the increased reflectance from Neptune's tropospheric clouds is unknown.

4.5 BIBLIOGRAPHY

Andrews, D.G. (2000) *An Introduction to Atmospheric Physics*. Cambridge University Press, Cambridge, U.K.

Atreya, S.K. (1986) *Atmospheres and Ionospheres of the Outer Planets and Their Satellites*. Springer-Verlag, Berlin.

Bagenal, F., T. Dowling, and W. McKinnon (Eds.) (2004) *Jupiter: The Planet, Satellites and Magnetosphere*. Cambridge University Press, Cambridge, U.K.

Beatty, J.K., C. Collins Petersen, and A. Chaikin (Eds.) (1999) *The New Solar System* (Fourth Edition). Cambridge University Press, Cambridge, MA.

Bergstrahl, J., E.D. Miner, and M.S. Matthews (Eds.) (1991) *Uranus*. University of Arizona Press, Tucson, AZ.

Cruikshank, D.P., M.S. Matthews, and A.M. Schumann (Eds.) (1995) *Neptune and Triton*. University of Arizona Press, Tucson, AZ.

Encrenaz, T., J.-P. Bibring, and M. Blanc (1995) *The Solar System* (Second Edition). Springer-Verlag, Berlin.

Gehrels, T. (Ed.) (1976) *Jupiter*. University of Arizona Press, Tucson, AZ.

Gehrels, T. and M.S. Matthews (Eds.) (1984) *Saturn*. University of Arizona Press, Tucson, AZ.

Houghton, J.T. (1986) *The Physics of Atmospheres* (Second Edition). Cambridge University Press, Cambridge, U.K.

Lewis, J.S. (1995) *Physics and Chemistry of the Solar System*. Academic Press, London.

Shirley, J.H. and R.W. Fairbridge (Eds.) (1997) *Encyclopaedia of the Planetary Sciences*. Chapman & Hall, London.

Weissman, P.R., L.-A. McFadden, and T.V. Johnson (Eds.) (1998) *Encyclopaedia of the Solar System*. Academic Press, London.

NASA planetary images in this chapter were mostly downloaded from

http://photojournal.jpl.nasa.gov/

ure 1.3. Jupiter as observed by *Cassini* in December 2000. The dark spot on the lower left is the shadow of Europa, of the Galilean satellites. The Great Red Spot (GRS) is clearly visible as is the prominent banding and the highly oulent cloud structure. Courtesy of NASA.

Figure 1.5. Saturn as observed by the Hubble Space Telescope (HST) in December, 1994. A large bright feature, known as a Great White Spot (GWS), is clearly visible at the center of Saturn's disk. Courtesy of NASA.

Figure 1.7. Saturn's night side as seen by *Cassini* ISS, with the Sun directly behind Saturn's disk.

Figure 1.8. False-color image of Saturn and its rings. Here, light emitted from the deep atmosphere at 5.1 μm is shown in red, light reflected at 3 μm (weak atmospheric absorption) is shown in green, and light reflected at 2.3 μm (strong atmospheric absorption) is shown in blue. At 2.3 μm the rings are highly reflecting, while at 3 μm they are highly absorbing due to water ice absorption. Hence, the rings appear blue in this image.

Figure 1.9. Uranus observed by *Voyager 2* in 1986. Courtesy of NASA.

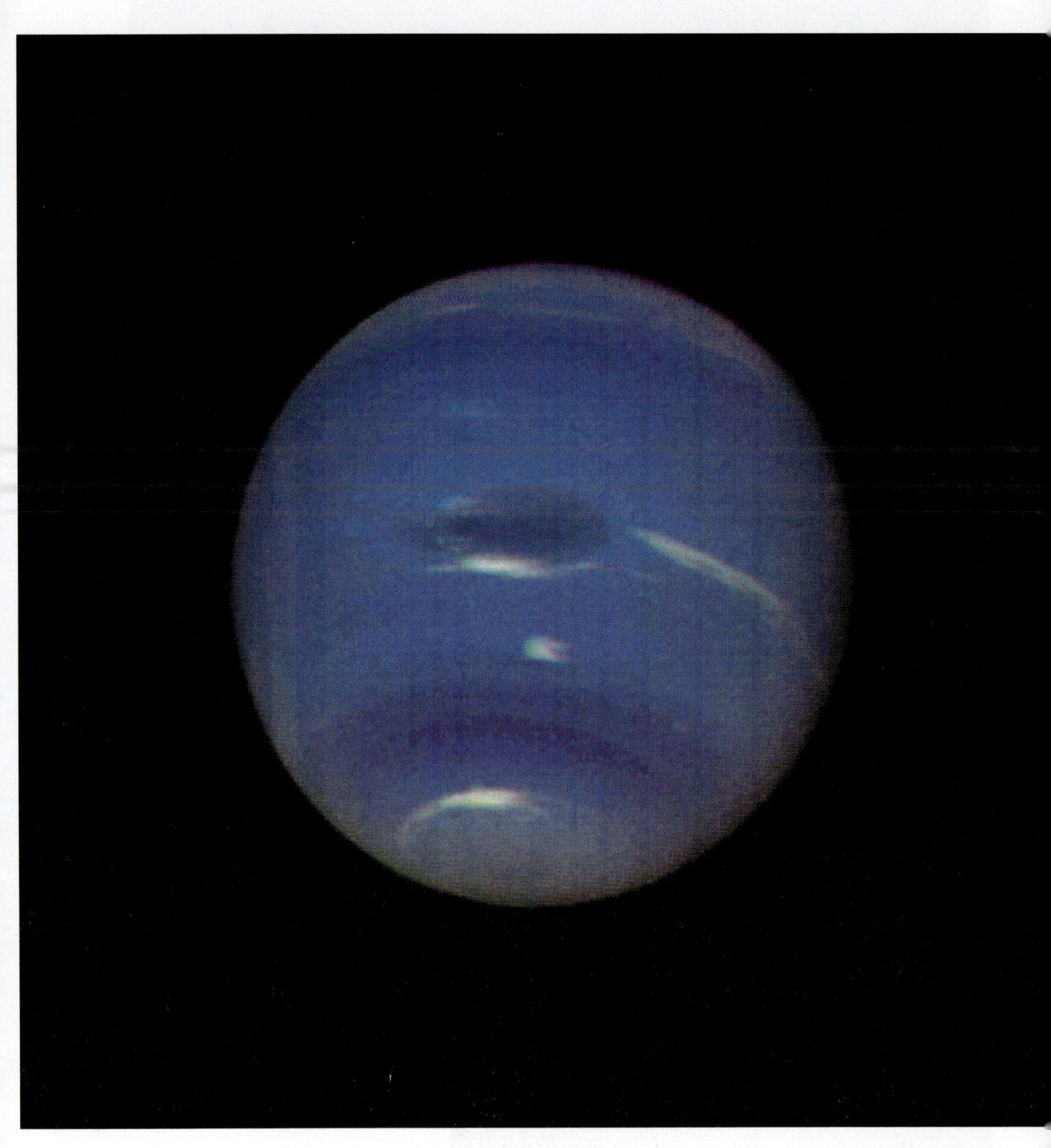

Figure 1.10. Neptune observed by *Voyager 2* in 1989. The Great Dark Spot (GDS) is clearly visible at the center of disk together with the darker midlatitude bands, and various small, convectively generated white methane clo Courtesy of NASA.

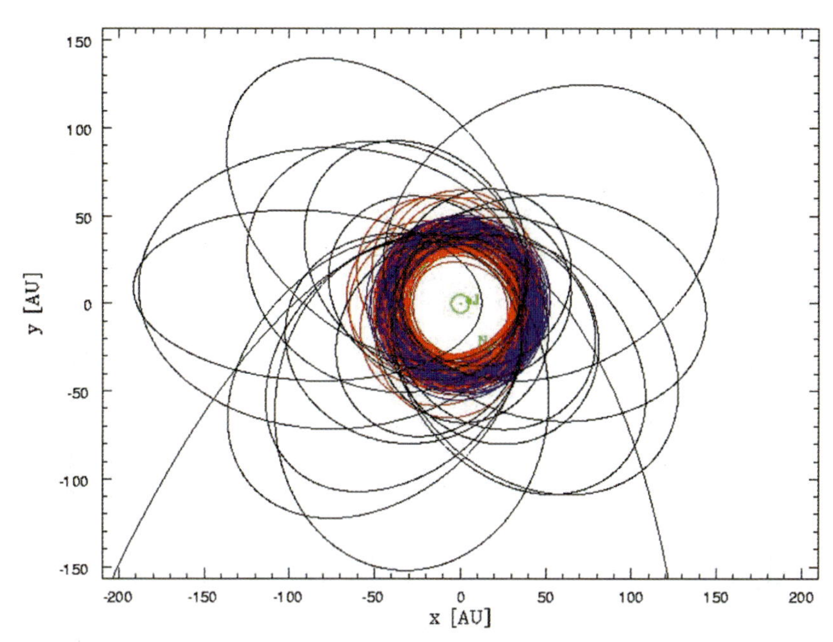

Figure 2.6. Plan view of known trans-Neptunian object orbits in the Kuiper–Edgeworth Belt. Red orbits relate to "Plutinos", objects which are in a 3:2 orbital resonance with Neptune. Blue orbits relate to classical Kuiper Belt Objects (KBOs), which do not have an orbital resonance with Neptune and typically orbit slightly farther from the Sun. Black orbits relate to scattered KBOs whose orbits have high eccentricity. Courtesy of David Jewitt, Institute for Astronomy, University of Hawaii.

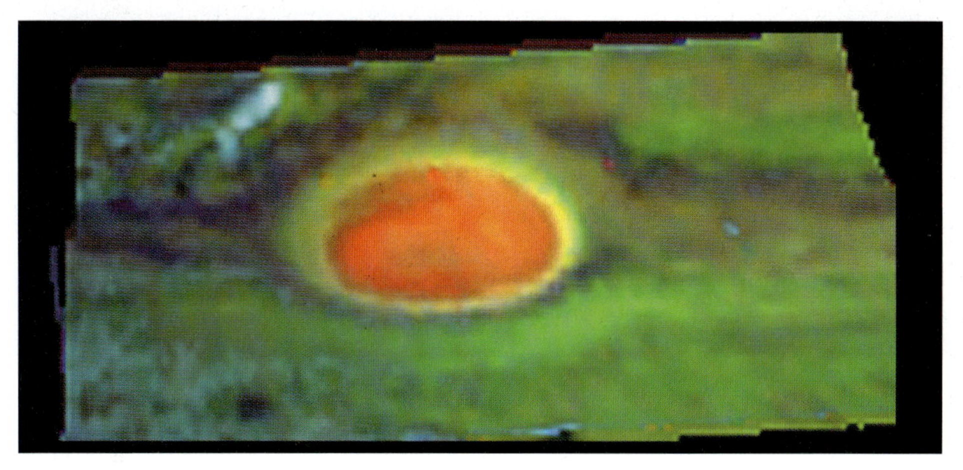

Figure 4.10. False-color image of the GRS constructed from near-IR data recorded in 1996 by *Galileo* NIMS. Reddish-orange areas show regions of high-level clouds, yellow areas depict mid-level clouds, and green areas depict lower level clouds. The darker areas are cloud-free regions. The light blue region to the northwest of the GRS has been identified as middle-to-high-altitude-level spectrally identifiable ammonia clouds (SIACs). From Baines *et al.* (2002). Courtesy of NASA.

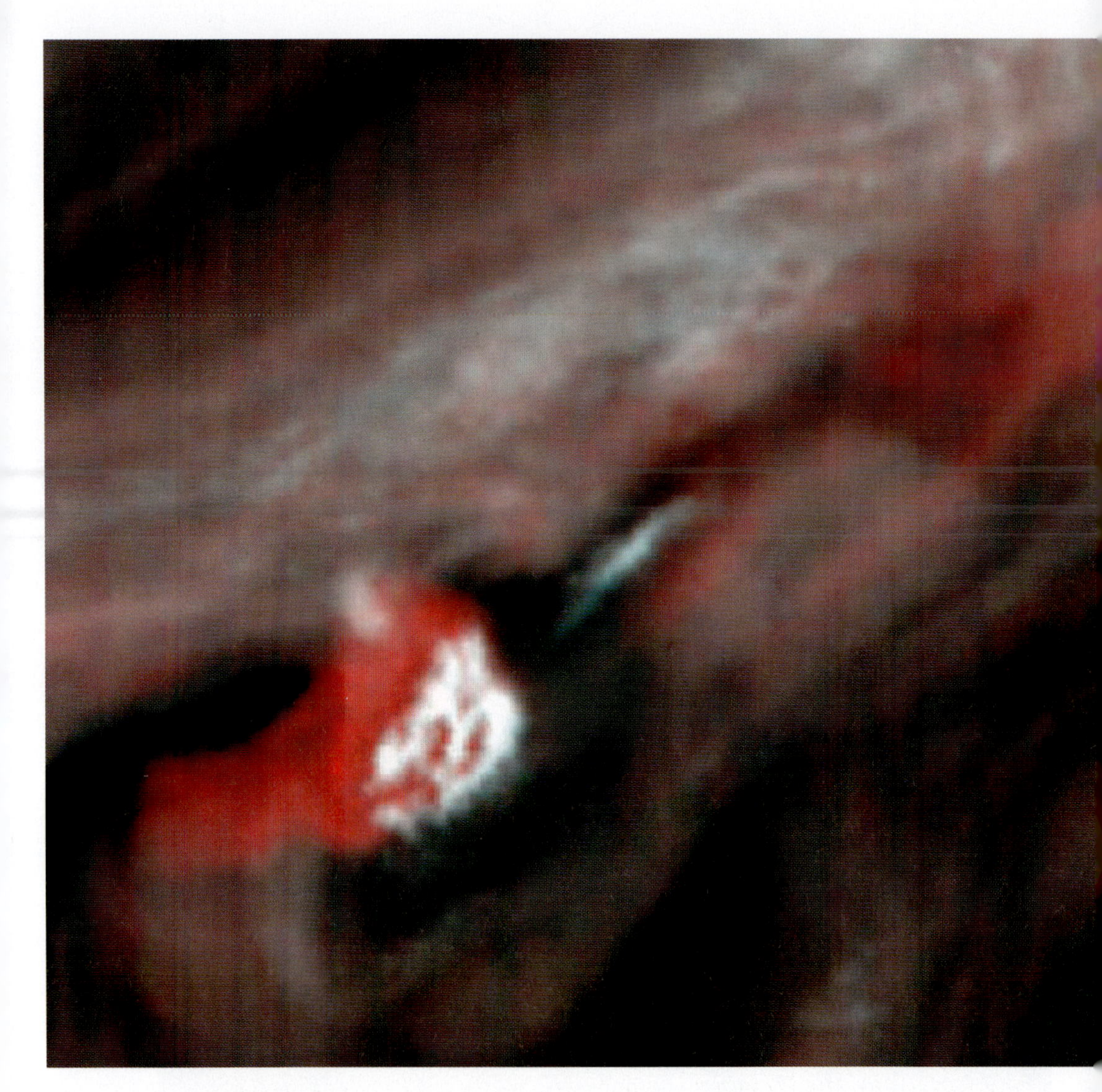

Figure 4.11. False-color picture of a convective thunderstorm 10,000 km (6,218 miles) northwest of the recorded by *Galileo* SSI in June 1996 (Banfield *et al.*, 1998). The picture is constructed of images record wavelengths of 756 nm (red), 889 nm (blue), and 727 nm (green). The central thick white cloud is over 1,000 km ac and is estimated to be standing almost 25 km higher than the surrounding clouds. Its base, extending off to th appears red in this representation, indicating a pressure of almost 4 bar. Hence the base of this cloud is al certainly to be composed of water. Courtesy of NASA.

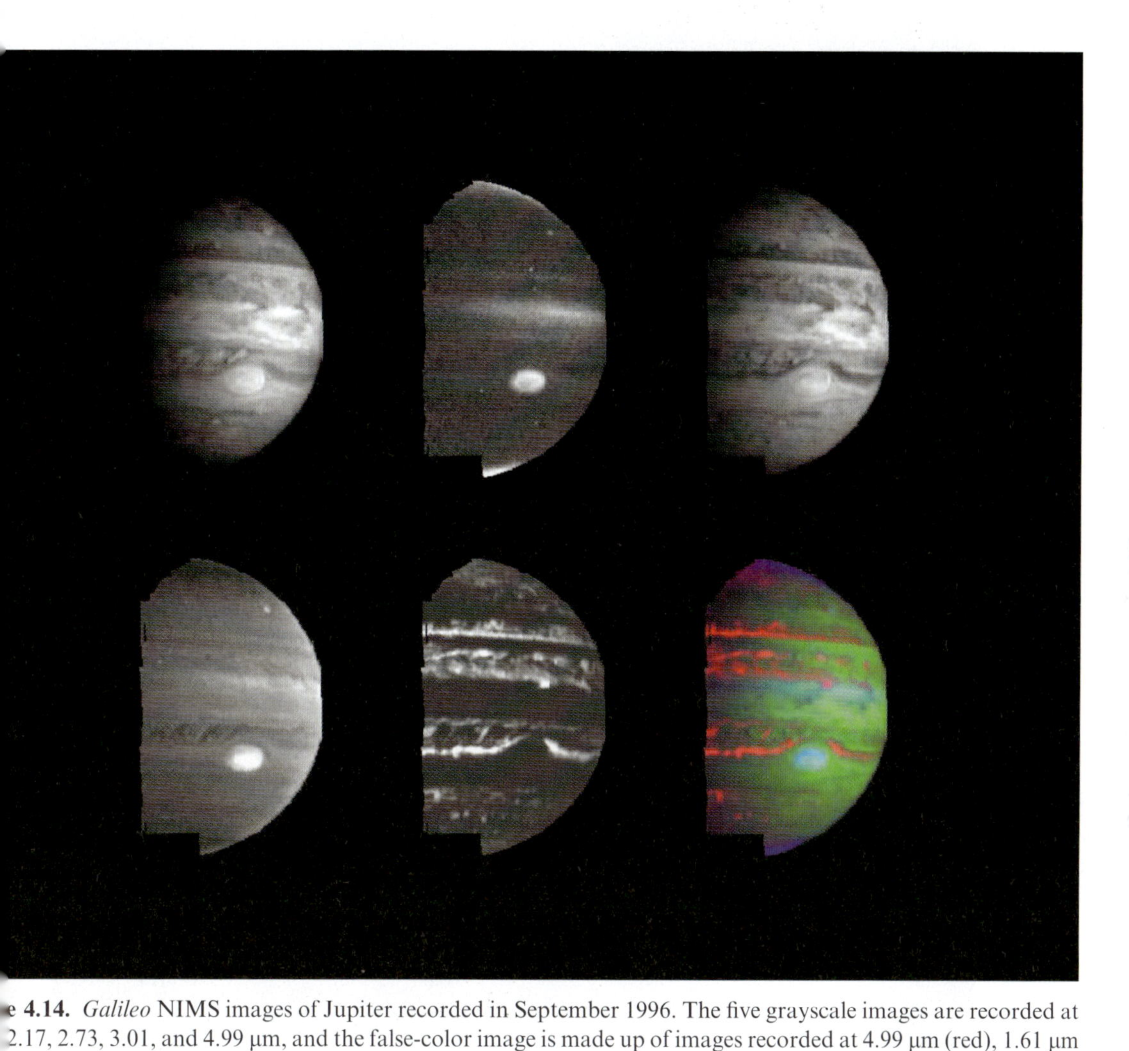

e 4.14. *Galileo* NIMS images of Jupiter recorded in September 1996. The five grayscale images are recorded at 2.17, 2.73, 3.01, and 4.99 μm, and the false-color image is made up of images recorded at 4.99 μm (red), 1.61 μm ı), and 2.17 μm (red), respectively. The Jovian atmosphere is relatively transparent at 1.61 and 2.73 μm and thus ıttern of belts and zones, familiar from visible wavelength images, is seen. At 2.17 μm, strong absorptions of ıne and hydrogen mean that only light reflected from the upper haze layers is visible and these are clearly seen he GRS and the NEB, while at 3.01 μm, where ammonia gas also absorbs, the features seen are due to both haze mmonia. The 4.99 μm image records thermal emission from the 5–8 bar region and the general anti-correlation en visible/near-IR albedo and thermal emission is clear. The GRS is clearly visible in this image, and its height means that it appears bright in all of the near-IR reflected sunlight images, and is correspondingly dark 5 μm image where it can be seen to be surrounded by a cloud-free annulus. Courtesy of NASA.

Figure 4.18. False-color image of Saturn composed of images recorded at 1.0 μm (low gaseous absorption, blue), 1.8 μm (medium gaseous absorption, green), and 2.1 μm (high gaseous absorption, red), in 1998 by the HST/Near-Infrared Camera and Multi-Object Spectrometer (NICMOS) instrument. In this representation the blue colors indicate a clear atmosphere down to the main, presumed ammonia cloud deck. The dark region around the south pole indicates a large hole in the main cloud layer here. The green and yellow regions indicate haze layers above the main cloud deck with thin hazes appearing green and thick hazes yellow. The red and orange regions indicate clouds reaching high up into the atmosphere. Thick clouds are seen at equatorial latitudes, while the southern hemisphere at this time appears to have higher abundances of upper tropospheric haze than the northern hemisphere. Courtesy of NASA.

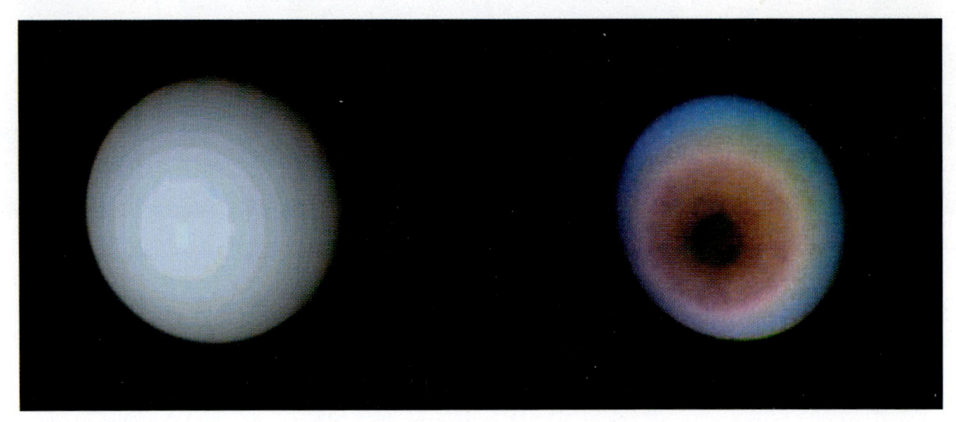

Figure 4.19. Uranus observed by *Voyager 2* in 1986. The left-hand frame shows a true-color image while the right-hand frame shows a false-color image where the original ultraviolet-, violet-, and orange-filtered images are displayed, respectively, as blue, green, and red (greatly stretched to improve contrast). Ultraviolet wavelengths are sensitive to the abundance of upper tropospheric and stratospheric haze, and during the *Voyager 2* flyby, when the South Pole of Uranus was pointed almost directly towards the Sun, the increased abundance of haze over the South Pole was clearly visible. Courtesy of NASA.

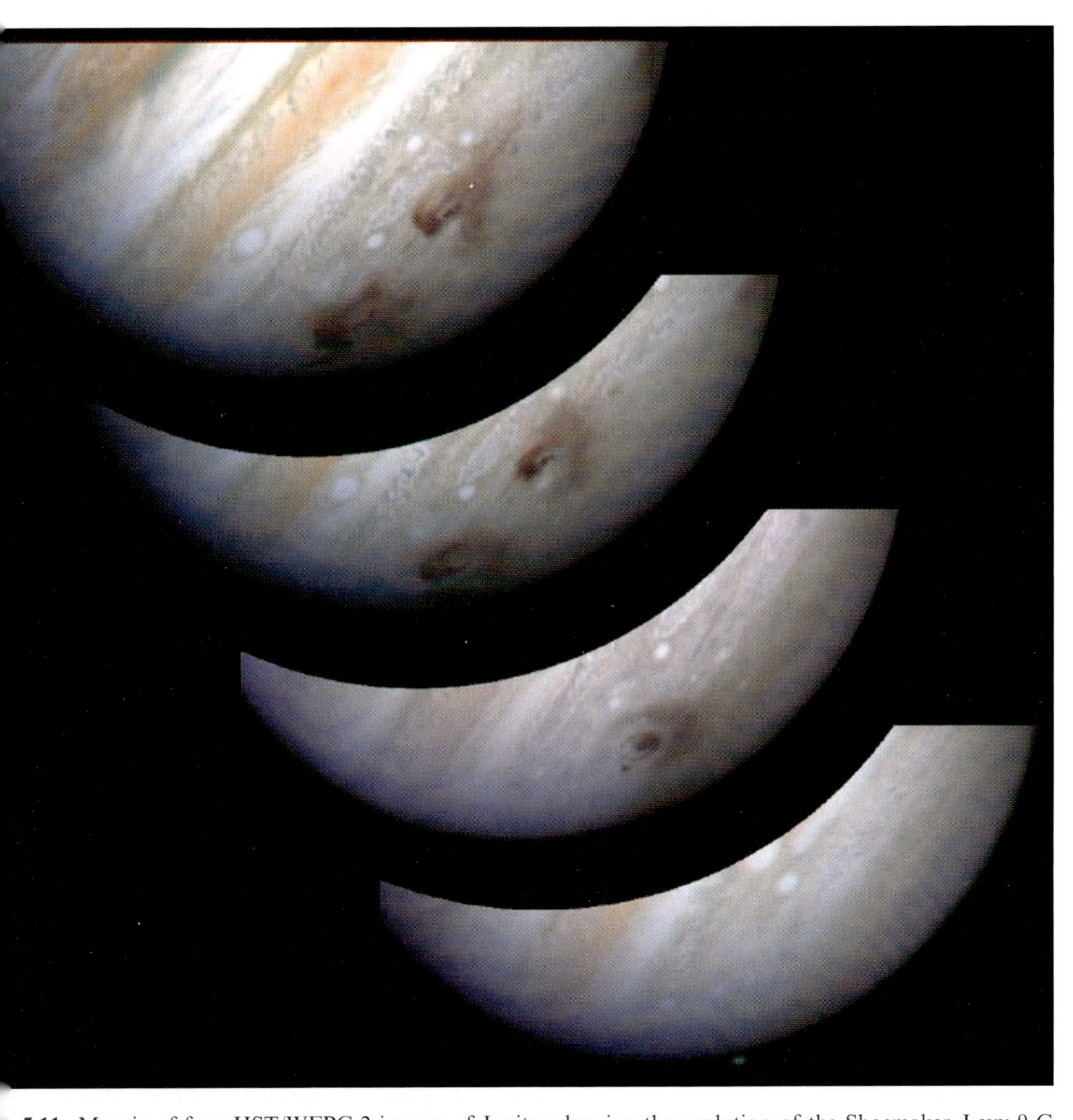

• **5.11.** Mosaic of four HST/WFPC-2 images of Jupiter showing the evolution of the Shoemaker–Levy 9 G
t site in 1994. The images from lower right to upper left show: (1) the impact plume at 7:38 UT on 18 July (about 5
es after the impact); (2) the fresh impact site an hour and a half later at 9:19 UT; (3) the impact site after evolution
winds of Jupiter (left), along with the L impact (right), taken at 6:22 UT on 21 July (3 days after the G impact and
ys after the L impact); and (4) further evolution of the G and L sites due to winds and an additional impact (S) in
vicinity, taken at 8:08 UT on 23 July (5 days after the G impact). Courtesy of NASA.

Figure 5.12. *Voyager 1* image of Jupiter's Great Red Spot and one of the STBs white ovals in 1979. Court
NASA.

e **5.13.** False-color *Galileo* SSI image of Jupiter's Great Red Spot observed in 1996. Light reflected at 886 nm g methane absorption) is shown in red, light reflected at 732 nm (medium methane absorption) is shown in , and light reflected at 757 nm (weak methane absorption) is shown in blue. Hence, in this color scheme: blue ck areas are deep clouds; pink areas are high, thin hazes; and white areas are high, thick clouds. Courtesy of A.

5.14. Cloud features on Jupiter. Cylindrical map of Jupiter observed by *Cassini* ISS extending from 60°S to The GRS is clearly visible towards the center of the image, as is the single remaining white oval to the south, proximately 180° to the east. Several smaller white ovals are seen farther to the south. The equatorial plumes northern edge of the equatorial zone are apparent, interspersed by darker regions, which appear bright at 5 μm e thus known as the 5 μm hotspots. The dark cyclonic ovals at the northern edge of the NEB are the brown . Several small thunderstorm clouds can be seen erupting in the NEB. Courtesy of NASA.

Figure 5.19. HST image of the GRS and Oval BA, which turned red in 2005 and is now known to many as the Little Red Spot (LRS) or Red Spot Jr. Courtesy of NASA.

Figure 5.22. Mosaic of five HST images of Saturn recorded between 1996 and 2000, showing Saturn's rings opening up from just past edge-on to nearly fully open as it moved from autumn towards winter in its northern hemisphere during its orbit about the Sun. Courtesy of NASA.

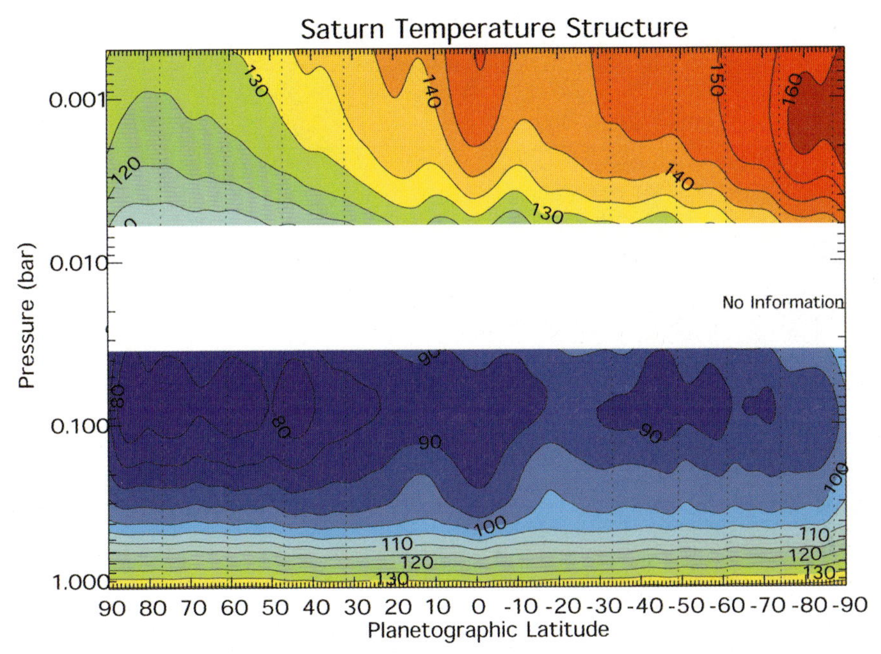

Figure 5.23. Temperature contour map of Saturn's atmosphere retrieved from *Cassini* CIRS observations (Fletcher *et al.* 2007b, 2008a)

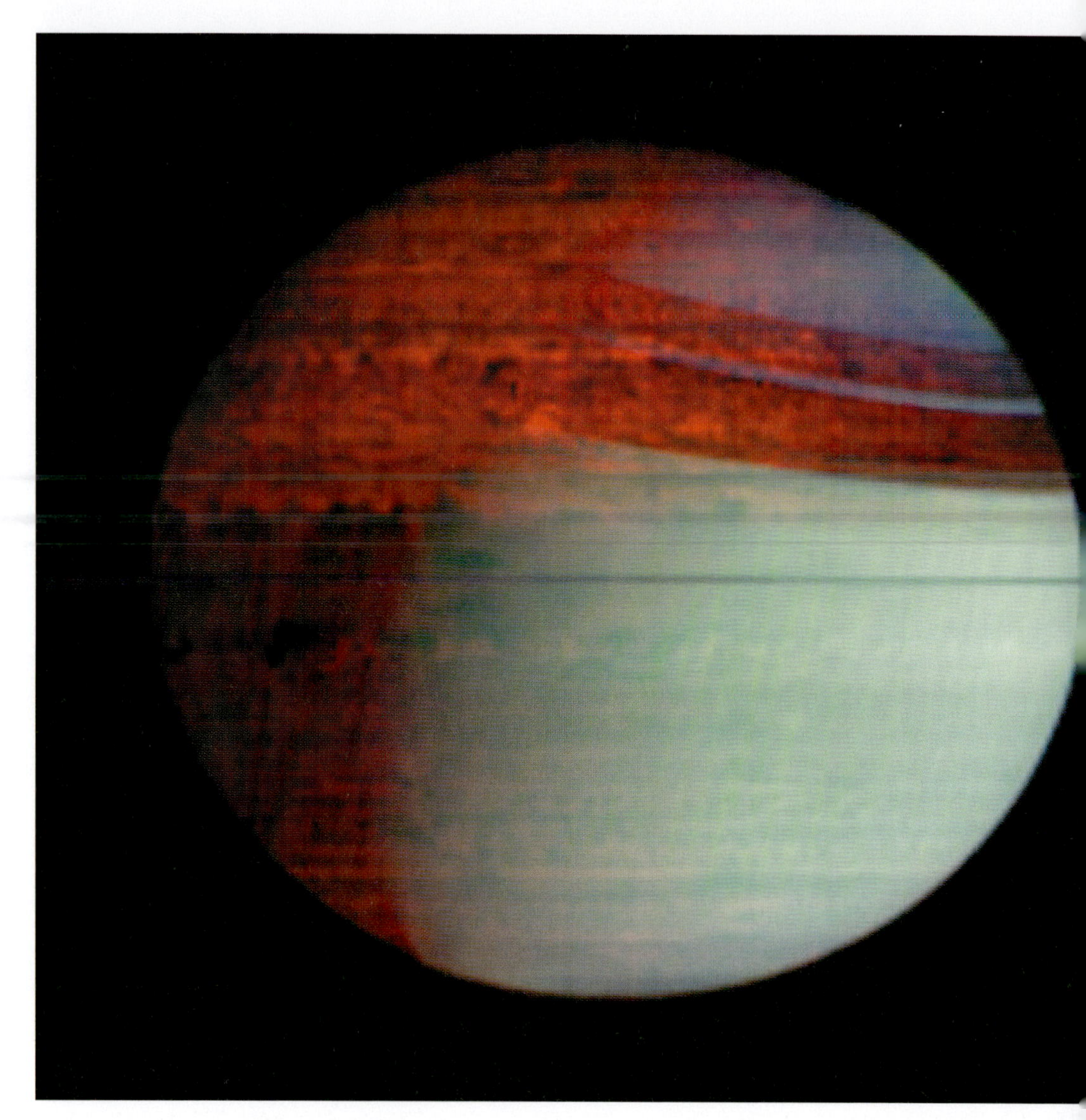

Figure 5.24. False-color mosaic of Saturn observed by *Cassini* VIMS in February 2006 from edge-on to th which appear as a thin blue line over the equator. The image is constructed from images at 1.07 μm (blue), (green), and 5.02 μm (red). The blue–green color (lower right) is sunlight scattered off clouds high in S atmosphere and the red color (upper left) is the glow of thermal radiation from Saturn's warm interior, easily Saturn's night side, within the shadow of the rings, and with somewhat less contrast on Saturn's day side. Cou NASA.

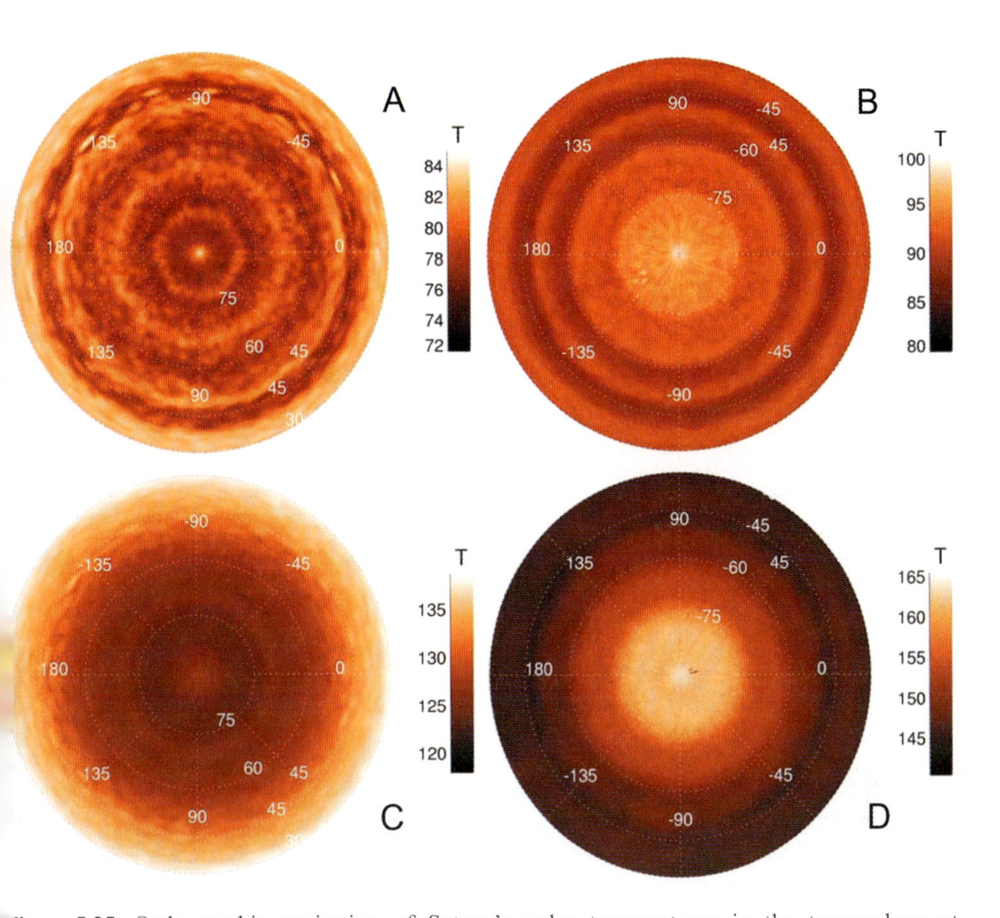

igure 5.25. Orthographic projection of Saturn's polar temperatures in the troposphere at 00 mbar (A and B) and the stratosphere at 1 mbar (C and D), after fig. 1 of Fletcher *et al.* 2008a). The northern hemisphere is shown on the left [(A) and (C)], and the southern hemisphere s shown on the right [(B) and (D)]. Temperatures were retrieved by fitting the 600–680 cm^{-1} and ,250–1,350 cm^{-1} regions of *Cassini* CIRS spectra at 15.0 cm^{-1} spectral resolution. The polar otspots are clearly visible in the troposphere, which are surrounded by warm polar belts at 79°N nd 76°S, with the former showing a coherent hexagonal structure. Roughly three warm belts are hown concentric around each pole in this latitude range. The stratosphere shows a large seasonal ifference in temperature and in the nature of the polar hoods, but reveals similarities in the warm olar vortices.

Figure 5.27. False-color image of Saturn recorded by *Voyager 1* in 1980 showing the unique red oval cloud located at 55°S, sometimes known as "Anne's Spot", which has since disappeared. Courtesy of NASA.

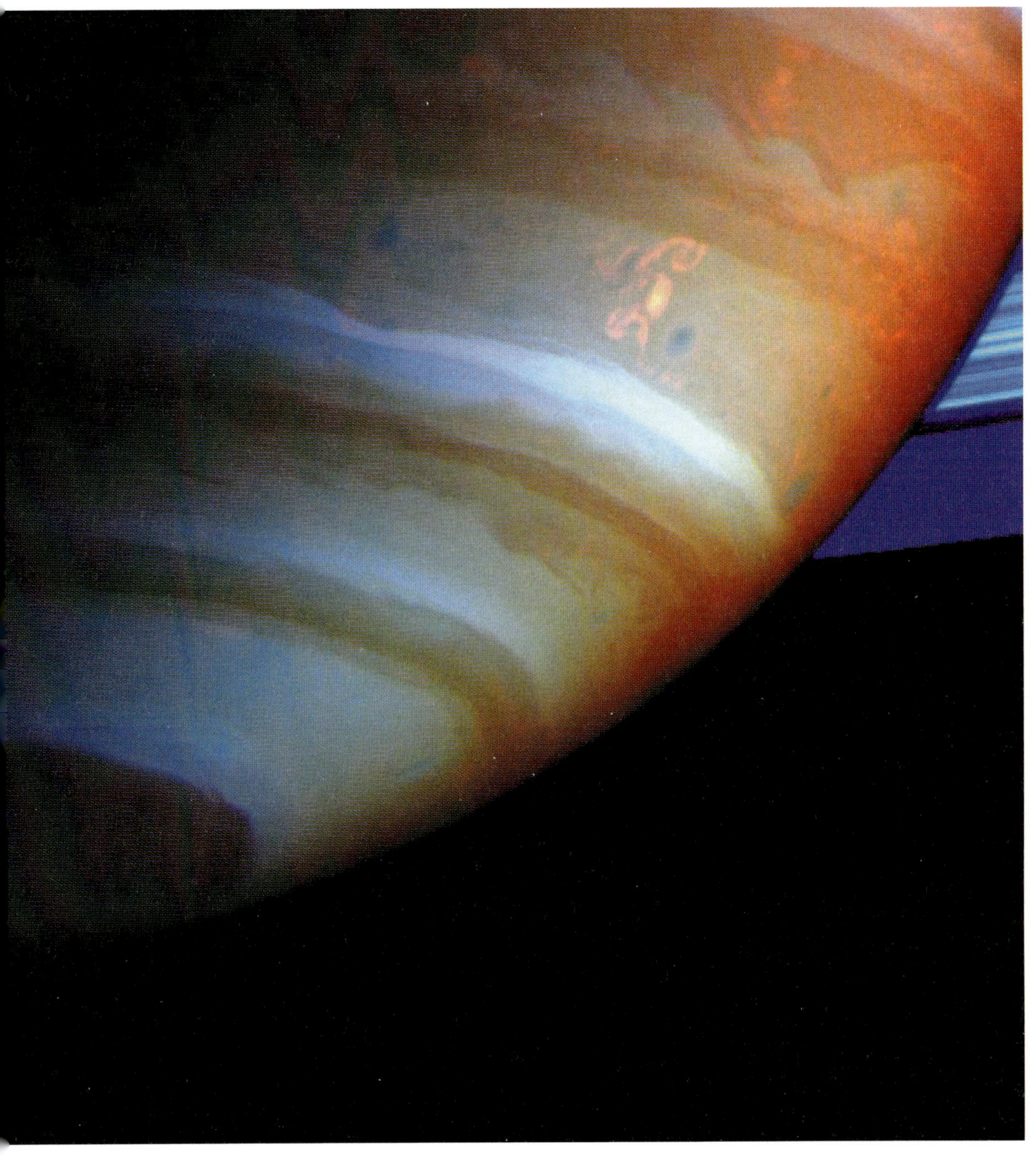

.28. False-color composite of Saturn, showing the "Dragon Storm" in Saturn's "Storm Alley" at 35°S, ▮ by *Cassini* ISS in September 2004. The Dragon Storm is linked to Saturn Electrostatic Discharges (SEDs) ▯ seen in July and September that year. The SEDs were seen to coincide with the Dragon Storm rising over ▯ side of the planet as seen from *Cassini*. This strong correlation between the position of the Dragon Storm ▯SEDs is a strong indication that the SEDs are caused by lightning generated by storms such as the Dragon ▯ourtesy of NASA.

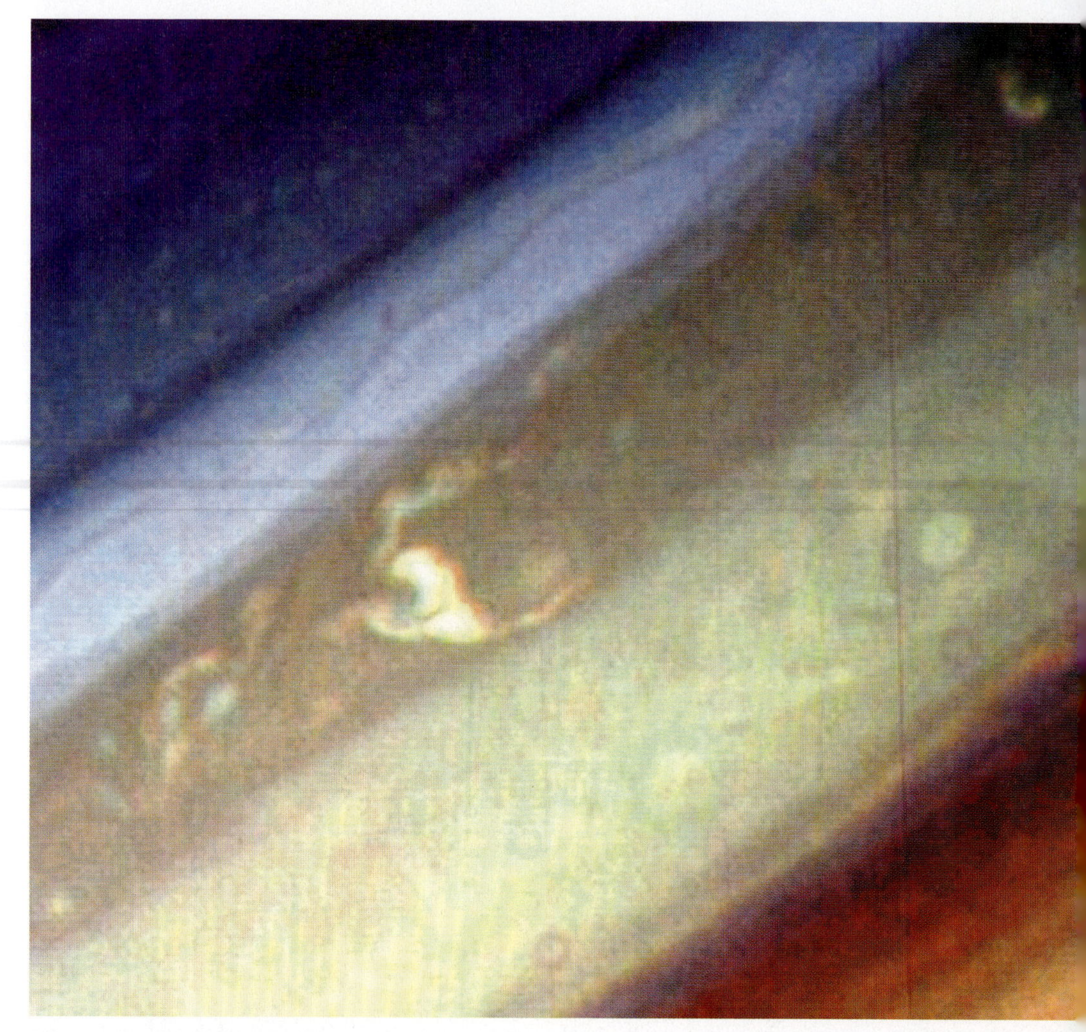

Figure 5.31. Highly enhanced image of Saturn's cloud features observed by *Voyager 2*. The Ribbon Wave is the top left in the bright zone centered at 47°N in the North Polar Zone (NPZ). A bright convective cloud is se south at latitude ~38°N in the North Temperate Belt. Farther south is the North Temperate Zone, th Equatorial Belt, and the edge of the Equatorial Zone at the bottom right of the image. These convective cloud to erupt and then shear apart in the zonal wind flow, much like very similar features in Jupiter's NEB. Co NASA.

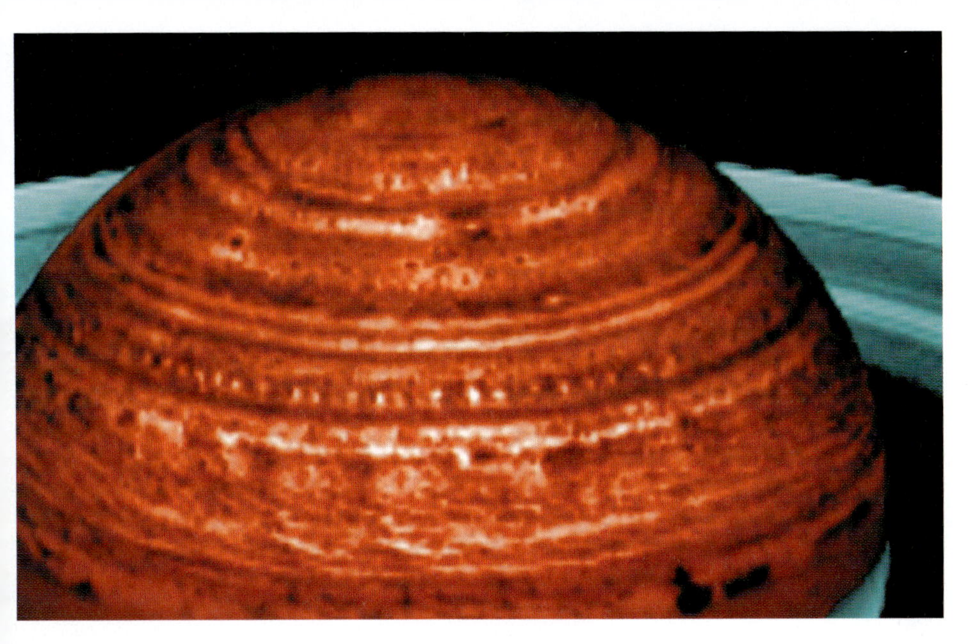

Figure 5.33. The "String of Pearls", viewed by *Cassini* VIMS in October 2006. The red part of this image is recorded at a wavelength of 5 μm and a series of bright spots can be seen in the center of the image at a latitude of ~40°N. Courtesy of NASA.

Figure 5.36. HST/NICMOS false-color image of Uranus, where the blue, green, and red components correspond to the real near-IR wavelengths of 0.9, 1.1, and 1.7 μm respectively. The increased haze opacity at 40–50°S is clearly seen as is the generally increased reflection over the South Pole. Individual methane clouds appear red in this image since they are most apparent at 1.7 μm (shown here as red) where the methane absorption is strong. The ring system plus several satellites are also clearly visible. Courtesy of NASA.

Figure 5.37. Color composite images of Uranus as observed by the Keck Observatory in July 2004 (Sromov Fry, 2005). The two images were recorded on successive nights and show opposite sides of Uranus. The brig at 45°S is clearly visible as are the ring system and several bright midlatitude convective features in both hemi with vigorous activity seen at northern midlatitudes. It is interesting to compare the appearance of Uranus with its appearance in 1998 (Figure 5.23) and 1986 (Figure 4.19). Courtesy of the Keck Observatory an Sromovsky.

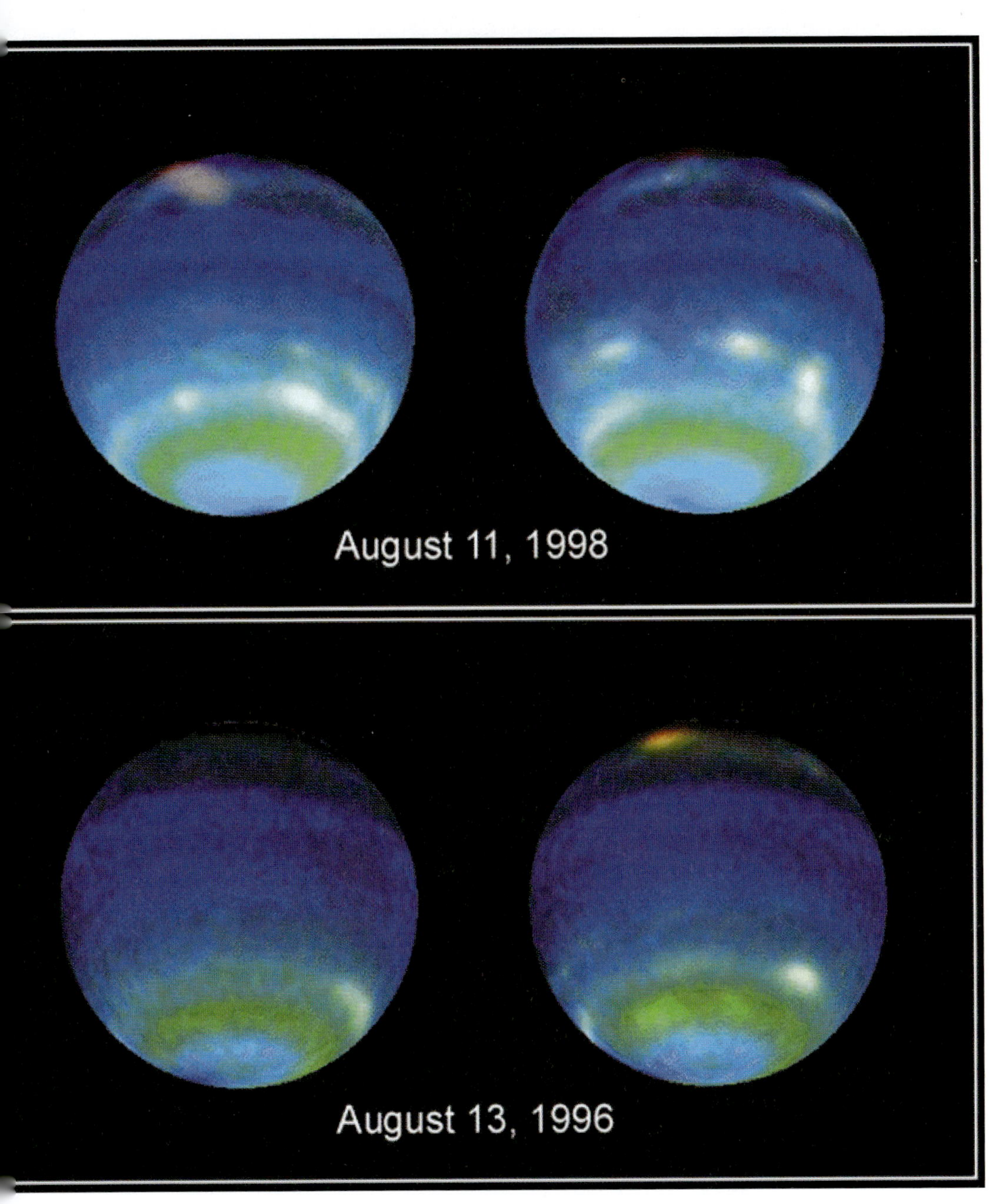

e **5.41.** False-color images of Neptune recorded by HST/WFPC-2 in 1996 and 1998. Courtesy of NASA.

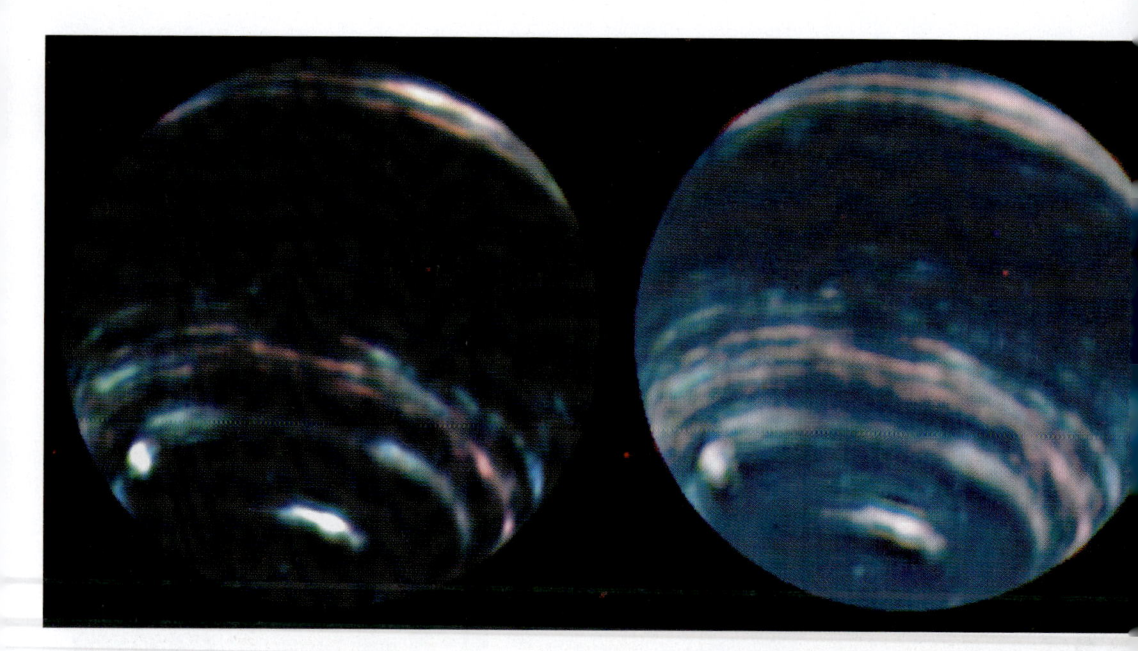

Figure 5.42. Color composite images of Neptune as observed by the Keck Observatory in August 2003 (Fı Sromovsky, 2004). The left-hand image has been stretched linearly, while the right-hand image has been str logarithmically to enhance dimmer cloud features. In this color scheme pink clouds are of high altitude, b opacity, and white clouds are high and optically thick. Courtesy of the Keck Observatory and Larry Srom

Figure 5.43. Cylindrical map of Neptune between 90°S and 90°N based on *Voyager 2* data, showing the GDS SPF, and a trace of the South Polar Wave (SPW). Courtesy of NASA and James Hastings-Trew.

7.15. False-color image of Jupiter obtained with UKIRT/UIST in July 2008 shortly after the close encounter :n the GRS, Red Spot Jr., and Baby Red Spot. Here, red = Fe II filter (low atmospheric absorption), green- ckett-gamma filter (medium atmospheric absorption), and blue = 2.27 μm (high atmospheric absorption). ve can see that polar stratospheric hazes are very high as they appear blue in this false-color image. Also, we can ıt the clouds over the Great Red Spot are higher than those over Red Spot Jr. to the southeast as the GRS rs whiter. To the east of the GRS can be seen scattered white/yellow high clouds. These are believed to be the nts of Baby Red Spot, which has since dissipated altogether.

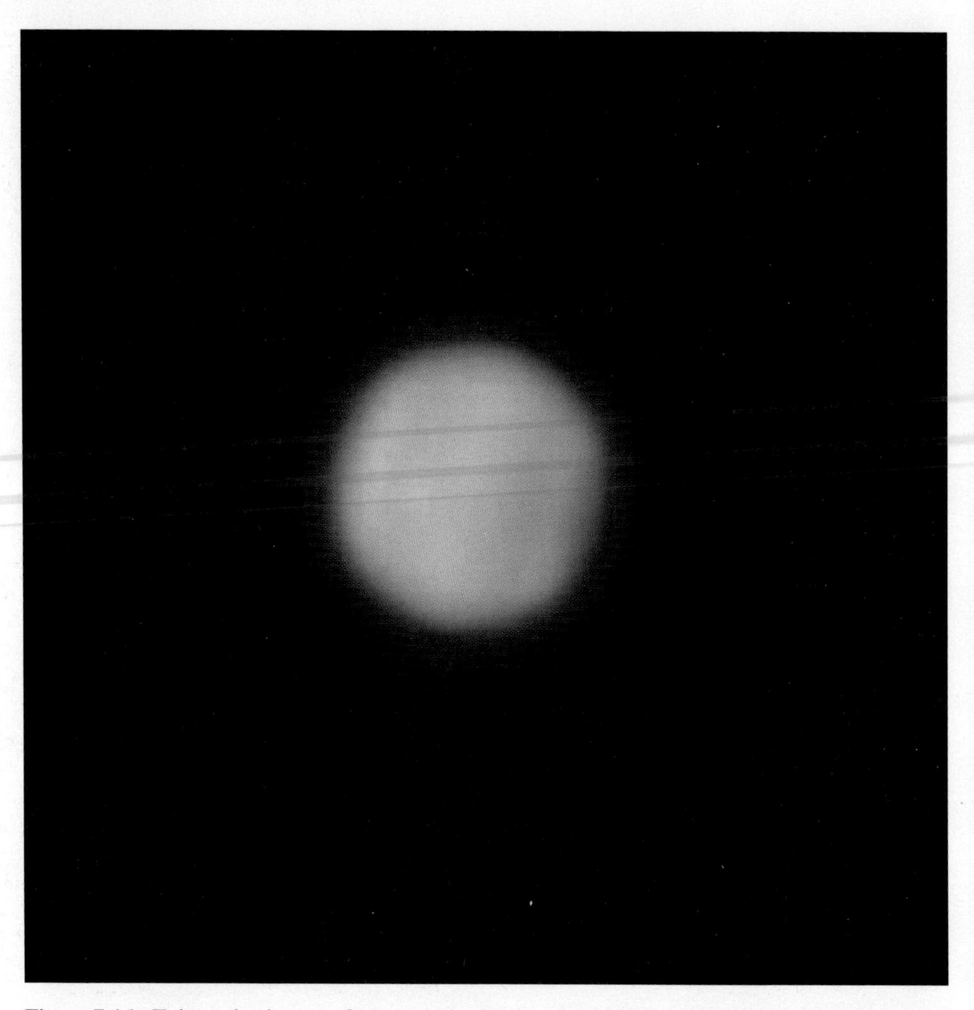

Figure 7.16. False-color image of Uranus obtained with UKIRT/UIST in July 2008. Here, red is the J98 filter at 1.3 μm (weak methane absorption), green is the 1.58 μm filter (weak methane absorption, less haze/air scattering), and blue the 1.69 μm filter (strong methane absorption). The circumpolar belts at 45°N and 45°S can be seen as the white/purple stripes to the left and right.

5

Dynamical processes

5.1 INTRODUCTION

In Chapter 4 we discussed the vertical profiles of temperature, pressure, composition, and clouds of the giant planet atmospheres, and also briefly discussed transport mechanisms such as convection and eddy mixing. Of course the real atmospheres of planets are three-dimensional and thus to understand fully the observations of ground-based telescopes and spacecraft missions, we need to understand how planetary atmospheres move and how heat and material are transported, not just vertically, but horizontally as well. In this chapter we will consider the basic equations of fluid motion in rotating planetary atmospheres and see how these theoretical considerations relate to the observed winds, clouds, and storm systems observed in the atmospheres of the giant planets.

5.2 MEAN CIRCULATION OF THE GIANT PLANET ATMOSPHERES

The deep interiors of the giant planets are fluid, as was discussed in Chapter 2. As the planets gradually contract, heat is released via the Kelvin–Helmholtz mechanism and this heat is believed to be transported outwards mainly via convection currents, although at certain depths, conduction and radiation may become more important. In the absence of heat being removed from the top of the convective cells, parcels of air cool adiabatically as they rise and then heat adiabatically as they descend by *exactly the same amount* and the temperature profile should thus be a perfect adiabat. Of course in the upper, optically thin, observable atmosphere, some heat is lost by infrared radiation, cooling the air more than it would do by adiabatic expansion alone. Hence, in a convective atmosphere where heat is being transported radially outwards, the temperature profile must on average be slightly subadiabatic, although the actual difference is in reality almost undetectable since the heat flow is negligible compared with the atmospheric heat capacity. Hence, the deep interiors of the giant planets to all intents and purposes may be considered to be perfectly adiabatic and thus perfectly *barotropic* (i.e., there are no temperature gradients on constant pressure surfaces).

In the upper layers of planetary atmospheres, where the atmosphere starts to become optically thin, the heating of the atmosphere, both through absorption of sunlight and by absorption of internal energy, provides a source of kinetic energy for atmospheric motion through a number of cycles akin to that of a heat engine shown in Figure 5.1. Consider local heating of the atmosphere. Air near the bottom of the atmosphere is heated quasi-isothermally which (in regions where free convection or forced convection occur) causes the air to rise and cool adiabatically. The air then radiates energy to space in the IR at the top of the atmosphere (where it is optically thin) causing it to cool further before it descends again and heats adiabatically until it reaches the bottom of the atmosphere and the cycle repeats. The integrated area of the p/V cycle is the work done per cycle, or equivalently is proportional to the rate at which thermal energy is converted into kinetic energy on a local scale. On a larger scale, solar heating is strongest at subsolar latitudes and least near the poles (except for Uranus, which spins on its side and where the subsolar latitude varies between the equator and poles during the course of a Uranian year). Hence, in the absence of atmospheric motion, the tropics would become warmer than the poles. Such differential heating creates pressure gradients, and atmospheres in general respond to this by moving in such a way as to minimize the temperature differences over the planet and become barotropic. Atmospheric motions are found to efficiently counterbalance differential solar heating on all of the giant planets. Even Uranus, which for large fractions of its orbit receives sunlight at either the North or South Pole only, has negligible equator-to-pole temperature difference at levels where $p > 0.5$ bar. In

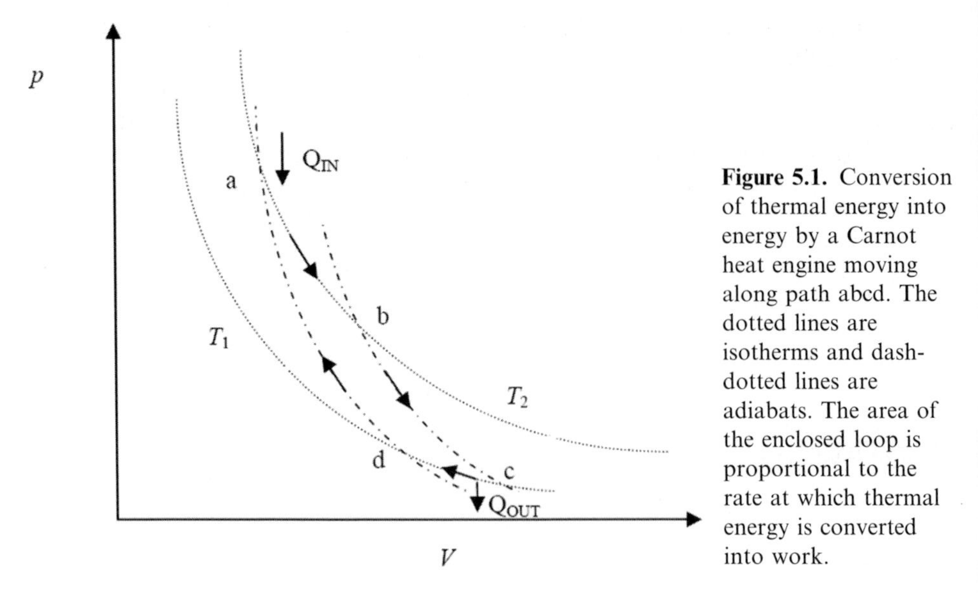

Figure 5.1. Conversion of thermal energy into energy by a Carnot heat engine moving along path abcd. The dotted lines are isotherms and dash-dotted lines are adiabats. The area of the enclosed loop is proportional to the rate at which thermal energy is converted into work.

addition to this differential solar heating, all of the giant planets, with the exception of Uranus, have large internal heat sources resulting from Kelvin–Helmholtz contractions and internal differentiation. It would appear that the circulations of the atmospheres are arranged such that more of this internal energy is released away from the subsolar latitudes since the net thermal emission to space is found to be almost independent of latitude (as was discussed in Chapter 1).

Before we can go on to describe the general circulations of the giant planets further we need to outline some basic atmospheric dynamics.

5.2.1 Equations of motion

Navier–Stokes equation

The fundamental equation governing the motion of air in a planetary atmosphere is the *Navier–Stokes* equation. Consider a parcel of air of volume $\delta V = \delta x\, \delta y\, \delta z$ and density ρ. From Newton's Second Law, $\mathbf{F} = m\mathbf{a}$ and thus

$$\mathbf{F} = (\rho\, \delta V)\frac{d\mathbf{V}}{dt} \tag{5.1}$$

where $\mathbf{V}$ is the velocity vector. There are three main forces that may act on air parcels in a non-rotating planetary atmosphere: gravity, the pressure gradient force, and friction. Incorporating these forces leads to the Navier–Stokes equation in an inertial frame

$$\frac{d\mathbf{V}}{dt} = \mathbf{g}_i - \frac{1}{\rho}\nabla p + \frac{\eta}{\rho}\nabla^2\mathbf{V} \tag{5.2}$$

where η is the viscosity; and $\mathbf{g}_i$ is the gravitational acceleration, excluding centrifugal forces. In reality, of course, planets rotate and thus to describe atmospheric motion we need to re-express this equation in a form suitable for a rotating frame of reference. A vector $\mathbf{A}$ in frame S_R, which is rotating at an angular velocity $\mathbf{\Omega}$ with respect to a stationary inertial frame S_I will have a component of motion $\mathbf{\Omega} \times \mathbf{A}$ in the stationary frame S_I. Hence, differentiating $\mathbf{A}$ with respect to time in the two frames leads to

$$\left(\frac{d\mathbf{A}}{dt}\right)_{S_I} = \left(\frac{d\mathbf{A}}{dt}\right)_{S_R} + \mathbf{\Omega} \times \mathbf{A}. \tag{5.3}$$

Putting $\mathbf{A} = \mathbf{r}$ we find

$$\mathbf{V}_I = \mathbf{V}_R + \mathbf{\Omega} \times \mathbf{r}, \tag{5.4}$$

and putting $\mathbf{A} = d\mathbf{r}/dt$ we have

$$\left(\frac{d\mathbf{V}_I}{dt}\right)_{S_I} = \left(\frac{d\mathbf{V}_R}{dt}\right)_{S_R} + \mathbf{\Omega} \times \mathbf{V}_R. \tag{5.5}$$

Combining the two we obtain the expression

$$\left(\frac{d\mathbf{V}_I}{dt}\right)_{S_I} = \left(\frac{d\mathbf{V}_R}{dt}\right)_{S_R} + 2\boldsymbol{\Omega} \times \mathbf{V}_R + \boldsymbol{\Omega} \times (\boldsymbol{\Omega} \times \mathbf{r}) \tag{5.6}$$

which may be substituted into the Navier–Stokes equation to give

$$\left(\frac{d\mathbf{V}_R}{dt}\right)_{S_R} = -\frac{1}{\rho}\nabla p + 2\mathbf{V}_R \times \boldsymbol{\Omega} + \mathbf{g}_i - \boldsymbol{\Omega} \times (\boldsymbol{\Omega} \times \mathbf{r}) + \frac{\pi}{\rho}\nabla^2\mathbf{V}_R \tag{5.7}$$

or more conveniently

$$\frac{d\mathbf{V}}{dt} = \mathbf{g} - \frac{1}{\rho}\nabla p + 2\mathbf{V} \times \boldsymbol{\Omega} + \frac{\eta}{\rho}\nabla^2\mathbf{V} \tag{5.8}$$

where the velocity and differentiation are assumed to be with respect to the rotating frame, and where $\mathbf{g} = \mathbf{g}_i - \boldsymbol{\Omega} \times (\boldsymbol{\Omega} \times \mathbf{r})$ is the local effective gravity, which includes the centrifugal force.

Equation (5.8) can be seen to be almost identical to the equation in the inertial frame with the exception that $\mathbf{g}$ is modified as just described and also the new equation has an additional term $2(\mathbf{V} \times \boldsymbol{\Omega})$ that arises due to the so-called *Coriolis force*, an apparent force experienced by objects moving in a rotating frame, which since it acts at right-angles to the velocity gives rise to circular motion. Near the equator, or at all latitudes on a slowly rotating planet such as Venus, the Coriolis force is small and in such conditions the primary circulation for a planet with low obliquity responding to solar heating tends to be a simple *Hadley cell* circulation whereby air rises at the equator, moves polewards at high altitude before descending and then returning to the equator at low altitude. For slowly rotating planets such as Venus the descending part of the circulation is at the poles themselves and the Hadley cell is thus global in extent. However, for the Earth and the giant planets, the Coriolis force becomes significant quite close to the equator and thus air is deflected eastwards as it moves towards the poles at high altitude, and deflected westwards as it returns to the equator at low altitudes. Coriolis force modification to the terrestrial equatorial Hadley cell, which extends from the equator to latitudes of approximately $\pm 30°$, thus forces the low-altitude, equatorward air to be deflected towards the west, giving rise to the easterly "trade winds" at subequatorial latitudes. While the terrestrial Hadley cell breaks down at latitudes of approximately $\pm 30°$, the corresponding critical latitude limit for the far more rapidly rotating giant planets is substantially smaller.

To solve the Navier–Stokes equation for planetary atmospheres where the Coriolis force is important, which is clearly the case in the rapidly rotating giant planet atmospheres, we first need to split up $\mathbf{V}$ into its three components (u, v, w), where u is the velocity east/west in the x-direction, v is the velocity north/south in the y-direction, and w is the vertical velocity in the z-direction. Using these components we find that for a spherical planet, following the approach of Houghton (1986),

$$\frac{d\mathbf{V}}{dt} = \left(\frac{du}{dt} - \frac{uv\tan\phi}{R} + \frac{uw}{R}\right)\mathbf{i} + \left(\frac{dv}{dt} - \frac{u^2\tan\phi}{R} + \frac{vw}{R}\right)\mathbf{j} + \left(\frac{dw}{dt} - \frac{u^2+v^2}{R}\right)\mathbf{k} \tag{5.9}$$

where ϕ is the latitude; and R is the planetary radius. Similarly, the Coriolis force term may be rewritten as

$$2\mathbf{V} \times \mathbf{\Omega} = 2\Omega(v \sin \phi - w \cos \phi)\mathbf{i} - 2\Omega u \sin \phi \mathbf{j} + 2\Omega u \cos \phi \mathbf{k}. \tag{5.10}$$

For the significantly oblate giant planets, Equation (5.10) must be expressed in terms of the planetographic latitude ϕ_g (introduced earlier in Chapter 2), not the planetocentric latitude. Modifications to Equation (5.9) for the oblate giant planets are more complicated, but fortunately the differences are in terms that are usually neglected since it is found to be very difficult to solve the Navier–Stokes equation explicitly. Instead, a number of reasonable and justifiable approximations are made. First of all, it is found to be a good approximation to assume that the magnitude of the winds is much less than the radius of the planet, and thus that the $1/R$ terms in Equation (5.9) may be ignored. Second, it is assumed that the vertical wind speeds are very much less than the horizontal wind speeds. This is found to be a good approximation since the action of gravity in the vertical direction keeps departures from hydrostatic equilibrium small and frictional forces are usually negligible meaning that horizontal winds, once initiated, tend to blow relatively freely. Finally, it is found that the vertical component of $2\mathbf{V} \times \mathbf{\Omega}$, and the vertical component of friction may both be neglected compared with the gravitational acceleration $\mathbf{g}$. Making these approximations to the Navier–Stokes equation and resolving into components we derive the *momentum equations*:

$$\frac{du}{dt} - fv + \frac{1}{\rho}\frac{\partial p}{\partial x} = F^{(x)} \tag{5.11}$$

$$\frac{dv}{dt} + fu + \frac{1}{\rho}\frac{\partial p}{\partial y} = F^{(y)} \tag{5.12}$$

$$\frac{dw}{dt} + \frac{1}{\rho}\frac{\partial p}{\partial z} + g = 0 \tag{5.13}$$

where $f = 2\Omega \sin \phi_g$ is called the Coriolis parameter and where we have written the components of friction acting in the x-direction and y-direction explicitly. In most cases the vertical accelerations of the winds are also negligible compared with local gravitational acceleration and thus Equation (5.13) may be well approximated by the hydrostatic equation

$$\frac{\partial p}{\partial z} = -g\rho. \tag{5.14}$$

Further approximations may be made to these equations, including the *geostrophic* approximation, which is relevant to giant planet dynamics.

Geostrophic approximation

For large-scale motion away from the planetary surface (always obeyed by the giant planets!) the frictional forces are to a first approximation negligible. Furthermore, for steady motion with small curvature, $d\mathbf{V}/dt$ is also negligible and hence the horizontal momentum equations reduce to:

$$fv = \frac{1}{\rho}\frac{\partial p}{\partial x} \tag{5.15}$$

and

$$fu = -\frac{1}{\rho}\frac{\partial p}{\partial y}. \tag{5.16}$$

These are the *geostrophic equations* and lead to the familiar situation for the Earth's atmosphere of winds blowing along isobars, with (for prograde-spinning planets) anticlockwise motion about centers of low pressure (cyclones) in the northern hemisphere, and clockwise motion about cyclones in the southern hemisphere. To make this geostrophic approximation we need to ensure that the acceleration terms in the momentum equations are much less than the Coriolis terms and we define the *Rossby number Ro* to be the ratio of the acceleration to Coriolis terms, for which an approximate expression is

$$Ro \approx \frac{U^2/L}{fU} = \frac{U}{fL} \tag{5.17}$$

where U is the mean wind speed; and L is the typical horizontal dimension of motion. The geostrophic approximation is then valid, and thus Coriolis forces are dominant, if $Ro \ll 1$. For the giant planets, the Rossby number is estimated to be of the order of 10^{-2} (Chamberlain and Hunten, 1987) and thus to a first approximation the atmospheric flows of these planets should be geostrophic.

Thermal wind equation

To use the geostrophic equation to analyze flow in a planetary atmosphere, we need to know the three-dimensional pressure field. While this can be measured for the Earth's atmosphere, it is very difficult to derive from remote-sensing observations of the giant planets. However, remote-sensing observations can determine the three-dimensional temperature field and, if we know the wind speeds at a certain pressure level (e.g., by tracking discrete cloud features at the cloud tops), then the wind speeds at all other levels can be deduced from the *thermal wind equation.*

To derive the thermal wind equation, the hydrostatic equation $dp = -\rho g\, dz$, must first be rearranged to give

$$\frac{1}{\rho} = -g\left(\frac{\partial z}{\partial p}\right)_{x,y,t} \tag{5.18}$$

which may be substituted in the geostrophic equation. Hence, for the meridional component of the geostrophic equation (Equation 5.15) we have, using the above expression for $1/\rho$ together with the reciprocal and cyclical relations,

$$fv = \frac{1}{\rho}\frac{\partial p}{\partial x} = -g\left(\frac{\partial p}{\partial x}\right)\Big/\left(\frac{\partial p}{\partial z}\right) = -g\left(\frac{\partial p}{\partial x}\right)\left(\frac{\partial z}{\partial p}\right) = g\left(\frac{\partial z}{\partial x}\right)_{p,y,t}. \tag{5.19}$$

Similarly, the zonal component of the geostrophic equation may be rearranged to give

$$fu = -g\left(\frac{\partial z}{\partial y}\right)_{p,x,t}. \tag{5.20}$$

The density of the atmosphere is given by $\rho = \bar{m}p/RT$, where $\bar{m}$ is the mean molecular weight of the air and R is the molar gas constant. Substituting this into the hydrostatic equation we find

$$g\left(\frac{\partial z}{\partial p}\right)_{x,y,t} = -\frac{1}{\rho} = -\frac{RT}{\bar{m}p} \tag{5.21}$$

which may be differentiated with respect to x, holding p constant, to give

$$g\frac{\partial^2 z}{\partial x\,\partial p} = \frac{R}{\bar{m}p}\left(\frac{\partial T}{\partial x}\right)_p. \tag{5.22}$$

Similarly, if we differentiate the meridional component of the geostrophic equation (Equation 5.19) with respect to p, holding x constant we find

$$f\left(\frac{\partial v}{\partial p}\right)_x = g\frac{\partial^2 z}{\partial p\,\partial x} \tag{5.23}$$

and thus equating Equation (5.22) and Equation (5.23) we derive

$$f\left(\frac{\partial v}{\partial p}\right)_x = \frac{R}{\bar{m}p}\left(\frac{\partial T}{\partial x}\right)_p \tag{5.24}$$

or

$$f\left(\frac{\partial v}{\partial p}\right)_x = \frac{1}{\rho T}\left(\frac{\partial T}{\partial x}\right)_p. \tag{5.25}$$

A similar manipulation of the zonal wind component of the geostrophic wind equation (Equation 5.16) leads to

$$f\left(\frac{\partial u}{\partial p}\right)_x = \frac{1}{\rho T}\left(\frac{\partial T}{\partial x}\right)_p. \tag{5.26}$$

These are the *thermal wind equations*. It should be noted that in deriving these equations we assumed that only temperature varied with horizontal location and that the molecular weight of the atmosphere remained constant. This is not in general true, especially for planets that have large fractions of condensable gases such as Uranus and Neptune and in these situations the thermal wind equation needs to be slightly modified.

Vorticity equation

The final concept needed to understand the general horizontal motions in giant planet atmospheres is the concept of vorticity $\boldsymbol{\omega}$, which is defined simply enough as the curl of the velocity vector

$$\boldsymbol{\omega} = \nabla \times \mathbf{V}. \tag{5.27}$$

Vorticity is important since it (and properties derived from it) is conserved under certain conditions. The physical meaning of vorticity is that it is a measure of the spin of a fluid flow at any particular point and as such tells us something about the angular momentum of the fluid flow. For velocities measured in a frame rotating with constant angular velocity $\boldsymbol{\Omega}$ we know that

$$\mathbf{V}_I = \mathbf{V}_R + \boldsymbol{\Omega} \times \mathbf{r}. \tag{5.28}$$

Taking the curl of both sides we find

$$\boldsymbol{\omega}_I = \boldsymbol{\omega}_R + 2\boldsymbol{\Omega}. \tag{5.29}$$

For the special case of horizontal flow only, which very well approximates to the conditions in giant planet atmospheres where zonal and meridional velocities greatly exceed vertical velocities $\mathbf{V} \sim (u, v, 0)$ and $\omega_R \sim (0, 0, \zeta)$, where ζ, known as *relative vorticity*, is equal to

$$\zeta = \frac{\partial v}{\partial x} - \frac{\partial u}{\partial y}. \tag{5.30}$$

Substituting for $\boldsymbol{\omega}_R$ in Equation (5.29) we then find

$$\boldsymbol{\omega}_I = (0, 2\Omega \cos \phi_g, \zeta + 2\Omega \sin \phi_g) \tag{5.31}$$

or

$$\boldsymbol{\omega}_I = (0, 2\Omega \cos \phi_g, \zeta + f). \tag{5.32}$$

A *vorticity equation* may be derived from the x and y components of the momentum equations expressed in whatever coordinates are most convenient and taking into account any suitable approximations. For example, differentiating Equation (5.11) with respect to y and differentiating Equation (5.12) with respect to x and subtracting (ignoring frictional forces) we find that, assuming pressure and density do not vary with horizontal position and that there are no vertical motions (Houghton, 1986),

$$\frac{d_h}{dt}(\zeta + f) = -(\zeta + f)\left(\frac{\partial u}{\partial x} + \frac{\partial v}{\partial y}\right) \tag{5.33}$$

where

$$\frac{d_h}{dt} = \frac{\partial}{\partial t} + u\frac{\partial}{\partial x} + v\frac{\partial}{\partial y}. \tag{5.34}$$

If the flow can be considered to be two-dimensional and non-divergent (i.e., vertical velocities are negligible) it can be seen from Equation (5.33) that the quantity $\zeta + f$, known as *absolute vorticity*, is conserved. We will see in Section 5.3.2 that this conservation law is the central restoring force that supports *Rossby* or *planetary waves*. A less stringent approximation is to assume that the air has roughly constant density and temperature for which the mass continuity equation

$$\frac{\partial \rho}{\partial t} + \nabla \cdot (\rho \mathbf{V}) = 0 \tag{5.35}$$

becomes simply $\nabla \cdot \mathbf{V} = 0$. This is an example of the *Boussinesq* approximation, where density and temperature variations are neglected, except where they appear in buoyancy terms. Substituting Equation (5.35) into Equation (5.33) we find

$$\frac{d_h}{dt}(\zeta + f) = -(\zeta + f)\frac{\partial w}{\partial z} \tag{5.36}$$

from which it may be deduced that (Houghton, 1986)

$$\frac{d_h}{dt}\left(\frac{\zeta + f}{h}\right) = 0 \tag{5.37}$$

where h is the separation between material levels in the fluid. The quantity $(\zeta + f)/h$ is a simplified form of *potential vorticity*, and its conservation is analogous to the conservation of angular momentum. When a parcel is vertically stretched (and hence made narrower by conservation of mass) it rotates faster to maintain its angular momentum in the same way as ice skaters spin faster when they draw their arms inwards. The most general form of potential vorticity is Ertel's potential vorticity which is defined as

$$q_E = \frac{\zeta}{\rho} \cdot \nabla\theta \tag{5.38}$$

where θ is the potential temperature, defined later in Section 5.3.1. This quantity may be calculated from temperature and pressure maps fitted to measured data and since q_E is conserved (under frictionless, non-diabatic heating cases) maps of this quantity may also be used as a tracer for atmospheric motion. However, recent observations of Jupiter's atmosphere by *Cassini* have shown that Ertel's potential vorticity is not actually conserved in Jovian atmospheres (Gierasch *et al.*, 2004) due to variations in the water vapor abundance and the ortho-hydrogen:para-hydrogen ratio, which mean that the potential temperature is no longer a function of just pressure and temperature. Gierasch *et al.* (2004) note that it is likely that a modified form of potential vorticity is conserved in these cases.

Another more approximate expression for potential vorticity is the quasi-geostrophic expression described, for example, by Andrews (2000).

Taylor–Proudman theorem

Another form of the vorticity equation is derived by Chamberlain and Hunten (1987, p. 106, Eq. 2.6.4) in isobaric coordinates for steady, slow motions in a frictionless atmosphere with acceleration terms and quadratic terms being negligible:

$$\frac{\partial(\zeta + f)}{\partial t} \approx -f\nabla_p \cdot \mathbf{V} = 0, \tag{5.39}$$

where ∇_p is the gradient along isobars; and $\mathbf{V} = (u, v, 0)$. Again, applying the continuity equation, this relation implies that $\partial\omega/\partial p = 0$, where $\omega = dp/dt$ is the equivalent of vertical velocity in pressure coordinates. This implies that $\omega = dp/dt$ equals a constant and the only physical solution is that $\omega = 0$ and thus that, to a first approximation, all steady motions in a rotating atmosphere with zero viscosity must be two-dimensional (i.e., barotropic). This is a fundamental conclusion and is equivalent to the *Taylor–Proudman* theorem in hydrodynamics that we shall come across again later.

5.2.2 Mean zonal motions in the giant planet atmospheres

All the giant planets are found to have a very stable, zonal banded structure of cloud opacity and winds, with the winds blowing strongly in the east–west direction with very little mean meridional motion. The fact that these planets have such different solar flux and internal energy forcings suggests that the zonal structure derives from the nature of these planetary atmospheres themselves, together with their rapid rotation, rather than due to the method of forcing. The measured zonal wind speeds at the cloud tops for the four giant planets are compared in Figure 5.2, while Figure 5.3 shows the zonal wind speeds separately for each planet, with regions of cyclonic vorticity highlighted in gray. In addition, Figure 5.4 shows the wind speeds in terms of degrees longitude per rotation of the planet superimposed on the planets' mean visible appearance. It can be seen that there is very close north/south symmetry in the wind structure for all four planets. For Jupiter, the wind variations are closely

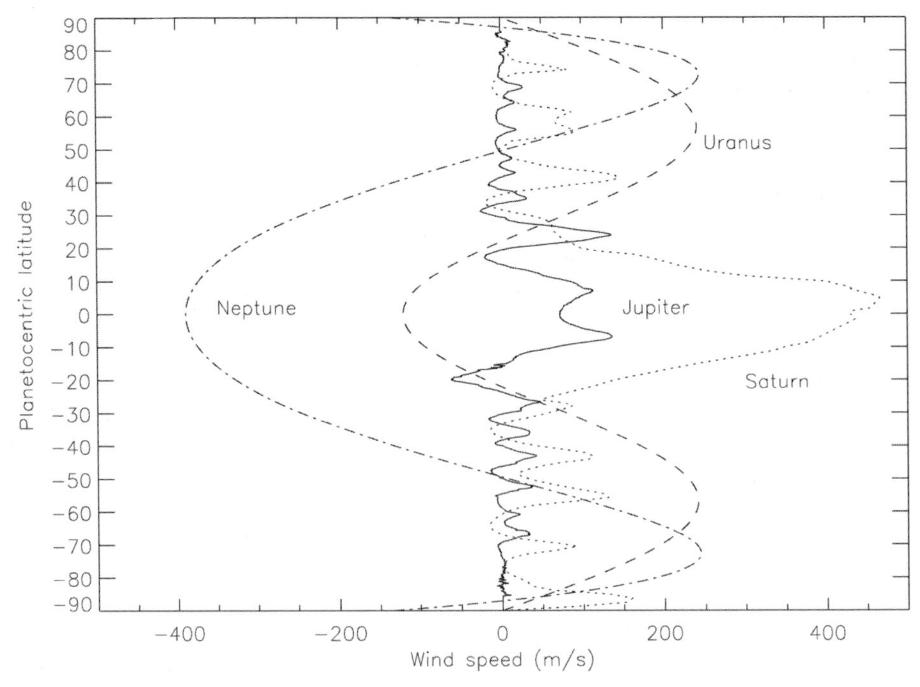

Figure 5.2. Zonal wind structure of the giant planets. Solid line: Jupiter; dotted line: Saturn; dashed line: Uranus; dot-dashed line: Neptune. The sources of these wind speeds are: Jupiter, Vasavada (2002), Porco *et al.* (2003); Saturn, Sánchez-Lavega *et al.* (2000); Uranus, Allison *et al.*, 1990 (unweighted *Voyager* fit); Hammel *et al.* (2001); Neptune, Sromovsky *et al.* (1993).

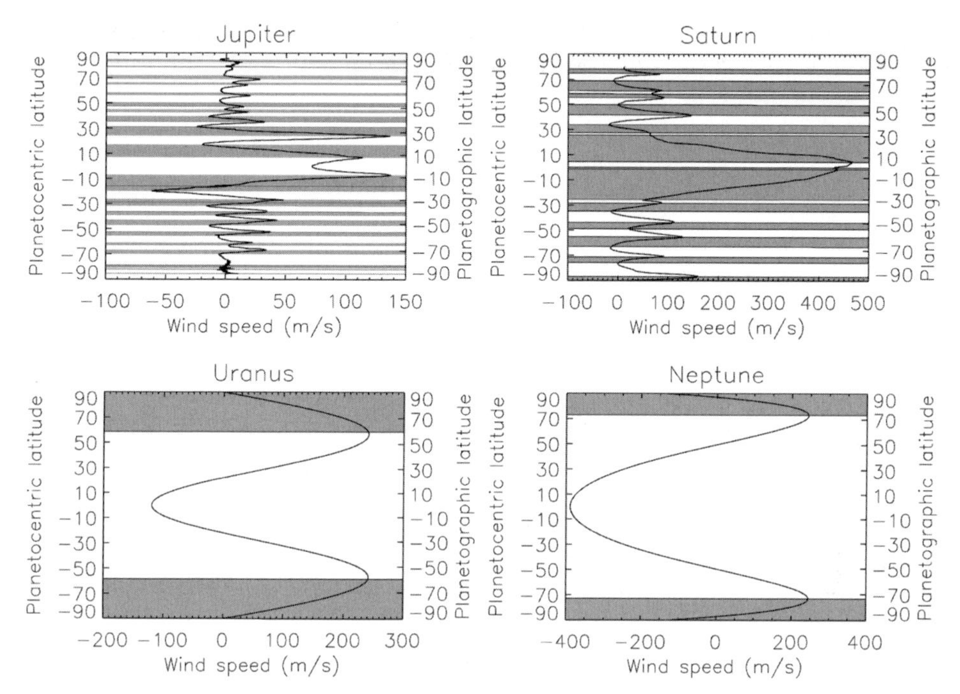

Figure 5.3. Zonal wind structure of the giant planets plotted separately and with regions of cyclonic vorticity shaded in gray.

associated with visible cloud features, where cyclonic vorticity shear regions (i.e., regions rotating in the same sense as the planet, resulting in anti-clockwise relative rotation in the northern hemisphere ($du/dy < 0$) and anti-clockwise rotation in the southern hemisphere ($du/dy > 0$) are seen as visibly dark, relatively cloud-free "belts", while anticyclonic vorticity shear regions (i.e., regions rotating in the opposite sense to the planet, resulting in clockwise relative rotation in the northern hemisphere ($du/dy > 0$) and anti-clockwise rotation in the southern hemisphere ($du/dy < 0$) are seen as visibly bright cloudy "zones". The mean cloudiness of the belts/zones is confirmed by observations of the thermal emission at 5 μm, where cloud-free belts allow radiance to escape from the warm 5 bar to 8 bar pressure levels underneath, while the cloudy zones appear dark. Hence, the visible and near-infrared albedo is closely anti-correlated with the 5 μm brightness (e.g., Irwin *et al.*, 2001). The belt/zone structure of Saturn appears much blander than that of Jupiter, due to the greater obscuration by tropospheric haze (Chapter 4), but there appears to be a more fundamental difference in that there is less correspondence between wind shear and albedo. In fact, the Saturn belts/zones appear better correlated with mean zonal wind than they do with wind shear. The appearance of Saturn in the 5 μm has now been recorded from ground-based observations (Yanamandra-Fisher *et al.*, 2001) and by *Cassini* VIMS (Baines *et al.*, 2005). While there is a greater component of reflected sunlight in Saturn's 5 μm spectrum, the observed radiance is, like Jupiter, principally

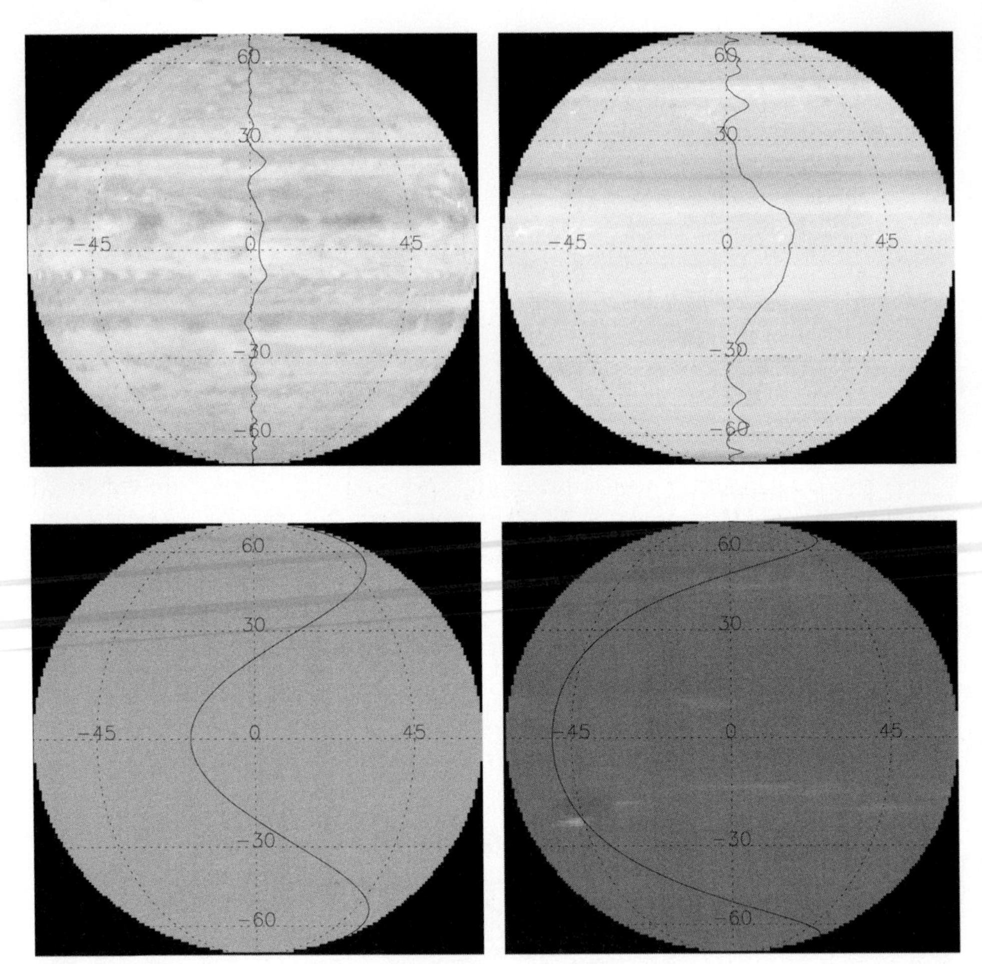

Figure 5.4. Zonal wind structure of the giant planets superimposed onto representations of their visible appearance. The wind speeds have been converted into degrees longitude/rotation of the planet. Note that the figures in the disks refer to degrees latitude and longitude—not to wind speed. Planetary images based on NASA data and processed by James Hastings-Trew and Björn Jónsson.

due to thermal emission of the deep atmosphere, modulated by cloud opacity in the 2 bar to 3 bar region, especially towards the longwave end of the 5 μm window. However, unlike Jupiter, there is observed to be little correlation between the 5.2 μm brightness and the visible albedo, although the main bright 5 μm band seen between 38°S and 49°S (planetocentric) does seem to coincide with an eastward jet, although puzzlingly with the anticyclonic side, rather than the cyclonic side, as is usually seen for Jupiter. Since the 5 μm observations map the total cloud opacity above the 5 bar level on Saturn there again seems to be less, or perhaps a different correlation between cloud opacity above 5 bar and zonal wind flow at the cloud tops.

Measurements of mean thermal emission as a function of latitude by spacecraft such as *Voyager* and *Cassini* may be used to calculate the zonal thermal vertical wind shears which, combined with measured cloud top winds (assuming the pressure level of this is known accurately) can be used to generate the variation of zonal wind speed with height. On all four giant planets, the zonal winds are found to decay with height to zero within about 3–4 scale heights of the cloud tops. The nature of the frictional force that this braking implies is not clear, although it is believed that the breaking of gravity waves or eddy motions may play a role. The latitude dependence of temperature in the upper troposphere may also be used to deduce meridional motion. Latitude bands with warmer than average temperatures imply horizontal convergence and thus presumably subsidence, while cooler latitudes imply divergence and thus upwelling from below. On Jupiter it is found that anticyclonic, more cloudy latitudes are cooler in the upper troposphere while cyclonic, less cloudy latitudes are warmer, leading to the canonical view of the flow in the Jovian atmosphere (shown in Figure 5.5) that zones are regions of moist, upwelling air and belts are regions of subsiding, dry air. However, *Cassini* observations suggest that this canonical view may be incorrect since convective clouds and lightning are found to occur exclusively in cyclonically sheared belts (Ingersoll *et al.*, 2000; Porco *et al.*, 2003). Similarly for Saturn, although convective storm activity is less vigorous than for Jupiter, three major storms have been observed in the cyclonically sheared region at 35°S, and these

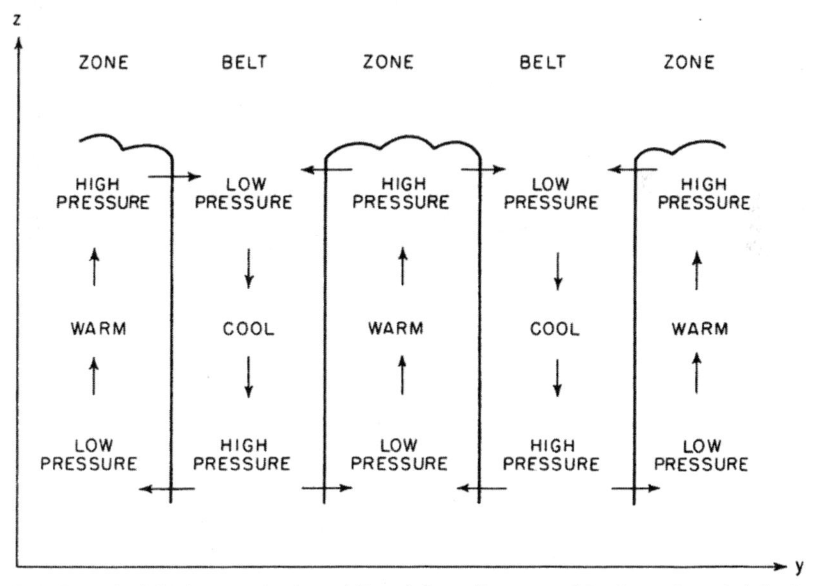

Figure 5.5. Stone's (1976) canonical meridional flow diagram of Jupiter where bright "zones" are regions of upwelling and thus divergence at the top of the atmosphere, and "belts" are regions of convergence and subduction. While this model fits well with the Jovian cloud structure and zonal winds, it less successfully models the flows in the other giant planet atmospheres. From Stone (1976). Reprinted by permission of the University of Arizona Press. © Arizona Board of Regents.

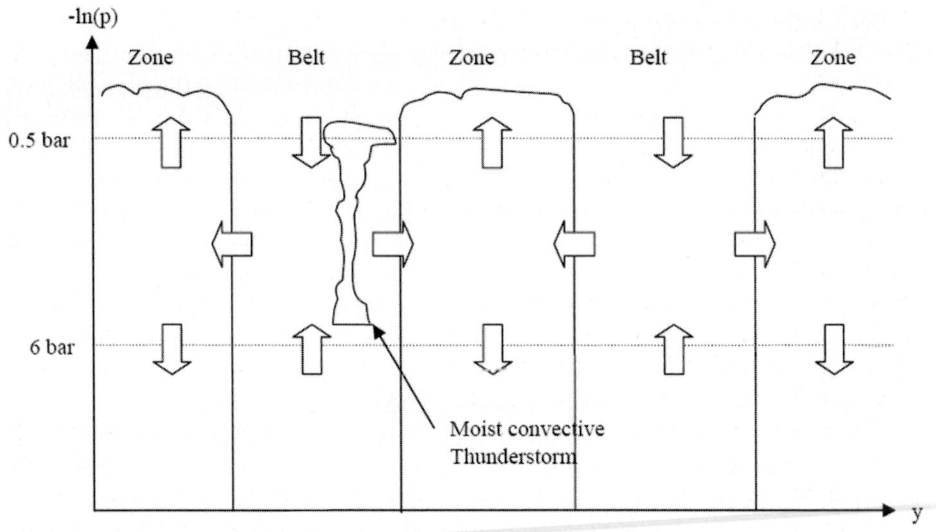

Figure 5.6. Modified meridional flow diagram of Jupiter suggested by Ingersoll *et al.* (2000).

outbreaks are correlated with electrostatic discharges thought to be due to lightning. These observations led Ingersoll *et al.* (2000) to suggest that the meridional flow we see at the cloud tops is not that of the deeper atmosphere below the water clouds and that, in fact, the deep flow is contrary to the upper flow (as can be seen in Figure 5.6). Moist convection in the form of thunderstorms requires the existence of convective available potential energy (CAPE). Positive CAPE exists when the background temperature profile is stable with respect to dry convection, but unstable with respect to moist convection. Essentially this requires that the lapse rate in the atmosphere lies somewhere between the DALR and SALR, such that if a parcel rises without condensation it is denser than the surrounding air and sinks back to its original level whereas if condensation occurs, the parcel is less dense than the surroundings and continues rising, condensing volatiles as it goes. For this scenario to be able to explain the conditions seen in the atmospheres of Jupiter and Saturn requires the existence of a stable layer underlying the condensation region, which inhibits dry convection below the water condensation level, but allows occasional vigorous events to penetrate to the level of positive CAPE leading to sudden, massive convective events (Showman and de Pater, 2005). Without such a stable layer, the CAPE would be quickly dissipated. In such a model the confinement of thunderstorms to cyclonic regions of Jupiter and Saturn can then be explained by the fact that anticyclonic regions are high-pressure areas that will press down on the underlying layers, increasing the thickness of the underlying stable layer, whereas cyclonic regions are low-pressure areas that will have a thin underlying stable layer and thus be more susceptible to convective outbursts. Such a scenario then explains why the abundance of ammonia at altitudes above the 5 bar level is so low on Jupiter (and Saturn): with convection in this region limited to small moist convective thunderstorms, the abundance of ammonia in the 1 bar to 2 bar region is much lower than expected.

Through horizontal mixing, some of this air reaches the zonal regions and is then upwelled to condense at levels considerably higher than expected from the deep abundances.

Jupiter and Saturn have similar zonal wind patterns, with strong eastward, or *prograde*, jets at the equator and rapidly varying wind direction towards both poles. At the equator, the air is effectively rotating faster than the bulk of the planet and thus is described as "super-rotating". How such a state might be driven is extremely puzzling since it can be proven that no axisymmetric (i.e., zonal mean) process can lead to such a state. Instead, nonlinear wave interactions are required such as diffusive small-scale eddies, or gravity/Kelvin/Rossby waves (discussed in Section 5.3.2). The nature of the super-rotation of the equatorial atmospheres of Jupiter and Saturn, and indeed the observed super-rotation of the slowly rotating worlds of Venus and Titan is an area of very active research.

The zonal wind profiles of Uranus and Neptune are completely different from those of Jupiter and Saturn with a very smooth, slow variation of zonal wind speed with latitude, and *retrograde* equatorial jets. Uranus had a particularly bland appearance at visible wavelengths during the *Voyager 2* flyby in 1986, with a hint of brighter albedos at the equator and poles and darker albedos in midlatitudes where the winds are eastward. Although the appearance of Uranus has become more dynamic as it passed through its Northern Spring Equinox in 2007, the generally darker appearance of the midlatitudes still holds true. Neptune also seems to show the same correspondence. Hence, the wind shear is anticyclonic at midlatitudes and cyclonic at the poles suggesting upwelling at midlatitudes in the upper tropospheres and subsidence at the equator and poles. This view is supported by *Voyager*, Hubble Space Telescope (HST), and ground-based observations of these planets that show the main convective activity in the upper methane cloud layers to occur at just these darker midlatitudes. In addition, measurements of zonal temperature structure show the upper troposphere to be warm at the equator and poles, but cooler at midlatitudes, which fits nicely with this picture. Hence, the brighter albedos observed at the equator and poles are believed to be due partly to the concentration of upper tropospheric haze in the convergence zones above the poles and equator and also partly due to either the cloud top pressure, or color of the main presumed H_2S cloud. Why the atmospheric circulations of Uranus and Neptune should be so different from those of Jupiter and Saturn is not known.

For many years, the jets were assumed to be confined to the upper weather layer of the atmosphere and not extend into the interior. The accepted explanation of the observed belts and zones was as mentioned earlier and shown in Figure 5.5 that, due to unspecified frictional drag processes, in anticyclonic shear regions where the Coriolis force is balanced by high pressures (assuming the geostrophic approximation) the flow is divergent within the clouds and at the cloud tops, but convergent beneath and so anticyclonic regions were preferentially heated from below by latent heat release (Ingersoll, 1990). This heating would act to sustain the anticyclone by heating the core, and from the thermal wind equation increasing the anticyclonicity with altitude, thus causing more divergence in the clouds and so on. This explanation is certainly consistent with the zonal cloud structure on Jupiter, but is far less successful in explaining the belt/zone structure of the other planets and also does not explain why convective thunderstorms on both Jupiter and Saturn occur only in cyclonic regions as we saw earlier. It also suffers from further flaws: (1) it does not actually explain the banded appearance of the planets since it works equally well for isolated anticyclonic spots as it does for extended anticyclonically sheared zones; and (2) it relies on difficult-to-observe (and estimate) processes such as frictional drag and latent heat release. One aspect of this picture that can easily confuse is that on the Earth we are familiar with anticyclones (high-pressure regions) being regions of downwelling, cloud-free air while cyclones (low-pressure regions) are regions of upwelling, cloudy air. This is clearly the opposite of what is found on the giant planets! However, it must be remembered that the Earth has a definite lower boundary and thus it is the frictional forces near the ground that cause convergence at the base of low pressures and divergence at the base of high pressures. On the giant planets there is no such lower boundary and the nature of any deep frictional forces that may be responsible for any meridional flow is unclear. However, the giant planets do appear to have an upper boundary in that the zonal winds are found to decay with height, presumably due to some kind of frictional damping. If this is also the main source of the friction driving the meridional circulations then the "highs" and "lows" on the giant planets have friction at their tops rather than at their bottoms, which would account for their inverted properties relative to their terrestrial cousins.

For all the giant planets, although there is evidence for significant turbulence and wave activity, the zonal winds appear to be extremely stable, and have not altered greatly since observations of these planets began (with the possible exception of Saturn, whose equatorial jet has slowed by some 100 m s^{-1} since the *Voyager* observations). In order to investigate the stability, driving, and dissipation of zonal mean circulation, we need to introduce the basic theory of eddy motion in the giant planet atmospheres.

5.3 EDDY MOTION IN THE GIANT PLANET ATMOSPHERES

The basic motion of the giant planet interiors is almost certainly barotropic up to pressure levels where the visible and thermal-infrared optical depth of the atmosphere becomes small enough to allow both solar heating and thermal radiation to space. Hence, while horizontal temperature variations are observed on the giant planets at pressures less than about 1 bar, temperature variations are observed to decrease at higher pressures and presumably rapidly diminish to zero in the deep atmosphere. The dominating barotropic nature of the flow, combined with the large Coriolis forces on the giant planets, low viscosity, and the absence of significant surface friction, seems to result in a simple zonal circulation for all four giant planets. However, while molecular viscosity is low, turbulence in the atmosphere leads to considerable eddy viscosity. This, together with differential solar heating, leads to time-varying eddy motions (where by eddy motion we technically mean anything departing from the mean flow) including turbulence, large-scale vortices, and waves, which will now be discussed.

5.3.1 Turbulence in the giant planet atmospheres

Although the atmospheres of these planets are very deep, the magnitude of the gravitational force greatly exceeds anything else and so the flow is to a first approximation two-dimensional. Under such conditions, any turbulence that is generated in the flow has the counter-intuitive property of being converted into larger and larger scale eddies. This is exactly the opposite of what occurs in more familiar three-dimensional turbulence, where turbulence cascades into smaller and smaller scales, a classic example being the break-up and dispersion of a smoke ring. The process is called the "backwards energy cascade" (Charney, 1971) and has fundamental consequences for planetary atmospheres. The energy spectrum of two-dimensional turbulence is found to follow the classic *Kolmogorov* scaling $E(k) \propto k^{-5/3}$ (Read, 2001). Turbulence in a planetary atmosphere may arise through a number of mechanisms such as static instability, where the atmosphere is unstable to convective overturning, or in cases where there is excessive horizontal or vertical wind shear. In general, it is found that once turbulence is initiated, the associated energy may be dissipated either due to friction, or transfer of energy to the mean flow. It can be shown (Andrews *et al.*, 1987) that many forms of turbulence will persist provided that a quantity known as the *Richardson number*, *Ri*, is less than approximately 1, where

$$Ri = \frac{gS}{T(\partial u/\partial z)^2}, \tag{5.40}$$

and where g is gravitational acceleration; S is the static stability described below in Equation (5.42); T is temperature; and u is the zonal (east–west) wind. In addition to assessing turbulence, the Richardson number provides a very useful measure of the dominant heat-transporting modes on planets as described by Stone (1976). For example, negative values of *Ri* (hence persistent turbulence) are associated with static instability and free convection, while large positive numbers are associated with Hadley cell circulations. A summary diagram of the different heat transfer modes associated with different *Ri* on the giant planets is shown in Figure 5.7 (after Stone, 1976).

For the giant planets, assuming geostrophic conditions and barotropic flow, turbulent eddies are predicted to grow and absorb smaller disturbances until their size reaches the *Rhines length* L_β (Rhines, 1973, 1975)

$$L_\beta = 2\pi\sqrt{\frac{2U}{\beta}} \tag{5.41}$$

(where U is mean wind speed and the β-parameter is defined below in Equation 5.46) when the disturbance is converted to planetary waves and the backwards energy cascade terminates. Thus, eddies have a maximum scale and this is likely to be related to the zonal structure scale (see Section 5.4.1). This property provides one link

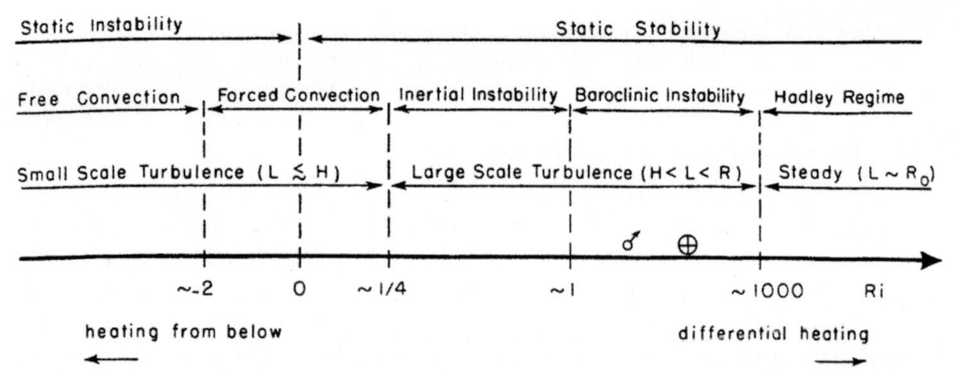

Figure 5.7. Stone's (1976) regime diagram of the main characteristic atmospheric motion as a function of the Richardson number (Ri). From Stone (1976). Reprinted by permission of the University of Arizona Press. © Arizona Board of Regents.

between U and L for the giant planets, but another condition is required to uniquely determine both the length and velocity scales. The other condition may be provided by the *radius of deformation*, which is the length scale at which Coriolis effects become significant (described in Section 5.3.2) and which is found to be similar to the width of the zones on Jupiter (Gierasch and Conrath, 1993), although this conclusion depends on the assumed static stability.

We will now outline some of the instability mechanisms which may, or may not, be responsible for turbulence in the giant planet atmospheres, and which are also summarized for convenience in Table 5.1 (after Stone, 1976).

Static instability

The simplest form of instability is static instability, whereby temperature decreases with height more rapidly than the adiabatic lapse rate. The static stability S of an atmosphere is defined as the rate of change of potential temperature with height

$$S = \frac{\partial\theta}{\partial z}. \tag{5.42}$$

The potential temperature is the equivalent temperature of an air parcel if it is compressed or expanded adiabatically (along a dry adiabat) to a reference pressure p_0, usually 1 bar, and may be shown to be equal to

$$\theta = T\left(\frac{p_0}{p}\right)^{R/C_p} \tag{5.43}$$

where R is the universal molar gas constant; and C_p is the molar heat capacity at constant pressure. In an atmosphere where temperature T falls with height at the dry adiabatic lapse rate, the potential temperature is constant with height and thus S is zero. For an atmosphere that is diabatically heated by, say, direct absorption of sunlight through gas or aerosol absorption, the temperature falls less quickly with height and may, as in the stratospheres of the giant planets, actually rise. In such cases

Table 5.1. Instability criteria (from Stone, 1976).

Dynamical mode (source)	*Criteria for instability*	*Structure*	*Characteristic horizontal scale*	*Characteristic timescale*
Barotropic instability	$\frac{d^2u}{dy^2} > \beta$	3-D	$\pi\left(\frac{u}{\beta}\right)^{1/2}$	$\frac{10}{(u\beta)^{1/2}}$
Baroclinic instability	$Ri > 0.84$	3-D	$2\frac{u}{f}(1+Ri)^{1/2}$	$\frac{3}{f}(1+Ri)^{1/2}$
Inertial instability	$0 < Ri < 1$ or $f - \frac{\partial u_g}{\partial y} < 0$	Axially symmetric	$2\left(\frac{u}{\beta}\right)^{1/2}\left(\frac{Ri^2}{1-Ri}\right)^{1/4}$	$\frac{1}{f}\left(\frac{Ri^2}{1-Ri}\right)^{1/2}$
Radiative instability	See text	Axially symmetric	$2\frac{(\alpha g H^2 S)^{1/4}}{\beta^{1/2}}$	$\frac{\tau}{2}\alpha HS$
Free convection	$S < 0$	3-D	~H	$\sim(\alpha g\lvert S\rvert)^{-1/2}$

α is the thermal expansion coefficient ($= 1/T$) and τ is the radiative relaxation time (from Stone, 1976).

S is positive, and the atmosphere is convectively stable—a parcel displaced vertically and adiabatically will be cooler, and so less buoyant than the surrounding air, and will thus sink back to its original level (in fact, it may oscillate about its original level as we shall see in Section 5.3.2). If S is negative, then the temperature falls with height more quickly than the lapse rate. Such a situation is unstable since a parcel displaced vertically and adiabatically will be warmer, and so more buoyant, than the surrounding air and will thus continue to rise. Whether the atmosphere actually turns over or not depends on the magnitude of the instability and the viscosity of the air (Chamberlain and Hunten, 1987). Clearly static instability dominates for $Ri < 0$.

Kelvin–Helmholtz instability

Another source of turbulence is vertical wind shear and one example is the Kelvin–Helmholtz instability which occurs for $0 < Ri < 0.25$. Such instability accounts for the formation of waves on the surface of the sea, and also for the so-called "mackerel skies" seen in upper cloud layers of the Earth during the approach of bad weather. Cloud patterns due to such instabilities have also been seen in Saturn's atmosphere at belt/zone boundaries.

Inertial instability

Inertial instability arises from mismatches between the pressure gradient force and centrifugal forces for a parcel displaced radially in an axisymmetric (i.e., independent of longitude) vortex (Andrews *et al.*, 1987) and are found to be important for atmospheres where $0.25 < Ri < 1$. They are driven primarily by the vertical eddy stresses associated with the vertical wind shear of thermal winds and transport heat both down the horizontal temperature gradient and vertically upwards, thereby increasing both the static stability of the atmosphere and Ri.

Inertial instability has been proposed as a mechanism to account for the formation of banded structures in giant planet atmospheres since it leads to axisymmetric motions consistent with the banded structure (Stone, 1976). Inertial instability may also explain the cloud structure on smaller scales such as in the Great Red Spot (GRS) since a second condition for inertial instability is that absolute vorticity is negative within a certain latitude band: that is,

$$f - \frac{\partial u_g}{\partial y} < 0. \tag{5.44}$$

It can be seen that the second condition for inertial instability favors regions of anticyclonic vorticity. Hence, in such anticyclonic eddies, inertial instability may create turbulence and thus enhanced cloudiness as is observed.

Barotropic instability

The word *barotropic* refers to cases where there are no temperature gradients on constant pressure surfaces (i.e., there is no differential heating) and thus the flow does not vary with height. This means that the basic zonal winds do not vary with height as can be understood from thermal wind equations. Under such conditions, instabilities arise mainly through excessive horizontal curvature of the basic flow profile (Andrews *et al.*, 1987). For barotropic waves in a shallow-atmospheric flow, where the height of the lower boundary is fixed, it may be shown (Houghton, 1986) that the condition for instability is the *Rayleigh–Kuo criterion*, which states that barotropic waves confined within a certain latitude band are stable unless the quantity

$$\beta - \frac{d^2\bar{u}}{dy^2} \tag{5.45}$$

changes sign somewhere within the latitude band, where the β-parameter is the latitudinal gradient of planetary vorticity defined as

$$\beta = \frac{df}{dy} \approx \frac{2\Omega \cos\phi_g}{R} \tag{5.46}$$

and where R is the planetary radius at the latitude in question. The barotropic instability condition is equivalent to saying that the meridional gradient of potential vorticity must change sign somewhere in a domain. While this may appear similar to inertial instability, barotropic instability is fundamentally different, not least because the Richardson number for barotropic conditions is poorly defined since the vertical wind shear in Equation (5.40) is zero.

Baroclinic instability

The word *baroclinic* refers to cases where temperature *does* vary on constant pressure surfaces and thus, from the thermal wind equation, zonal winds *do* vary with height. Hence, baroclinic instabilities depend, broadly speaking, on the vertical curvature of the flow. However, the flow under baroclinic conditions may also have large horizontal curvature, and thus under these conditions the relevant instability criterion is the *Charney–Stern* criterion, which states that a shallow-atmosphere baroclinic wave confined within a certain latitude band is stable if the gradient of the potential vorticity does not change sign (e.g., Dowling, 1995). Baroclinic instabilities give rise to the midlatitude storms seen in the Earth's atmosphere, but their importance in the Jovian atmospheres is unclear. Baroclinic instabilities are important for $Ri > 0.84$.

Radiative instability

This is a possible mechanism that has been proposed to account for the banded structure of the giant planets. If, in a condensing region of the atmosphere, the condensate enhances the greenhouse effect, then the increased thermal blanketing heats the atmosphere further thus enhancing vertical convection via positive feedback. Latitudinal temperature differences would then drive strong zonal winds from the thermal wind equation.

5.3.2 Waves in the giant planet atmospheres

We saw in Section 5.3.1 that turbulent motion arises when parcels that are deflected from their original positions feel forces that pull them further from their equilibrium positions. In some circumstances, however, displaced parcels may feel forces that force them to return to their equilibrium positions. Such forces give rise to a wide range of wave motions that are clearly observed in all planetary atmospheres, including the giant planets.

The two main properties of planetary atmospheres that may lead to waves (ignoring sound waves) are vertical stratification and the rotation of the atmosphere (Andrews *et al.*, 1987). *Internal gravity* waves result directly from stable stratification, while larger scale *inertia-gravity* waves result from a combination of stratification and Coriolis effects. *Planetary*, or *Rossby* waves, result from the polar gradient in planetary vorticity (the β-effect) and the conservation of potential vorticity. To deal with the whole spectrum of waves fully is beyond the scope of this book and the reader is referred to books such as Andrews (2000), Andrews *et al.* (1987), Holton (1992), or Houghton (1986). Historically, the equations of motion have first been linearized by separating parameters such as pressure and wind into their mean and perturbation components and then looking for waves with small enough amplitude that the linearizing approximation still holds. This *linear wave theory* is explored in detail by Andrews *et al.* (1987). A wide spectrum of waves is predicted under different assumed conditions, including waves that are free to travel in all directions and others that are trapped in certain latitude bands. A particularly important example of the latter for study of giant planet dynamics, are the so-called *equatorially trapped* waves

(Allison, 1990; Andrews *et al.*, 1987). More recently, advances in computation have made it possible to search for wave motion in the full nonlinear primitive equations and thus look for waves with large amplitudes (e.g., Dowling *et al.*, 1998). However, in this section we will summarize the main features of linear waves that are predicted to occur under different conditions in giant planet atmospheres.

Gravity waves and radius of deformation

Consider a parcel of air at a certain height and at the same temperature as the surroundings T_0, which is moved vertically and adiabatically from its equilibrium position by a small distance δz. If no condensation occurs then the new temperature of the parcel will be $T_1 = T_0 - \Gamma_d\,\delta z$ where Γ_d is the dry adiabatic lapse rate. The temperature of the surrounding atmosphere, however, will be $T_{e1} = T_0 + (dT/dz)\,\delta z$, where dT/dz is the background lapse rate, and if the two are different the parcel will feel a buoyancy force of

$$F = \rho_1 V \frac{d^2(\delta z)}{dt^2} = gV(\rho_{e1} - \rho_1) \tag{5.47}$$

where ρ_1 is the density of the parcel; ρ_{e1} is the density of environmental air at the same altitude; and V is the parcel's volume. Hence, the parcel's acceleration is given by

$$\frac{d^2(\delta z)}{dt^2} = g\left(\frac{\rho_{e1}}{\rho_1} - 1\right) = g\left(\frac{T_1}{T_{e1}} - 1\right) \tag{5.48}$$

where on the right-hand side we have substituted the expression for density in terms of temperature and pressure, assuming that the gas is ideal. Substituting for T_1 and T_{e1} in terms of lapse rates and displacements, and rearranging, we find

$$\frac{d^2(\delta z)}{dt^2} = -\frac{g}{T_0}\left(\frac{dT}{dz} + \Gamma_d\right)\delta z \tag{5.49}$$

or

$$\frac{d^2(\delta z)}{dt^2} + N_B^2\,\delta z = 0 \tag{5.50}$$

where

$$N_B^2 = \frac{g}{T_0}\left(\frac{dT}{dz} + \Gamma_d\right). \tag{5.51}$$

In a statically stable atmosphere, N_B^2 is positive and thus Equation (5.50) represents simple harmonic oscillation of the parcel about its equilibrium position. The oscillation angular frequency, N_B, is known as the *buoyancy*, or *Brunt–Väisälä* frequency. If N_B^2 is negative, then the solution of Equation (5.50) is exponential in z and thus the atmosphere is unstable. It can be shown (Andrews, 2000) that Equation (5.51) may be further simplified to a form incorporating potential temperature and static stability

$$N_B^2 = \frac{g}{\theta_0}\left(\frac{d\theta}{dz}\right) = \frac{gS}{\theta_0}. \tag{5.52}$$

If the horizontal direction is also considered, it is easily shown that these gravity waves propagate horizontally as well as vertically (e.g., Andrews, 2000; Houghton, 1986). Vertical perturbations in the lower troposphere, such as those generated by convection, may initiate such oscillations in the stable part of the atmosphere above the radiative–convective boundary. The resulting waves propagate both horizontally and vertically, and their amplitude is found to increase exponentially with height since their energy flux, which is conserved, is proportional to ρv_0^2, where ρ is the density and v_0 is the velocity amplitude of the wave. Small-scale, high-frequency waves do not feel the effects of the Coriolis force and are called internal gravity waves. Larger scale, lower frequency waves, however, are affected by the planet's rotation and are called inertia-gravity waves. The length scale at which Coriolis effects become important is defined by the *radius of deformation*, $a = c/f$ (sometimes called the *Rossby radius*), where c is the phase speed of the wave and f is the Coriolis parameter (Gill, 1982). It should be noted that there is not a single radius of deformation, but in fact an individual deformation radius for every particular class and mode of wave since the phase speed c depends on wavenumber, and also on vertical stratification through the Brunt–Väisälä frequency. For atmospheres where the geostrophic approximation applies it is found that the energy of shortwavelength disturbances is mainly in the form of kinetic energy, while the energy of longwavelength disturbances is mainly in the form of potential energy. Both are equal at the scale of the radius of deformation. This scale is also sometimes called the "synoptic scale". The radius of deformation may also be thought of as the "preferred" horizontal length of disturbances in the sense that waves arising from baroclinic instabilities (Section 5.3.1) grow most rapidly (and are thus most visible) when their horizontal dimension is of this radius. For a continuously stratified atmosphere, a useful expression for the mean radius of deformation (Gierasch and Conrath, 1993) is $a = N_B H/f$, where H is the scale height of the atmosphere.

Gravity waves are thought to be the dominant source of vertical eddy mixing in the stratospheres of the giant planets. At very high altitudes the waves 'break' as described in Chapter 4, much as surface water waves on the lakes and seas break when they reach beaches. At altitudes where the gravity waves break, the momentum of the wave is transferred to the momentum of the mean flow.

Kelvin waves

Kelvin waves are a special class of gravity waves that are found to move eastwards relative to the mean zonal flow, but only at latitudes close to the equator. Air moving in the atmospheres of rotating planets is affected by the Coriolis force, which for prograde-spinning planets deflects the air to the right in the northern hemisphere and to the left in the southern hemisphere. Consider a pressure disturbance moving eastwards relative to the mean zonal flow, which is symmetric about the equator. The Coriolis force is zero at the equator, but increases with latitude. Hence, the part of the disturbance to the north of the equator is deflected south, while that to the south of the equator is deflected north, which by conservation of mass increases the pressure at the equator. Eventually, the equatorial pressure rises to a level sufficient to balance the Coriolis "compression" and the disturbance spreads latitudinally again before the cycle repeats. These equatorially trapped *Kelvin* waves are observed in the Earth's atmosphere and may also be important at equatorial latitudes in the giant planets. This description is very simplistic and more detailed treatments are summarized by Allison (1990) and Andrews *et al.* (1987). The amplitude of these waves is calculated to diminish away from the equator with an exponential decay length L_{eq}, known as the *equatorial deformation radius*

$$L_{\text{eq}} = \left(\frac{2c}{\beta_{\text{eq}}}\right) \tag{5.53}$$

where c is the phase speed. A key diagnostic feature of equatorially trapped Kelvin waves is that the meridional velocity across the equator is zero.

Rossby or planetary waves

We saw in Section 5.2.1 that potential vorticity is conserved by an atmosphere where friction and diabatic heating are negligible. Consider a parcel of air of height h at some latitude ϕ in the northern hemisphere, which initially has zero relative vorticity ζ. Suppose the parcel is displaced northwards. From Section 5.2.1 we know that its potential vorticity $(\zeta + f)/h$ must be conserved. However, the Coriolis parameter f increases as the parcel moves north and thus the absolute vorticity ζ must decrease in order to compensate. Hence, the parcel gains negative (clockwise) *relative* vorticity. Similarly a parcel displaced southwards gains positive (anti-clockwise) relative vorticity. This conservation mechanism gives rise to *Rossby*, or *planetary*, waves at latitudes where the air stream is moving in the eastward direction. Air deflected to the south will gain positive relative vorticity and thus turn towards the left and then northwards. After crossing the starting latitude, the air will gain negative relative vorticity and thus turn towards the right and then southwards and so on, setting up a stable wave. To outline how we may show this analytically, consider the momentum equations in pressure coordinates for frictionless flow, where vertical motion is neglected (Houghton, 1986):

$$\left(\frac{\partial}{\partial t} + u\frac{\partial}{\partial x} + v\frac{\partial}{\partial y}\right)u - fv = -g\frac{\partial z}{\partial x}, \tag{5.54}$$

$$\left(\frac{\partial}{\partial t} + u\frac{\partial}{\partial x} + v\frac{\partial}{\partial y}\right)v - fu = g\frac{\partial z}{\partial y}. \tag{5.55}$$

Differentiating Equation (5.55) with respect to x and differentiating Equation (5.54) with respect to y and then subtracting gives another form of the vorticity equation

$$\left(\frac{\partial}{\partial t} + u\frac{\partial}{\partial x} + v\frac{\partial}{\partial y}\right)\zeta + v\frac{df}{dy} = 0. \tag{5.56}$$

We now substitute the following mean and perturbation values: $u = \bar{u} + u'$, $v = v'$, $\zeta = \zeta'$ and define a perturbation stream function $u' = -\dfrac{\partial\psi}{\partial y}$ and $v' = -\dfrac{\partial\psi}{\partial x}$

such that $\zeta' = \nabla^2\psi$. The *perturbation form* of Equation (5.56) is then

$$\left(\frac{\partial}{\partial t} + \bar{u}\frac{\partial}{\partial x}\right)\nabla^2\psi + \beta\frac{\partial\psi}{\partial x} = 0. \tag{5.57}$$

This is a wave equation and assuming a wave solution of the form $\psi = Ae^{i(\omega t + kx + ly)}$ and substituting this into Equation (5.57) we may then derive the dispersion relation

$$c = -\frac{\omega}{k} = \bar{u} - \frac{\beta}{k^2 + l^2}. \tag{5.58}$$

This expression shows that wave motion is indeed possible, and that the waves move westward relative to the zonal flow. Clearly if the zonal flow is eastward with a speed equal to this phase speed, then the Rossby wave appears stationary. The analysis may be extended to the vertical dimension also, but this is again beyond the scope of this book. For Rossby waves it can be seen that the meridional velocity variations are non-zero.

A periodic combination of westward-moving Rossby waves, and eastward-moving gravity waves, including equatorially trapped Kelvin waves is believed to be responsible for the *quasi-biennial oscillation* (QBO) observed in the Earth's stratosphere at equatorial latitudes. Here, mean zonal winds (between 15 mbar and 200 mbar) are found to vary quasi-periodically between eastward and westward flow with a period of about 28 months. In periods of eastward flow near the bottom of the stratosphere, equatorially trapped mixed Rossby gravity waves may propagate vertically upwards and dump their energy into an upper westward flow. As a result of this energy dumping, the region of westward flow gradually moves down. When the westward flow region reaches the bottom of the stratosphere, only Kelvin waves may propagate vertically upwards. These establish an eastward flow in the upper stratosphere where they break, which itself slowly moves down over time as more and more energy is dumped there. Eventually the eastward flow reaches the bottom of the stratosphere and mixed Rossby gravity waves begin to propagate vertically again and thus the cycle repeats itself. A similar process may be responsible for the quasi-quadrennial oscillation (QQO) observed in the equatorial stratosphere of Jupiter (discussed further in Section 5.5.3) and/or a similar semi-annual oscillation seen in Saturn's equatorial stratosphere (discussed in Section 5.6.1).

5.3.3 Vortices in the giant planet atmospheres

In addition to turbulence and waves, the giant planet atmospheres exhibit one more, very distinctive planetary-scale motion peculiar to them: long-lived vortices, or oval circulations such as the GRS on Jupiter, and the dark spots on Neptune. Such systems typically appear at latitudes of large horizontal wind shear and all vortices seen in the giant planet atmospheres are seen to have a relative vorticity of the same sign as the jets within which they lie. In one respect their presence is expected from two-dimensional turbulence since any small vortices generated in such shear regions between the belts and zones are expected to merge with other vortices and grow.

However, what is not clear is by what instability mechanism these vortices are initiated, and how they are maintained once they are established.

A fundamental clue to the nature of the large ovals is to examine the distribution of cyclones and anticyclones across the different planets. The geostrophic equations derived in Section 5.2.1 are completely symmetric with respect to the sign of the vorticity and thus if the atmospheres of the giant planets were purely geostrophic, we might expect that cyclonic ovals were equally numerous as anticyclonic ones. However, almost all observed ovals on the giant planets are anticyclonic (e.g., on Jupiter 90% of all ovals are anticyclones). Where might such an asymmetry arise? One way of achieving such an asymmetry is to include centrifugal forces. Consider the acceleration of air moving with speed V in a circle of radius R. Starting with the horizontal momentum equations, and ignoring friction, we find that (e.g., Chamberlain and Hunten, 1987; Holton, 1992; Houghton, 1986)

$$\frac{V^2}{R} = -fV + \frac{1}{\rho}\frac{\partial p}{\partial R}. \tag{5.59}$$

Equation (5.59) may be solved for V to give

$$V = -\frac{fR}{2} \pm \left(\frac{f^2R^2}{4} + \frac{R}{\rho}\frac{\partial p}{\partial R}\right)^{1/2}. \tag{5.60}$$

This is the *gradient wind approximation* and by convention V is taken as positive for cyclonic motion, negative for anticyclonic motion, and R is positive and measured from the center of curvature. This equation has one cyclonic solution (low pressure in the center) and thus $\partial p/\partial R$ is positive, but three anticyclonic solutions, two of which have high pressures in the center of the cyclone, but one that has low pressure at the center. Hence, by including the acceleration term, the symmetry between cyclones and anticyclones is broken, although it is not known if this also accounts for the observed cyclonic/anticyclonic asymmetry on the giant planets. Another reason for the greater number of anticyclones than cyclones might be that anticyclones are more stable under Jovian conditions and thus that cyclones rapidly become disrupted and break up. One reason for this may be that cyclones are more susceptible than anticyclones to moist convection (Dowling, 1997). As discussed in Section 5.2.2, the extra mass of an anticyclone depresses the atmospheric layers beneath it, whereas cyclones have the opposite effect and can raise deep moist air beyond its lifting condensation level. Hence, moist convection may be triggered which, if vigorous enough, may disrupt organized cyclonic circulation.

Much modeling has been done on long-lived eddies and, in addition, laboratory studies have been conducted with rotating annulus experiments (Section 5.4.1) to simulate a range of driving conditions (Read, 1986; Read and Hide, 1983, 1984). The GRS on Jupiter is the largest of the long-lived anticyclones observed on the giant planets and the way it is driven and sustained is the source of much debate. There are at least four possible driving mechanisms that have been considered: (1) barotropic shear; (2) baroclinic shear; (3) local forcing (e.g., moist convection, ortho–para conversion); and (4) capture and absorption of smaller eddies. Unfortunately the precise forcing mechanism is unclear, although the capture of smaller eddies would lead to the deposition of their momentum in the outer annulus of the GRS, whose observed width of roughly 300 km to 500 km is consistent with the smallest scale of observed eddies, and may be equal to the radius of deformation for Jupiter. While many studies concentrate on how such flows may be maintained against dissipation, another possibility is that the GRS is a "free mode" of the Jovian circulation system and thus needs very little driving against dissipative effects. If frictional forces on the giant planets really are as low as they appear to be then such vortices may appear spontaneously and be naturally long-lived (Lewis, 1988). Hence, for example, the GRS can be considered to be a giant "flywheel" which, rather than being difficult to drive, is actually rather difficult to stop!

The GRS may be a special case, but in a number of numerical simulations (e.g., Vasavada and Showman, 2005), jet instabilities can lead to the formation of up to ~10 mid-size vortices, which then undergo mergers to form fewer larger vortices. Such behavior is remarkably consistent with how Jupiter's white ovals were observed to form in 1939 (Section 5.5.2).

Once initiated, because of the β-effect isolated vortices should be dispersed by Rossby waves, but this may be halted by their position between jets, from which some models suggest they may draw their energy (Achterberg and Ingersoll, 1994). Near the equator, where β is greatest, the dispersion is so strong that vortices generally cannot survive (Showman and Dowling, 2000) and no vortices are seen within 10° of the equators of any of the giant planets.

In theoretical studies (e.g., Achterberg and Ingersoll, 1994; Yamazaki *et al.*, 2004), isolated vortices tend to drift westwards, and in the absence of strong background winds, also drift equatorwards. Such equatorwards drifting is seen on Neptune (Section 5.8.2), but not on Jupiter, where such motion may be inhibited by the jet system. Furthermore, some studies show that thin vortices migrate more slowly than thick vortices and are more easily confined by jets (Williams, 1996). A small thickness would be more consistent with the observed dynamical lifetimes of most large vortices, and some studies (Dritschel *et al.*, 1999) suggest that vortices are baroclinically unstable if their thickness exceeds their width by more than the order of f/N_B, where f is the Coriolis parameter and N_B is the Brunt–Väisälä frequency (Section 5.3.1). Putting in typical values for the Jovian atmosphere suggests that the GRS and white ovals extend no more than ~500 km below the clouds.

While some vortices such as the GRS certainly appear to be isolated, other examples of ovals, such as the brown barges of Jupiter (discussed in Section 5.5.2), appear in regular chains suggesting a link with planetary-scale Rossby waves, perhaps through the Rhines effect (Section 5.3.1). Another example was the North Polar Spot (NPS) observed by *Voyager 2* on Saturn and its associated North Polar Hexagon (NPH) wave. While it initially appeared as though the wave arose through deflection of the mean flow around the NPS, followed by subsequent oscillation, it may be that the NPS was just a manifestation of a global series of cyclones and anticyclones at this latitude with an accompanying, apparently wave-like flow around them. While the NPH has been seen again by *Cassini* (Baines *et al.*, 2007b; Fletcher *et al.*, 2008a), *Cassini* has not detected any associated North Polar spot during the current epoch. We will return to this topic is Section 5.6.3.

5.4 MEAN AND EDDY CIRCULATION OF THE GIANT PLANET ATMOSPHERES

5.4.1 Tropospheric circulation and jets

We have seen that the zonal wind circulation of the giant planets is very vigorous. What is not so clear, however, is how these jets are initiated, maintained, and how deep into the interior these zonal winds extend. In this section we will review some the modeling work that has been done to understand the mean circulation of the giant planet atmospheres.

The Rhines length L_β was introduced in Section 5.3.1, in the context that vortices in a two-dimensional turbulent flow will grow via the inverse energy cascade until they reach a size comparable with the Rhines length, when they are then dissipated by Rossby waves. Rhines (1975) also realized that the variation of f with latitude would lead to an elongation of such vortices in the east–west direction and that under certain circumstances the flow reorganizes itself into jets spaced by L_β. This idea has been extended by Vallis and Maltrud (1993).

Theories for the vertical structure of the jets range between two limiting scenarios (Vasavada and Showman, 2005). In one, the jets are modeled as being confined to a shallow "weather layer" near the visible cloud level, where absorption of sunlight and latent heat release from cloud formation would lead to thermal contrasts of the order of 5 K to 10 K, setting up vertical wind shears from the thermal wind equation. In such a model, cyclonic regions must be cold at depth in order to reduce the winds to zero and anticyclonic regions must be warm. This contrasts with the observed behavior *above* the clouds that cyclonic regions are warm and anticyclonic regions cold. The other limiting scenario for the jets is that they extend thoroughout the entire molecular-hydrogen interior.

To describe fully the various models of the global circulation of Jupiter and the other giant planets is beyond the scope of this book. While the main theories are introduced in the following sections, for a more complete treatment the reader is referred to more detailed works such as Vasavada and Showman (2005).

Shallow-layer models

If the interior of a giant planet rotates at the same rotation rate at all levels then the deep atmosphere may be reasonably approximated as a fixed lower surface since the interior is adiabatic and has a huge mass. Any "weather" arising from differential heating and cooling is likely to be confined to the surface layers. Such shallow-layer models, adapted from terrestrial models, have provided a reasonable first analysis model at interpreting the dynamics of the giant planet atmospheres (e.g., Huang and Robinson, 1998; Williams, 1978, 1979, 1985, 1996, 2002, 2003a, b; Williams and Robinson, 1973). In such models, belts and zones appear spontaneously and there are examples of the kind of vortices found on the giant planets. However, a problem is that such models constantly need "pumping" with energy in order to keep them going, and soon disappear if the forcing is removed. In addition, the calculated outward thermal flux greatly exceeds that actually observed and jets resulting from such models are found to be stable with respect to the barotropic (or Rayleigh–Kuo) stability criterion, whereas the observed jets on Jupiter and Saturn have curvatures exceeding β by a factor of 2–3. A further shortcoming of shallow-layer models is that they predict equatorial jets with a similar width to midlatitude jets, while the observed equatorial jets of Jupiter and Saturn are approximately twice as wide. Furthermore, the equatorial jets of Jupiter and Saturn are eastwards, while most shallow-layer models produce westward jets.

With all these shortcomings, shallow-layer models are the only models that have an asymmetry in that they favor organized anticyclonic vortices and disorganized cyclonic regions, much as is observed on Jupiter and the other giant planets. Hence, while there is evidence for a deep component to the zonal flow of the giant planets, as outlined in the following sections, some shallow-layer aspects appear to remain and the true flow probably lies somewhere between the shallow-layer and deep models.

Deep models

While shallow-layer models are reasonably simple and are clearly applicable to the atmospheres of the terrestrial planets, the shallow weather layer theory of Jovian dynamics has suffered two setbacks since space age observation of the planetary atmospheres began. First of all, it is observed that although they experience differential solar heating, the giant planets have very little temperature variation with latitude, and any latitudinal variation that is present rapidly diminishes as the pressure increases. If zonal winds really were confined to the surface weather layer, then there must be large vertical wind shear below the cloud tops, and from the thermal wind equation an accompanying large variation in temperature with latitude, particularly for Saturn and Neptune, which have such high zonal winds. The low-temperature variation actually observed clearly suggests that the zonal wind structure is deep. For Saturn, zonal winds are estimated to extend to pressures of at least 10 bar (Smith *et al.*, 1981). Similar low-temperature variations are found at Jupiter and, in addition, radio tracking of the *Galileo* entry probe (Young, 2003) allowed the direct determination of the deep wind structure at the edge of the 5 μm hotspot it entered. Rather than decrease with depth, the winds were found to initially increase with depth and then tend to a constant value. However, this result must be qualified with the fact that the *Galileo* entry probe entered a somewhat anomalous region of the planet and thus conditions there may not be generally representative. The second major problem with the shallow weather layer model is that for Jupiter and Saturn (but not Uranus and Neptune) the rapidly varying zonal wind structure gives wind curvatures at the eastward jets that violate the barotropic instability (or Rayleigh–Kuo) criterion that $\beta - u_{yy}$ should not change sign. Since the zonal wind structure in fact appears very stable, this violation would suggest either that the physical assumptions used in

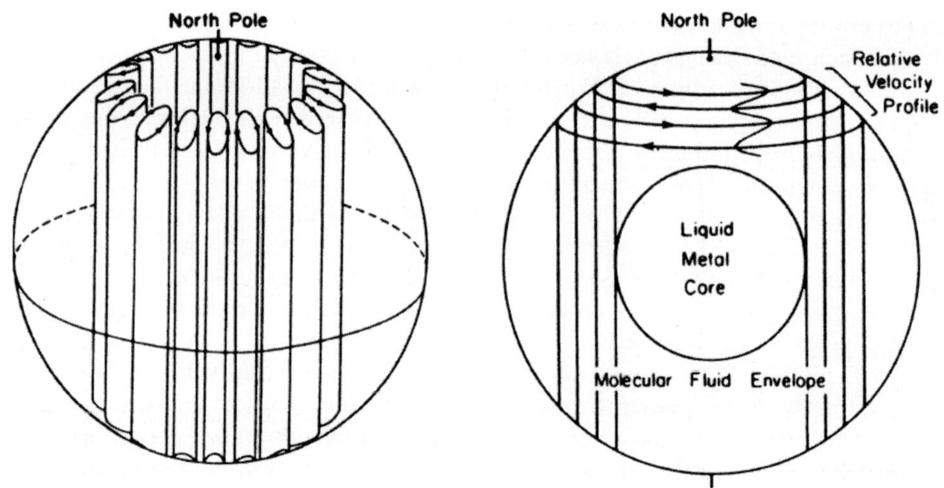

Figure 5.8. Taylor–Proudman columns and differential cylinders. From Ingersoll and Pollard (1982). Reprinted with permission from Elsevier.

deriving the criterion are invalid for Jovian atmospheres or that the shallow-layer model is not applicable. Hence, the zonal winds of the giant planets would appear to be deep—not shallow—and several other features in the giant planet atmospheres, discussed later, suggest significant coupling between the surface weather layer and the deep interior. Hence, a more complete picture of the giant planet atmospheres would appear to require the consideration of deep interior flows also.

One effect of giant planet rapid rotation on the interior fluid dynamics is the suppression of motion parallel to its rotation axis, known as the Taylor–Proudman effect (Busse, 1976, 1994, 2002; Ingersoll and Pollard, 1982) introduced earlier (Section 5.2.1). This tends to force the fluid to move as semi-rigid columns that are aligned with the rotation axis as shown in Figure 5.8. A remarkable experiment was performed on *Spacelab 3* in 1985 (Hart *et al.*, 1986), where a liquid confined between two hemispherical surfaces was spun about its own axis, with an electrostatic field used to simulate gravity. Under certain conditions a clear "banana cell" convection flow was seen. The oblate–spheroidal shape of giant planets cause such columns to stretch as they move towards or away from the rotation axis and, via the conservation of angular momentum, this vortex tube stretching effect is suggested to give rise to Rossby waves. An obvious consequence of this model is that atmospheric motions should be symmetric about the equator, which to a very good approximation they are. Thus to explain the zonal structure of the giant planet atmospheres it is possible to imagine Taylor–Proudman columns organizing themselves into a number of concentric cylinders, all rotating at slightly different rates. This theory elegantly explained the symmetric zonal structure of the giant planets and was also consistent with the findings of the *Voyager* missions that the zonal structure broke down at high latitudes and was replaced by chaotic overturning. The latitude where this occurred was found to be close to that where a cylinder just touching the metallic–molecular

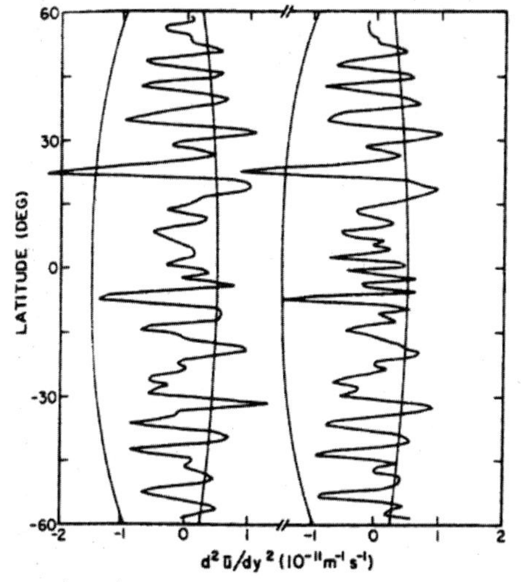

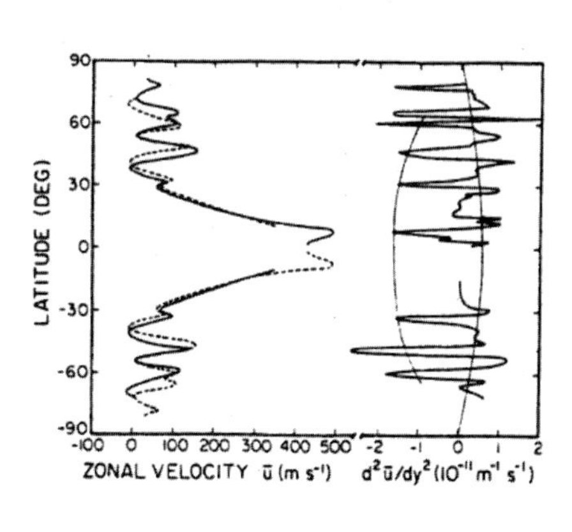

Figure 5.9. Stability of zonal structure of Jupiter and Saturn assuming: (1) a shallow layer and barotropic instability; and (2) assuming deep Taylor–Proudman columns (Ingersoll and Pollard, 1982). The left-hand figure shows the curvature of the zonal winds of Jupiter derived from both *Voyager 1* and *2* data. The barotropic stability curves are the smooth curves for positive values of d^2u/dy^2, and it can be seen that the winds on Jupiter clearly exceed this curve at several points. The winds do seem stable with respect to the deep Taylor–Proudman model, however, since the curvature is rarely more negative than the second stability curve shown, derived by Ingersoll and Pollard (1982). Similar results are shown on the right-hand side of the figure for the zonal winds measured on Saturn by *Voyagers 1* and *2*. Reprinted with permission of Elsevier.

boundary was predicted to intersect the surface spheroid. This critical latitude is at approximately 40°–45° for Jupiter and 65° for Saturn and it is interesting to note from Figures 5.2–5.4 that both Jupiter and Saturn have three eastward jets between the equator and these latitudes (Smith *et al.*, 1982). Polewards of this latitude, less organized motion would be expected since Taylor–Proudman columns could not pass right through the planet, but instead would intersect the metallic-hydrogen/molecular-hydrogen phase boundary. An additional advantage of this model is that the stability criterion of zonal flow is different from the barotropic instability criterion mentioned earlier and is found to be better satisfied by the zonal flows of all the giant planets (Ingersoll and Pollard, 1982) as can be seen in Figure 5.9 for the case of Jupiter and Saturn. It should also be mentioned that the deep zonal flow need not be forced by deep convection. Showman *et al.* (2006) show that surface forcing can also lead to deep zonal flow if the interior has low static stability, which it should do. Hence, eddy pumping can accelerate the jets, which can accelerate the interior also.

However, this model, while very elegant, has suffered a setback following the recent Jupiter flyby by *Cassini/Huygens*. The *Cassini* ISS camera has now found organized, long-lived zonal motion extending all the way to Jupiter's poles and high-latitude organized polar motion has also been observed on Saturn by *Voyager*,

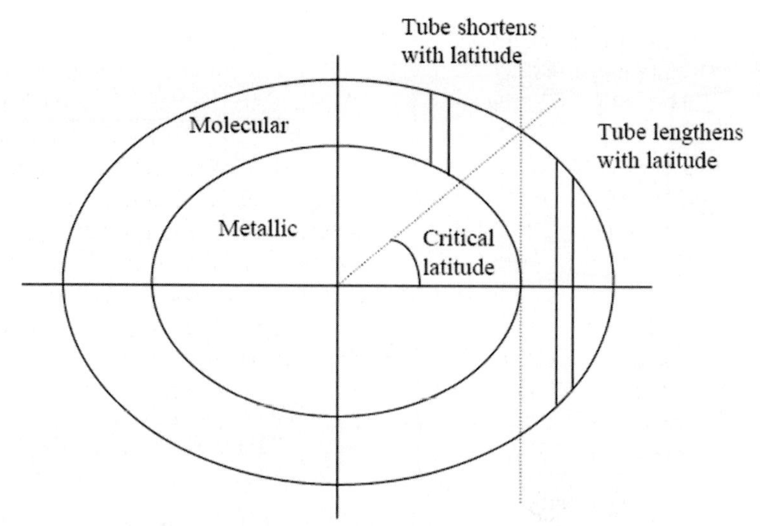

Figure 5.10. Vortex tube–stretching associated with Taylor–Proudman columns. Columns intersecting at latitudes less than the critical latitude pass right through from north to south. Motion of the column towards the rotation axis stretches the column. Columns at latitudes poleward of the critical latitude do not pass through to the other hemisphere. Furthermore, motion of the column towards the rotation axis *compresses* the column. Hence, this model would suggest very different atmospheric flow on either side of the critical latitude, which is not actually seen on any of the giant planets.

Cassini, and ground-based observations. There is also a problem that very different behavior is expected either side of the critical latitude in that on the equatorward side, northern motion stretches the columns, but on the poleward side, northern motion compresses the columns (Figure 5.10). Such differential behavior in atmospheric flow is not observed.

To address these problems, computer modeling of the dynamics in the molecular-hydrogen envelopes in giant planets has been performed (Aurnou and Olson, 2001; Christensen, 2001, 2002; Grote *et al.*, 2000). Such models solve the Navier–Stokes momentum equations, subject to the Boussinesq approximation that density variations are negligible except where coupled with buoyancy terms. Key parameters in the resulting equations are the Ekman number (ratio of frictional to Coriolis forces), the Prandtl number (ratio of kinematic viscosity to thermal diffusivity), and the Rayleigh number, which is a measure of convective vigor (Vasavada and Showman, 2005). Many solutions exist, depending on these parameters, but a general feature is that convective overturning in the molecular-hydrogen region gives rise to deep, symmetrical zonal flows from the equator to the pole. Although such models do produce an eastward-flowing jet at the interior, as in seen on Jupiter and Saturn, a major drawback is that the jets are generally too wide with just the equatorial eastward and half of the adjacent westward jets appearing at latitudes less than the critical latitude, where a cylinder just touching the metallic-molecular boundary intersects the surface spheroid. If it can be argued that the metallic-hydrogen region is actually rather larger than has previously been estimated, or that vigorous convection is limited to the outer layers of the molecular-hydrogen region then the critical latitude moves closer to the equator and the flow resembles the gas giants much more closely (Heimpel and Aurnou, 2007; Heimpel *et al.*, 2005), with wider jets near the equator and narrower ones at latitudes greater than the critical latitude. Recent estimates of Guillot *et al.* (2004) put the metallic-hydrogen/molecular-hydrogen boundary at a pressure level of 1 Mbar to 3 Mbar, placing it somewhere in the range 0.7 to 0.9 Jupiter radii, which would be consistent with these models, although it is also possible that a barrier to convection lies within the molecular-hydrogen mantle itself, since so little is currently known about the interior structures of these planets. However, a drawback of these models is their use of the Boussinesq approximation, which leads to an incompressible continuity equation. Since the actual density of the Jovian interior increases by a factor of $\sim$1,000 from the cloud decks down to the molecular–metallic boundary, the real circulation in Jupiter's interior is likely to be more complicated than these models suggest. Two-dimensional studies solving the *anelastic* equations, which are similar to the Boussinesq Navier–Stokes equations, but which have a more realistic continuity function containing an assumed variation of density with radius, have shown that convection is indeed altered when density variations are included (Evonuk and Glatzmaier, 2004). Another drawback of these deep models is that none can produce the multitude of vortices seen in giant planet atmospheres, which are easily produced by shallow-layer models.

Although there are many problems with the Taylor–Proudman column theory, it remains an attractive and intriguing possibility and thus it is of great interest to determine how deep the winds in the giant planets actually extend. Coherently organized internal motion of the kind just described will create perturbations in the gravitational equipotential "surface" of the planets, which will affect gravitational J-coefficients. For Neptune, which has a very broad and rapidly rotating zonal structure, the measured J_4-coefficients have been found to be inconsistent with Taylor–Proudman columns extending throughout the planet and instead the winds are concluded to be limited at most to the outer hydrogen–helium shell (Hubbard, 1997d). However, this does not discount the Taylor–Proudman theory since for Neptune (and Uranus), the transition to the icy interior, which should present the same boundary to Taylor–Proudman columns as the transition from molecular hydrogen to metallic hydrogen does in Jupiter and Saturn, occurs just a few thousand kilometers below the cloud tops. Planets such as Jupiter and Saturn have a much finer zonal structure and thus the effects of such columns (if they exist) will only be apparent in the higher order J-coefficients, which have not yet been accurately measured. However, the *Juno* mission (Section 8.6.1), due for launch in 2011, will be placed in a close polar orbit to Jupiter and close tracking of its motion should allow the determination of Jupiter's higher order J-coefficients.

One-and-a-half-layer models

We have seen that the apparently deep nature of the zonal flow argues against shallow-layer models of the giant planets in favor of perhaps a system of co-rotating cylinders. However, this deep-atmosphere theory is apparently cast into doubt by the absence of a difference in atmospheric flow at the latitude of the cylinder tangential to the metallic-hydrogen/molecular-hydrogen boundary and the observation of zonal jets extending polewards of the critical latitudes. Hence, the real flow of these planets would appear to be more complicated than either of these more simple approaches suggests and ideally a fully three-dimensional model of the atmospheric flow needs to be constructed to investigate the dynamics of the giant planets. While there has been some progress in this area, as we have seen in the previous section ("Deep-layer models"), simplifying assumptions have to be made in order to make the calculations computationally achievable, whose validity calls into question the predictions of such models. However, since the deep interiors of the giant planets are almost certainly barotropic, and since the surface layers exhibit many similarities with shallow-layer models, it may well be that these planets can be represented by models that have deep barotropic flows up to near the observable levels, capped by a statically stable layer that is driven by turbulent energy, injected from below (Leovy, 1986). One way of representing this idea is to use "one-and-a-half"-layer models, where deep atmospheric flow is represented as latitudinal variations in the height of a lower boundary layer. These height variations may be determined by tracking the vorticity of features on the edge of anticyclones such as the GRS, which have significant north/south motion and assuming that the potential vorticity $q = (\zeta + f)/h$ is conserved. Hence, the effective height h of the weather layer may be determined by observing variation in the vorticity of the flow (Dowling and Ingersoll, 1988, 1989). One-and-a-half-layer models with lower topography determined in this way are able to simulate the zonal flow of Jupiter (and other planets) and mimic the spontaneous formation and growth of large oval circulations. However, like shallow-layer models, they have the disadvantage that they need to be continuously forced in order to maintain the flow and thus seem to lack the principal energy source that maintains the zonal circulations of these planets.

The idea of a thin stable surface layer on top of barotropic, but differentially rotating interior cylinders also provides one explanation for how Jupiter, Saturn, and Neptune, which have significant internal heat sources, but differential solar heating all radiate approximately equally in all directions. The solar heating will be greatest at equatorial latitudes and this will tend to increase the static stability of the air and thus reduce the amount of internal heat that is convectively transported. Conversely, solar heating is least at the poles, which will tend to make the atmosphere neutrally statically stable and thus convection is uninhibited. Hence, more internal heat is radiated at the poles than the equator, counteracting the differential absorption of sunlight. However, the fact that Uranus, which also has significant differential solar heating, but negligible internal heat, also radiates equally in all directions suggests that reality may once again be more complicated!

Eddy–mean interactions

The interaction between the zonal mean flow and waves/vortices is currently unclear and opinion is divided between two main points of view. Some scientists believe that the zonal motions are, in effect, a free mode of a low-viscosity atmospheric circulation driven directly by internal energy and absorbed sunlight and hence that the observed eddy motion is as a result of turbulence at the belt/zone boundaries, with the eddies drawing their energy from the mean flow. The alternative point of view is that it is the eddies themselves which primarily draw their energy from internal sources and absorbed sunlight, and that it is these eddies that then drive the zonal flow and the large vortices.

One feature of the atmospheres of Jupiter and Saturn that argues in favor of zonal winds being driven by eddies (which includes wave motion) is the super-rotation of the equatorial zones of these planets. Another argument in favor of eddy driving comes from the analysis of the motion of eddies observed in *Voyager 1* and *Voyager 2* images of Jupiter by Beebe *et al.* (1980) and Ingersoll *et al.* (1981). In these studies, observation of the motion of individual clouds allowed estimation of Reynolds stress $\overline{u'v'}$ which is the average northwards transport of momentum by eddies. [NB: Here zonal and meridional winds have been split into their zonal mean and transient, or eddy, components: $u = \bar{u} + u'$, $v = \bar{v} + v'$.] Ingersoll *et al.* (1981) found that the eddies were pumping momentum into the jets and thus sustaining them. This scenario was cast into doubt by Sromovsky *et al.* (1982) who suggested that Ingersoll's conclusion was probably caused by a biased sampling of prominent eddy cloud features and that a more uniform spatial sampling showed no evidence for eddy pumping. However, more recent cloud-tracking observations by *Cassini* at Jupiter (Salyk *et al.*, 2006) and Saturn (Del Genio *et al.*, 2007) supports the theory that it is the eddies that drive the jets, not the jets that drive the eddies. However, a note of caution is sounded by Vasavada and Showman (2005), who point out that the real situation might again be more complicated than has been previously considered and note that Read (1986) showed that momentum transfer depends not only on terms such as $-\partial(\overline{u'v'})/\partial y$, but also many other terms, all of which need to be known before the rate of change in mean zonal velocity $\bar{u}$ with respect to time can be calculated. In some circumstances it may be sufficient to consider only the Reynolds stress $\overline{u'v'}$, but this has not yet been proven to be the case in giant planet atmospheres.

Of course, if the eddies drive the jets then that begs the question what drives the eddies! One possibility is that eddies may be produced by baroclinic instability, which releases stored potential energy set up by horizontal temperature gradients (Ingersoll, 1990). However, the only temperature variations that have been observed are associated with the jets and it is not tenable to believe that the jets sustain the eddies, which then sustain the jets! Another possibility is that the smallest eddies derive their energy by moist convection and latent heat release as suggested by Ingersoll *et al.* (2000).

Laboratory simulations

Modeling the dynamics of giant planet atmospheres with computer models remains a challenging exercise, even with the most modern and advanced computers. An alternative approach to computer modeling is to conduct laboratory experiments with forcings and conditions matching as closely as possible those found in giant planet atmospheres. Using a rotating annulus of fluid that was heated on the inside and cooled on the outside, Read and Hide (1983, 1984) were able to generate baroclinic eddies with many of the features of Jupiter's Great Red Spot. More recently, Read *et al.* (2004) and Aubert *et al.* (2002) have shown that in a shallow rotating tank with a sloping bottom to represent the β effect, multiple jets can form with widths that scale with the Rhines length.

5.4.2 Stratospheric and upper-tropospheric circulation

Temperatures in the upper tropospheres of the giant planets may be estimated from thermal-infrared observations at approximately 18 μm (550 cm^{-1}), both by spacecraft and by ground-based observations. Zonal contrast is clearly seen, and applying the thermal wind equation the zonal wind structure is predicted to decay to zero at approximately 3–4 scale heights above the cloud tops. The source of the friction implied is probably due to eddy motions or gravity-wave breaking (as was mentioned earlier). The temperatures in the stratosphere at ~20 mbar may be estimated from observations in the methane ν_4 vibration–rotation band at 7.7 μm (1,300 cm^{-1}). A number of two-dimensional radiative–dynamical models have been constructed to estimate how the atmosphere responds to solar irradiation and thermal cooling to space. These models have been used to calculate the long-term meridional flow structures in the stratospheres of giant planets, which match the estimated stratospheric temperatures. Conrath *et al.* (1990) considered direct heating of the stratosphere through absorption of visible and near-infrared solar irradiation by methane gas alone and predicted that the *residual mean* circulation (or *diabatic* circulation) in the stratospheres of all giant planets had air rising near the subsolar latitude (where the solar flux is highest) and descending near the poles. A similar residual circulation is observed in the Earth's stratosphere and is known as "Brewer–Dobson circulation". Air rising at the subsolar latitude means that Conrath *et al.*'s calculations are seasonally dependent for Saturn, Uranus, and Neptune, but less so for Jupiter whose obliquity is close to zero. West *et al.* (1992) challenged Conrath *et al.*'s findings for the case of Jupiter since the absorption of UV sunlight by stratospheric hazes near the pole had been neglected in Conrath's model. In West's model, air rose over the poles above the ~10 mbar level and descended at the equator! However, at lower altitudes, air descended over both poles as in Conrath's model. Hence, air drifted equatorwards above the 10 mbar level and polewards below it. However, these models are highly dependent on the assumed gas and haze absorption coefficients, and a later study of the Jovian atmosphere by Moreno and Sedano (1997), based upon West's model, but with revised haze and methane absorption characteristics, has a meridional flow structure closer to Conrath *et al.*'s calculations. For Saturn, the role of UV-absorbing polar stratospheric hazes may also affect the calculations of Conrath *et al.*, but for Uranus and Neptune, which do not have UV-absorbing polar stratospheric hazes, Conrath *et al.*'s model would seem to be reliable.

These residual mean calculations are useful in understanding mean stratospheric meridional flow, but they represent time averages over long periods and do not necessarily model how tracers are actually transported in the stratosphere. In particular, they neglect horizontal eddy diffusion processes that can transport material meridionally in much shorter time periods. This was well demonstrated by the collision of Comet Shoemaker–Levy 9 with Jupiter's atmosphere in 1994 (Figure 5.11, see color section). All the models of Jovian stratospheric circulation mentioned predict that air moves poleward between the 100 mbar and 10 mbar pressure levels and thus the sooty debris of the impact deposited at these altitudes at 45°S was expected to drift towards the South Pole (Friedson *et al.*, 1999). Instead, the debris (observed at 230 nm by HST) drifted towards the equator and had reached a latitude of 20°S by 1997. In addition, trace constituents introduced by the comet such as HCN and CO_2 were observed to cross the equator into the northern hemisphere in the four years after the Comet Shoemaker–Levy 9 impact (Lellouch *et al.*, 2002; Moreno *et al.*, 2003). Observations by *Cassini* CIRS in 2000 (Kunde *et al.*, 2004) found that both CO_2 and HCN were more concentrated in the southern hemisphere than the northern hemisphere, but that the latitude of maximum HCN abundance was 45°S, while that of CO_2 was 60°S. The abundance of HCN was found to drop at the South Pole and also fall away polewards of 40°N.

5.5 METEOROLOGY OF JUPITER

5.5.1 General circulation and zonal structure

Jupiter emits 1.67× more radiation than it receives from the Sun indicating a substantial internal heat source and hence, presumably, vigorous convection. The zonal structure of Jupiter appears to be neutrally stable with well-defined belts and zones (summarized in Figure 1.4 and reviewed in great detail by Rogers, 1995). In fact, this canonical stable belt/zone structure is a little misleading since the dark belts occasionally brighten and the bright zones occasionally darken with typical timescales ranging from days to years. However, since the advent of space missions it has become clear that the zonal wind structure associated with the bands is far more invariant and that atmospheric motion may best be referred to the fast-moving "jet-streams" seen in the zonal wind flow. Belts and zones occur in pairs, or domains, to the north and south of the equatorial zone, with belts bounded by an eastward-flowing jet-stream on the side closest to the equator, and a westward-flowing jet-stream on the poleward side. Thus, the belts are regions of cyclonic vorticity and the zones regions of anticyclonic vorticity (as described earlier). Sandwiched within this general zonal flow structure are several short-lived and long-lived ovals, of which the largest and most long-lived are the Great Red Spot (GRS) shown in Figures 5.12 and 5.13 (see color section for both), which lies at 22°S (planetographic) between the South Equatorial Belt (SEB) and South Tropical Zone (STrZ), and the South Temperate Belt-south (STBs) white oval, which lies at 32.6°S (planetographic) between the South Temperate Belt and the South Temperate Zone (STZ). Jupiter's visible cloud features are shown in Figures 5.14 (see color section) and 5.15 and its appearance at visible, UV, and near-infrared wavelengths is shown in Figure 5.16.

Figure 5.15. Southern hemisphere of Jupiter observed by *Cassini* ISS in December 2000. The Galilean satellite Io is visible in the middle right together with its shadow. The image shows several of the main cloud features of Jupiter. The GRS is clearly visible, together with small SSTB white ovals to the south. The turbulence in the EZ is clear, as are two of the dark plumes on its northern edge. In the NEB, a bright white transitory convective cloud is clearly seen in the process of being torn apart by the horizontal wind shear. Courtesy of NASA.

As mentioned previously, there is a clear correlation between cloud opacity and latitudinal wind shear with anticyclonic latitudes appearing bright at visible wavelengths and dark at 5 μm, and cyclonic latitudes appearing dark at visible wavelengths, but bright at 5 μm. This correlation becomes less clear at wavelengths where the atmosphere is more opaque and thus where most of the observed reflection comes from the upper troposphere. This is probably because the small aerosols detected near the tropopause are transported horizontally from zone to belt on a timescale short compared with the precipitation time.

igure 5.16. Three images of Jupiter observed by *Cassini* ISS on 8 October 2000, as it approached Jupiter uring its flyby. The image on the left was taken through the blue filter and appears similar to Jupiter's sible appearance. The middle image is recorded in the ultraviolet. At this wavelength, light is Rayleigh-attered from the upper atmosphere and the disk appears bright unless there are high abundances of upper-opospheric and stratospheric hazes. The strong haze absorption near the poles is clearly visible as is the creased haze abundance over the EZ. A wavelike pattern is also clear at the northern edge of this haze, hich appears correlated with the plumes/hotspots seen in the blue image. The image to the right is recorded the near-infrared at 0.89 μm where methane absorption is strong and so only light scattered by high-titude hazes is visible. This image is almost the negative of the UV image at equatorial latitudes although e anti-correlation breaks down at polar latitudes indicating the peculiar properties of the aerosols near the oles. The high haze opacity over the GRS is clearly visible. Courtesy of NASA.

The zonal structure of Jupiter was observed by the *Voyager* spacecraft up to latitudes of ±60° (Limaye, 1986; Smith *et al.*, 1979a, b), and the organized, symmetric zonal structure appeared to diminish towards the pole and be replaced by more chaotic motion. Such an observation was consistent with the model of Ingersoll and Pollard (1982) that the zonal flow of Jupiter arises from the flow of the interior organizing itself into a series of differentially rotating concentric cylinders, which would cease at a critical latitude where a cylinder would be tangential to the molecular-hydrogen/metallic-hydrogen boundary. Observations of the zonal winds by *Cassini* (Porco *et al.*, 2003; Vasavada, 2002) revealed little change to the winds between ±60°, although the speed of the eastward jet at 24°N had slowed by 40 m s^{-1} to 50 m s^{-1}. However, *Cassini* showed organized zonal flow to extend to at least to ±70°, and still possess north/south symmetry, which presents a considerable problem for the cylinder model. Although the planet takes on a mottled disorganized appearance polewards of 70°, cloud tracking by *Cassini* revealed that the features at these latitudes are still organized into regular eastward and westward jets that extend all the way to the poles. Thermal measurements by both *Voyager* (Hanel *et al.*, 1979a, b) and *Cassini* indicate that these winds decay with height and tend to zero within 3–4 scale heights of the cloud tops. An unusual feature of the zonal wind structure of Jupiter is that the equatorial jet appears to have "horns" in that the wind speed initially increases away from the equator before rapidly diminishing. Such a structure is consistent with a small Hadley cell centered at the equator, where air rises at the equator (thus forming the zone's bright clouds), then moves polewards at the cloud tops, and picks up zonal speed due to the conservation of angular momentum, or equivalently the conservation of vorticity, before descending at the edges of the zone.

Long-term imaging of Jupiter at 18 μm (Orton *et al.*, 1994) and 7.4 μm (Orton *et al.*, 1991), sounding the 250 mbar and 20 mbar levels, respectively, have yielded unique information on seasonal variability, albedo correlation, and wave motion in the Jovian atmosphere. Although Jupiter has very small obliquity, clear seasonal variations are seen at both pressure levels, especially at high latitudes. At the 250 mbar level the seasonal maxima/minima occur roughly 2 years after the solstices. Such a time lag is expected since the atmosphere has a finite heat capacity and solar heating is balanced by increased radiation to space at a later date. The radiative time constant (Chapter 6) at 250 mbar is estimated (Orton *et al.*, 1994) to be 6×10^7 s or 1.9 years, which is consistent with observations. A seasonal cycle is also seen at 20 mbar, but at this altitude there appears to be no lag between the solstice and maximum temperature, which is inconsistent with radiative equilibrium models and suggests that additional factors affect the stratospheric temperatures. Periods of upper-tropospheric equatorial cooling in 1980 and 1992 coincided with a visible whitening of the Equatorial Zone (EZ), consistent with an episode of increased upwelling and condensation of cloud particles. However, the equatorial cooling observed in 1988 did not correspond to any albedo change. Interesting variability in the strong prograde jet at 20°N was observed between 1984 and 1990 where, from the thermal wind equation, it appeared that the jet went from a condition where it decayed with height, to one where it remained almost constant with height! During this period the North Temperate Belt (NTB) brightened in the middle of 1987 and then darkened in 1990 during a major outbreak of white and dark spots in the STB. In addition, the North Equatorial Belt (NEB) broadened to the north in 1988 and then receded in 1999 leaving an array of brown barges (Orton *et al.*, 1994). Whether or not these changes were caused or influenced by the apparent change of vertical structure of the 20°N jet is not known.

5.5.2 Storms and vortices

The atmosphere of Jupiter contains numerous examples of large, long-lived ovals, of which almost 90% of are anticyclonic (as was discussed earlier in Section 5.3.3). Stable cyclones do exist, however, and the most prominent are the brown barges which appear at the northern edge of the NEB at 16°N (Figure 5.17). Brown barges were very prominent during the *Voyager* epoch, but have become less prominent during the *Galileo* and *Cassini* epochs. The strong cyclonic shear found at this latitude may help to stabilize the brown barges, which are observed to be dark at visible wavelengths, but bright at 5 μm, suggesting that they are regions of reduced cloud cover and subsidence.

In addition to brown barges, features called "equatorial plumes" appear at the southern edge of the NEB and appear (at visible wavelengths) as a sequence of bright regions, separated by darker, hook-shaped features, which extend from the NEB into the EZ. These dark features coincide with 5 μm "hotspots", regions of very low total

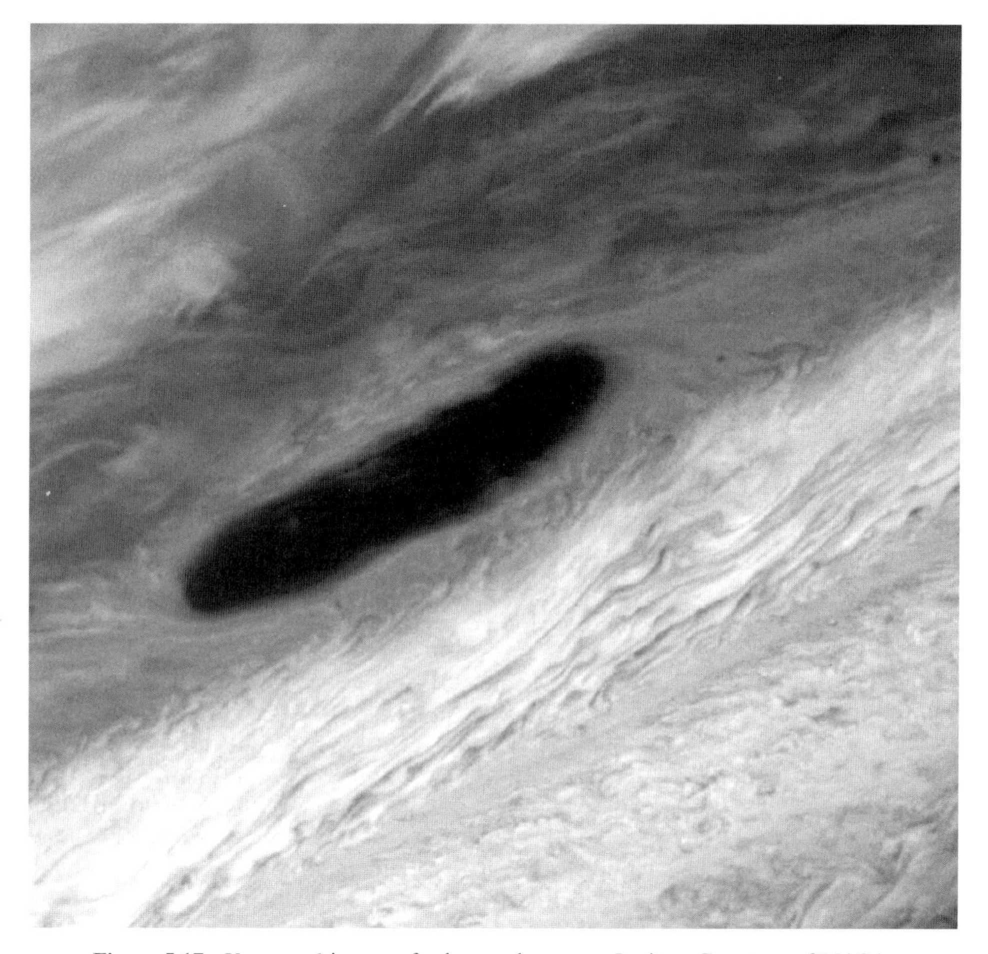

Figure 5.17. *Voyager 1* image of a brown barge on Jupiter. Courtesy of NASA.

cloud cover allowing thermal radiation from the 3 bar to 8 bar region to be observed at 5 μm. Like brown barges, equatorial plumes were more prominent in the *Voyager* epoch than in more recent observations.

Analysis of cloud top wind vectors observed by *Voyager* (noted earlier) suggested that small-scale eddies, in general, pump energy into the mean zonal flow and large-scale eddies (Beebe *et al.*, 1980; Ingersoll *et al.*, 1981), and analysis of *Cassini* observations by Salyk *et al.* (2006) seems to confirm this scenario, although with 2–4× smaller eddy–zonal kinetic energy conversion than previously estimated. Salyk *et al.* (2006) estimated that the power transfer between eddies and jets is 48% of the total thermal energy emitted by Jupiter.

The largest visible Jovian oval is the GRS, which is a huge anticyclonic vortex centered at $22.4^\circ \pm 0.5^\circ$S (planetographic) and has a constant latitudinal extent of 11° or 12,000 km, and a longitudinal extent of 17° or 20,000 km in 2002. Winds in the vortex rise to over 100 m s^{-1} in the outer annulus, but the center is found to be quiescent and also roughly 8 K cooler than the surroundings at the cloud tops. Applying the thermal wind equation, this implies that the wind speed should decrease with depth into the atmosphere and thus that the GRS is probably only 200 km thick, which is tiny compared with its horizontal dimensions. Clouds in the center of the spot are found to be very thick and very high and also tilt slightly from the north to south and more subtly from east to west. The GRS thus resembles a "tilted pancake" (Simon-Miller *et al.*, 2002; West, 1999) and is quite unlike a terrestrial hurricane with which it is often compared (West, 1999), whose breadths are only 20–30× their heights. The thick clouds and low upper-tropospheric temperatures imply upwelling in the center of the GRS, and at the edges, high 5 μm emissions indicate low cloud opacity and thus subsidence, again unlike a hurricane where subsidence takes place in the central eye. The GRS appears to be very long-lived, although the visibility of the spot changes greatly with time and was particularly clear during the *Pioneer* encounters. However, at other times, such as during SEB disturbances the spot almost disappears. A large spot at the current GRS latitude was first observed in 1665 by Robert Hooke, and a year later by Jean-Dominique Cassini (Rogers, 1995; Simon-Miller *et al.*, 2002). However, it is not clear whether the current GRS is actually the same spot since continuous observations can only be traced back to 1830, 120 years after the last sighting of Hooke's spot. Today's GRS is observed to be gradually shrinking in the longitudinal direction at a rate of 0.193° per year (Simon-Miller *et al.*, 2002), which translates to approximately 4,000 km between the time of the *Voyager* and *Cassini* flybys. In addition, the winds in the collar have increased since the *Voyager* flyby. If the current rate of shrinking continues then by around the year 2040 the GRS will be perfectly circular. A circular aspect ratio is believed to be an unstable configuration for such a large anticyclone and hence it is possible that the GRS may actually disappear 30 years from now! Rogers (1995) has proposed that this may be what happened to Hooke's spot (which was reported to be roughly circular) in around 1700 and that the current GRS formed from a belt-wide disturbance (rather like the formation of the STBs white ovals in 1939) at about the same time and has been continuously shrinking in the longitudinal direction ever since. Its formation may even have been fed by its predecessor, Hooke's spot! The quiescent conditions at the center of the GRS may mean that air is trapped inside it for substantial periods of time, which may lead to the production of the characteristic red chromophore that gives the GRS its apparent reddish color. An alternative explanation is that the clouds seen in the center of the GRS have much higher cloud tops than anywhere else on the planet, suggesting vigorous convection. Hence, high levels of gases such as phosphine may be present whose photolysis may lead to the production of triclinic red phosphorous $P_4(s)$.

The STBs white ovals are the most prominent storm systems after the GRS and the current oval system first appeared in 1939 when a wavy disturbance appeared in the STB, although similar ovals had previously been seen at this latitude. This disturbance developed into six pinched regions that were labeled A to F which eventually coalesced into the three white ovals labeled BC, DE, FA sandwiched between the STZ and STB and which were observed during the *Voyager* flybys in

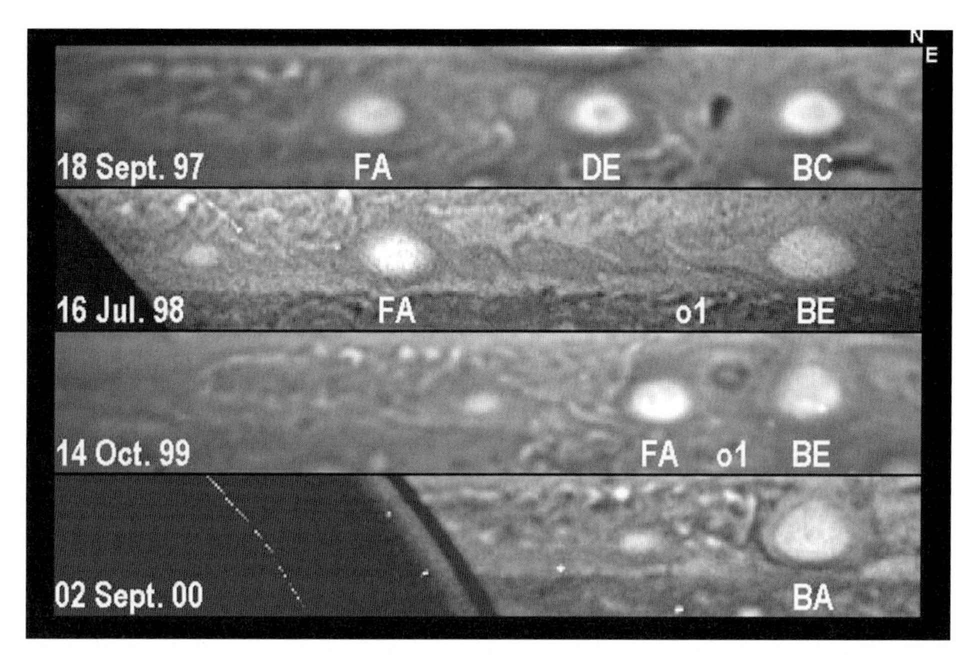

Figure 5.18. The merger of the white ovals from 1997 to 2000 observed by HST. Courtesy of NASA.

1979. Two of these ovals were observed to merge together in 1998, and in March 2000 the resultant two remaining ovals coalesced to form a single white oval (Figure 5.18), known as Oval BA. More recently, Oval BA was observed by amateur astronomers in 2005 to turn the same color of red as the GRS and the storm is now known to many as the "Little Red Spot" (LRS) or "Red Spot Jr.". Oval BA has been extensively observed in its new red state by HST and also by the *New Horizons* spacecraft during its flyby in February 2007. HST observations (Simon-Miller *et al.*, 2006) have shown that Oval BA appears to be getting stronger, with wind speeds reaching 400 mph (645 km/h), similar to those seen in the GRS (Figure 5.19, see color section). Its current size is about the diameter of Earth. Approximately every 2 years, Oval BA passes close by the GRS; previous encounters in 1998 and 2000 were accompanied by dramatic changes in the white ovals. The most recent encounter occurred in June 2008 and was widely monitored with ground-based telescopes. During the encounter between the GRS and Red Spot Jr. a third, smaller red oval, called the "Baby Red Spot", became involved, which was caught between the two larger ovals and was torn apart and absorbed by the GRS.

The region to the northwest of the GRS is an area of cyclonic vorticity and usually appears to be particularly chaotic and rapidly changing. Small bright clouds regularly appear, which have been widely interpreted as thunderstorm clouds and the base of these clouds appear to be at pressures greater than 4 bar (Banfield *et al.*, 1998 and Figure 4.11) suggesting a moist convective cumulus cloud rising from the base of the expected water cloud. In addition, the spectral signature of ammonia ice has been detected in these bright white clouds indicating rapid updraft and formation of pure white ammonia crystals (Baines *et al.*, 2002 and Figure 4.10). The absence of clear ammonia ice features elsewhere, except in the NEB plumes, indicates that these crystals are rapidly modified or perhaps coated in some way as to hide their pure spectral signature (Atreya *et al.*, 2005; Irwin and Dyudina, 2002; Irwin *et al.*, 2005; Kalogerakis *et al.*, 2008), as was described in Chapter 4.

A number of thunderstorms have now been observed in Jupiter's atmosphere. The first thunderstorms were observed by *Galileo* at latitudes of cyclonic shear (Dyudina and Ingersoll, 2002; Gierasch *et al.*, 2000; Little *et al.*, 1999) by observing their flashes on the night side of Jupiter (Figure 4.12). These storms have a lifetime of ~4 days and the size of lightning spots in images suggest they result from point sources within or below the expected water cloud, which appear to be much more energetic than the average terrestrial lightning bolt. Lightning flashes on Jupiter's night side were also observed by the *Cassini* ISS instrument using an H_α filter (Dyudina *et al.*, 2004). Dyudina *et al.* found that the flashes were 10× less bright than expected compared with the earlier clear-filter *Galileo* SSI observations, indicating that the flashes are generated at levels deeper than 5 bar, consistent with supersolar abundances of water vapor. Most recently, during the flyby of the *New Horizons* spacecraft in February 2007, the LORRI instrument reported many lightning flashes in Jupiter's polar regions (latitude >60°) (Baines *et al.*, 2007a; Weaver *et al.*, 2007). Until this observation, polar lightning had only ever been observed in Earth's atmosphere. The most poleward lightning flashes were seen at 80°N and 74°S. The energies of these lightning strikes are estimated to be between 0.2 GJ and 13 GJ, which are comparable with values seen by *Galileo* and *Cassini* at midlatitudes, but much larger than values estimated in the equatorial region, which is consistent with the hypothesis that the release of internal heat is mainly driving Jupiter's convection. The spatial extent of the flashes is again consistent with them originating in the 5 bar to 8 bar water-rich region of Jupiter's atmosphere. Although cyclonic shear latitudes are generally cloud-free regions, higher occurrence of lightning flashes is not thought to be purely an observational effect in that the flashes are simply more visible where there are fewer overlying clouds. Instead, modeling of the scattering properties of the clouds indicates that deep flashes occurring in zones would also be clearly visible. Hence, the correlation suggests that regions of cyclonic shear are simply more susceptible to moist convection (as was mentioned in Section 5.3.3). The moist convection scenario is also supported by an indication of increased water humidity in these areas (Roos-Serote *et al.*, 2000). Although the *Galileo* entry probe descended in just such a cyclonic shear zone at 6.5°N, no lightning was detected within 10,000 km and indeed the nearest lightning strike detected in images was observed at 8.6°N.

At higher altitudes, in the stratosphere, other transient spot-like features have been noted in the polar regions of Jupiter by HST, *Galileo*, and *Cassini* at UV wavelengths, sounding approximately the 1 mbar level. These wavelengths are sensitive to the abundances of stratospheric hazes and the ovals appear with sizes comparable with the GRS. An example of such a UV spot can be seen in the middle UV image of Figure 5.16 (on the top, right-hand limb). What these spots are is not yet fully known. However, during the *Cassini* flyby, the *Cassini* ISS instrument recorded the birth, development, and subsequent decay of a large dark UV spot at 60°N with a total lifetime of approximately two months (Porco *et al.*, 2003). This spot was found to lie within the main auroral oval, strongly suggesting a link with auroral processes. How auroral processes might affect stratospheric hazes is unclear, although the *Cassini* CIRS instrument found the upper stratosphere within the auroral oval to have anomalously high temperatures at pressures less than 4 mbar (Flasar *et al.*, 2004b; Kunde *et al.*, 2004; Nixon *et al.*, 2007). This region is also coincident with an area of enhanced X-ray emission observed in December 2000 by the *Chandra X-Ray Telescope*, launched into Earth orbit in July 1999.

In 2007 Jupiter went through a period of massive global upheaval, last seen in 1990 (Sánchez-Lavega *et al.*, 1991). The changes (Go *et al.*, 2007) started in mid-2006 with a darkening of the central and southern EZ, and the detachment of the GRS from the SEB, leading to cessation of the usually strong convective activity seen to the northwest of the GRS (Baines *et al.*, 2007a). At the start of 2007, long-running activity in the SEB had ceased and two South Tropical Zone (STrZ) disturbances had formed, which appeared as dark hooks emanating from the southern edge of the SEB. Small dark spots running along the northern edge of the STrZ were observed to be deflected by these "hooks" and then run back in the opposite direction along the southern edge of the STrZ. In March 2007, two brilliant white spots emerged in Jupiter's North Tropical Zone (NTrZ), reaching an altitude 30 km above the surrounding clouds (Sánchez-Lavega *et al.*, 2008) and initially separated by 55° longitude, which then took about 6 weeks to travel around the planet at a high speed relative to the System III longitude system, before they merged and subsided, leaving behind a trail of dark material, which slowly darkened, reviving the North Temperate Belt. Modeling of the three-dimensional form of the disturbance suggests that the plumes originated from great depth, and their observed high speed suggests that Jupiter's zonal wind system extends well below the depths that can be penetrated by sunlight, adding weight to the theory that Jupiter's dynamics are driven mostly by the release of internal heat. During the NTB disturbance, the SEB slowly faded, but was then revived by an SEB outbreak that appeared in mid-May. Unlike in 1990, this upheaval was observed by astronomers all over the world and also by the HST and partially imaged during the flyby of the *New Horizons* spacecraft in February 2007.

Although warm cyclonic polar vortices have been found on Saturn and Neptune, it is not yet clear if the same happens on Jupiter as we rarely see the poles. However, the only spacecraft to have flown over one of Jupiter's poles, *Pioneer 11*, reported higher temperatures at the cloud tops at the pole than at the equator (Ingersoll, 1990), suggesting such polar vortices may also exist on Jupiter.

5.5.3 Waves

The Jovian atmosphere contains numerous examples of waves on a wide range of lengthscales. At the smallest scale, Flasar and Gierasch (1986) discovered waves in the equatorial region in *Voyager* images traveling east–west at the cloud tops with wavelengths of ~300 km, gathered together in wave packets of length ~1,300 km in the meridional direction and 3,000 km to 13,000 km in the zonal direction. These were interpreted as equatorially trapped modes with smaller gravity waves superimposed on them, perhaps generated by Kelvin–Helmholtz instabilities (Bosak and Ingersoll, 2002), and were most apparent at the edges of the equatorial jet at $\pm 8^\circ$. A later *Galileo* observation of these waves (Belton *et al.*, 1996) is shown in Figure 5.20. The interpretation of these features led to the suggestion of a statically stable duct beneath the NH_3 cloud deck and this hypothesis was later supported by *Galileo* probe temperature measurements (Seiff *et al.*, 1998), although the *Galileo* probe entered a 5 μm hotspot, which may not be very representative of mean near-equatorial conditions. These mesoscale gravity waves were observed again during the *New Horizons* flyby throughout the equatorial region within about 5° of the equator, during a time when the equatorial region was unusually cloud-free. By assessing their visibility in different spectral channels it is possible to conclude that they are formed at altitudes above the 600 mbar level. Their wavelength was measured to be 330 km, consistent with earlier determinations and they have a phase speed of $\sim$250 m s^{-1}. Unfortunately this speed is inconsistent with current theories of how they are formed!

At larger, planetary scales, waves have been detected in both thermal maps of Jupiter (Deming *et al.*, 1989, 1997; Magalhães *et al.*, 1989, 1990; Orton *et al.*, 1991, 1994) and in the planetary variation of cloud opacity as determined by emission at 5 μm (Harrington *et al.*, 1996; Ortiz *et al.*, 1998). A near-stationary wavenumber-9 wave was discovered by Magalhães *et al.* (1989, 1990) from *Voyager* IRIS data in the upper troposphere at 270 mbar near 15°N, and a similar wavenumber-11 wave was observed at 20°N at 45 μm (which sounds down to 1 bar in the absence of clouds). Similar wavenumber-10 waves were observed by Deming *et al.* (1989) from ground-based observations (8–13 μm), again at 20°N and also at the equator. Ground-based observations at 7.8 μm (Orton *et al.*, 1991), sounding temperatures at $\sim$20 mbar, found near-stationary waves at $\pm 20^\circ$, which were interpreted as planetary waves generated by instabilities in the strong cloud top prograde jets at $\pm 18^\circ$. The near-stationary appearance of these waves with respect to System III implies some sort of dynamical link with the interior bulk rotation of the planet. Observations by Orton *et al.* (1994) at 18 μm, which sounds the 250 mbar temperatures, found waves at 13°N in the NEB at the same time as the wave previously mentioned in the stratosphere near 20°N. Both disturbances appeared to have a zonal group velocity of -5.5 m s^{-1} with respect to System III longitude and there appeared to be some correlation with height implying vertical propagation of a Rossby wave. A later study by Deming *et al.* (1997) found such waves to be ubiquitous at near-equatorial latitudes with wavenumbers anywhere between 2 and 15. The amplitude of thermal waves in the lower stratosphere (20 mbar) sounded at 7.8 μm was found to be roughly 3× greater than thermal waves in the upper troposphere (250 mbar), sounded at 18 μm. The waves at these two altitudes appear to be correlated and the amplitude growth is consistent with a $\rho^{-1/2}$-dependence expected for vertically propagating Rossby gravity waves. By analyzing the amplitude of these stationary Rossby waves, Deming *et al.* (1997) infer latitudinal deflections of only 1°, which may arise from interaction with the interior "banana cell" (Section 5.4.1) convective structure or through interaction with

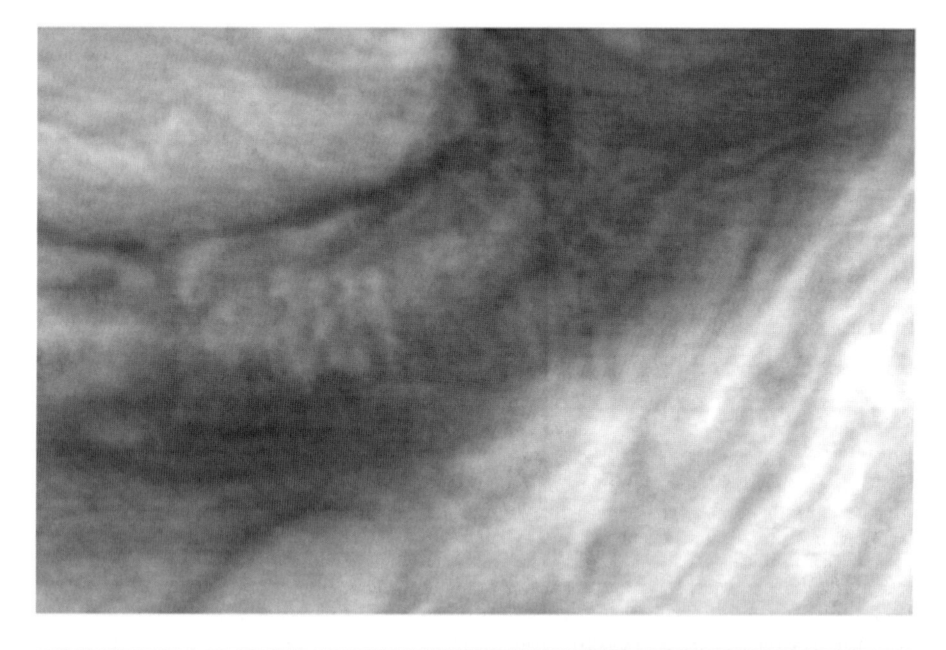

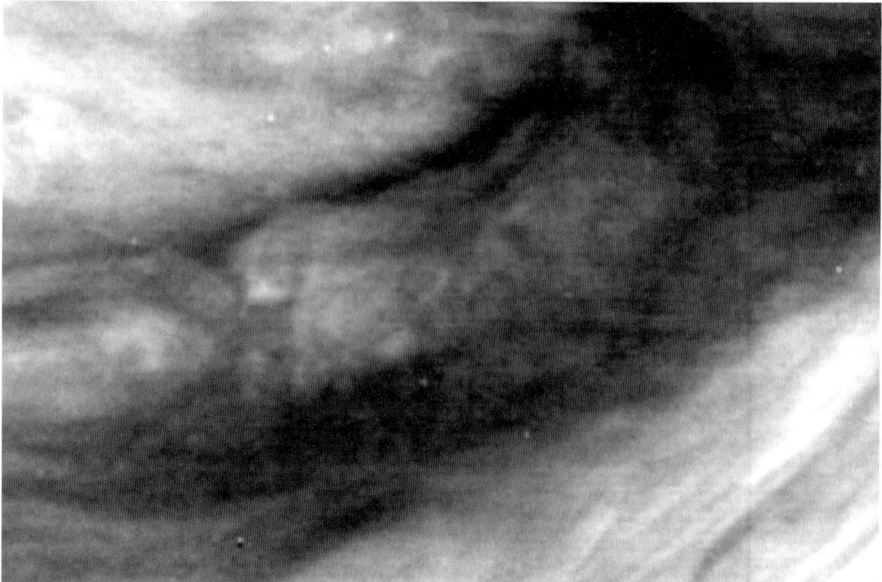

Figure 5.20. Two images of Jupiter's atmosphere recorded by *Galileo* SSI with the "violet" filter in 1996, centered at 15°S and 307°W. The pixel resolution is approximately 30 km. Mesoscale Kelvin–Helmholtz gravity waves can just be seen in the center of the upper image where they appear as a series of about 15 nearly vertical north–south stripes. The combined wave packet is about 300 km long and is aligned in the east–west direction. In the lower image, recorded 9 hours later, there is no indication of the waves, though the clouds appear to have been disturbed. Such waves were also seen by the *Voyager* spacecraft in 1979. Courtesy of NASA.

vortices, which are themselves moving slowly with respect to System III, sandwiched between easterly and westerly flow regions.

Waves have also been detected in ground-based images of Jupiter recorded at 5 μm, a wavelength sensitive to the total cloud opacity above the warm 5 bar to 8 bar pressure regions and thus sensitive to the opacity of the expected ammonium hydrosulfide and ammonia cloud decks. A number of waves were discovered at many latitudes (Harrington *et al.*, 1996) including near-stationary Wavenumber-10 features at 7°N–8°N and eastward-propagating Wavenumber-4 waves at the equator. The former wave appears to be associated with the 5 μm hotspots at the southern edge of the NEB, which appear to be distributed semi-uniformly with longitude, and are interspersed by highly reflective equatorial plumes lying between the hotspots and slightly to the south. These hotspots were studied by Ortiz *et al.* (1998) who concluded that they were manifestations of equatorially trapped Rossby waves. The 5 μm hotspots are regions of very low cloud cover, which makes them appear very bright at IR wavelengths and dark at visible wavelengths. For some time these features were interpreted as being regions of rapid downdraft, which would explain their low cloud cover and also the observed depletion of volatiles such as ammonia and water. However, such models required excessive downdraft velocities and also predicted that the abundance of volatiles such as ammonia, water vapor, and hydrogen sulfide should all return to their "deep" abundances at roughly the same pressure level. Instead, the *Galileo* entry probe, which sampled just such a 5 μm hotspot, found that the abundance of ammonia increased first as the probe descended, then H_2S, then H_2O which was still increasing at 20 bar when communication with the probe was lost. Hence, an alternative theory, that these hotspots are associated with a planetary-wave system that alternately compresses and expands the vertical air column, has been developed and extended with nonlinear modeling (Friedson, 1999; Showman and Dowling, 2000) and is currently the favored explanation. In this model, the bright anticyclonic regions (equatorial plumes) appear on the upward portion of the planetary wave concentrating volatiles at high altitudes, where they condense to form bright white clouds. The 5 μm hotspots then appear on the downward portion of the wave, where the statically stable air column is vertically stretched, increasing the base pressure by a factor of almost 2. The accompanying adiabatic heating forces the clouds to sublimate and reduces the apparent volatile abundances, but retains the relative abundances of H_2O, H_2S, and NH_3, which thus increase towards their deep values at different rates as observed by the *Galileo* probe. Showman and Dowling (2000) found that their modeled waves were only stable if large initial pressure perturbations were assumed, suggesting that nonlinear effects are central to the stability of this wave. Furthermore, the authors found that the zonal wind profile measured by the *Galileo* entry probe (Atkinson *et al.*, 1998), which implies static stability in the troposphere consistent with the probe's atmospheric temperature experiment (Seiff *et al.*, 1998), may in fact be a local effect. The vertical negative wind shear (winds decreasing with height) was reproduced by Showman and Dowling's model at the southern edge of their simulated hotspots (where the probe entered), but was zero in the center and positive at the northern edge implying that the measured wind profile may owe more to local dynamical effects than to the general zonal wind structure. The spectral signature of pure ammonia ice has been observed in the equatorial plumes (and in localized thunderstorm clouds seen in other cyclonic regions; Baines *et al.*, 2002), which is not generally seen elsewhere. This is interpreted as being due to the rapid condensation of new "fresh" ammonia crystals in these regions, which are then subsequently degraded or coated with chromophores after a few days. A link between these waves and upper tropospheric properties was observed by *Cassini* ISS and can be seen in Figure 5.16 where a wavenumber-16 wave is clearly seen at 9°N–19°N in both the strong methane-absorbing band at 0.89 μm and also in the accompanying UV image. Hence, the wave clearly extends into the upper troposphere and has an effect on either the abundance or reflecting properties of the haze particles (Porco *et al.*, 2003). Several wavelike features were also visible in *Cassini* CIRS observations of atmospheric temperature (Flasar *et al.*, 2004b). In the troposphere, within 15° of the equator, a weak variation of temperature with longitude has been seen, which may be associated with the waves thought to be driving the QQO, introduced in Section 5.3.2. Farther away from the equator, numerous chains of warm features are seen at a variety of latitudes. These features, below the tropopause, seem to be almost stationary with respect to the interior, and their cause is not known. At higher altitudes, in the stratosphere, many thermal waves were seen, some in the southern hemisphere with westward zonal velocity, suggesting that these are caused by planetary Rossby waves. Stratospheric waves are probably initiated by disturbances in the troposphere, but waves propagating vertically upwards should, to a first order, retain their zonal velocity. Hence, how westward-moving stratospheric waves are related to tropospheric stationary disturbances is mysterious.

One of the most striking features of Jovian stratospheric temperatures observed by Orton *et al.* (1991) was a periodic, approximately 4-year variation of the zonal temperature at 20 mbar, known as the *Quasi-Quadrennial Oscillation* or QQO. This feature, which has been continuously observed since 1978, takes the form of a periodic warming of the equator and simultaneous cooling of latitudes between ±(15–30°), followed by a cooling of the equator and warming of the ±(15–30°) latitude band, with a period of between 2 and 5 years and an amplitude of 1 K to 2 K. During the cool equatorial phase, longitudinal structure has been observed in the northern band, which are the thermal waves mentioned earlier. The oscillation has also been observed in the upper troposphere (250 mbar) at equatorial latitudes by Orton *et al.* (1994), where the temperature oscillation is found to be roughly 180° out of phase with the stratosphere. Leovy *et al.* (1991) likened this variability to variation in the Earth's equatorial stratosphere known as the QBO (discussed in Section 5.3.2). However, Friedson (1999) found that this analysis underestimated the vertical and horizontal averaging of ground-based observations and that Leovy's identification of the driving waves as alternating equatorially trapped Kelvin and mixed Rossby gravity waves was not able to account for the amplitude of the oscillation observed. Instead, Friedson (1999) considered a wide range of equatorially trapped modes and found that forcing of the upper-tropospheric flow and lower-stratospheric flow by smaller scale internal gravity waves produced temperature variations much closer to those seen. However, Li and Read (2000) also modeled the QQO and found that the effect was sufficiently well modeled by alternating wavenumber 8–11 waves with an equatorial Rossby mode moving eastwards at around $100\,\mathrm{m\,s^{-1}}$ (or perhaps a Kelvin mode) and a mixed Rossby gravity wave stationary with respect to System III, apparently excited by a wave source moving with the zonal wind in the deep atmosphere. While these studies appear inconsistent, Li and Read (2000) did find that the identification of wave mode depended substantially on model assumptions and thus that more detailed nonlinear modeling might be necessary. Hence, while it seems likely that Jupiter's QQO is substantially like the Earth's QBO, the precise identification of the wave motions forcing it remains elusive.

Cassini CIRS made temperature measurements in the upper troposphere and stratosphere (Flasar *et al.*, 2004b), which can be used with the thermal wind equation to probe winds in the stratosphere, providing we fix winds at the cloud top level to those determined by cloud tracking. *Cassini* CIRS found that the winds decrease with altitude in the upper troposphere as was previously determined by *Voyager*. However, at some latitudes (23°N, the equator, and 18°S), wind speeds were found to pick up again, especially at the equator, where a jet of $140\,\mathrm{m\,s^{-1}}$ was deduced at a pressure of 3 mbar to 4 mbar. It is thought that this stratospheric jet is related to the QQO.

5.6 METEOROLOGY OF SATURN

5.6.1 General circulation and zonal structure

Saturn emits 1.78× more energy than it receives from Sun, but compared with Jupiter the overall energy emission is much less (only 1/3 that of Jupiter). Hence, it might be expected that Saturn should have a commensurately less vigorously overturning atmosphere. However, the atmosphere appears just as energetic as Jupiter's with strong zonal winds that reach speeds of $400\,\mathrm{m\,s^{-1}}$ in the eastward-flowing equatorial jet. Although the atmosphere is very dynamic, the appearance of Saturn is generally much more subdued than that of Jupiter, with the belt/zone structure being much less clear. The tropospheric cloud structure appears to be much more masked by tropospheric and stratospheric haze layers due both to the expected ammonia ice cloud deck condensing at deeper levels than in Jupiter's atmosphere and also due to the greater scale height of the Saturnian atmosphere (as outlined in Chapter 4). The belts and zones have similar names to those of Jupiter and the universally accepted naming convention is shown in Figure 5.21.

The obliquity of Saturn (26.7°) means that it is much more prone to seasonal effects (Figure 5.22, see color section) than Jupiter and the *Voyager* spacecraft found in 1980 and 1981 (at a time corresponding roughly to the Northern Spring Equinox) that the upper tropospheric temperature at the 210 mbar level was approximately 10 K warmer in the southern hemisphere than the northern hemisphere (Bézard *et al.*, 1984; Conrath and Pirraglia 1983; Hanel *et al.*, 1981, 1982), decreasing at higher pressures. The thermal response of the atmosphere at this level is estimated from the radiative time constant (Gierasch and Goody, 1969) to be roughly 5 years or equiva-

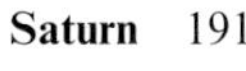

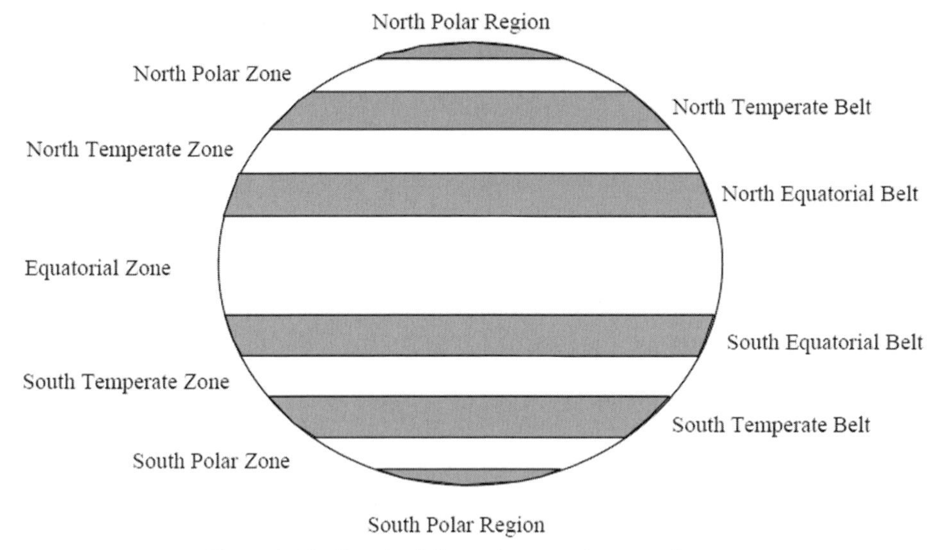

Figure 5.21. Standard Saturnian zonal nomenclature.

lently $\pi/2$ out of phase with solar forcing, which corresponds well to the observations. Like Jupiter, the variations in temperature with latitude were used to calculate vertical wind shear, which again was well correlated with cloud top zonal winds, indicating that the jets decay with height (Pirraglia *et al.*, 1981; Smith *et al.*, 1981, 1982). Prior to the arrival of the *Cassini* mission a similar north–south asymmetry in stratospheric temperature in 1995 was reported by Karkoschka (1998a) by analyzing the 890 nm methane absorption feature, whose width depends on temperature. Stratospheric temperature asymmetries consistent with the 5-year time lag have also been observed in the approach to and during the *Cassini* mission with ground-based observations (Greathouse *et al.*, 2005; Orton and Yanamandra-Fisher, 2005; Yanamandra-Fisher *et al.*, 2001) and by *Cassini* observations themselves (Flasar *et al.*, 2005; Fletcher *et al.*, 2007a, b) (Figure 5.23, see color section).

During its observations *Cassini* has found (Fletcher *et al.*, 2007b) that the northern hemisphere is consistently cooler than the southern hemisphere, though the seasonal contrast again becomes smaller for larger pressures, where the radiative time constant for the thermal response of the atmosphere to changes in seasonal forcing is longer (Bézard *et al.*, 1984; Conrath and Pirraglia, 1983). At the top of the upper troposphere at 100 mbar, the South Polar region is 10 K warmer than the North Polar region, which is consistent with the findings of *Voyager* (Hanel *et al.*, 1982), bearing in mind that *Cassini* started observing prior to Saturn's Northern Spring Equinox in 2009, while the *Voyager* observations were made just after Saturn's previous Northern Spring Equinox. *Cassini* has observed smaller latitudinal structure superimposed onto this seasonal asymmetry, and zonal temperature minima occur symmetrically about the equator at planetographic latitudes of 58–60°, 46–47°, 30–31°, and possibly near 71°S, 83°S, and 74°N (Fletcher *et al.*, 2007b), which coincide well with the zonal wind structure determined from cloud tracking (Porco *et al.*, 2005), with eastward prograde jets appearing on the poleward side of temperature minima.

In the stratosphere at 1 mbar, no belt/zone variations are observed by *Cassini* at midlatitudes, but strong hemispherical asymmetry exists with the South Pole appearing 10 K to 12 K warmer than the equator and 25 K to 30 K warmer than the North Pole (Figure 5.23, see color section). In addition, although a temperature minimum is seen at the equator in the upper troposphere, in the stratosphere there is a local temperature maximum at the equator, which is 8 K to 10 K warmer than at 20°N or 20°S. Furthermore, Orton *et al.* (2008) find, from analyzing a long time series of ground-based observations, that the temperature of the equatorial stratosphere changes regularly with a period of ~15 years, or half a Saturnian year. This suggests that the equatorial stratospheric temperature is forced by seasonal variations and has similarities with Earth's QBO and the QQO in Jupiter's equatorial stratosphere. During the *Voyager* observations, the stratosphere at ±20° was at its coldest relative to the equatorial belts, but during the early *Cassini* measurements it was only slightly colder.

Although seasonal effects are observed in stratospheric temperatures, the north–south symmetry of the zonal wind pattern observed at the cloud tops suggests that tropospheric circulation is most influenced by rotational forces and perhaps internal energy sources. Estimates of the zonal wind speeds near the equator of Saturn were determined via cloud top tracking by the *Voyager* spacecraft (Ingersoll *et al.*, 1984; Smith *et al.*, 1981, 1982) at two wavelengths (479 nm, 556 nm) and later reanalyzed by Sánchez-Lavega *et al.* (2000). Using ground-based observations in 1995–1997, Sánchez-Lavega *et al.* (1999) found that the winds in the Equatorial Zone (EZ) had slowed significantly since the *Voyager* epoch and the authors suggested that this might have been due to the Great White Spot (GWS) activity of 1990 and 1994 injecting significant westward momentum into the equatorial jet (Hueso and Sánchez-Lavega, 2004). However, since *Voyager* and later *Cassini* thermal measurements (Flasar *et al.*, 2005) show that the equatorial winds slow with height, some uncertainty remained as to whether a real change in wind speeds had been seen or whether additional cloud opacity in the equatorial region in 1995–1997 and different filter characteristics meant that the later measurements were tracking clouds at a higher altitude, where the winds were slower. Hence, Sánchez-Lavega *et al.* (2004) reanalyzed HST observations at three different wavelengths (439 nm, 814 nm, 890 nm) to confirm their conclusion that the winds in the EZ had indeed slowed by about 150 m s^{-1} since the *Voyager* observations. The *Cassini* mission made new observations of the cloud top speeds upon *Cassini*'s arrival in the Saturnian system in 2004 at 727 nm and 752 nm, and Porco *et al.* (2005) also found that winds in the equatorial region had slowed since the *Voyager* observations, although at other latitudes the winds appeared to have changed very little. Using the vertical wind shears determined by the *Cassini* CIRS experiment (Flasar *et al.*, 2005) at ±5° latitude, Pérez-Hoyos and Sánchez-Lavega (2006a) have shown that, by matching the vertical cloud structure to the limb darkening in the EZ observed by *Voyager* and later observations, all cloud-tracking observations since the *Voyager* mission are consistent with each other. Hence, they conclude that the equatorial winds at the 350 mbar level have indeed physically dropped by almost 100 m s^{-1} since the *Voyager* epoch.

The periodic variations in temperature in the equatorial stratosphere reported by Orton *et al.* (2008) should, from thermal wind shear arguments, give rise to periodically alternating prograde and retrograde jets in the stratosphere either side of the equatorial band. At the beginning of the *Cassini* observations the winds should thus have increased with height above the tropopause to create prograde stratospheric jets. However, these prograde stratospheric jets should reverse as we approach the Northern Spring Equinox in 2009 (Fletcher, pers. commun.). Curiously, the *Voyager* observations (during which equatorial tropospheric winds were very strong) coincided with a period of maximum temperature contrast between stratospheric equatorial and near-equatorial temperatures, while HST and early *Cassini* determinations (when equatorial tropospheric winds were seen to be slower) coincided with a period of minimal temperature contrast. However, no regular periodic variation in equatorial upper-tropospheric temperatures has been seen and there is no known mechanism for variations in the thin stratosphere to affect conditions in the deeper, denser troposphere. Hence, the apparent correlation is probably just a coincidence.

One of the big surprises of the *Cassini* mission was the appearance of Saturn at 5 μm, measured by the VIMS instrument. Jupiter's appearance at 5 μm is to a good approximation closely anti-correlated with its visible appearance, with zonal regions which are bright at visible and near-IR wavelengths appearing dark at 5 μm and belt regions, which are dark at visible and near-IR wavelengths, appearing bright at 5 μm. At visible and near-IR wavelengths the large optical depths of tropospheric and stratospheric hazes in Saturn's atmosphere mask the structure of the planet's main clouds and thus the 5 μm images were expected to show more detail, but *Cassini* VIMS has shown the deep belt/zone structure of Saturn to be far more complex than that of Jupiter with the latitudinal scale of the belts/zones appearing to be much finer (as can be seen in Figure 5.24, see color section).

The circulation near the poles of planets often reveals interesting dynamics that help to probe the general circulation. A notable example is the warm polar dipole seen near Venus' poles (Piccioni *et al.*, 2007) and the poles of Saturn have proven to be equally intriguing regions. Sánchez-Lavega *et al.* (2002) used HST observations to show that a dark "cap" of radius 1,000 km to 2,000 km existed over the South Pole from 1997 to 2004 and that a prograde jet with winds of 90 m s^{-1} existed at 73°S. Keck imaging observations in 2004 at wavelengths between 8 μm and 24 μm (Orton and Yanamandra-Fisher, 2005) showed that Saturn had a warm South Polar cap and a compact hot point within 3° of the South Pole. These observations were consistent with a temperature rise at the 100 mbar level of ~2 K between 69°S and 74°S (planetocentric) and an increase of ~5 K at 3 mbar. The temperature was estimated to rise still further between ~87°S and 90°S by ~2.5 K at the 100 mbar level, correlated with a rise of ~1 K at 3 mbar. While the asymmetry in Saturn's stratospheric temperatures is a known seasonal effect, due in part to solar heating of aerosols, photodissociation of methane, and thermal blanketing by dissociation products such as acetylene and ethane, Flasar *et al.* (2005) showed that the radiative time constant in the stratosphere is 9–10 years and thus the temperatures should lag the solar forcing by roughly a sixth of an orbital period or 5 years. The fact that *Cassini* CIRS found the South Polar stratosphere to be much warmer than would be expected with a simple radiative forcing model, suggests that dynamics also plays a part.

The zonal wind speeds determined by *Voyager* extended from ~70°S to 82°N. Measurements of zonal wind speeds were extended to nearly 80°S by HST (Sánchez-Lavega *et al.*, 2004) and have been extended all the way to the South Pole by *Cassini* ISS (Sánchez-Lavega *et al.*, 2006) to further explore the South Polar Vortex (SPV). A prograde jet with speeds of $90\,\mathrm{m\,s^{-1}}$ was seen at 74°S and a second, much stronger prograde jet was seen at 87°S, with speeds of $160 \pm 10\,\mathrm{m\,s^{-1}}$. Similar jets would be expected about the North Pole, but *Cassini* will be unable to extend wind measurements towards the North Pole until sunlight returns in the Northern Spring Equinox in 2009 to allow cloud tracking. Like Flasar *et al.* (2005), Sánchez-Lavega *et al.* (2006) found that the temperature inside the warm, cyclonic SPV was higher than would be expected by solar heating alone, implying a dynamical component that could be explained by the region downwelling at a rate of $1.4\,\mathrm{m\,s^{-1}}$. In addition, from the thermal wind equation it was concluded that the wind speeds in the jets should decrease with altitude.

Fletcher *et al.* (2008a) have analyzed almost all of the *Cassini* CIRS observations made during *Cassini*'s prime mission to compare and contrast the conditions at Saturn's North and South Poles. Such a study is only possible with *Cassini* CIRS data since ground-based observations can only ever observe the summer pole, while CIRS can observe both poles in the thermal-infrared and is not reliant on reflected sunlight. Fletcher *et al.* (2008a) find that warm cyclonic vortices are actually present at both poles (Figure 5.25, see color section) and both are surrounded by polar collars that are cool in the 70 mbar to 300 mbar pressure range, which themselves are surrounded by warmer polar belts (again between 70 mbar and 300 mbar) at 79°N and 76°S. Cyclonic warm features, such as the polar belts and hotspots, are related to horizontal convergence and subsidence of tropospheric air, whereas the cold polar zones are related to divergence of upwelling, suggesting upwelling all around the polar vortices and downwelling within. Since Fletcher *et al.* (2008a) found that the abundance of PH_3 (whose mole fraction decreases with height due to photodissociation) was depleted within both vortices, this interpretation would seem sound. Differences between the poles are apparent, however, as the North Polar hotspot and cold collar seem more tightly confined than in the south.

Dyudina *et al.* (2008) used *Cassini* ISS observations to further characterize the SPV. Analyzing images at different wavelengths, Dyudina *et al.* conclude that the SPV is effectively clear of hazes in the upper troposphere down to the tops of tropospheric clouds at 2 bar to 3 bar, again consistent with the view that this is a region of downwelling. Furthermore, the region around the vortex is seen to have high levels of tropospheric haze, consistent with upwelling. Dyudina *et al.* find that the central haze-free "eye" (Figure 5.26) is surrounded by two cloud walls, one roughly 1° away from the pole with an oval shape and a second circular cloud wall

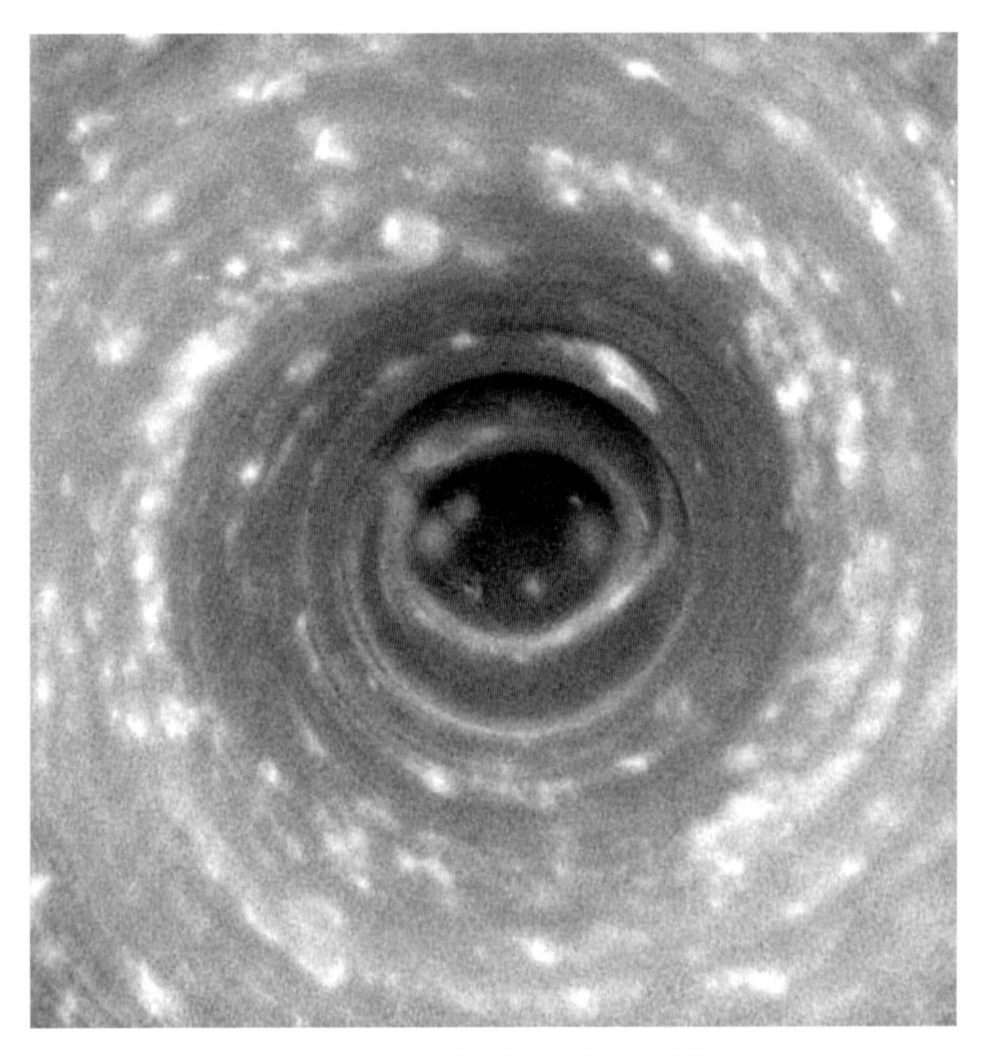

Figure 5.26. *Cassini* ISS image of Saturn's South Polar Vortex (SPV), recorded in November 2006 (after Dyudina *et al.*, 2008). In this image the Sun is shining from the top and the shadows cast by the SPV cloud eye walls can be seen on the underlying clouds. Courtesy of NASA.

roughly 2° away from the pole. By measuring the length of the shadows cast by these walls, Dyudina *et al.* estimate that the outer eye cloud wall rises to ~40 km above the main cloud deck, while the inner cloud wall rises to a height of ~70 km, implying that it extends all the way to the tropopause at 100 mbar. No equatorwards or polewards mean motion was seen. Although stratospheric haze is seen to extend over the South Pole and become particularly UV absorbing (Chapter 4), the subsidence in the region leads to a clearing of the tropospheric haze layer allowing remarkably clear images of the tops of tropospheric clouds at 2 bar to 3 bar. This clearing can be seen in both

Cassini VIMS and ISS observations (Dyudina *et al.*, 2008). Another curious feature in the South Polar region is that the small eddies surrounding the vortex core are seen to be dark in 5 μm images, pointing to their large optical depths, but appear to be grouped into two classes according to their visible and near-IR reflectivities with one type appearing 2/3 the brightness of the other.

While the warm cyclonic vortices of Saturn's atmosphere are particularly striking, it may be that such phenomena are found in most giant planet atmospheres, and are related to the release of internal heat and convective overturning. As mentioned in Section 5.5.2, *Pioneer 11* observations of Jupiter (Ingersoll, 1990) found the temperature above the cloud tops to be slightly higher at the North Pole than the equator, and a similar cyclonic polar hotspot has been observed at Neptune's South Pole (Orton *et al.*, 2005), where the temperature at the 100 mbar level is seen to be warm polewards of 70°S.

5.6.2 Storms and vortices

The cloud features that are visible on Saturn appear to be the tops of active convection systems that push their way up into the overlying semitransparent region. While generally less active than Jupiter's atmosphere, a number of spots have been observed in Saturn's atmosphere from ground-based observations over many years (Sánchez-Lavega, 1982).

Like Jupiter, cloud tracking of the motion of eddies in Saturn's atmosphere suggests that it is the small eddies that drive the zonal jets and not *vice versa*, though Del Genio *et al.* (2007) estimate that Saturn's jets are driven with a smaller rate of energy conversion than Jupiter's. In addition, Del Genio *et al.* (2007) find that convection occurs preferentially in cyclonic regions, but unlike Jupiter, some convection is also seen in eastward jet regions.

While most features are small, large white spots occasionally form in one of the planet's zones, which rapidly expand (on timescales of days) in the east–west direction to girdle the whole planet before gradually subsiding (on timescales of months). Such Great White Spots (GWSs) have now been observed from the ground in all zones except the STZ. Although no such storm was observed during the *Voyager* encounters, an equatorial storm (or "Equatorial Disturbance") was observed by the Hubble Space Telescope (HST) in September 1990 at 4°N (Westphal *et al.* 1992), which had almost completely disappeared by June 1991. Subsequently, a new storm (Figure 1.5) was observed in September 1994 (Sánchez-Lavega *et al.*, 1996), which lasted for more than a year (Sánchez-Lavega *et al.*, 1999), disappearing in 1996. It is thought that these storms may affect the zonal winds of the equatorial zone (as was discussed in Section 5.6.1). Although the origin of these storms remains unknown, detailed observations suggest that they result from a sudden outburst of convective activity, presumably originating from disturbances deep below the visible cloud tops, which trigger rapid vertical convection and resultant ammonia cloud condensation (with possible thunderstorm-style deep vertical convection). These localized, thick, bright, high clouds then spread latitudinally and are subsequently torn apart by the strong latitudinal wind shear observed in Saturn's atmosphere and spread right around the planet before eventually settling. Previous storms in the EZ were observed in 1876 and 1933, and at first glance these major equatorial storms would seem separated by approx 57 years (2 Saturn years) and appear correlated with the northern hemisphere summer suggesting a link with solar forcing. If this is true then the next equatorial disturbance may be expected around the year 2047 (Beebe, 1997).

While the *Voyager* spacecraft did not observe a GWS, numerous anticyclonic ovals were observed in Saturn's atmosphere including, among others, Brown Spots 1, 2, and 3 at 42°N, "Anne's Spot" (which had a reddish color) at 55°S, shown in Figure 5.27 (see color section), and the "UV Spot" at 27°N. A North Polar spot (sometimes called "Big Bertha") was also observed at 75°N, whose then apparent interaction with the North Polar Hexagon feature will be discussed in Section 5.6.3. These features were generally found to have the highest contrast in green-filtered images, although as the name suggests, the UV spot was most prominent at UV wavelengths. In addition to these regular ovals, a number of convective regions were also seen near 39°N (Figure 5.20), a region of cyclonic vorticity, which were seen to have a similar appearance to the plumes that appear in Jupiter's NEB (Smith *et al.*, 1981, 1982; Sromovsky *et al.* 1983). It is possible that the convective events observed were triggered by the passage of a cyclonic white spot immediately to the south of the outbreaks.

Prior to the *Cassini* observation period, ground-based imaging of Saturn in the 5 μm window (Yanamandra-Fisher *et al.*, 2001) showed the main thermal emission to originate between 38°S and 49°S (planetocentric), coinciding with an eastward jet. Discrete dark (cold) features were apparent in this latitude band, with lengths of 30,000 km to 50,000 km, which are quite unlike anything that has been observed in Jupiter's atmosphere. However, such large features have not been seen in *Cassini* VIMS (Baines *et al.*, 2005) observations, nor in more recent ground-based observations (Orton and Yanamandra-Fisher, 2005) so they would appear to be ephemeral.

Since *Cassini* started observing Saturn, many eddies and vortices have been seen, but none matches those previously seen by *Voyager*. Hence, such systems would appear to come and go in Saturn's atmosphere and are not long-lived as some of the vortices in Jupiter's atmosphere appear to be. *Cassini* has observed a number of sudden, convectively active storms near 35°S in Saturn's so-called "Storm Alley". One, the "dragon" storm, is shown in Figure 5.28 (see color section), while another example, on Saturn's night side, but illuminated by "ringshine" (i.e., light reflected from Saturn's rings) is shown in Figure 5.29. Unlike Jupiter, lightning has not been directly detected on Saturn through observing lightning flashes on Saturn's night side. This is partly due to the fact that Saturn's night side is often not very dark due to the significant levels of light reflected from Saturn's rings (as seen in Figure 5.29) and is also due to the expected Saturnian water clouds lying at much greater depths (~20 bar) and thus being obscured by much greater optical depths of overlying clouds and hazes. However, lightning is indicated by radio wave emissions. *Voyager 1* first detected such radio emissions known as Saturn Electrostatic Discharges (SEDs) (Warwick *et al.*, 1981) in 1980. More recently, the *Cassini* RPWS instrument observed a sudden burst of SEDs in 2004 (Fischer *et al.*, 2006), and a particularly large

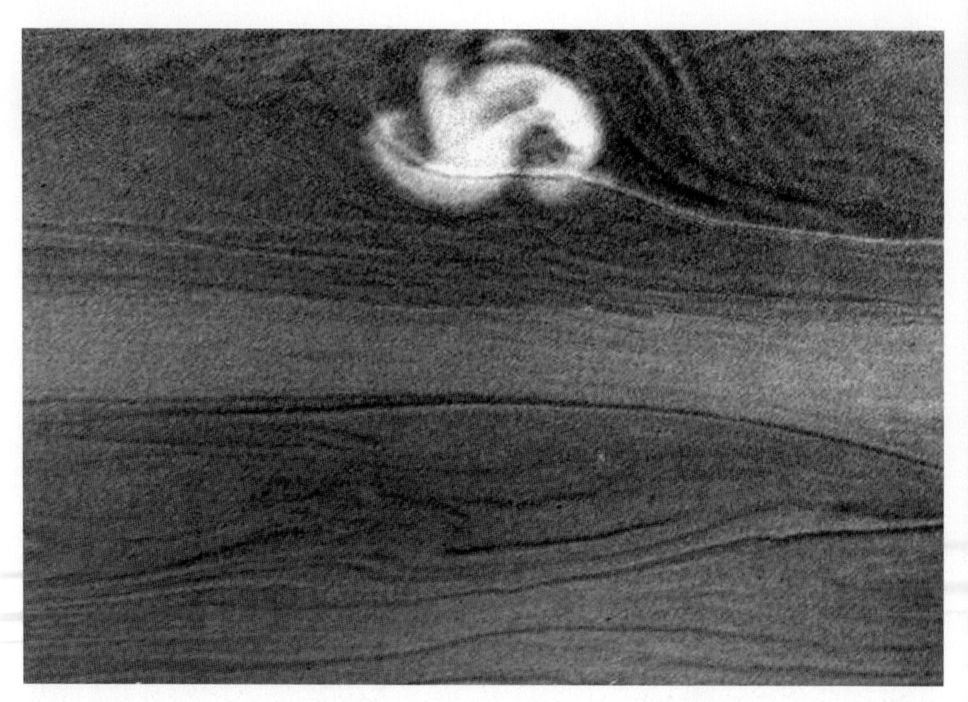

Figure 5.29. A convective thunderstorm observed by *Cassini* ISS in Saturn's Storm Alley in January 2006. The storm is here seen on Saturn's night side and is illuminated by "ringshine" (i.e., sunlight reflected off Saturn's rings and onto Saturn's cloud tops). Courtesy of NASA.

outburst was detected in 2006 (Fischer *et al.*, 2007). Comparing these observations with *Cassini* ISS observations of Saturnian storm systems, Dyudina *et al.* (2007) find that the SEDs are correlated with the massive storms seen at 35°S in Saturn's Storm Alley, which suggests strongly that these observations do indicate lighting activity in Saturn's atmosphere. Intriguingly, while optical lightning flashes are seen on Jupiter, there is no equivalent to Saturn's high-frequency SEDs, although whistler signals are seen, which are not detected on Saturn. Zarka (1985) suggests that the lack of high-frequency electrostatic disturbances for Jupiter is due to strong absorption of the radiowaves propagating through Jupiter's lower-ionospheric layers.

5.6.3 Waves

Prior to the arrival of the *Cassini* spacecraft in 2004, a number of waves had been seen in Saturn's atmosphere. Achterberg and Flasar (1996) analyzed *Voyager* IRIS mid-IR observations to find wavenumber-2 waves near the tropopause at northern midlatitudes between 20°N and 40°N, which were quasi-stationary with respect to Saturn's interior (as defined by System III). However, these waves were dwarfed by two major planetary-scale waves observed at visible wavelengths by *Voyager* ISS: the

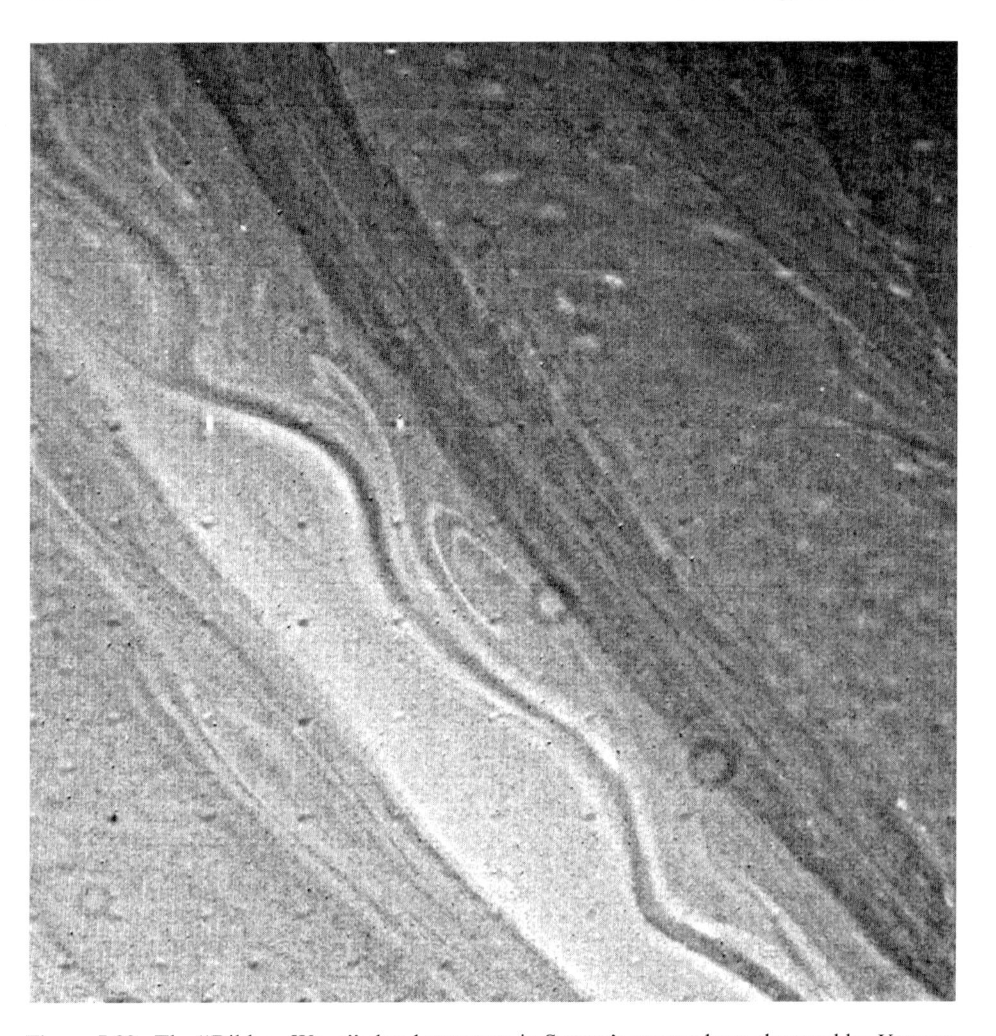

Figure 5.30. The "Ribbon Wave" cloud structure in Saturn's atmosphere observed by *Voyager 2* with a "green" filter. Courtesy of NASA.

"Ribbon Wave" at 41°N (planetocentric, 46° planetographic), and the North Polar Hexagon associated with the jet at 74°N (planetocentric, 77° planetographic).

The Ribbon Wave was first detected by the *Voyager 2* spacecraft in 1981 (Smith *et al.*, 1982) and appeared at visible wavelengths as a thin wavy line within the bright zone at 41°N (planetocentric, 46°N planetographic), coincident with the first major eastward jet in Saturn's atmosphere north of the equatorial jet (Figures 5.30 and 5.31, see color section for latter). The wavelength of the Ribbon Wave was of the order of 5,000 km and the phase speed was approximately 140 m s^{-1}. The wave feature was Fourier-analyzed by Sromovsky *et al.* (1983) and later Godfrey and Moore (1986), who found that the dominant wavenumbers were 8 and 20. Although both sides of the zone were equally bright in green and red filters, the southern side appeared brighter in violet images. However, this contrast was completely reversed in the UV, where the southern side appeared almost black. These observations pointed to the existence of a high-altitude haze of small particles to the south of the wave and low haze opacity to the north. This suggested that air was rising to the south of the Ribbon Wave and sinking to the north, consistent with cloud tracking of features that showed the northern edge to be cyclonic and the southern edge to be anticyclonic. Sromovsky *et al.* (1983) derived a dispersion relation for the Ribbon Wave, which they argued showed that the wave was fundamentally a barotropic Rossby wave. However, Godfrey and Moore (1986) pointed out that the zonal wave curvature d^2u/dy^2 (for which high values are expected to lead to barotropic instability) was at a minimum at the central latitude of the wave and this appeared to be inconsistent with the barotropic wave hypothesis. Instead, they favored a baroclinic Rossby wave explanation since the *Voyager* IRIS measurements showed the temperature at 150 mbar (Hanel *et al.*, 1982) to have a sharp gradient at this latitude with temperatures to the south being roughly 6 K cooler than the north. This is again consistent with the picture of rising air to the south and descending air to the north. The position of the wave at UV wavelengths (where we only see high in the atmosphere) was identical to that in visible wavelengths, indicating little vertical shear in the feature.

The Ribbon Wave was observed again in 1995 by HST (Sánchez-Lavega, 2002) with the same wavelength and phase speed and thus appeared at the time to be a long-lived feature. However, *Cassini* has found no trace of the Ribbon Wave, although numerous other small wave features, due to Kelvin–Helmholtz instabilities, have been detected on many belt/zone boundaries (Figure 5.32). These are regions of strong thermal gradient, and thus strong wind shear, where instabilities would appear to lead to the formation of waves. Hence, the Ribbon Wave does not now appear to be a special feature, but rather a particularly strong and prominent example of a class of waves that occur all the time in Saturn's atmosphere, which happened to be particularly visible during the *Voyager* observations and in the early 1990s. In fact, ground-based monitoring reveals that at the start of the *Cassini* observations in 2004, Saturn was going through a period of peculiarly negligible zonal wave activity, although this seemed to be increasing again in 2007. In addition, prior to *Cassini* no wave features had been observed in the southern hemisphere, but wave features have now been observed in both hemispheres.

Apart from waves seen at visible wavelengths on belt/zone boundaries, *Cassini* has made observations of other types of waves. One of these is the "String of Pearls" seen by *Cassini* VIMS at ~40°N, which is due to variations in optical depth of the main (presumably ammonia) cloud deck and thus reveals motions in the 2 bar to 3 bar region (Figure 5.33, see color section). More than 24 bright spots appeared at this latitude in April 2006 separated by about 3.5° longitude. This is the first time that such a long and regular train of cloud clearings has been seen and would appear to be a manifestation of a large planetary wave. One of the goals of the *Cassini* mission was to probe the thermal structure of planetary waves to get a better understanding of their dynamics. Since wave activity has been relatively light during *Cassini*'s mission so far there has been limited opportunity to do this. However, the *Cassini* CIRS instrument has seen a wave-like feature at 45°N to 50°N, which appears as a series of

Figure 5.32. Kelvin–Helmholtz instability waves seen at a belt/zone boundary of Saturn by *Cassini* ISS in 2004. Courtesy of NASA.

hotspots, spaced by 30°, originating from the 100 mbar to 500 mbar level. This feature is still being investigated. Orton and Yanamandra-Fisher (2005) also saw zonal temperature oscillations near 32°S (planetocentric), consistent with a wave of wavenumber-9 or wavenumber-10.

Although the *Voyager* flybys of Saturn occurred in 1980 and 1981, the second major wave system after the Ribbon Wave, the "North Polar Hexagon" was not discovered until 1988, when *Voyager 2* images were re-projected to produce polar maps (Godfrey, 1988). The feature revealed by this re-projection, shown in Figure 5.34, was a very regular wavenumber-6 wave centered in a 100 $m s^{-1}$ eastward jet at 76°N (planetographic). On the southern edge of one of the Hexagon's faces was the North Polar Spot (NPS). What is most remarkable about these features is that they were found to remain almost static with respect to the System III rotation frame, which has a period close to Saturn's bulk rotation rate (see Section 2.7.3). It seems unlikely that the magnetic field itself could be exerting such an influence on the motion of the observable troposphere and it thus appeared that both the NPS and the Hexagon were linked in some way to the deep interior. Allison *et al.* (1990) suggested that the Hexagon was a stationary Rossby wave forced by interaction between the eastward jet and the adjacent NPS and meridionally trapped by the strong relative vorticity gradient of the flow itself. Both the NPS and NPH were observed again in 1990 with ground-based observations (Sánchez-Lavega *et al.*, 1993) and HST (Caldwell *et al.*, 1993). Sánchez-Lavega *et al.* (1993) found that the appearance of the NPS had changed in the intervening 10 years. *Voyager* observed that the NPS had greatest contrast at 419 nm, but was barely visible at 566 nm. However, in 1990 the NPS was observable from yellow to red wavelengths and

Figure 5.34. Saturn's North Polar Hexagon, North Polar Spot, and Ribbon Wave as observed by *Voyager 2*. Polar stereographic projection after Godfrey and Moore (1986). The Ribbon Wave at 47°N is clearly visible. The Hexagon Wave is not so clear, but is visible at latitude ~78°N. The North Polar Spot is at latitude 75°N and longitude (planetographic) of 320°W.

was particularly bright in the near-IR methane bands, indicating a high cloud top of ~90 mbar. An additional finding was that while the *Voyager* and 1990 observations estimated the NPS to be drifting with longitude at similar rates, extrapolating the position of the *Voyager* NPS forward to 1990 predicted a position approximately 60° away from where the NPS was actually observed. Sánchez-Lavega *et al.* suggested that an explanation for the observed differences might be that there were in fact two spots, on two adjacent sides of the Hexagon, which alternately brightened and faded. Since the NPS was seen to drift so slowly with respect to System III, it was suggested that the NPS might in some way be fixed to the deep interior and its presence deflected the jet stream around it, setting up the NPH wave system. However, the internal rotation rate of Saturn is the source of some speculation, as we saw earlier in Section 2.7.3, since we now know that the System III rotation rate varies with time and is not rigidly linked to interior bulk rotation (Anderson and Schubert, 2007; Gurnett *et al.*, 2005). Hence, the relationship between the NPH and the interior is not clear. An alternative and probably more plausible explanation of the NPH (Aguiar *et al.*, 2008) is that the zonal winds at the NPH latitude of 77°N violate the *Rayleigh–Kuo criterion* for barotropic instability (i.e., the latitudinal gradient of potential vorticity changes sign at the latitude of the jet; Section 5.3.1) and thus the NPH results from a finite-amplitude, nonlinear equilibration of a barotropic instability of the jet.

The *Cassini* mission has made extensive observations of both poles of Saturn and the warm cyclonic vortices found there were discussed in Section 5.6.1. Since Saturn's North Pole has been in darkness since *Cassini* arrived it has not been possible to establish whether the NPH and NPS are still detectable at visible wavelengths. However, it has been possible to observe the 5 μm thermal emission from the deep atmosphere (which maps the variation in optical thickness of the main cloud deck at 2–3 bar) with the VIMS instrument and also the thermal emission from the upper troposphere up to the stratosphere with *Cassini* CIRS. Remarkably the NPH has been observed by both *Cassini* instruments. Fletcher *et al.* (2008a) found a warm cyclonic hexagon-shaped belt in the upper troposphere (800–100 mbar) at 79°N (planetographic) in CIRS observations (Figure 5.25), which coincides with a 5 μm bright hexagon seen by VIMS (Figure 5.35). Surrounding this feature, a cold anti-cyclonic hexagon-shaped zone is seen by CIRS at 76°N (planetographic), which coincides with a 5 μm dark hexagon seen by VIMS. Fletcher *et al.* (2008a) suggest that upwelling on the equatorial side of the polar jet and subsidence on the poleward side might be responsible for both the warming at 100 mbar to 800 mbar and also the cloud clearing at 2 bar to 3 bar leading to the 5 μm bright hexagon seen in VIMS images. The NPH thus appears to be an extraordinarily long-lived feature that appears to extend from the deep atmosphere right up to the tropopause. The fact that this feature exists over such a great altitude range supports the theory that it is due to barotropic instability, but why the wave should be wavenumber-6 and not any other is unclear, although it is interesting to note that the wavelength is very similar to the mean distance between Saturn's belts and zones and may be a "preferred length" of the atmosphere. However, although the NPH has been rediscovered by *Cassini*, neither CIRS nor VIMS detected the NPS, although VIMS did observe numerous vortices outside the NPH, one of which might perhaps be associated with the NPS. Further mapping of the area at visible and near-IR wavelengths will be an important objective once sunlight returns to the North Pole in 2009.

5.7 METEOROLOGY OF URANUS

5.7.1 General circulation and zonal structure

The amount of energy emitted by Uranus is at most only 1.06× that received from the Sun indicating that Uranus has a very low internal heat source and, one might

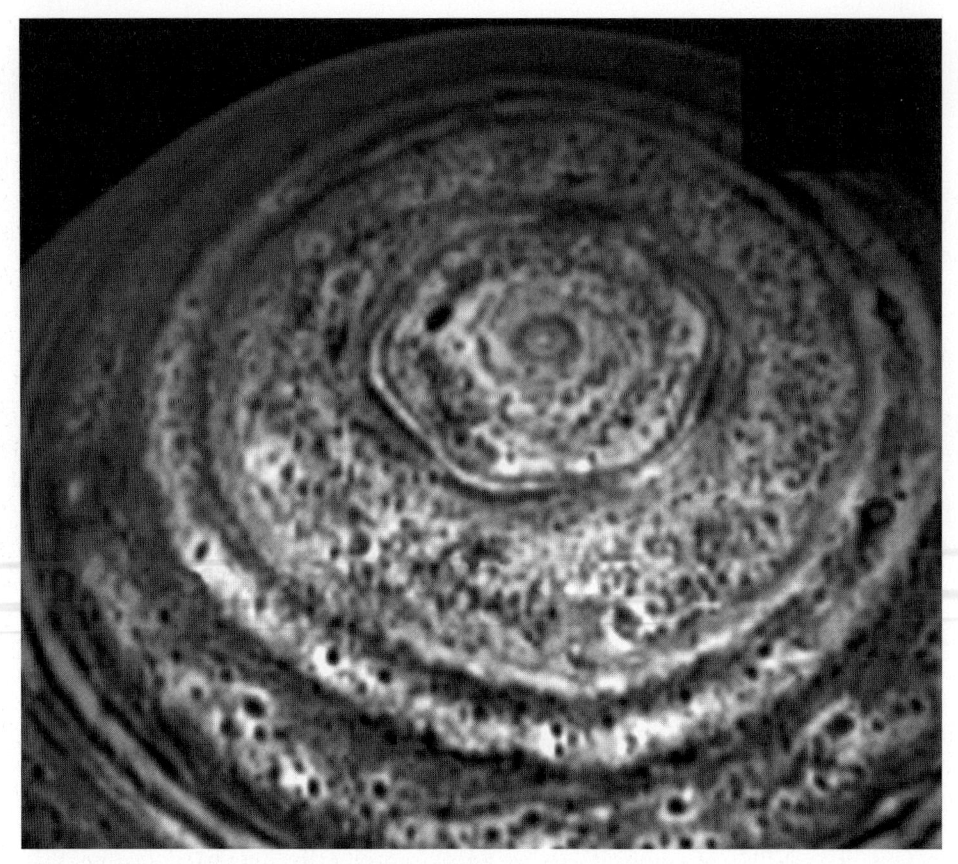

Figure 5.35. Saturn's North Polar Hexagon (NPH) as viewed by *Cassini* VIMS in March 2007 at 5 μm, showing thermal emission from the deep atmosphere, modulated by variable cloud opacity in the 2–3 bar region. Courtesy of NASA.

suspect, a sluggish circulation system driven primarily by latitudinal variations of the solar flux, which at Uranus' distance from the Sun is a meager 3.7 W m^{-2}. Another difference between Uranus and Jupiter/Saturn is Uranus' extremely large obliquity, which means that Uranus receives direct sunlight over both poles as well as the equator during the course of a Uranian year. [NB: Even though both poles experience a night lasting half a Uranian year long, they receive annually 50% more sunlight per unit area than the equator.] The final difference between Uranus and Jupiter/Saturn is that the visible hydrogen–helium atmosphere is only a small fraction of the total planetary mass and thus the Taylor–Proudman column hypothesis is unlikely to apply due to the boundary between the hydrogen–helium outer atmosphere and the denser icy–rock interior occurring at a depth of just 5,000 km below the visible cloud tops. During the *Voyager 2* flyby mission, when Uranus' South Pole was facing almost directly towards the Sun, the dominant circulation must have been one that efficiently redistributed relatively warm polar air over the planet. Modeling the planet as a black sphere in radiative equilibrium with the received insolation, the expected equator-to-pole temperature difference would be of the order of 10 K. However, very little thermal variation was found in the 0.5 bar to 1 bar pressure levels by *Voyager 2* (Flasar *et al.*, 1987; Hanel *et al.*, 1986). This lack of thermal variation is surprising given that estimates of the para-H_2 fraction and the eddy-mixing coefficients suggest that the vertical and meridional circulation of Uranus is very sluggish. How such an apparently sluggish atmosphere can have such small temperature gradients, given its differential radiative forcing, is very unclear.

Deeper in the atmosphere, microwave observations reveal a clear zonal structure, with latitudes polewards of 45°N or 45°S appearing significantly brighter than midlatitudes (Hofstadter and Butler, 2003). While these differences could just conceivably be due to temperature variations at the 50 bar level, it seems more plausible that the variations in brightness are due to a higher abundance of ammonia gas at latitudes less than ±45°, suggesting either meridional circulation with air rising at midlatitudes and descending at the poles, or alternatively that convective overturning polewards of ±45° is somehow inhibited. Long-term monitoring of Uranus' microwave brightness (Klein and Hofstadter, 2006) shows seasonal variations consistent with this broad latitudinal structure. Interestingly, VLA observations in 2006 at microwave wavelengths detect some banded structure at latitudes less than ±45° (Orton *et al.*, 2007c), which may point to a latitudinal variation in convective overturning rather than more general Hadley-like overturning of the atmosphere. The apparent special significance of the ±45° latitude boundary is further emphasized by the fact that long-term monitoring of Uranus' disk-averaged visible albedo is also well matched by a model where latitudes of less than ±45° are darker than polar latitudes by a factor that varies from 0.72 to 0.98 (Hammel and Lockwood, 2007), as was discussed in Chapter 4.

Although very little temperature variation was seen by *Voyager 2* in the 0.5 bar to 1 bar pressure range, significant variation was seen in the upper troposphere/lower stratosphere (60–200 mbar), where the equator was found to be actually warmer than the pole and where midlatitudes were found to be colder than the equator by ~5 K (Flasar *et al.*, 1987). Ground-based VLT observations in 2006 of the temperature near 100 mbar (Orton *et al.*, 2007c), found the same variation in brightness temperature even though almost an entire Uranian season had elapsed since the *Voyager 2* observations. This upper-tropospheric temperature structure is also seen in Neptune's atmosphere and is strongly indicative of upwelling and divergence at midlatitudes (causing cooling) followed by meridional circulation to the poles and equator, where the air then converges (causing heating) and descends.

Although Uranus had very little visible belt/zone structure during the *Voyager* epoch (Smith *et al.*, 1986, and Figure 1.9), with only a very slight north–south asymmetry apparent in broadband visible images (with South Polar latitudes appearing slightly brighter due to thicker methane haze), enough variable features existed to allow cloud tracking and thus determination of cloud top winds by the *Voyager 2* cameras, which was discussed earlier in Section 5.2.2. Thermal wind shears calculated from the retrieved temperature fields of *Voyager 2* suggested that zonal winds decayed with height with a vertical scale of ~10 scale heights or 300 km.

Convective activity has become significantly more vigorous in the lead-up to Uranus' Northern Spring Equinox in 2007 and further wind speed data have been deduced from more recent HST (Hammel *et al.*, 2001) and Keck observations (Hammel *et al.*, 2005a). However, even though Uranus has become more convectively active than it was during the *Voyager 2* flyby, the zonal winds do not appear to have changed significantly. Images of Uranus recorded by the HST at the end of the 1990s (Karkoschka, 1998b, 2001; Rages *et al.*, 2004) (Figure 5.36, see color section) revealed that the brightness of the South Polar latitudes had decreased since the *Voyager* epoch and a noticeably bright "zone" had appeared at 40°S to 50°S, together with a less bright zone appearing, for a time, at 70°S. Since then, as Uranus approached its Northern Spring Equinox in 2007 (Figure 5.37, see color section), the contrast of the zone at 45°S has been seen to decrease and the beginnings of a zone at 45°N has been detected. Observations of Uranus during the previous equinox revealed a very different zonal appearance (Alexander, 1965; Karkoschka, 2001) and there was much anticipation that rapidly changing solar forcing would lead to dramatic changes in Uranus' atmosphere during the 2007 equinox. Hence, many observations of Uranus have taken place during 2006–2008 and it appears that the atmosphere has indeed changed significantly during this event, but this time the changes have been very well observed and are currently being analyzed.

5.7.2 Storms and vortices

Although Uranus exhibits a clear zonal wind structure, no regular cyclonic or anticyclonic eddies have been observed. Small white clouds are occasionally observed at midlatitudes (at approximately $\pm 30°$), which have been interpreted as localized methane clouds forming in restricted areas of rapid upwelling, rather like the equatorial brightenings seen on Saturn, and these have become particularly numerous during the run-up to the Northern Spring Equinox in 2007. Observations of these clouds (Figure 5.38) with HST are reported by Karkoschka (1998b, 2001), Sromovsky *et al.* (2000), and more recently by Hammel *et al.* (2005a, b) and Sromovsky *et al.* (2007). A particularly bright cloud was observed by HST in 2004 by Hammel *et al.* (2005b) at 36°S. Such clouds, if associated with vigorous enough convection may help to transport methane through the cold trap of the tropopause and on into the stratosphere. However, an observation of an even brighter, higher cloud by the Keck Observatory in 2006 (Sromovsky and Fry, 2007) put the cloud top at 200 mbar, which is still below the tropopause and estimates of the abundance of methane in the stratosphere lie close to the "cold trap" value. HST observations in 2000 found that the northern midlatitudes just coming into view as Uranus approached equinox were considerably more convectively active then southern mid-latitudes. However, during the equinox in 2007 both hemispheres appeared equally convectively active.

5.7.3 Waves

Up until the end of the 1990s no wave structures had ever been observed in the Uranian atmosphere. However, near-IR Keck observations in 2003 (Hammel *et al.*,

Figure 5.38. Three HST/WFPC-2 images of Uranus, recorded in 1994 in a methane absorption band, revealing the motion of a pair of bright clouds in the planet's southern hemisphere, and a high-altitude haze that forms a "cap" above the planet's South Pole. The two high-altitude clouds are 4,300 and 3,100 km across, respectively. Three hours have elapsed between the first two images, and five hours have elapsed between the second pair of observations. Courtesy of NASA.

2005a) found a subtle wave feature at the equator, with diffuse patches appearing every 30° in longitude, which is suggestive of an equatorially trapped Kelvin wave. Furthermore, in 2005, Keck observations of a very bright cloud at 30°N (Sromovsky *et al.*, 2007) found that it oscillated its position in both latitude and longitude. The oscillation was modeled as being due to a slow-period Rossby wave combined with a shorter period inertial oscillation. Hence, it would seem that wave processes are likely to be as prevalent in Uranus' atmosphere as in the atmospheres of all the other giant planets, but have not been possible to observe until recently, when bright, trackable features have become observable.

5.8 METEOROLOGY OF NEPTUNE

5.8.1 General circulation and zonal structure

The atmosphere of Neptune is powered by extremely low energy fluxes. Internal heat energy flux is estimated to be $0.45\,\mathrm{W\,m^{-2}}$ (Table 3.2) while the absorbed solar flux is estimated to be $0.27\,\mathrm{W\,m^{-2}}$, compared with values of $0\,\mathrm{W\,m^{-2}}$ and $205.5\,\mathrm{W\,m^{-2}}$, respectively, for the Earth. However, the cloud top zonal winds on Neptune are found to be very high with an extremely fast westward retrograde equatorial jet reaching speeds of $400\,\mathrm{m\,s^{-1}}$, gradually decreasing in the poleward direction and becoming eastward and prograde at latitudes poleward of 50°. It has been postulated that such high winds are allowed because the atmosphere of Neptune has low turbulence and thus low eddy viscosity. The general zonal wind structure is similar to that of Uranus, as we saw in Section 5.2.2, and why the equatorial jets of both planets should be blowing in the opposite direction to that of Jupiter and Saturn is unclear. In some ways it is easier to see how a retrograde equatorial jet is driven than a prograde, super-rotating jet. Air rising from deep levels at the equator, and initially rotating at the internal rotation rate would be expected to slow at higher levels (and thus greater distance from the center) due simply to conservation of angular momentum, giving rise to a westward-blowing air stream. Similarly, fast-flowing air at the equator which moves polewards would be expected to acquire additional eastward momentum via the same mechanism, giving prograde jets near the poles. Alternatively, air rising at midlatitudes and then traveling both polewards and equatorwards would give rise to a similar wind structure. This view is supported by *Voyager 2* thermal-IR measurements (Conrath *et al.*, 1989), which found that in the 0.3 bar to 1 bar pressure region, midlatitudes were ~2 K cooler than the equator and poles, which had roughly equal temperatures. This temperature difference was found to increase to ~5 K in the 30 mbar to 120 mbar pressure region. While *Voyager 2* found the equator and poles to have similar temperatures, VLT observations in 2006 (Orton *et al.*, 2007c) found that at the 100 mbar level the temperature polewards of 70°S was significantly greater than the temperature at both the equator and midlatitudes. This observation is suggestive of Neptune possessing a warm cyclonic vortex at its South Pole in 2006, similar to the cyclonic vortices found at both poles in Saturn's atmosphere.

Thermal wind shears calculated from *Voyager 2* thermal-IR measurements (Conrath *et al.*, 1989) show that, like all the other giant planets, zonal winds decay with height. It is also of great interest, as we have seen, to determine how deep the zonal winds extend into the interior. This may be answered unequivocally for both Neptune and Uranus by determination of the J_4 gravitational constant. If the zonal currents were superficial such that virtually all the planetary mass rotates with the same deep period, then calculations suggest that J_4 should be negative and have a value close to that given in Table 2.4. If, however, the zonal currents are surface expressions of internal differentially rotating cylinders then J_4 is predicted to be positive and have an absolute value roughly twice as big as that observed (Hubbard, 1997d). Hence, at least for the case of Uranus and Neptune, the zonal winds do appear to be restricted to the upper levels of the atmosphere, although the thermal wind equation suggests they must still be reasonably deep given the small latitudinal variation in temperature seen at pressures greater than approximately 1 bar.

Another curious feature of Neptune's atmosphere is that, while some banding is observed (more so than Uranus) the relationship between visible albedo and vorticity appeared from *Voyager* observations at some latitudes to be the opposite of that observed for Jupiter and Saturn with the anticyclonic, midlatitude regions appearing generally darker, and the equator and poles appearing generally bright (as was mentioned in Section 4.4.4). Unfortunately the flux at 5 μm is too low to determine if these albedo variations are due to total cloud opacity changes or to some other effect. In addition to these general banded features, transitory white clouds are often observed and these clouds remain bright in the near-IR methane absorption bands, indicating high cloud tops. The most plausible explanation of these features is that they are convectively produced methane clouds, although how they are initiated is unclear since the zonal wind flow is everywhere apparently stable to baroclinic and barotropic instabilities according to the Charney–Stern criterion. These clouds are observed at a number of latitudes, but they usually appear at midlatitudes (i.e., in anticyclonic vorticity regions just as they do for all of the other giant planets). This interpretation was strengthened by *Voyager 2* IRIS observations that the coolest tropopause temperatures, indicating divergence at the tropopause, and thus convection from below, are found at midlatitudes. The rapid overturning of Neptune's atmosphere is also indicated by the detection of disequilibrium species such as CO and the modeled possible presence of N_2. The background banded appearance of Neptune is thus probably due to variations in either the depth or color of the lower H_2S cloud and it would appear that this cloud deck responds counter-intuitively to the convective motion of the atmosphere above by appearing darker or being depressed to deeper levels where upper-level convection occurs. This picture is remarkably similar to the model proposed by Ingersoll *et al.* (2000) to account for the dynamics and cloud formation in Jupiter's atmosphere (Section 5.2.2 and Figure 5.6). Since the *Voyager 2* flyby in 1989, Neptune has been observed with HST and from the ground. The disk-averaged visible albedo of Neptune has been seen to increase steadily since the 1950s (Section 4.4.4) and may possibly be correlated with the latitude of the subsolar point. This increase appears to be due, in part, to an increase in convective activity in the two main "cloud belts" seen from 25°S to 60°S and 20°N to 40°N.

On an historical note, the high equatorial wind speeds on Neptune led to early confusion about its internal structure. As we discussed in Chapter 2, for a body in hydrostatic equilibrium rotating with a single period P, the oblateness, J_2, and P are uniquely related. Prior to the arrival of *Voyager 2* at Neptune in 1989, the rotational period was estimated from observation of the variation of Neptune's disk-averaged albedo caused by the transition of distinct cloud features which occur at midlatitudes and equatorial latitudes. Since these clouds lie in a strong retrograde zonal flow, the rotational period was estimated to be 18 hours which was found to be incompatible with the observed oblateness and J_2. However, *Voyager 2* measured the rotational period of the magnetic field (and thus presumably the interior) to be just over 16 hours, which is consistent with the other data.

5.8.2 Storms and vortices

The atmosphere of Neptune is observed to be highly dynamic and convective, with numerous storm and eddy features that allowed early estimates of the planet's rotation rate to be made due to diurnal variations of the planet's observed reflectivity. As telescope technology improved, ground-based images in the 0.89 μm methane band began to resolve distinct high-altitude cloud features on scales of 10,000 km or more (Baines, 1997b) and the distribution of these features was found to be highly variable.

The arrival of *Voyager 2* at Neptune in 1989 heralded an enormous development in our understanding of clouds and storm systems in Neptune's atmosphere (Smith *et al.*, 1989). *Voyager 2* observed four large features that persisted for the duration of the *Voyager* observations (from January to August 1989). The largest of these was the Great Dark Spot (GDS) and its white companion immediately to the south, which together drifted from 26°S to 17°S during the period of observation, or equivalently

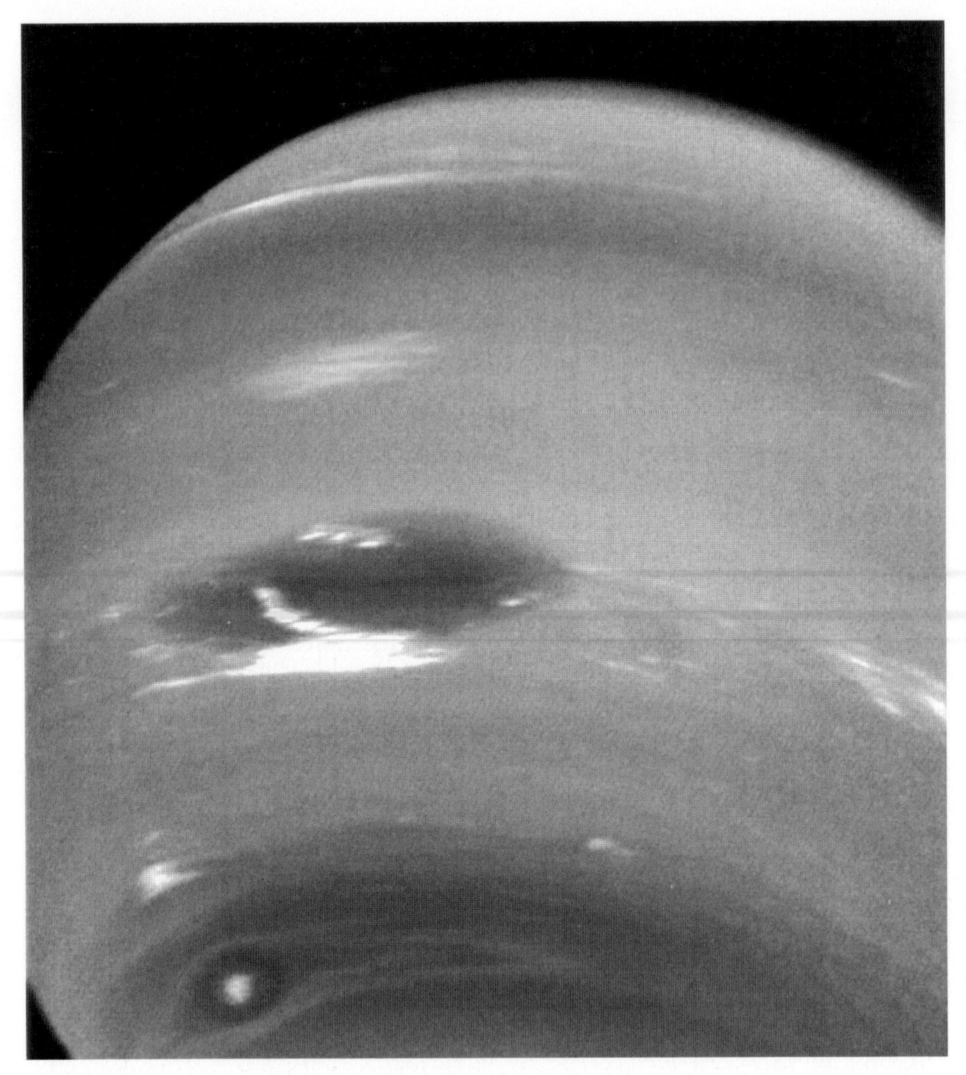

Figure 5.39. Detail of Neptune's GDS and DS2 observed by *Voyager 2* in 1989. Courtesy: NASA.

drifted equatorwards at a rate of 15° yr^{-1} (Figure 5.39). Such a drift of these features down the potential vorticity gradient towards the equator is expected from dynamical modeling (LeBeau and Dowling, 1998). Furthermore, once such features get too close to the equator, they are expected to rapidly dissipate through conversion into planetary waves (Sromovsky *et al.*, 2001c), as was discussed in Section 5.3.3. At first glance, the GDS appeared similar to the Great Red Spot of Jupiter since they both rotated anticyclonically and were of comparable size. However, in other respects they were very different. For a start the GDS appears to have been a short-lived disturbance, which by the time of later HST observations in 1994 had completely disappeared. Second, the GRS is believed to be a region of rapid updraft and the cloud cover is particularly thick and high, whereas the dark color of the GDS may be attributed either to a low methane ice and stratospheric haze abundance above the GDS or, more likely, suggests that that the main 3.8 bar cloud itself is somehow darker, or deeper. The wispy white clouds associated with the companion to the GDS were observed to move at a different speed to the GDS and companion, suggesting that these features form and evaporate high above the GDS as they pass through a local pressure anomaly (West, 1999). Stratman *et al.* (2001) performed dynamical modeling that supported this view and further determined that the cloud tops must be just below the tropopause. The second dark spot (DS2) was observed at 55°S in the dark circumpolar band and appeared to be roughly 180° in longitude away from the transient bright clouds seen at 70°S, called the South Polar Feature (SPF) (Sromovsky *et al.*, 1993). An additional bright cloud was seen near 42°S and acquired the name "Scooter".

A striking feature of Neptune's dark spots, discovered by *Voyager 2*, was that they oscillated. The longitudinal and latitudinal widths of the GDS were found to vary sinusoidally and in antiphase, with a period of approximately 200 hours and with amplitudes of 7.4° longitude and 1.5° latitude (Sromovsky *et al.*, 1993). While the shape of DS2 did not undergo such oscillations, its position was found to vary sinusoidally with a period of 36 days and amplitude of 2.4° latitude and 47.5° longitude (relative to longitudinal mean drift). These positional variations were accompanied by a variation in the area of the bright core of the DS2 which had maximum area when the DS2 was farthest north.

After the *Voyager 2* encounter, high-resolution imaging of Neptune did not begin again until HST observed the planet in 1994. Remarkably it was discovered that all the discrete atmospheric features observed by *Voyager 2*, with the exception of the SPF, had completely disappeared (Hammel *et al.*, 1995)! In their place a new Great Dark Spot had formed in the northern hemisphere at 32°N, which acquired the name NGDS-32 (Figure 5.40). Unlike the *Voyager* GDS, NGDS-32 did not drift equatorwards (Sromovsky *et al.*, 2002), but was seen at the same latitude in 1996 HST observations (Sromovsky *et al.*, 2001c), together with a new dark spot at 15°N (NGDS-15). However, NGDS-15 had completely disappeared by the next HST observations in 1998 (Sromovsky *et al.*, 2001d) and only the bright companions of NGDS-32 were still visible. The remains of NGDS-32 have now completely disappeared and no further dark spots have since been seen on Neptune.

In addition to ongoing HST observations, advances in the deconvolution of telescope images (Sromovsky *et al.*, 2001a) and the development of adaptive optics have greatly improved the spatial resolution of ground-based observations at visible and near-IR wavelengths, and have thus greatly improved the time sampling and spectral sampling of Neptunian clouds. Numerous clouds have now been recorded in Neptune's atmosphere over a number of years (Figures 5.41 and 5.42, see color section for both). Transient clouds generally appear between 29°S and 45°S, and 29°N and 39°N [NB: 39°N is the farthest north that can currently be seen due to Neptune's obliquity and season] and convective activity has been seen to increase

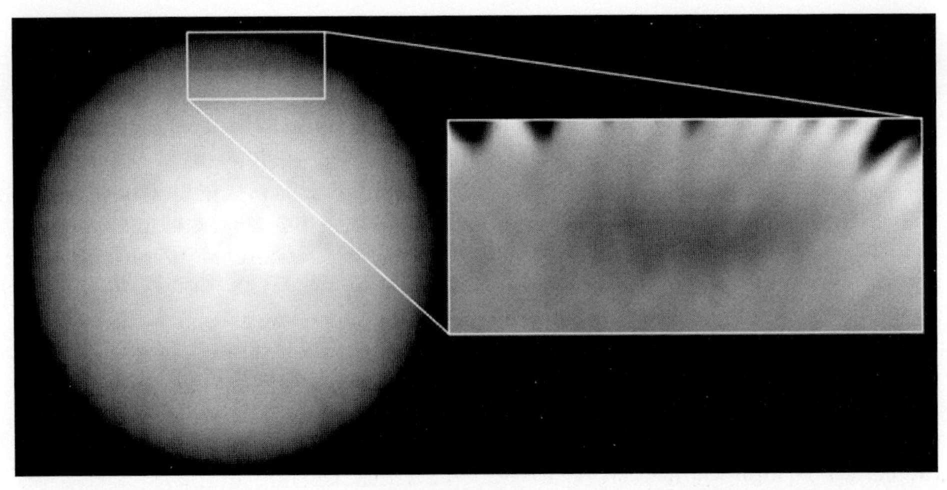

Figure 5.40. HST image of a new "Great Dark Spot" on Neptune, NGDS-32, located at 32°N, recorded in 1994. By this time the *Voyager* GDS found in the southern hemisphere had completely disappeared. Courtesy of NASA.

steadily since the 1990s, which seems to account for the steady rise in Neptune's disk-averaged visible albedo since the start of continuous measurements in 1950 (Lockwood and Jerzykiewicz, 2006). These latitudes correspond to the cooler tropopause temperatures indicative of upwelling and divergence at high altitudes discussed earlier.

5.8.3 Waves

The main apparent wave features observable in Neptune's atmosphere are the South Polar Wave (SPW), and possibly transient cloud features known as "outbursts". Other examples include the observed oscillation of discrete cloud features that was discussed in Section 5.8.2.

A dark, apparently axisymmetric, feature was detected near Neptune's South Pole by *Voyager 2* in 1989 (Figure 5.43, see color section). The band extended between planetocentric latitudes of 65°S and 40°S. However, within this band was a narrower, darker region that appeared as a wavenumber-1 planetary wave. The poleward excursion of the inner band was bordered on the north by a dark anticyclone (DS2) centered at 55°S, while the longitude of maximum equatorial excursion was marked by an outbreak of bright clouds at 70°S, the SPF. This wave appears in many respects to be analogous to the North Polar Hexagon in Saturn's atmosphere, which is believed by some to be a planetary wave forced by interaction with its own equator-side anticyclone, the NPS (although the NPS has now disappeared). However, while the North Polar Hexagon is quasi-stationary with respect to System III, suggesting a link with the rotation of the bulk interior, Neptune's SPW was estimated to drift significantly with respect to interior rotation (Sromovsky *et al.*, 1993). Since the *Voyager* encounter, HST and ground-based telescopes between 1994 and 1998 (Sromovsky *et al.*, 2001c) showed that while the SPW was still clearly visible, the latitudinal extent of the band had shrunk to between 55°S and 65°S. In addition, the SPF was rarely apparent and DS2 had completely disappeared. Dynamically, Sromovsky *et al.* (2001c) suggest that it is hard to see how such a wave could still be present in the absence of a forcing anticyclone and suggest that DS2 may have in fact still been present, but was too faint to observe. However, it may also be possible that the SPW is simply caused by barotropic instability and requires no forcing anticyclone. A dark band continues to exist at this latitude and still had a Wavenumber-1 characteristic at least until 2002 (Larry Sromovsky, pers. commun.).

As mentioned in Section 5.8.2, while the *Voyager* GDS had disappeared by 1994–1996, a new dark feature NGDS-32 had appeared at 32°N, which remained fixed in latitude during the observation campaign. In 1994 many bright clouds were observed extending from 30°N to the equator at roughly the same longitude as NGDS-32 (Sromovsky *et al.*, 2001c). Furthermore, the zonal velocity of these clouds was substantially less than the average zonal velocity at the equator and instead the speed seemed locked to the zonal velocity at the NGDS-32 latitude. This sudden appearance of bright equatorial clouds (Sromovsky *et al.*, 2001b) would seem to be similar to the "outbursts" postulated to explain the increased disk-averaged albedo observed with ground-based telescopes between 1986 and 1987 by Hammel *et al.* (1992) and likened to the GWS of Saturn. A similar link between near-equatorial dark spots and equatorial clouds was observed by *Voyager 2* where a bright "smudge" was observed just past the equator at the same longitude as the GDS and apparently co-rotating with it. This co-rotation suggests some sort of wave interaction and since the phase speed of the wave would appear to be eastward the most likely candidate is an equatorially trapped Kelvin wave (Sromovsky *et al.*, 2001c). However, the equatorial confinement of such a wave with an estimated phase speed of $210\,\mathrm{m\,s^{-1}}$ is calculated from the equatorial deformation radius to be 17° of latitude while the observed clouds actually extended to 30°N. Hence, the nature of this interaction clearly warrants further study.

5.9 BIBLIOGRAPHY

Andrews, D.G. (2000) *An Introduction to Atmospheric Physics*. Cambridge University Press, Cambridge, U.K.

Andrews, D.G., J.R. Holton, and C.B. Leovy (1987) *Middle Atmosphere Dynamics*. Academic Press, San Diego, CA.

Bagenal, F., T. Dowling, and W. McKinnon (Eds.) (2004) *Jupiter: The Planet, Satellites and Magnetosphere*. Cambridge University Press, Cambridge, U.K.

Beatty, J.K., C. Collins Petersen, and A. Chaikin (Eds.) (1999) *The New Solar System* (Fourth Edition). Cambridge University Press, Cambridge, MA.

Bergstrahl, J., E.D. Miner, and M.S. Matthews (Eds.) (1991) *Uranus*. University of Arizona Press, Tucson, AZ.

Cruikshank, D.P., M.S. Matthews, and A.M. Schumann (Eds.) (1995) *Neptune and Triton*. University of Arizona Press, Tucson, AZ.

Gehrels, T. (Ed.) (1976) *Jupiter*. University of Arizona Press, Tucson, AZ.

Gehrels, T. and M.S. Matthews (Eds.) (1984) *Saturn*. University of Arizona Press, Tucson, AZ.

Holton, J.R. (1992) *An Introduction to Dynamical Meteorology* (Third Edition). Academic Press, San Diego, CA.

Houghton, J.T. (1986) *The Physics of Atmospheres* (Second Edition). Cambridge University Press, Cambridge, U.K.

Jones, B.W. (2007) *Discovering the Solar System* (Second Edition). John Wiley & Sons, Chichester, U.K.

Rogers, J.H. (1995) *The Giant Planet Jupiter*. Cambridge University Press, Cambridge, U.K.

Shirley, J.H. and R.W. Fairbridge (Eds.) (1997) *Encyclopaedia of the Planetary Sciences*. Chapman & Hall, London.

Weissman, P.R., L.-A. McFadden, and T.V. Johnson (Eds.) (1998) *Encyclopaedia of the Solar System*. Academic Press, San Diego, CA.

NASA planetary images in this chapter were mostly downloaded from *http://photojournal.jpl.nasa.gov/*

Current images of the outer planets may be viewed at the International Outer Planet Watch (Atmospheres Division) *http://pds-atmospheres.nmsu.edu/ijw/ijw.html* The main International Outer Planet Watch website is at *http://www-ssc.igpp.ucla.edu/IJW/*

6

Radiative transfer processes in outer planetary atmospheres

6.1 INTRODUCTION

With the exception of Jupiter, none of the atmospheres of the giant planets has been directly sampled. Even with Jupiter the only *in situ* measurements that are available were made from the *Galileo* entry probe, which sampled a single, probably not very representative, region of the planet, namely a 5 μm hotspot. Hence, the bulk of our knowledge of the composition, cloud structure, and dynamics of these planets has not come from direct measurements, but instead has come indirectly from analyzing features in their electromagnetic spectrum, measured by ground-based telescopes, Earth-orbiting telescopes, and from specific flyby and orbiting spacecraft missions. There are two main components of the observed spectra that provide atmospheric information: reflected sunlight from the cloud and haze layers, and thermal emission from the atmosphere itself.

In this chapter we will examine how electromagnetic radiation interacts with molecules and thus how the spectra of these planets are formed through the process of *radiative transfer*. Once a satisfactory radiative transfer model has been developed (sometimes called a *forward model*), synthetic spectra of the planets may be generated from initial assumptions of the mean atmospheric vertical profile and compared with observations. Differences between observed and modeled spectra then may be used to revise the atmospheric profile assumptions used to generate the synthetic spectra and thus improve the fit. The process of revising the atmospheric profiles to improve the spectral fit is known as the *inverse model* or *retrieval model* and will be discussed in Chapter 7.

6.2 INTERACTION BETWEEN ELECTROMAGNETIC RADIATION AND PARTICLES

6.2.1 Fermi's golden rule

From quantum mechanics, the time-dependent Schrödinger equation governing the wavefunction Ψ of a particle is

$$i\hbar\frac{\partial\Psi}{\partial t} = \hat{H}\Psi \tag{6.1}$$

where $\hat{H} = -\frac{\hbar^2}{2m}\nabla^2 + V(\mathbf{r}, t)$ is the Hamiltonian operator representing the total energy of the system. If the potential energy V is time-independent, then Schrödinger's equation may be separated into time and space factors with the wavefunction written as $\Psi(\mathbf{r}, t) = \psi(\mathbf{r})e^{-iEt/\hbar}$, and where the separation constant E (which has units of energy) and time-independent wavefunction ψ satisfy the time-independent Schrödinger equation

$$-\frac{\hbar^2}{2m}\nabla^2\psi + V(\mathbf{r})\psi = E\psi. \tag{6.2}$$

It may be shown that there are a number of possible time-independent wavefunctions, or stationary states ψ_m, satisfying Equation (6.2) and thus a number of possible separation constants (or energies) E_m. Furthermore, it may be shown that these stationary states form an orthonormal basis. Hence, the general time-dependent wavefunction Ψ may be represented as a linear combination of the stationary states as

$$\Psi(\mathbf{r}, t) = \sum_m c_m(t)\psi_m(\mathbf{r})e^{-iE_m t/\hbar} \tag{6.3}$$

where the c_m are coefficients to be determined and where the probability of finding a particle in the mth stationary state is $|c_m|^2$.

Suppose now that a photon interacts with the particle to promote a change between two such stationary states. Let us assume that the transition between two energy levels is caused by a time-dependent influence, which we represent by a small additive potential $v(\mathbf{r}, t)$. Suppose also that at time $t = 0$ we know that the particle is in a certain stationary state ψ_n of energy E_n. The coefficients c_m some time later (where $m \neq n$) may be shown from perturbation theory (e.g., Rae, 1985) to be

$$c_m \approx \frac{1}{i\hbar}\int_0^t H'_{mn}e^{i\omega_{mn}t} \tag{6.4}$$

where $H'_{mn}(t) = \int \psi_m^* v(\mathbf{r}, t)\psi_n \, d\tau$ is the integral over volume (with $d\tau$ representing an element of volume), and $\omega_{mn} = (E_m - E_n)/\hbar$. For an interaction between a particle and an electromagnetic field of angular frequency ω, the time-dependent perturbation varies, to a first approximation, sinusoidally with time as

$$v(\mathbf{r}, t) = v'(\mathbf{r}) \cos\omega t. \tag{6.5}$$

Substituting this into the previous equations and defining $H''_{mn} = \int \psi_m^* v'(\mathbf{r})\psi_n \, d\tau$ we find that the probability that the particle will be in state m some time t after being in state n is given by

$$|c_m|^2 \approx \frac{|H''_{mn}|^2}{4\hbar^2} \frac{\sin^2[(\omega_{mn} - \omega)t/2]}{[(\omega_{mn} - \omega)/2]^2}. \tag{6.6}$$

This probability is negligible except when ω is close to ω_{mn}, provided that the time t is long compared with the period of the perturbation, and thus it can be seen that the absorption or emission of a photon of appropriate frequency may change the state of a particle. In reality, various broadening effects such as Doppler broadening mean that there are a number of pairs of energy levels which have an energy difference $\hbar\omega_{mn}$ and thus we need an additional term, known as the density of states $g(\omega_{mn})$, where $g(\omega_{mn})\,d\omega_{mn}$ is defined as the number of pairs of energy levels with energy difference between $\hbar\omega_{mn}$ and $\hbar(\omega_{mn} + d\omega_{mn})$. The total probability for the transition to take place is then

$$P(t) = \frac{|H''_{mn}|^2}{4\hbar^2} \int \frac{\sin^2[(\omega_{mn} - \omega)t/2]}{[(\omega_{mn} - \omega)/2]^2} g(\omega_{mn})\,d\omega_{mn}. \tag{6.7}$$

In general the density of states function $g(\omega_{mn})$ is much more slowly varying than the $\sin^2$ term in Equation (6.7), which may thus be well approximated by

$$P(t) \approx \frac{\pi |H''_{mn}|^2}{2\hbar^2} g(\omega)t. \tag{6.8}$$

Hence the observed *transition rate* W is given by

$$W = \frac{dP}{dt} = \frac{\pi |H''_{mn}|^2}{2\hbar^2} g(\omega). \tag{6.9}$$

This formula is known as Fermi's golden rule.

6.2.2 Electric and magnetic moments

In Section 6.2.1 we considered a general sinusoidal perturbation of the Hamiltonian of a particle introduced by interaction with an electromagnetic wave. To be more specific about spectral transitions we must consider the detailed interaction of an atom or molecule with an electromagnetic wave.

An electromagnetic wave has an electric field whose strength varies with time and position as $\mathbf{E} = \mathbf{E}_0 \cos(\omega t - \mathbf{k}\cdot\mathbf{r})$, where ω is the angular frequency and $\mathbf{k}$ is the wavevector ($|\mathbf{k}| = 2\pi/\lambda$). The wave also has an associated magnetic field $\mathbf{B} = (\mathbf{E}_0/c)\cos(\omega t - \mathbf{k}\cdot\mathbf{r})$ and thus the total force acting on each electron and nucleus in the molecule is given by $\mathbf{F}_i = q_i(\mathbf{E} + \mathbf{v}_i \times \mathbf{B})$, where q_i is the electric charge of the ith particle and $\mathbf{v}_i$ is its velocity. The energy of interaction between the wave and the molecule is then defined as $v(\mathbf{r}, t) = \sum_i \mathbf{F}_i \cdot \mathbf{r}_i$.

Since the electric field strength is greater than the magnetic field strength by a factor of c, and since particle velocities are in general small compared with c, the electric field terms in the perturbation potential tend to completely dominate the magnetic ones. However, since the electric field varies with position, there are different ways in which the electric interaction can take place. The electric field of the wave may be expanded as the following power series

$$\mathbf{E} = \mathbf{E}_0 \cos(\omega t - \mathbf{k}\cdot\mathbf{r}) = \mathbf{E}_0[\cos\omega t + (\mathbf{k}\cdot\mathbf{r})\sin\omega t - \tfrac{1}{2}(\mathbf{k}\cdot\mathbf{r})^2\cos\omega t + \cdots] \quad (6.10)$$

and the molecule may be considered to interact with different terms independently. Interaction with the first term, where the amplitude of the field is constant over the molecule, leads to *electric dipole* transitions. Interaction with the second, much weaker term leads to *electric quadrupole* transitions, interaction with the third even weaker term leads to *electric octopole* transitions, and so on. Although small, magnetic field interactions may be similarly resolved as *magnetic dipole* transitions, *magnetic quadrupole* transitions, etc. For most molecules, electric dipole transitions completely dominate everything else and thus it is these transitions that are usually considered. However, some molecules such as H_2 do not have a dipole moment and thus may not engage in electric dipole transitions although they do have an electric quadrupole moment and hence may engage in electric quadrupole transitions. Such transitions are clearly observable in giant planet spectra. Magnetic transitions are very weak and difficult to observe in giant planet spectra and may be neglected except for the effect of magnetic dipole O_2 absorption in Earth's atmosphere discussed in Section 7.3.1.

6.3 MOLECULAR SPECTROSCOPY: VIBRATIONAL–ROTATIONAL TRANSITIONS

At visible wavelengths, most transitions that are observed in the laboratory are due to electronic transitions within atoms, where electrons move between different energy levels in atoms. However, at infrared (IR) wavelengths, photons have insufficient energy to interact directly with atoms. Instead, at these lower energies molecular absorption mechanisms come into play, which are the core processes in forming the observed IR spectra of planets. Depending on temperature and molecular structure, molecules possess both rotational and vibrational degrees of freedom and absorption of IR photons can promote transitions between these different rotational–vibrational states, as will now be described.

6.3.1 Molecular vibrational energy levels

The bonds between the atoms in a molecule are subject to stretching and (sometimes) bending degrees of freedom that may contribute to IR spectra. The simplest case is for a diatomic molecule composed of atoms of mass m_1 and m_2. To a first approximation we may consider the binding forces between the atoms to be similar to a spring of stiffness k. Such a system, for which the restoring force is proportional to the displacement, will vibrate with simple harmonic motion and we may use quantum mechanics to show that the allowed energy levels of such an oscillator are (e.g., Rae, 1985)

$$E_\nu = (\nu + \tfrac{1}{2})\hbar\sqrt{\frac{k}{\mu}} \tag{6.11}$$

where μ is the reduced mass $\mu = \frac{m_1 m_2}{m_1 + m_2}$; and ν is the vibrational integer quantum number.

6.3.2 Molecular rotational energy levels

In addition to the vibrational degrees of freedom, molecules also have rotational degrees of freedom, which lead to discrete energy levels. Consider a molecule which has moments of inertia I_a, I_b, I_c about its three *principal axes* a, b, c, which are aligned along lines of rotational or reflectional symmetry of the molecule, where the moment of inertia is defined as

$$I = \sum_i m_i r_i^2. \tag{6.12}$$

By convention the axes are ordered such that $I_a \leq I_b \leq I_c$. For the molecule to have a rotational degree of freedom about a certain axis, the moment of inertia must be nonzero. Hence, linear molecules, for which the moment of inertia is zero along one axis, have two degrees of rotational freedom, while nonlinear molecules, for which the moment of inertia is nonzero along all three axes, have three degrees of rotational energy. Depending on their moments of inertia, molecules may be categorized as *linear rotors*, *symmetric rotors* (oblate or prolate), *asymmetric rotors*, and *spherical tops*, as defined in Table 6.1 and discussed below. From quantum mechanics, the square of the total angular momentum is quantized as

$$L^2 = L_a^2 + L_b^2 + L_c^2 = \hbar^2 J(J+1) \tag{6.13}$$

where J is an integer quantum number and the allowed components of angular momentum along a symmetry axis are given by

$$L_\alpha = \hbar K \tag{6.14}$$

where K is a quantum number that satisfies $|K| \leq J$.

Table 6.1. Symmetry classifications of molecules relevant to giant planets.

Name	*Definition*	*Examples*
Linear rotor	$I_a = 0$, $I_b = I_c$	CO_2, C_2H_2 (acetylene)
Symmetric rotor or top	Prolate: $I_a = I_A$, $I_b = I_c = I_B$ Oblate: $I_a = I_C$, $I_a = I_b = I_B$	CH_3C_2H (methyl acetylene), C_2H_6 (ethane), C_2H_4 (ethylene), CH_3D NH_3, AsH_3, PH_3
Asymmetric rotor or top	$I_a < I_b < I_c$	H_2O, O_3, C_3H_8 (propane), H_2S
Spherical top	$I_a = I_b = I_c$	CH_4, GeH_4

NB: Diatomic molecules are by definition linear rotors.

Classical rotational energy is defined as

$$E = \frac{L_a^2}{2I_a} + \frac{L_b^2}{2I_b} + \frac{L_c^2}{2I_c} \tag{6.15}$$

and from this expression and the quantum mechanical expressions for the angular momenta, we may calculate the rotational energy levels for the four different molecule types as follows:

(i) *Linear rotors*. For linear rotors $I_a = 0$, $I_b = I_c = I$, and the energy levels may be found from simple quantum theory to be

$$E_J = \frac{\hbar^2}{2I} J(J+1) \tag{6.16}$$

where, since $-J \leq K \leq J$, the energy levels have a degeneracy of $2J+1$.

(ii) *Spherical tops*. For these molecules $I_a = I_b = I_c = I$, and the energy levels are found to be essentially the same as Equation (6.16), although in this case the degeneracy of the rotational energy levels is found to be partially lifted.

(iii) *Symmetric rotors*. For prolate symmetric rotors where $I_a = I_A$, $I_b = I_c = I_B$, classical rotational energy may be written as

$$E = \frac{L_a^2}{2I_A} + \frac{L_b^2 + L_c^2}{2I_b} = \frac{L^2}{2I_B} + \left(\frac{1}{2I_A} - \frac{1}{2I_B}\right) L_a^2 \tag{6.17}$$

from which the quantum mechanical expression may be derived

$$E_J = \frac{\hbar^2}{2I_B} J(J+1) + \frac{\hbar^2}{2}\left(\frac{1}{I_A} - \frac{1}{I_B}\right) K^2. \tag{6.18}$$

Similarly for the oblate case where $I_a = I_b = I_B$, $I_c = I_C$, the energy levels are found to be

$$E_J = \frac{\hbar^2}{2I_B} J(J+1) + \frac{\hbar^2}{2}\left(\frac{1}{I_C} - \frac{1}{I_B}\right) K^2. \tag{6.19}$$

Hence the rotational energy levels of symmetric rotors are considerably more complex than those of linear rotors and spherical tops since the energies depend on both J and K.

(iv) *Asymmetric rotors*. Although the total angular momentum for molecules of this type is well defined, there is no principal axis along which the component L_α may be defined, which greatly complicates the calculation of energy levels, as is discussed further by Hanel *et al.* (2003). Water is a typical example of an asymmetric rotor, and rotational energy levels are found to be so complex that they have an apparently random distribution of energy levels.

At higher moments of inertia, the bonds between the atoms of all molecular types become stretched due, essentially, to centrifugal forces. This stretching increases the moment of inertia and thus lowers the rotational energy levels. Including centrifugal effects, the rotational energy levels of, for example, a simple linear rotor are modified as

$$E_J = \frac{\hbar^2}{2I} J(J+1) - DJ^2(J+1)^2 \tag{6.20}$$

where D is a small constant. The rotational energy levels of other molecular types are similarly adjusted.

6.3.3 Rotational transitions

The spacing of rotational energy levels is very much less than that between vibrational energy levels. At long wavelengths, electric dipole rotational transitions may be promoted by the absorption or emission of a photon provided, as mentioned earlier, that the molecule has an electric dipole moment defined as

$$\mathbf{M} = \sum_i q_i \mathbf{r}_i. \tag{6.21}$$

For homonuclear diatomic molecules such as H_2, and symmetric linear polyatomic molecules such as CO_2, the electric dipole moment $\mathbf{M}$ is zero and thus pure rotational transitions are forbidden. However, heteronuclear diatomic molecules and all polyatomic molecules other than symmetric linear ones do have an electric dipole moment and thus may absorb photons subject to the selection rule $\Delta J = 1$ and may emit a photon provided $\Delta J = -1$ (Hanel *et al.* 2003). Hence, a molecule in the Jth rotational energy level may only absorb a photon with an energy needed to promote it to the $(J+1)$th energy level. This required energy, known as the *transition energy*, is given by the difference between the $(J+1)$th and Jth rotational energy levels: that is,

$$\Delta E = E_{J+1} - E_J = \frac{\hbar^2}{2I}[(J+1)(J+2) - J(J+1)] = \frac{\hbar^2}{I}(J+1). \tag{6.22}$$

Thus, since the transition energy is proportional to the total angular momentum J of the lower state, the frequencies of the rotational absorption lines are to a first approximation all equally spaced. In reality, effects such as centrifugal distortion lead to rotational energy levels becoming more closely spaced at higher J.

The strength of a rotational absorption line, or the *line strength*, depends both on the transition probability derived earlier from quantum mechanics and also on the number of molecules which are in the Jth rotational energy level at any particular time. For *rotation bands* it is found that the dominant factor affecting line strengths is the population of the lower rotational state. Assuming thermodynamic equilibrium, the population of states varies with energy according to the Boltzmann distribution

$$N = N_0 g_J \exp\left(-\frac{E_J}{k_B T}\right) \tag{6.23}$$

where k_B is the Boltzmann constant; E_J is the energy of the Jth energy level; and g_J is the degeneracy of the Jth level (i.e., the number if individual states with the same energy E_J). Hence, for linear rotors, substituting for the energy E_J from Equation

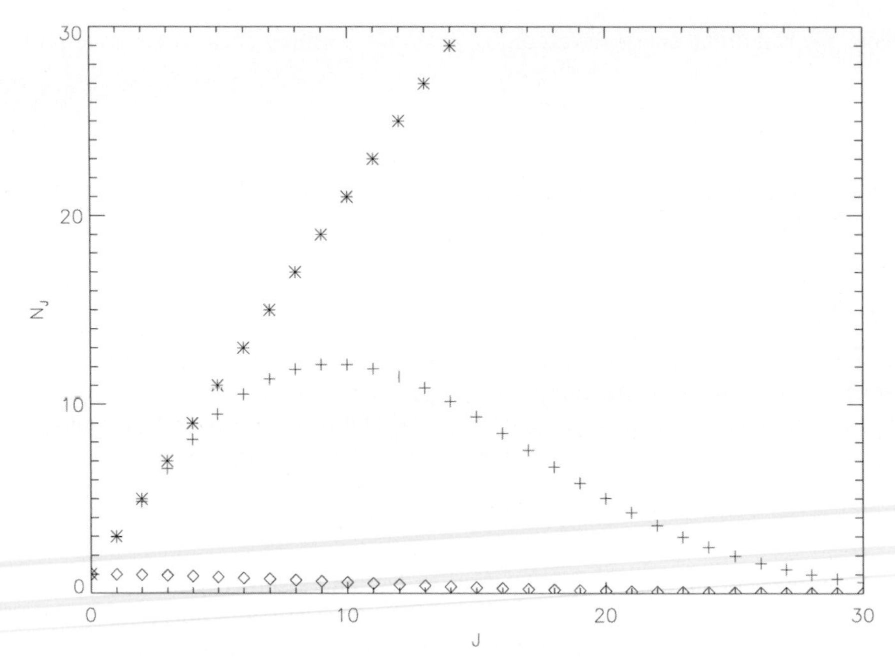

Figure 6.1. Population of rotational energy states. Diamond symbols indicate a typical Boltzmann distribution, while the asterisk symbols indicate the degeneracy $2J + 1$. The product of the two is shown by the cross symbols.

(6.16) and setting the degeneracy g_J equal to $2J + 1$, the number of molecules actually in the K th rotational energy state is given by:

$$N_J = (2J + 1)N_0 \exp(-J(J + 1)\hbar^2/2Ik_BT) \tag{6.24}$$

which has a distribution as shown in Figure 6.1. The measured line strengths of the rotation band of the heteronuclear linear rotor CO are shown in Figure 6.2 and it can be seen that the variation in strength closely resembles the population curve of Figure 6.1 and that the lines are equally spaced as expected.

6.3.4 Vibration–rotation bands

At shorter wavelengths, individual photons carry more energy and may thus excite transitions between vibrational states. The transition rule for molecules changing their vibrational state through an electric dipole transition is simply $\Delta\nu = \pm 1$ (Hanel *et al.*, 2003). Now since the vibrational and rotational degrees of freedom of a molecule are (more or less) independent, at wavelengths where vibrational transitions are excited, they may also be associated with rotational level changes giving rise to combined rotation–vibration spectra. The form of these *vibration–rotation bands*

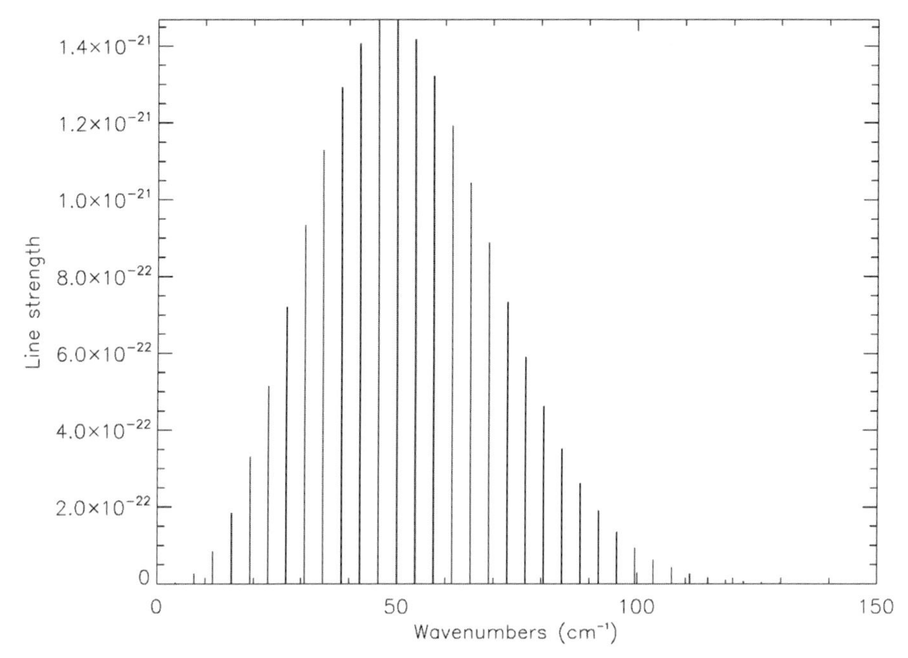

Figure 6.2. Measured line strengths at 296 K in the rotation band of CO (line strengths measured in units of $cm^{-1}(mol\,cm^{-2})^{-1}$).

depends on the symmetry of the molecule as follows:

(i) *Diatomic molecules.* Diatomic molecules have a single vibrational mode and thus a single set of vibration–rotation bands centered at frequencies $n\nu_1$ where n is an integer and ν_1 is the fundamental frequency given (from Equation 6.11) by

$$\nu_1 = \frac{1}{2\pi}\sqrt{\frac{k}{\mu}}. \tag{6.25}$$

Just as for rotational transitions, only heteronuclear diatomic molecules may engage in electric dipole transitions and during each vibrational transition, the rotational energy level may also change, for which the selection rule is $\Delta J = \pm 1$. Hence, each vibration rotation band actually consists of two branches, the "P-branch" at frequencies below the central vibrational frequency for which $\Delta J = -1$ and the "R-branch" at frequencies above the central vibrational frequency for which $\Delta J = +1$. The shape of the R-branch is identical to the pure rotational band since the population and degeneracy of rotational states are not affected by the vibrational state. The shape of the P-branch is the mirror image of the R-branch.

(ii) *Linear polyatomic molecules and spherical tops.* For these molecules, the main vibration–rotation bands are actually composed of *three* bands which again

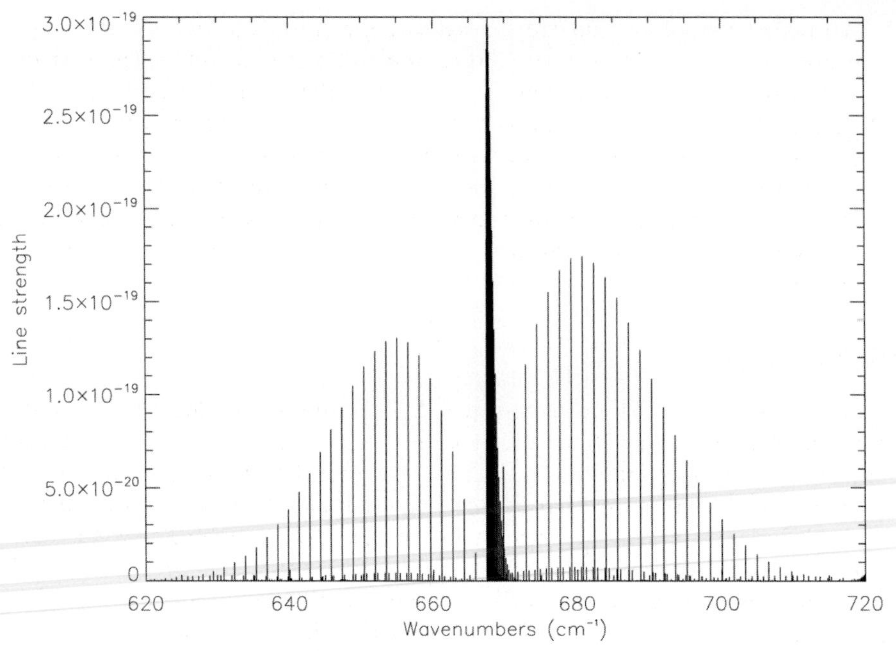

Figure 6.3. Measured line strengths at 296 K in the ν_2 vibration–rotation band of CO_2 (line strengths measured in units of $cm^{-1}(mol\,cm^{-2})^{-1}$).

incorporate possible simultaneous changes in rotational energy levels. Just as for heteronuclear diatomic molecules, rotational energy changes governed by $\Delta J = \pm 1$ give rise to the P-branches and R-branches. However, for linear polyatomic and spherical top molecules, an additional "Q-branch" arises since transitions with $\Delta J = 0$ are also allowed. For an ideal molecule all the possible transitions in the Q-branch would perfectly overlap, but since the rotational transitions are not perfectly equally spaced, a tight collection of transitions are seen at the central vibrational frequency. A good example of a classic P, Q, R vibration–rotation band is the 15 μm ν_2 band of CO_2 shown in Figure 6.3.

(iii) *Symmetric rotors*. The situation for symmetric rotors is a little more complicated since additional structure arises from possible variations in the K angular momentum quantum number. Two types of bands arise. Transitions with $\Delta K = 0$ have a very similar appearance to the classic P, Q, R structure of linear and symmetric top molecules just described and are called *parallel* bands. However, a second type of band arises from transitions for which $\Delta K = \pm 1$, known as *perpendicular* bands. These have the appearance of a number of P, Q, R branches superimposed on each other with a small frequency shift between each central Q-branch giving rise to a regularly spaced series of Q-branches and associated P-branches and R-branches. The vibration–rotation band of ethane at 800 cm^{-1} [NB: wavenumbers are often used to describe frequency in visible/

infrared spectroscopy and are defined as the reciprocal of wavelength, usually expressed in centimeters] is a good example of a perpendicular band which is clearly seen in the thermal emission spectra of the giant planets.

(iv) *Asymmetric rotors.* The vibration–rotation bands of asymmetric rotors are even more complex and appear similar to either the parallel or perpendicular bands of symmetric rotors depending on the values of the three principal moments of inertia. Water molecules are asymmetric rotors and the spectrum of each band is so complex that they have the appearance of a random jumble of line positions and strengths.

Vibrational modes

A molecule composed of N atoms has by definition $3N$ degrees of freedom comprising translation, rotation, and vibration. Of these, three degrees of freedom define the mean translational position of the molecule. In addition, linear molecules have two degrees of rotational freedom and all other molecules have three degrees of rotational freedom. Hence, a linear molecule has $3N - 5$ vibrational degrees of freedom, while a nonlinear molecule has $3N - 6$ vibrational degrees of freedom. Substituting for N it can be seen that a diatomic molecule has a single vibrational degree of freedom, a molecule with three atoms (such as CO_2, or H_2O) has either three or four vibrational modes depending on its linearity, a molecule with four atoms has six or seven vibrational modes, and so on. Clearly the number of possible vibrational modes increases rapidly with the number of atoms, although not all will contribute to the observable vibration–rotation spectra since not all will be able to interact with electric dipole radiation, as we shall now see.

Consider a simple tri-atomic molecule such as CO_2. Since this is a linear molecule, there are $3 \times 3 - 5 = 4$ vibrational modes, which are shown in Figure 6.4. The first mode (ν_1) is called the symmetric stretch mode and the spacing of the vibrational energy levels is equivalent to a photon with a wavenumber of approximately 1,100 cm^{-1}. However, the motion of the oxygen atoms associated with this mode do not change the center of charge of the molecule and thus this mode may not interact with electric dipole radiation. The next two vibrational modes (ν_2) are the bending modes, which have identical frequencies, and clearly vary the center of charge of the molecule giving rise to the fundamental vibration–rotation band centered at 667 cm^{-1} (shown in Figure 6.3). The last mode (ν_3) is the asymmetric stretch mode and may also interact with electric dipole radiation giving rise to the fundamental vibration–rotation band at 2,350 cm^{-1}.

Overtones and hot bands

The vibrational selection rule $\Delta\nu = \pm 1$ for electric dipole transitions is only absolute for a pure simple harmonic oscillator. The binding force between real molecules, while proportional to displacement for small oscillations, has non-negligible higher orders, or anharmonic elements for larger oscillations. These anharmonic elements relax the selection rules to $\Delta\nu = \pm 1, \pm 2, \pm 3, \ldots$ giving rise to "overtone" bands, for

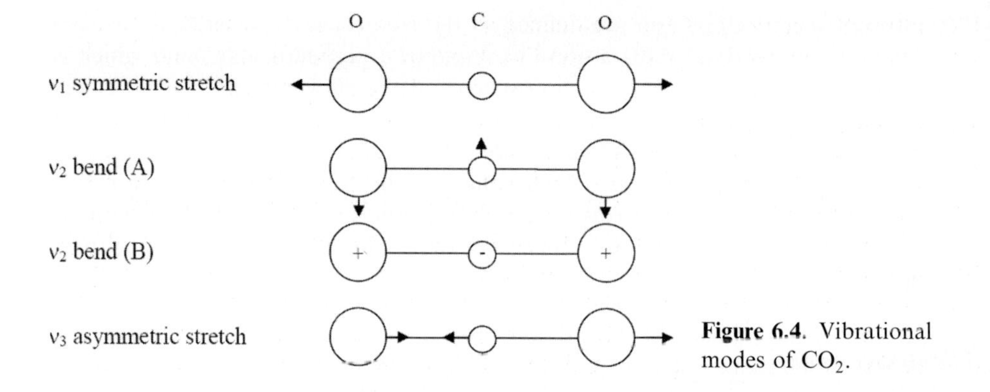

Figure 6.4. Vibrational modes of CO_2.

which $|\Delta\nu| > 1$. Overtone bands are always far less intense than fundamental bands, for which $\Delta\nu = \pm 1$.

The population of vibrational energy states is also covered by a Boltzmann distribution, and thus at low temperatures, most molecules are in their lowest vibrational state and the absorption spectra of low-temperature molecules are dominated by transitions from this ground state. However, as the temperature rises, so does the population of the higher vibrational states giving rise to so-called "hot bands".

6.3.5 Inversion bands and inversion doubling

Another type of band, relevant to microwave spectra, is the inversion band of ammonia. The arrangement of NH_3 has the N atom slightly above or below the plane of the three H atoms and the inversion band arises from symmetry considerations of whether the N atom is above or below this plane. This gives rise to a cluster of absorption lines centered at a wavelength of 1.3 cm, which is the dominant feature of microwave spectra of Jupiter and Saturn, although not Uranus and Neptune, which is how the low abundance of ammonia above ~80 bar in these atmospheres is inferred. In addition to the microwave inversion band, this molecular structure also gives rise to a splitting of the rotational energy levels of ammonia, known as inversion doubling (Herzberg, 1945).

6.3.6 Diatomic homonuclear molecules

Diatomic homonuclear molecules do not interact with electromagnetic radiation via electric dipole transitions at all since no vibrational or rotational change of state may induce an electric dipole in the molecule. Although such molecules may interact weakly with electromagnetic radiation in other ways, they are usually effectively considered to be radiatively inactive in terrestrial atmospheres since their weak absorptions are totally dominated by the electric dipole transitions of other molecules. However, the atmospheres of the giant planets are dominated by molecular hydrogen, which is just such a diatomic homonuclear molecule and in certain parts of the spectrum the more familiar electric dipole transitions of other molecules are themselves so negligible that it becomes necessary to consider in detail how such diatomic molecules may interact with electromagnetic radiation.

Electric quadrupole transitions

We saw earlier that, in addition to electric dipole transitions, there are a number of less likely, and thus weaker interaction mechanisms between a molecule and radiation. While diatomic homonuclear molecules may not engage in electric dipole transitions they may engage in the much weaker electric quadrupole transitions. For these electric quadrupole transitions the selection rules are $\Delta J = \pm 2$, and $\Delta \nu = \pm 1, \pm 2, \pm 3, \ldots$ Thus, instead of having a P, Q, R-branch structure, electric quadrupole transitions are confined to the O-branch ($\Delta J = -2$) and S-branch ($\Delta J = 2$).

In the atmospheres of the giant planets, which are dominated by molecular hydrogen, the electric quadrupole lines of H_2 are detectable for a number of vibrational transitions. In the far-infrared the S(0) and S(1) lines associated with the (1–0) vibrational transitions are observable where "S" means $\Delta J = 2$, and the numbers refer to the total angular momentum J of the lower state. At visible wavelengths the S(0) and S(1) transitions associated with the (3–0) and (4–0) vibrational energy changes are observable. Electric quadrupole transitions have similar linewidths to the electric dipole transitions described later in Section 6.3.7.

Collision-induced dipole transitions

In the gas phase, collisions and interactions with other molecules can lead to transitory dipole moments being induced on homonuclear diatomic molecules that may then interact with IR light. Although the absorption is weak, the abundance of H_2 in the giant planet atmospheres is so high that the far-IR spectra of the giant planets is dominated by H_2–H_2 and H_2–He pressure-induced or collision-induced absorptions (CIA). Another example is N_2–N_2 and N_2–CH_4 collision-induced absorption in Titan's atmosphere. These pressure-induced dipole absorptions have $\Delta J = 2$ (i.e., S(0), S(1)) and are thus found near the wavelengths of the pure H_2 quadrupole lines just discussed. However, temporarily induced dipoles have very short lifetimes, and thus from Heisenberg's uncertainty principle their line shape is extremely broad. For more detailed information on H_2–H_2 and H_2–He collision-induced absorptions, the reader is referred to Birnbaum *et al.* (1996).

6.3.7 Line broadening

We have now discussed a number of different absorption mechanisms of molecules and have also outlined that the strength of an absorption line depends both on the transition probability and on the population of the lower energy level. However, these absorption lines are not infinitely thin, but instead have a finite width as we have alluded to previously and which arise from a number of possible mechanisms that we shall now describe.

Natural broadening

In any transition, there is a natural linewidth that arises due to the finite time over which a photon is absorbed or emitted by an atom or molecule. From Heisenberg's uncertainty principal we know that $\Delta E\, \Delta t \approx h$, or $\Delta \nu\, \Delta t \approx 1$ since $E = h\nu$. This may be further refined to give the width of an absorption line in terms of wavenumbers (cm^{-1})

$$\Delta \tilde{\nu} \approx \frac{1}{c\Delta t}. \tag{6.26}$$

In practice natural broadening is usually negligible compared with the following mechanisms. As an order of magnitude, the natural-broadened linewidth is of the order of $10^{-7}\ \text{cm}^{-1}$ (Hanel *et al.*, 2003).

Collision broadening or pressure broadening

In a gas, molecules suffer repeated collisions with other molecules and there is a certain probability that such a collision may occur while molecules are absorbing or emitting photons. This effectively shortens the length of the absorbed or emitted wavetrains, and thus from the uncertainty principle increases the spread of wavelengths. This process is known as *collision broadening* or sometimes *pressure broadening* since this effect becomes dominant at high pressures. The absorption coefficient $k_{\tilde{\nu}}$ (usually defined in units of molecules per square meter) at wavenumber $\tilde{\nu}$ due to a collision-broadened line centered at $\tilde{\nu}_0$ is given by the Lorentz lineshape

$$k_{\tilde{\nu}} = \frac{s\gamma_L}{\pi((\tilde{\nu} - \tilde{\nu}_0)^2 + \gamma_L^2)} \tag{6.27}$$

where $s = \int_0^\infty k_{\tilde{\nu}}\, d\tilde{\nu}$ is the line strength and $\gamma_L = (2\pi tc)^{-1}$ is the linewidth (cm^{-1}). The parameter t is the mean time between collisions, which depends on density and thus for atmospheres mostly on pressure. We may thus rewrite γ as

$$\gamma_L = \gamma_{L_0} \frac{p}{p_0} \left(\frac{T_0}{T}\right)^n \tag{6.28}$$

where γ_{L_0} is the linewidth at a reference pressure of p_0. The temperature coefficient n is by simple theory equal to 0.5, although in reality it varies slightly from molecule to molecule. At a pressure of 1 bar, and room temperature, a typical collision-broadened linewidth is of the order of $0.1\ \text{cm}^{-1}$.

Doppler broadening

This arises due the line-of-sight motion of the emitting/absorbing molecules in the gas, which is due to the molecules moving with a Maxwell–Boltzmann distribution of speeds. This distribution depends upon the temperature of the gas and the molecular weight of the molecules. Molecules approaching the observer will absorb at slightly higher frequencies than receding molecules due to the Doppler effect and the absorption spectrum of a Doppler-broadened line is given by

$$k_{\tilde{\nu}} = \frac{s}{\gamma_D \pi^{1/2}} \exp\left(-\left(\frac{\tilde{\nu} - \tilde{\nu}_0}{\gamma_D}\right)^2\right) \tag{6.29}$$

where

$$\gamma_D = \frac{\tilde{\nu}_0}{c}\left(\frac{2RT}{M_r}\right)^{1/2} = \gamma_{D_0}\left(\frac{T}{T_0}\right)^{1/2} \tag{6.30}$$

is the Doppler linewidth; M_r is the molecular weight; and γ_{D_0} is the Doppler linewidth at a reference temperature. For example, for methane, for which $M_r = 16g$, the Doppler width at room temperature (293 K) and a wavenumber of 1,300 cm^{-1} is 0.002 cm^{-1}.

Voigt broadening

Considering the previous two mechanisms it is found that pressure broadening dominates at high pressures in an atmosphere, whereas Doppler broadening dominates at low pressures since pressure broadening is directly proportional to pressure that falls exponentially with height in an atmosphere, while Doppler broadening is proportional to $\sqrt{T}$ which decreases much less rapidly. At intermediate temperatures and pressures both mechanisms are significant and thus the lineshape of an observed absorption line is due to a combination of both pressure broadening and Doppler broadening giving rise to the Voigt lineshape

$$k_{\tilde{\nu}} = \frac{sy}{\gamma_D \pi^{3/2}} \int_{-\infty}^{\infty} \frac{\exp(-t^2)}{(x-t)^2 + y^2}\, dt \tag{6.31}$$

where $x = (\tilde{\nu} - \tilde{\nu}_0)/\gamma_D$ and $y = \gamma_L/\gamma_D$. Unfortunately this equation does not have an analytical solution and so must be integrated numerically.

6.3.8 Giant planet gas transmission spectra

The absorption of gases in planetary atmospheres has been seen to be due to both vibration–rotation bands and collision-induced absorptions, and to demonstrate the absorptions of different gases Figure 6.5 shows the transmission of a typical path in an atmosphere of approximately solar composition between 0 cm^{-1} and 2,500 cm^{-1}. Here we can see the basic properties outlined above. The various vibration–rotation bands, and pure rotation bands if allowed, can be seen for the main gases of interest in the tropospheres of the giant planets. Together with the line spectra, the importance of H_2–H_2 and H_2–He collision-induced absorption at long wavelengths is clearly seen. Figure 6.5 also shows similar transmission spectra for various hydrocarbons observed in the giant planet stratospheres.

At shorter wavelengths, Figure 6.6 shows the calculated transmission between 0.4 μm and 5.5 μm for the same solar composition path used in Figure 6.5, where further vibration–rotation bands can be seen. The absorption of methane, ammonia, and again H_2–H_2 and H_2–He CIA are most important in this region. Although water vapor also has strong absorptions, the abundance of water vapor at the cloud tops of

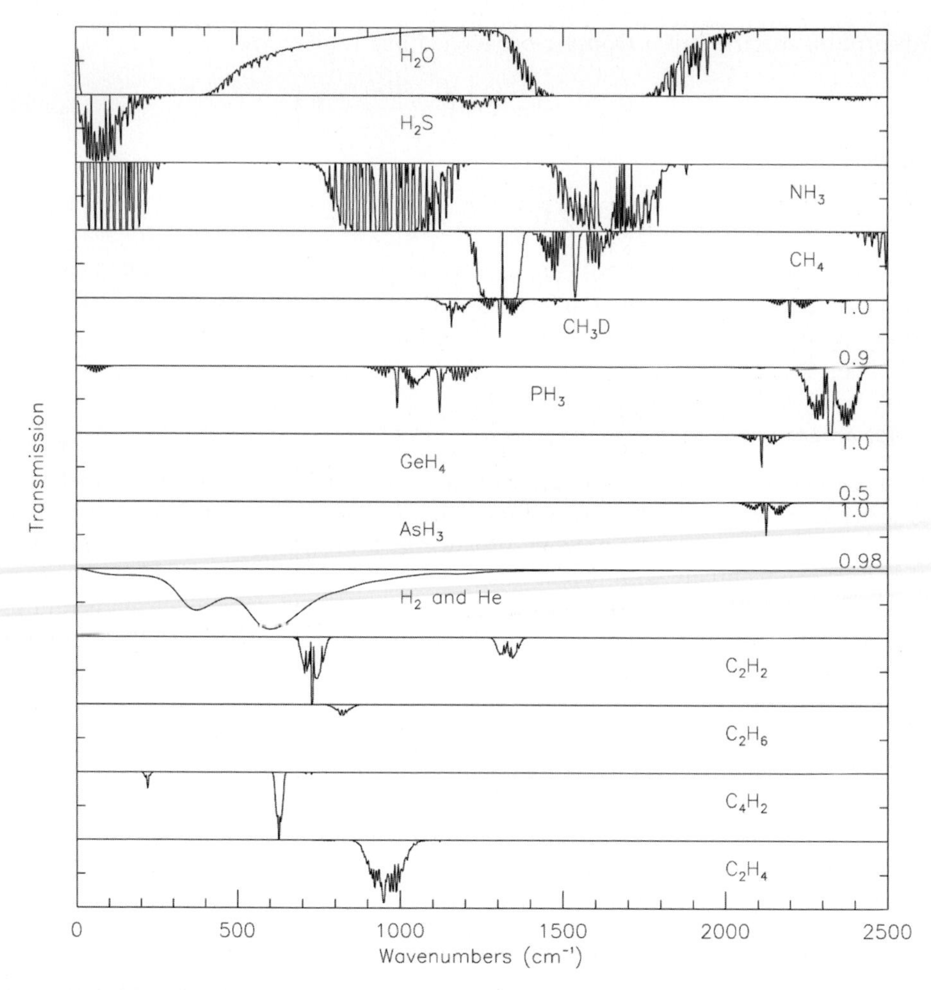

Figure 6.5. Mid-IR to far-IR transmission of tropospheric and stratospheric gases for a solar composition path with $p = 0.3$ bar, $T = 127$ K, and pathlength = 10 km. Calculated transmissions are between 0 and 1 unless indicated otherwise. [NB: The stratospheric hydrocarbon abundances have been increased to give clear absorption spectra.]

the giant planets is so low that water vapor is not detectable at these wavelengths, except for Jupiter.

6.4 RADIATIVE TRANSFER IN A GRAY ATMOSPHERE

Now we have discussed the mechanisms by which gases may absorb IR radiation we will consider how the thermal emission spectra of planets are generated.

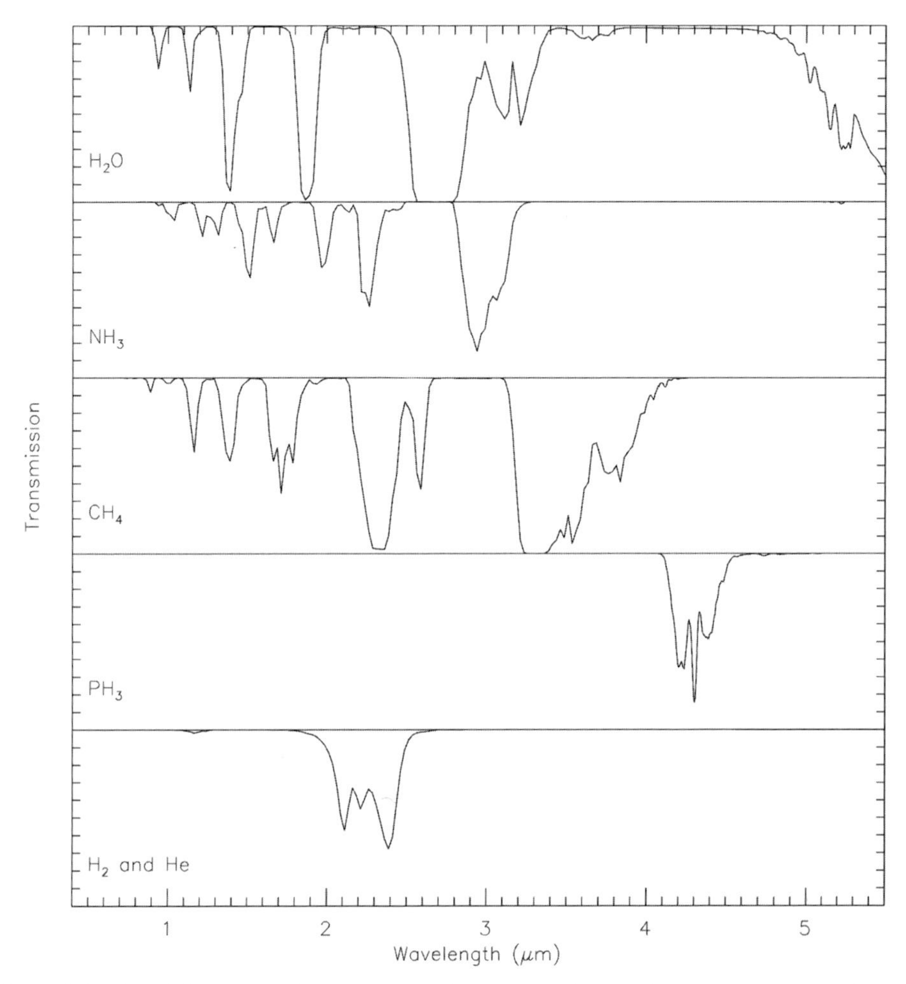

Figure 6.6. Near-IR transmission of tropospheric gases for a solar composition path with $p = 0.3$ bar, $T = 127$ K, and pathlength $= 10$ km.

6.4.1 Nadir viewing

When considering radiation that is emitted to space at angles reasonably close to the vertical, it is a very good approximation to consider the atmosphere as being effectively plane-parallel. This approximation breaks down at very high emission angles such as when the limb of the planet is observed (but even then the radiative transfer equations turn out to be rather similar). Consider such a plane-parallel atmosphere as shown in Figure 6.7. Lambert's law of absorption states that at wavenumber $\tilde{\nu}$ the absorption of radiation of *spectral radiance* $I_{\tilde{\nu}}$ (typically measured in units of W cm^{-2} sr^{-1} (cm^{-1})$^{-1}$) traveling at a *zenith angle* θ to the vertical at altitude z through a path of vertical thickness dz (meters), mean number density n (molecules per cubic meter)

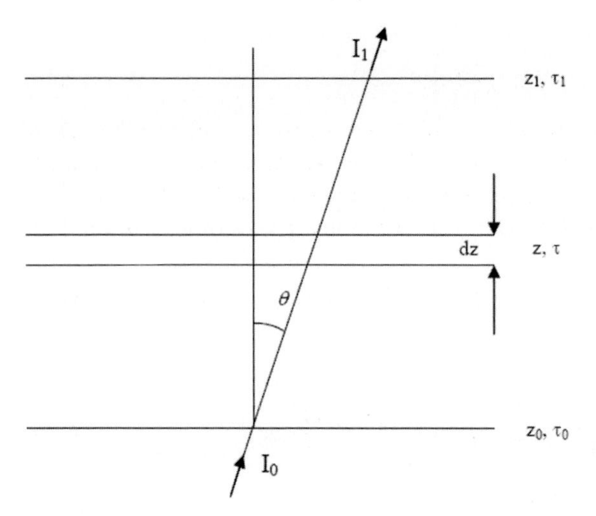

Figure 6.7. Radiative transfer in a gray plane-parallel atmosphere.

containing material with absorption coefficient $k_{\tilde{\nu}}$ (square meters per molecule) is given by

$$dI_{\tilde{\nu}} = -Ik_{\tilde{\nu}}n\,dz/\cos\theta. \tag{6.32}$$

In other words the layer element, of thickness dz, has absorption along the slant path equal to $\alpha = k_{\tilde{\nu}}n\,dz/\mu$, where $\mu = \cos\theta$. Consider a thick slab of atmosphere between levels z_0 and z_1, with spectral radiance $I_{\tilde{\nu}_0}$ incident upwards at z_0 at an angle θ. This radiation will be attenuated before reaching z_1 and thus upward radiation at the top of the atmosphere from the base is given by

$$I_{\tilde{\nu}_1} = I_{\tilde{\nu}_0} \exp\left(-\int_{z_0}^{z_1} \frac{k_{\tilde{\nu}}(z)n(z)}{\mu}\,dz\right) \tag{6.33}$$

which may be re-expressed as

$$I_{\tilde{\nu}_1} = I_{\tilde{\nu}_0} \exp(-\tau_{\tilde{\nu}}(z_0, z_1)/\mu) = I_{\tilde{\nu}_0} T_{\tilde{\nu}}(\mu, z_0, z_1) \tag{6.34}$$

where $\tau_{\tilde{\nu}}(z_0, z_1)$ is the *optical thickness* at wavenumber $\tilde{\nu}$ (or *optical depth* if measured downwards from the top of the atmosphere) between z_0 and z_1 for a vertical path and $T_{\tilde{\nu}}(\mu, z_0, z_1)$ is the transmission from z_0 to z_1 at angle θ.

Now the thin slab at altitude z of thickness dz will also emit thermal radiation and from Kirchoff's law the emissivity of the layer is equal to its absorptivity. Hence, the spectral radiance emitted by this layer at angle θ is equal to

$$dI_{\tilde{\nu}} = k_{\tilde{\nu}}(z)n(z)B_{\tilde{\nu}}(z)\,dz/\mu, \tag{6.35}$$

where $B_{\tilde{\nu}}(z)$ is the Planck function at wavenumber $\tilde{\nu}$ and altitude z, where the temperature is $T(z)$. This spectral radiance is itself attenuated by overlying layers before reaching the top of the atmosphere and thus the contribution to $I_{\tilde{\nu}_1}$ from this slab is

$$dI_{\tilde{\nu}_1} = \frac{k_{\tilde{\nu}}n(z)B_{\tilde{\nu}}(z)\,dz}{\mu} \exp\left(-\int_z^{z_1} k_{\tilde{\nu}}(z)n(z)\,dz'/\mu\right) \tag{6.36}$$

or

$$dI_{\tilde{\nu}_1} = B_{\tilde{\nu}}(z)\, dT_{\tilde{\nu}}(\mu, z, z_1) \tag{6.37}$$

where transmission is measured upwards from lower altitudes and thus increases from zero for z very small to 1.0 at the top of the atmosphere. Thus, summing the thermal emission from all the layers the total spectral radiance reaching the top of the atmosphere at z_1 (where the optical depth is zero and thus the transmission is one) is

$$I_{\tilde{\nu}_1}(\mu) = I_{\tilde{\nu}_0} T_{\tilde{\nu}}(\mu, z_0, z_1) + \int_{T_{\tilde{\nu}}(z_0, z_1)}^{1} B_{\tilde{\nu}}(z)\, dT_{\tilde{\nu}}(\mu, z, z_1) \tag{6.38}$$

or alternatively

$$I_{\tilde{\nu}_1}(\mu) = I_{\tilde{\nu}_0} T_{\tilde{\nu}}(\mu, z_0, z_1) + \int_{z_0}^{z_1} B_{\tilde{\nu}}(z) \frac{dT_{\tilde{\nu}}}{dz}\, dz \tag{6.39}$$

which can be written as

$$\begin{aligned} I_{\tilde{\nu}_1}(\mu) &= I_{\tilde{\nu}_0} T_{\tilde{\nu}}(\mu, z_0, z_1) + \int_{z_0}^{z_1} B_{\tilde{\nu}}(z) K_{\tilde{\nu}}(z)\, dz \\ &= I_{\tilde{\nu}_0} T_{\tilde{\nu}}(\mu, z_0, z_1) + \int_{z_0}^{z_1} C_{\tilde{\nu}}(z)\, dz \end{aligned} \tag{6.40}$$

where $K_{\tilde{\nu}}(z) = \dfrac{dT_{\tilde{\nu}}}{dz}$ is known as the *transmission weighting function*, and $C_{\tilde{\nu}}(z) = B_{\tilde{\nu}}(z) K_{\tilde{\nu}}(z)$ is commonly known as the *contribution function*. The transmission weighting function for nadir viewing (for which the zenith angle $\theta = 0$) is a smoothly varying function as can be seen from Figure 6.8, which shows the calculated weighting function at 600 cm^{-1} and 1,300 cm^{-1} for Jupiter's atmosphere. Also shown in Figure 6.8 is the variation of transmission with height and it can be seen that the weighting function peaks roughly where the optical depth is unity, or equivalently where the transmission to space is 0.368. Since transmission of the atmospheric gases varies with wavelength, as was shown in Figures 6.5 and 6.6, the altitude of the peak of the weighting function varies correspondingly. Hence, in spectral regions of high absorption (such as at 1,300 cm^{-1} in Figure 6.8), the weighting function peaks high in the atmosphere and thus most of the radiation observed is emitted from high levels, whereas in spectral regions of low absorption (such as at 600 cm^{-1}), the weighting functions peak at low altitudes and thus most of the radiation comes from deep levels. Figure 6.9 shows how the altitude of the peak of the calculated weighting function varies with wavelength for the Jovian atmosphere, assuming no clouds. At 2,000 cm^{-1} (or 5 μm) gas absorption is particularly low and thus radiation is mostly emitted from the deep troposphere. At 1,300 cm^{-1}, in the middle of a strong CH_4 absorption, most of the radiation is emitted from the stratosphere. Similar variation of the peak of the weighting function with wavelength is found for the other giant planets. Clearly if we have good models for the absorption spectra of gases we may use the observed thermal emission spectra to infer the variation of both temperature and composition with height in the giant planet atmospheres.

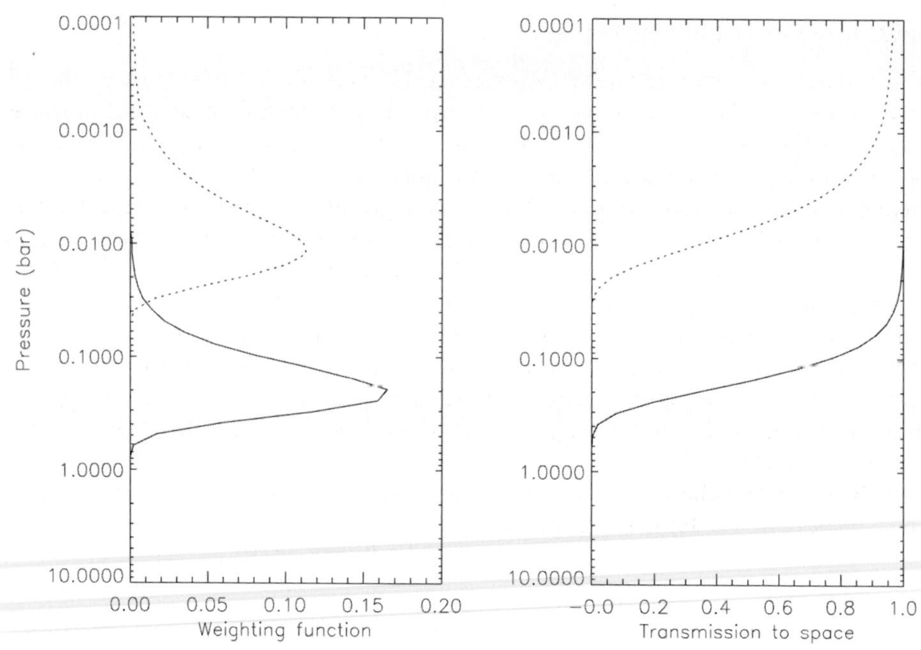

Figure 6.8. Calculated transmission weighting functions (nadir) for Jupiter at 600 cm^{-1} (solid line) and 1,300 cm^{-1} (dotted line).

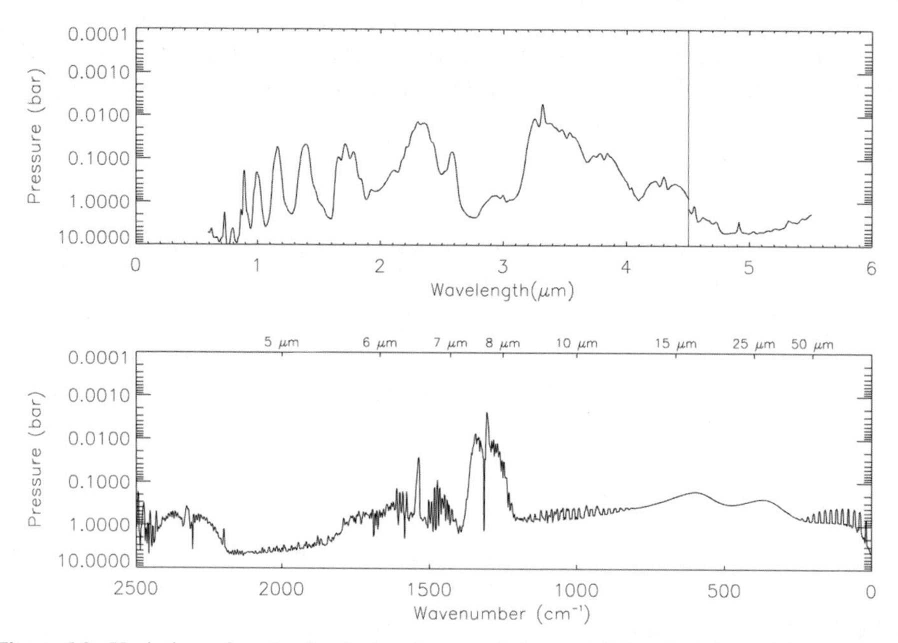

Figure 6.9. Variation of peak of calculated transmission weighting function with wavelength for Jupiter. Below 4.5 μm, the pressure level where the transmission to *and from* space is 0.5 has been plotted.

6.4.2 Net flux and disk averaging

The radiative transfer equation (Equation 6.38) very successfully models the IR observations of the giant planets made by passing spacecraft instruments such as *Voyager* IRIS which have a small field of view and thus sample only a small range of positions and zenith angles. Ground-based observations of the planets historically have poorer spatial resolution and thus often (especially for planets such as Uranus and Neptune which have a small angular size) the spectrum measured is that averaged over the whole planet and thus samples all zenith angles between 0° and 90°. If we assume that the temperature and composition profiles are the same all over the planet, then the disk-averaged spectrum is directly proportional to the averaged spectrum emitted by a single location on the planet into a hemisphere. Hence, to calculate the disk-averaged spectrum, all we need to do is integrate Equation (6.38) over all zenith angles and then scale this appropriately to get the disk-averaged spectral radiance.

The spectral radiance emitted upwards at the top of the atmosphere at zenith angle θ (Equation 6.38) may be rewritten as

$$I_{\tilde{\nu}_1}(\mu) = I_{\tilde{\nu}_0} e^{-\tau_0/\mu} + \int_0^{\tau_0} B_{\tilde{\nu}}(\tau) e^{-\tau/\mu} \frac{d\tau}{\mu} \tag{6.41}$$

where the vertical coordinate is now the optical depth of the atmosphere τ. This expression is valid for $0 < \mu \leq 1$ and gives us the spectral radiance emitted in any given direction. To calculate the total spectral *irradiance* (or *flux*) leaving the atmosphere, we need to integrate Equation (6.41) over the solid angle of a hemisphere. The total upward spectral flux ($\mathrm{W\,m^{-2}\,(cm^{-1})^{-1}}$) is then given by

$$F_{\tilde{\nu}_1} = \int I_{\tilde{\nu}_1} \mu \, d\Omega = 2\pi \int_0^1 I_{\tilde{\nu}_1}(\mu) \mu \, d\mu \tag{6.42}$$

where $d\Omega = 2\pi \sin\theta \, d\theta = -2\pi \, d\mu$ is the element of solid angle and where the extra factor of μ arises since it is assumed that the emitting surfaces are horizontal. Substituting for $I_{\tilde{\nu}_1}(\mu)$ in Equation (6.42) we find

$$F_{\tilde{\nu}_1} = 2\pi I_{\tilde{\nu}_0} \int_0^1 e^{-\tau_0/\mu} \mu \, d\mu + 2\pi \int_0^{\tau_0} B_{\tilde{\nu}}(\tau) \left(\int_0^1 e^{-\tau/\mu} \, d\mu \right) d\tau \tag{6.43}$$

or

$$F_{\tilde{\nu}_1} = 2\pi I_{\tilde{\nu}_0} E_3(\tau_0) + 2\pi \int_0^{\tau_0} B_{\tilde{\nu}}(\tau) E_2(\tau) \, d\tau \tag{6.44}$$

where $E_n(x) = \int_1^{\infty} \frac{e^{-wx}}{w^n} dw = \int_0^1 \mu^{n-2} e^{-x/\mu} \, d\mu$, known as an *exponential integral*, is a standard tabulated integral (e.g., Goody and Yung, 1989), which has a number of properties including $E_n'(x) = -E_{n-1}(x)$. Using this last property the total flux

(Equation 6.44) may be re-expressed as

$$F_{\tilde{\nu}_1} = 2\pi I_{\tilde{\nu}_0} E_3(\tau_0) + 2\pi \int_{E_3(\tau_0)}^{0.5} B_{\tilde{\nu}}(\tau)\, dE_3(\tau) \tag{6.45}$$

where we have substituted $E_3(0) = 0.5$. The total spectral power emitted by a planet is then $4\pi R^2 F_{\tilde{\nu}_1}$, where R is the radius of the planet and hence the measured spectral flux or irradiance of the disk-averaged spectrum seen at a distance D is given by

$$I_{\tilde{\nu}} = F_{\tilde{\nu}_1} \left(\frac{R}{D}\right)^2. \tag{6.46}$$

The spectral irradiance may be measured in units such as $\mathrm{W\,cm^{-2}\,(cm^{-1})^{-1}}$, $\mathrm{W\,m^{-2}\,\mu m^{-1}}$, or sometimes in *Janskys* (for Earth-based observations), which have units of $1\ \mathrm{Jy} = 10^{-26}\ \mathrm{W\,m^{-2}\,Hz^{-1}}$. To obtain the disk-averaged spectral radiance of a planet, we then divide this spectral irradiance by the solid angle projected by the planet which is equal to $\pi(R/D)^2$. Hence, the disk-averaged spectral radiance is given by

$$\bar{I}_{\tilde{\nu}_1} = 2 I_{\tilde{\nu}_0} E_3(\tau_0) + 2 \int_{E_3(\tau_0)}^{0.5} B_{\tilde{\nu}}(\tau)\, dE_3(\tau) \tag{6.47}$$

which is similar to the nadir radiative transfer equation, Equation (6.38). In Figure 6.10 the nadir transmission and transmission weighting function is compared with

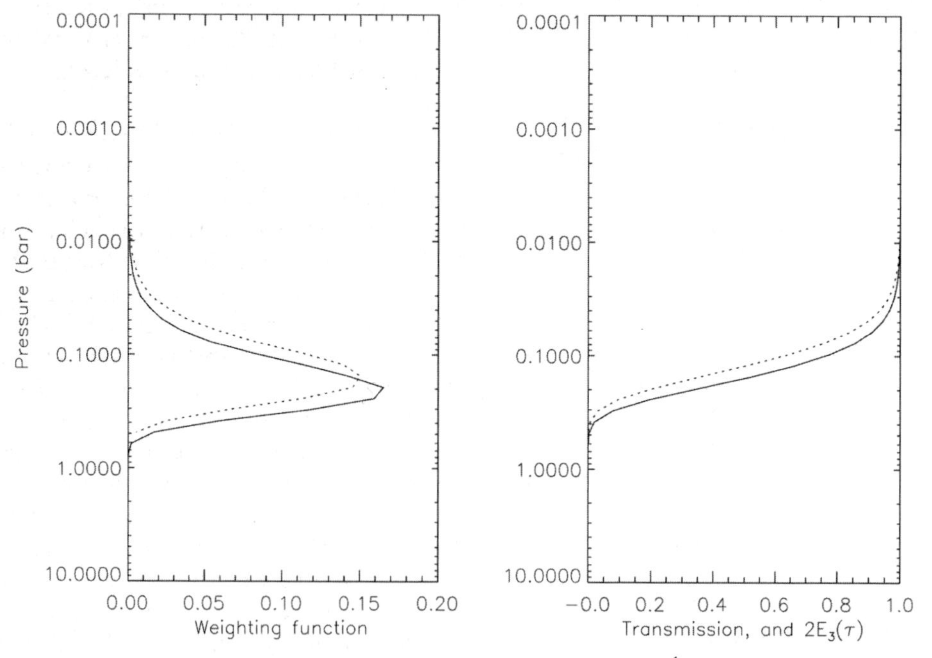

Figure 6.10. Calculated weighting functions for Jupiter at $600\,\mathrm{cm^{-1}}$ for nadir-viewing (solid line) and for disk-averaged (dotted line) conditions.

$2E_3(\tau)$ and $2dE_3(\tau)/dz$. As can be seen the curves are very similar with the only real difference being that the disk-averaged weighting function peaks at slightly higher altitudes, roughly at the altitude of the transmission weighting function calculated for a zenith angle of $\sim 50°$, but is slightly broader.

6.4.3 Limb viewing

We saw in Section 6.4.1 that for nadir, or near-nadir, viewing geometries the weighting functions are rather broad. This low vertical resolution can be a problem when the spectra are used to assess the vertical profiles of gases whose abundances vary rapidly with altitude. One way to increase the vertical resolution is to view the limb of the planet. The radiative transfer equation is very similar to Equation (6.40), but instead of integrating over height we instead integrate directly along the path l

$$I_{\tilde{\nu}_1} = \int_{l_0}^{l_1} B_{\tilde{\nu}}(l)\frac{dT_{\tilde{\nu}}}{dl}\,dl = \int_{z_0}^{z_1} B_{\tilde{\nu}}(z)K'_{\tilde{\nu}}\,dz \tag{6.48}$$

where

$$K'_{\tilde{\nu}} = \frac{dT_{\tilde{\nu}}}{dl}\frac{dl}{dz}. \tag{6.49}$$

Since density varies exponentially with height in an atmosphere, the density of the slant path is highest near the tangent altitude, and thus at wavelengths where the gas absorption is low the transmission weighting function is an exponential shape function with a sharp base at the tangent altitude z_0. As gas absorption becomes stronger, more and more emission comes from molecules between the observer and the tangent point and the weighting function becomes broader. Example limb-weighting functions are shown for Jupiter at 600 cm^{-1} in Figure 6.11. At a tangent pressure of $\sim$10 mbar, the weighting function can be seen to be very narrow with a sharp lower boundary. However, as the tangent pressure increases, and thus the absorption increases, the weighting function becomes broader, as can be seen in the second weighting function of Figure 6.11, where the pressure at the tangent altitude is $\sim$0.38 bar.

In addition to increased vertical resolution, limb sounding also provides greater sensitivity to the detection of trace atmospheric constituents by providing much longer pathlengths. For an atmosphere whose number density varies with height as $n = n_0 \exp(-(z - z_0)/H)$ above some reference altitude z_0, where H is the scale height, it may easily be shown that the column amount (i.e., molecules per square meter) of molecules above this altitude for nadir viewing is $A_N = n_0 H$. However, if limb viewing the atmosphere with a tangent altitude at z_0, the total amount of molecules in the path may be shown to be well approximated by $A_L = n_0(2\pi RH)^{1/2}$, where R is the planetary radius at the tangent altitude z_0. Since the radius of all planets is very large compared with H, the limb path contains much greater amounts, and thus absorption features of trace constituents are much easier to observe in limb paths than nadir ones. For the giant planets, the increase in path amount A_L/A_N is of the order of 100.

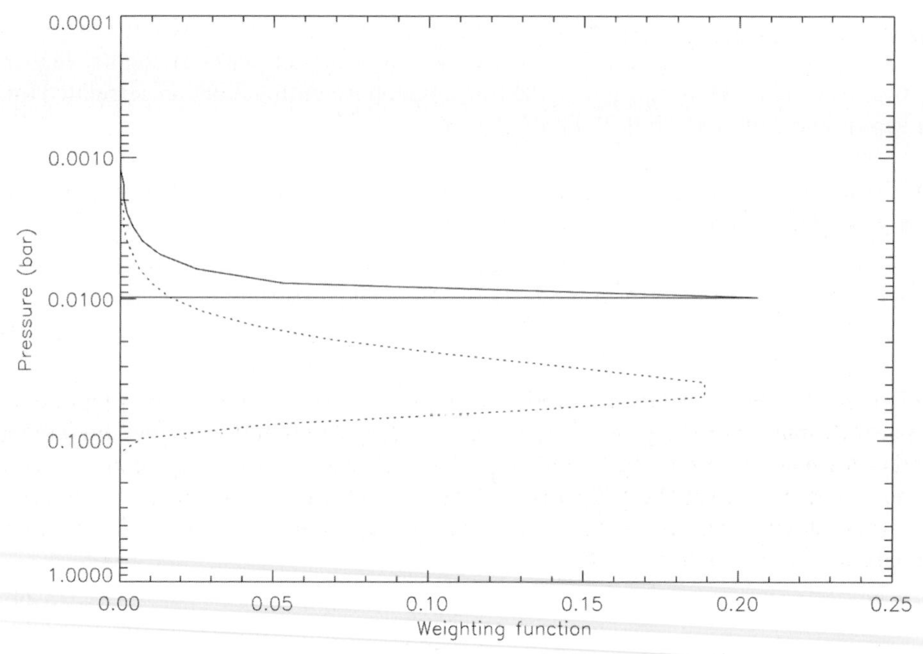

Figure 6.11. Calculated limb weighting functions for Jupiter at 600 cm^{-1} and tangent height pressures of 9.7 mbar (solid lines) and 387 mbar (dotted line).

The disadvantage of limb viewing it that it places considerably tighter constraints on telescope field-of-view and pointing accuracy and is hence much more difficult to achieve in practice. However, the possible improvement in vertical resolution and sensitivity to trace species makes it a very attractive technique and it is widely used in terrestrial remote sensing. The field of view of the *Voyager* IRIS spectrometers was too broad to allow limb sounding of the giant planets, although it was achieved for Titan. The *Cassini* CIRS instrument, however, was specifically designed with limb viewing in mind and has been able to make limb-sounding observations of both Titan and Saturn since it arrived in the Saturnian system in 2004.

6.4.4 Radiative balance

The balance between radiative heating and cooling affects the temperature profile in upper tropospheres and stratospheres of the giant planets and thus affects the circulation of air at these altitudes. At an altitude z, or optical depth τ, in the atmosphere, the upward spectral flux of radiation is from Equation (6.45) equal to

$$F_{\tilde{\nu}}(\uparrow) = 2\pi I_{\tilde{\nu}_0} E_3(\tau_0 - \tau) + 2\pi \int_{E_3(\tau_0-\tau)}^{0.5} B_{\tilde{\nu}}(\tau')\, dE_3(\tau' - \tau) \tag{6.50}$$

and the downward spectral flux is

$$F_{\tilde{\nu}}(\downarrow) = 2\pi \int_{E_3(\tau)}^{0.5} B_{\tilde{\nu}}(\tau')\, dE_3(\tau - \tau'). \tag{6.51}$$

If the fluxes are then integrated over all wavelengths, the difference between the two may be used to calculate the heating rate as

$$\rho c_p \frac{dT}{dt} = \frac{d\left(\int_0^\infty (F_{\tilde{\nu}}(\uparrow) - F_{\tilde{\nu}}(\downarrow))\, d\tilde{\nu}\right)}{dz} \tag{6.52}$$

where ρ is the mass density; and c_p is the specific heat capacity at constant pressure. In the stratospheres of the giant planets it is of particular interest to know how long it takes for temperature perturbations introduced by effects such as dynamics to relax back to zero (i.e., for the temperature profile to return to its equilibrium radiative balance state). Near the emitting level, a simple estimate of the radiative time constant is (Allison *et al.*, 1991)

$$t_R \approx \frac{c_p p T}{g \sigma T_e^4} \tag{6.53}$$

where T_e is the effective emitting temperature. This expression is basically the heat stored per unit area of the atmosphere above the level at pressure p divided by the outgoing thermal radiation flux. Once an estimate for the radiative time constant has been derived, the local heating rates $\dot{Q}$ implied by local departures of temperature from equilibrium can be estimated from (e.g., Conrath *et al.*, 1990)

$$\frac{\dot{Q}}{c_p T} = \frac{1}{T}\frac{T - T_{\text{eqm}}}{t_R}, \tag{6.54}$$

where T_{eqm} is the local radiative equilibrium temperature.

6.4.5 Local thermodynamic equilibrium

For radiative transfer calculations, we normally assume that the atmosphere is in local thermodynamic equilibrium (LTE). This means that we can use the Planck function $B_{\tilde{\nu}}(z)$, as we have done in our previous equations. However, at very low pressures the time between collisions becomes equal to, or greater than, the time of interaction of the molecules with photons, and thus the population density of states deviates from a simple Boltzmann distribution (Goody and Yung, 1989). Under these conditions of non-LTE, the calculation of the *source function*, used instead of the Planck function in the previous radiative transfer equations, becomes very complicated and is a whole research area in its own right (e.g., Lopez-Puertas and Taylor, 2001).

6.4.6 Transmission calculations

Given the absorption coefficients of the various gases and aerosols in a planetary atmosphere, it can be seen that the emerging radiance for a real atmosphere may be calculated. Since the line strengths and linewidths are functions of temperature and pressure, and since atmospheres are extremely inhomogeneous in both respects, the monochromatic transmission at wavenumber $\tilde{\nu}$ of a path through an atmosphere between two levels z_1 and z_2 at zenith angle θ is given by

$$T_{12}(\tilde{\nu}) = \exp\left(-\frac{1}{\mu}\int_{z_1}^{z_2} n(z)\left(\sum_j q_j(z)\sum_i k_{ij}(\tilde{\nu}, p(z), T(z))\right) dz\right) \tag{6.55}$$

where $n(z)$ is the number density of the atmosphere at altitude z (molecules per cubic meter); $p(z)$ and $T(z)$ are the atmospheric pressure and temperatures; $q_j(z)$ is the mole fraction of gas j at altitude z; and the summation over i is over all the absorption lines of gas j that contribute to the optical depth at this wavenumber. There are a number of databases available that list the molecular absorption lines of important gases, including HITRAN (Rothman *et al.*, 2005), which is extensively used for terrestrial atmospheric calculations, and GEISA (Jacquinet-Husson *et al.*, 2005), which also includes lines of exotic gases only found in the atmospheres of the giant planets. For each line of each gas, the absorption coefficient k_{ij} must be calculated from the line strengths and line shape, both of which are functions of temperature and pressure. Since thousands of absorption lines may contribute to the absorption at a particular wavelength, it can easily be seen that such *line-by-line* calculations are computationally expensive and thus slow, although they are clearly the most accurate method available.

In most cases, spectral calculations are made to compare with real measured spectra which have limited spectral resolution. Hence, the transmissions used in the radiative transfer equation must be integrated over the instrument function. While it is still most accurate to smooth line-by-line calculated spectra to the required spectral resolution, there are alternative methods of simulating finite-resolution spectra that are much faster and only slightly less accurate. Since the line parameters used to generate line-by-line spectra are themselves accurate to only 10% in some cases, the use of these lower accuracy models is very common and perfectly defendable. There are two main approaches, band models and correlated-k models, which will now be discussed.

Band model approximation

A number of possible band models exist, including the Goody–Lorentz, Godson–Lorentz, Malkmus–Lorentz, and Goody–Voigt band models. A full discussion of these models is beyond the scope of this book, but they are discussed in detail in a number of more general radiative transfer books such as Goody and Yung (1989). The basic idea of such models is that if we have an atmosphere of fixed temperature, pressure, and composition, and if we then measure (or calculate using a line-by-line model) the mean transmission over the wavenumber range $\tilde{\nu}$ to $\tilde{\nu} + \Delta\tilde{\nu}$ of a number of paths of different lengths through this atmosphere, then the mean transmission as a function of the path amount m (molecules per square meter), due to individual gases, may be well approximated by a smoothly varying analytical function of just a few parameters. For example, the *band transmission* as a function of the path amount for the Goody–Voigt approximation due to a certain gas of mole fraction q is given by

$$\bar{T}(m) = \exp\left(-2k_{\bar{\nu}}(T)m\int_0^\infty \frac{V(x,y)}{1+k_{\bar{\nu}}(T)mV(x,y)(\delta/\alpha_D^0)\sqrt{T}}dx\right) \tag{6.56}$$

where $V(x,y)$ is the Voigt function and where

$$y = \frac{\alpha_L}{\alpha_D} = \frac{\alpha_L^0}{\alpha_D^0}\frac{p}{p_0}\frac{\sqrt{T_0}}{T}\left(q+\frac{1-q}{SFB}\right) = y_0\frac{p}{p_0}\frac{\sqrt{T_0}}{T}\left(q+\frac{1-q}{SFB}\right) \tag{6.57}$$

where T_0 and p_0 are reference temperatures and pressures; and SFB is a factor to be fitted or calculated (see below). The mean absorption coefficient $k_{\bar{\nu}}$ may be expanded as

$$k_{\bar{\nu}}(T) = k_{\bar{\nu}}(T_0)\left(\frac{T_0}{T}\right)^{q_r}\exp\left[1.439E_l\left(\frac{1}{T_0}-\frac{1}{T}\right)\right] \tag{6.58}$$

where q_r is 1.0 for linear molecules and 1.5 for nonlinear molecules. There are thus five independent parameters in the Goody–Voigt model: $k_{\bar{\nu}}(T_0)$, y_0, δ/α_D^0, E_l, and SFB, which may either be fitted to laboratory transmission spectra, or alternatively derived from tabulated line listings such as HITRAN and GEISA. These parameters, while usually fitted directly to measured spectra, are related to the properties of the real absorption lines in the spectral band: $k_{\bar{\nu}}(T_0)$ is the integrated line strength of all the lines in the wavenumber range considered at the standard temperature; y_0 is the mean ratio of pressure-broadened to Doppler-broadened linewidths at STP (standard temperature and pressure); δ/α_D^0 is mean line spacing divided by the Doppler-broadened width at STP; E_l is the mean energy of lower states; and SFB is the mean self-broadening to foreign-broadening ratio of absorption lines.

Once band data have been tabulated, the transmission of homogeneous paths may be rapidly and accurately calculated. However, real atmospheres are inhomogeneous in that pressure, temperature, and composition vary rapidly with altitude. How then may band models be applied? It may be shown that the mean transmission of a path though an inhomogeneous atmosphere may be well approximated by the mean transmission of an equivalent homogeneous path, whose path amount, mean pressure, and mean temperature are given by:

$$m = \frac{1}{\mu}\int_{z_1}^{z_2} n(z)\,dz, \qquad \bar{p} = \frac{\frac{1}{\mu}\int_{z_1}^{z_2} n(z)p(z)\,dz}{m}, \qquad \bar{T} = \frac{\frac{1}{\mu}\int_{z_1}^{z_2} n(z)T(z)\,dz}{m}. \tag{6.59}$$

This is known as the *Curtis–Godson* approximation. Hence, to use band models to calculate thermal emission spectra, the inhomogeneous atmosphere is represented by a series of equivalent Curtis–Godson paths from space to progressively deeper levels in the atmosphere and the difference between the band-calculated mean transmissions used to find the mean transmission weighting function. In addition, since the absorption lines of different molecules are rarely correlated with each other, the total transmission of Curtis–Godson paths for all the gases concerned is simply found by multiplying the individual gas transmissions together, or equivalently summing the optical thicknesses. The band model approach is very fast, but is found to be useless for multiple scattering calculations and hence is mostly used in the mid-infrared to far-infrared where scattering is generally less important (as we shall see in Section 6.5.2).

Correlated-k approximation

An alternative approach to calculating finite resolution spectra is to use k-distributions. For a path of absorber amount m in an atmosphere of uniform pressure p and T, the mean transmission is given by

$$\bar{T}(m) = \frac{1}{\Delta\tilde{\nu}} \int_{\tilde{\nu}}^{\tilde{\nu}+\Delta\tilde{\nu}} \exp(-k(\tilde{\nu})m)\, d\tilde{\nu} \tag{6.60}$$

where the absorption coefficient at a particular wavelength is the summation of all the individual line contributions. Since the absorption coefficient $k(\tilde{\nu})$ is a rapidly varying function of a wavenumber, in order to numerically calculate the mean transmission accurately, a very fine wavenumber step must be chosen. However, when calculating the mean transmission in a spectral interval it does not matter which parts of the interval are actually highly or poorly absorbing. All we need to know is what fraction of the interval has low absorption, what fraction has high absorption, and so on. In other words, if we calculate a high-resolution absorption coefficient spectrum using a regularly spaced high-resolution grid, and then sort the absorption coefficients into order starting with the low absorption coefficients first and then working monotonically through to the high absorption coefficients, the resulting integral of the sorted spectrum is identical to that of the original. The advantage of this approach is that the sorted spectrum, known as the k-distribution, $k(g)$, is a *smoothly varying* function that is usually expressed in terms of the fraction of the interval g, which varies between 0 and 1. Since it is a smoothly varying function, the integral may be accurately integrated with far fewer quadrature points and thus calculation of mean transmission is very much faster. In practice ten to twenty quadrature points are usually found to be satisfactory and the mean transmission may be approximated by

$$\bar{T}(m) \equiv \int_0^1 \exp(-k(g)m)\, dg \simeq \sum_{i=1}^{N} \exp(-k_i m)\, \Delta g_i \tag{6.61}$$

where k_i is the k-distribution calculated at each of the N quadrature points; and Δg_i are the quadrature weights. The k-distributions may be pre-calculated for each gas for a range of temperatures and pressures found in real atmospheres and then stored in look-up tables for rapid interpolation and calculation of mean transmission. Since the absorption lines of different gases may be assumed to a good approximation to be uncorrelated, it is also reasonably straightforward to combine k-distributions together (Lacis and Oinas, 1991). The k-distribution look-up tables may be calculated either directly from line data or, if the available line data are of poor quality, indirectly from band data using the technique of exponential sums (e.g., Irwin *et al.*, 1996). While we can see that k-distributions can speed up transmission calculations for homogeneous paths, how can they help us for inhomogeneous paths? For monochromatic calculations, the transmission of an inhomogeneous path is found by splitting the path into small subpaths, calculating the transmission, and then multiplying all the transmissions together. However, for band-averaged transmissions, such as those used by band models, this multiplication is not possible and thus the Curtis–Godson approximation must be used. The Curtis–Godson approximation may also be used with k-distributions, but there is then no advantage over the band model approach. Instead, it is found that regions of high and low absorption within the spectral band are spectrally correlated between various subpaths within the inhomogeneous path. This correlation exists between the k-distributions also (Lacis and Oinas, 1991; Goody *et al.*, 1989). Hence, the k-distributions may effectively be multiplied together almost as though they were monochromatic to determine the mean transmission of the inhomogeneous path

$$\bar{T} \simeq \sum_{i=1}^{N} \exp\left(-\sum_{j=1}^{M} k_{ij} m_j\right) \Delta g_i \tag{6.62}$$

where the inhomogeneous path has been split into M subpaths. This is the correlated-k approximation and is found to have a similar accuracy to that of the Curtis–Godson approximation. The great advantage, however, lies in the fact that thermal emission and in particular scattering calculations (discussed in Section 6.6) may also be summed in exactly the same way. Hence, the technique of correlated-k allows for rapid calculation of spectra in multiply scattering atmospheres and is thus used extensively to simulate the near-IR reflectance spectra of the giant planets.

6.5 SCATTERING OF LIGHT BY PARTICLES

We have seen in Section 6.4 that the equations for radiative transfer in a nonscattering "gray" atmosphere are relatively simple. However, these equations are not applicable to the analysis of sunlight reflected by clouds in planetary atmospheres and are hence only of use in modeling the thermal-IR spectra of the planets. Even in the thermal-infrared, however, neglecting the scattering effects of atmospheric aerosols can sometimes lead to errors, especially if cloud particles are of a size approximately equal to or greater than the wavelength. The scattering effect of aerosols greatly complicates the equations of radiative transfer, as we shall see in Section 6.6. However, before we can investigate the effects of scattering, we must first introduce the basic definitions of scattering parameters and how the scattering properties of individual particles may be calculated.

Consider a single photon of wavelength λ incident on a particle that is then scattered forward at an angle θ to the original direction (Figure 6.12). This angle is defined as the *scattering angle* and in an experiment where light is incident upon

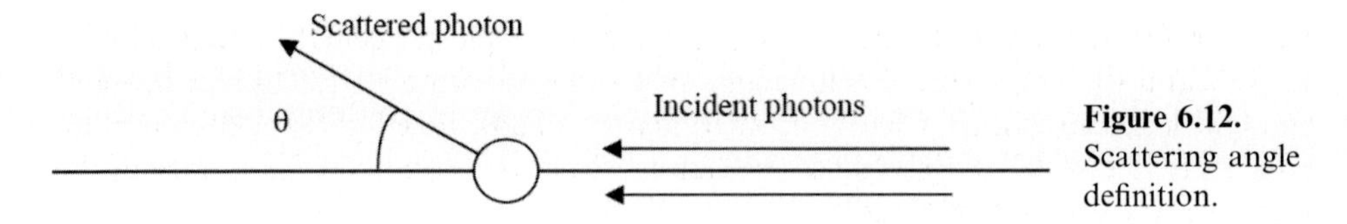

Figure 6.12. Scattering angle definition.

such a particle, the numbers of photons scattered into different directions will, for spherically symmetric particles, be a function of this scattering angle only. [NB: For nonspherical particles, discussed further in Section 6.5.3, the scattering efficiency will in general also depend on the orientation of the particle and the rotational angle of the vector of the scattered photon about its initial direction.] The function that gives the probability that a photon will be scattered into an element of solid angle $d\Omega$ and scattering angle θ is known as the *phase function* $p(\theta)$ and by convention is normalized such that

$$\int_0^{4\pi} p(\theta)\, d\Omega = \int_{\phi=0}^{2\pi} \int_{\theta=0}^{\pi} p(\theta) \sin\theta\, d\theta\, d\phi = 4\pi. \tag{6.63}$$

Note that in some schemes this integral is normalized to unity, which can lead to confusion! The phase function itself is a function of wavelength, particle composition, mean particle radius, and particle shape.

The probability that a photon will actually be absorbed or scattered by a particle, regardless of direction, depends upon the *absorption* and *scattering cross-sections* σ_{abs} and σ_{sca}, respectively. These cross-sections also depend on the wavelength, particle composition, and mean particle radius and shape. The *extinction cross-section* is defined as $\sigma_{\text{ext}} = \sigma_{\text{sca}} + \sigma_{\text{abs}}$, and the *single-scattering albedo* ϖ is defined as the ratio

$$\varpi = \frac{\sigma_{\text{sca}}}{\sigma_{\text{ext}}}. \tag{6.64}$$

The absorption, scattering, and extinction scattering *efficiencies* Q_{abs}, Q_{sca}, and Q_{ext} are defined as the ratios of the respective cross-sections to the geometric cross-sectional area of the particles.

6.5.1 Rayleigh or dipole scattering

When the wavelength is very much larger than the particle size, scattering particles tend to behave as simple dipoles and we have the condition for *Rayleigh scattering* that we referred to in Chapter 4. The phase function for Rayleigh scattering may be shown from standard electromagnetic theory to be (e.g., Goody and Yung, 1989)

$$p(\theta) = \tfrac{3}{4}(1 + \cos^2\theta) \tag{6.65}$$

and thus the probability that photons will be scattered at a scattering angle between θ and $\theta + d\theta$ is

$$P(\theta)\, d\theta = \tfrac{3}{8}(1 + \cos^2\theta) \sin\theta\, d\theta. \tag{6.66}$$

Such dipole scatterers are purely scattering and thus have zero absorption cross-sections. Their scattering cross-section varies with wavelength as $1/\lambda^4$ as described in

Chapter 4 (Equation 4.34) and the most familiar example of such scattering is in the Earth's atmosphere, where the molecules of N_2 and O_2 scatter a fraction of incident sunlight in all directions. Clearly, from Equation (4.34), blue light is scattered more effectively than red light leading to the familiar blue sky seen from the surface of the Earth (in the absence of cloud!).

Raman scattering and fluorescence

Quantum mechanically, an atom or molecule Rayleigh-scatters a photon by first absorbing it and becoming excited to an intermediate or virtual state, whereupon it immediately relaxes back to its initial state, releasing a photon with the same wavelength in a direction governed by Equation (6.65). However, it is also possible that the atom or molecule relaxes back to a different state, thus releasing a photon of either longer or shorter wavelength than the original photon. *Stokes Transitions* lead to scattered photons with longer wavelengths than the incident light, while *Anti-Stokes Transitions* lead to scattered photons with shorter wavelengths. The phenomenon is called *Raman* scattering and is usually rather weak compared with Rayleigh scattering and thus may usually be neglected. An exception to this is in the case for the giant planets (occurring in the UV spectra) of Uranus and Neptune where distinct solar spectrum features appear shifted to longer wavelengths in the observed albedos of these planets by Raman scattering associated mainly with the rotational S(0) transition of hydrogen molecules (giving a wavenumber shift of 354 cm^{-1}) and to a much lesser extent the S(1) and Q1(1) transitions (shifted by 587 cm^{-1} and 4,161 cm^{-1}, respectively). Raman scattering in the outer planet atmospheres is described in detail by Karkoschka (1994).

Should scattered photons have considerably less energy than incident photons, and thus significantly longer wavelengths, but be released quickly (within roughly 10^{-7} s), then the effect is sometimes also known as *fluorescence*. It is observed that many household materials glow, or fluoresce, under UV illumination (Hecht and Zajac, 1974). If there is an appreciable delay in the release of lower energy photons, sometimes several hours, then the effect is known as *phosphorescence*.

6.5.2 Mie theory

For particles that have a non-negligible size compared with the wavelength, Rayleigh scattering no longer applies, and calculation of the phase function and extinction cross-section becomes more complicated. However, provided the aerosol particles are spherical (and are thus liquid), and provided that the complex refractive index $(n_r + in_i)$ as a function of wavelength is known, Maxwell's equations may be solved analytically via a method known as Mie theory to calculate the scattering properties. This method deals with the classical case of a dielectric sphere interacting with a plane electromagnetic wave, and is too complex to be covered in detail here. The reader is referred to a number of more detailed references for further information: Goody and Yung (1989); Hanel *et al.* (2003); and Hansen and Travis (1974). Using Mie theory, Q_{ext}, ϖ, and $p(\theta)$ may all be calculated as a function of wavelength. The typical

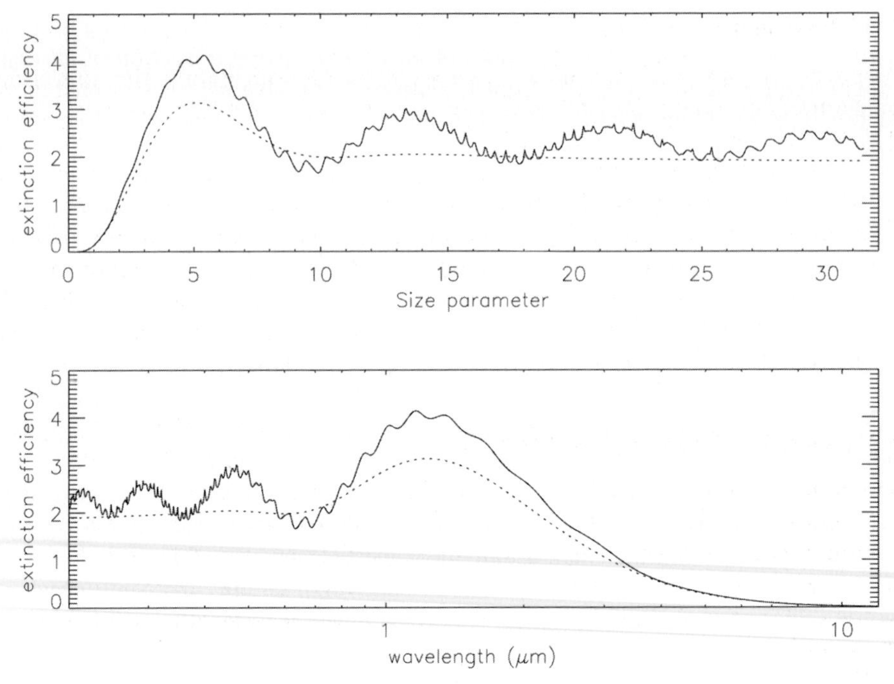

Figure 6.13. Mie scattering calculation of Q_{ext} as a function of wavelength for particles of refractive index $(1.4 + 0i)$. The solid line shows the properties of particles with a single radius of 1 μm while the dotted line shows the properties if there is a small distribution of particle size about 1 μm. The spectra are plotted as a function of both wavelength and size parameter, defined as $2\pi r/\lambda$, where r is the radius and λ the wavelength.

variation of Q_{ext} with wavelength is shown in Figure 6.13. From this figure, together with the calculated phase functions, the bulk scattering properties of spherical liquid aerosols are found to be:

(1) For particles that are small compared with the wavelength, the extinction cross-section tends to $1/\lambda^4$, and phase functions tend toward the Rayleigh or dipole scattering case discussed earlier.
(2) A particle is most efficient at scattering light with wavelength approximately equal to its own radius.
(3) For particles that are large compared with the wavelength, it is found that the amount of light diffracted is equal to the amount striking the particle (independent of particle shape and refractive index) and thus $Q_{ext} = 2$ (i.e., its extinction cross-section is twice its geometric cross-section). It is also found that the phase function becomes more and more forward scattering and in the limit where the particle is very much greater than the wavelength of the incident light, $p(\theta) = \delta(\theta)$, where $\delta(\theta)$ is the Dirac delta function.

6.5.3 Nonspherical particles

A number of analytical/numerical methods exist for calculating the scattering behaviour of nonspherical scatterers (i.e., ice crystals) with electromagnetic radiation. A particular example is for rotationally symmetric nonspherical scatterers in fixed or random orientations, which can be modeled using T-matrix theory (Mishchenko *et al.*, 1996, 2006). These methods add a further layer of complexity. Fortunately, the nonspherical nature of real ice crystals may often be ignored since a set of randomly orientated crystals is, to a first approximation, indistinguishable from a set of spheres with the same mean radius. However, if crystals become aligned in some way through dynamics or other effects, the difference can become significant. The general case of nonspherical particles is outlined by Goody and Yung (1989).

6.5.4 Analytical forms of phase functions

The phase functions calculated for both spherical particles (using Mie theory) and nonspherical particles do not have a simple analytical expression. However, it is sometimes useful to have a more simple parameterized form of the phase function to use in radiative transfer models that can reasonably well approximate real phase functions of particles. One such representation is the double Henyey–Greenstein representation

$$p(\theta) = \left\{ f \frac{1 - g_1^2}{(1 + g_1^2 - 2g_1 \cos\theta)^{3/2}} + (1 - f) \frac{1 - g_2^2}{(1 + g_2^2 - 2g_2 \cos\theta)^{3/2}} \right\} \qquad (6.67)$$

where g_1 is the asymmetry of the forward-scattering lobe (varying between 0 and 1); g_2 is the asymmetry of the back-scattering lobe (varying between 0 and −1); and f is the fractional contribution (between 0 and 1) of the forward-scattering part. Calculated Henyey–Greenstein phase functions for a range of parameters are shown in Figure 6.14. An example of the use of Henyey–Greenstein phase functions is that scattering radiative transfer models may be optimized to use them. Hence, by approximating real phase functions by Henyey–Greenstein functions, great reductions in computation times may be achieved.

6.6 RADIATIVE TRANSFER IN SCATTERING ATMOSPHERES

The radiative transfer equations of Section 6.4 for a gray atmosphere must be substantially modified to deal with atmospheres that contain significant abundances of scattering particles, as we shall shortly see. These modifications greatly increase the computation time and thus scattering calculations are notoriously difficult and slow. There are two main ways of approaching the problem. The simplest is to ignore the curvature of the planetary atmosphere and approximate the problem by a stack of plane-parallel layers. Using this approach, a number of good approximations may be made that considerably reduce the computation time. Most calculations are made with this *plane-parallel approximation*. However, under certain conditions, such as for

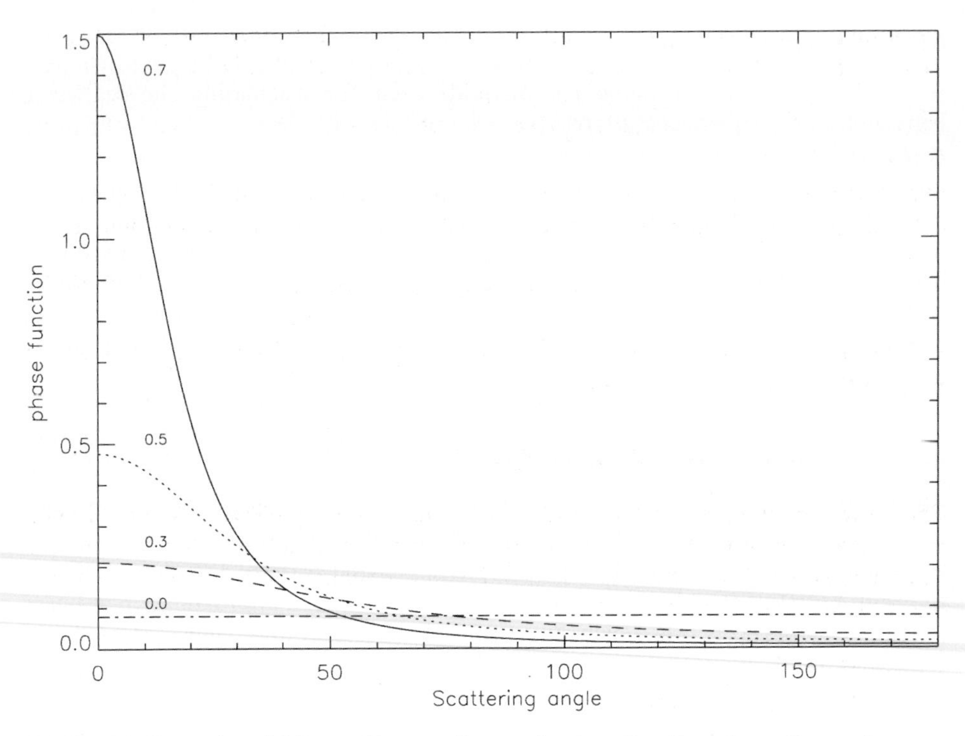

Figure 6.14. Examples of different Henyey–Greenstein phase functions depending on f, g_1, g_2. Here $f = 1$, $g_2 = 0$, and $g_1 = 0.7$ (solid), 0.5 (dots), 0.3 (dashes), and 0.0 (dot-dash).

limb observations, the plane-parallel approximation no longer applies, and thus much slower, but more general purpose techniques such as *Monte Carlo* calculations must be used. In this section we will outline how the scattering properties of particles, just discussed, may be used in radiative transfer models to calculate the synthetic spectra of the giant planets.

6.6.1 Plane-parallel approximation

In Section 6.4.1 we found that the thermal contribution of a thin slab of thickness dz at altitude z to the upwelling spectral radiance at wavenumber $\tilde{\nu}$ and zenith angle θ and azimuth angle ϕ at that level in a plane-parallel atmosphere is given by $dI_{\tilde{\nu}} = k_{\tilde{\nu}}(z)n(z)B_{\tilde{\nu}}(z)\,dz/\mu$ (Equation 6.35). Given that the optical depth of the atmosphere (which is zero at the top of the atmosphere, assumed here to be at $z = \infty$, and increases steadily as we move to deeper levels) at altitude z is defined in Equations (6.33) and (6.34) as

$$\tau_{\tilde{\nu}} = \int_z^{\infty} k_{\tilde{\nu}}(z)n(z)\,dz \tag{6.68}$$

the upwelling spectral radiance at altitude z (or equivalently optical depth $\tau_{\tilde{\nu}}$) in the direction of zenith angle θ and azimuth angle ϕ may equivalently be expressed as

$$dI_{\tilde{\nu}}(\tau_{\tilde{\nu}}, \mu, \phi) = -\frac{1}{\mu} B_{\tilde{\nu}}(\tau_{\tilde{\nu}})\, d\tau_{\tilde{\nu}} \tag{6.69}$$

where the minus sign arises since $\tau_{\tilde{\nu}}$ decreases as z increases. From Kirchoff's law, the slab will also absorb radiation already upwelling along this path by an amount

$$dI_{\tilde{\nu}}(\tau_{\tilde{\nu}}, \mu, \phi) = \frac{1}{\mu} I_{\tilde{\nu}}(\tau_{\tilde{\nu}}, \mu, \phi)\, d\tau_{\tilde{\nu}} \tag{6.70}$$

and thus the total radiative transfer equation for spectral radiance in the atmosphere at an optical depth $\tau_{\tilde{\nu}}$ may be expressed as

$$\mu \frac{dI_{\tilde{\nu}}(\tau_{\tilde{\nu}}, \mu, \phi)}{d\tau_{\tilde{\nu}}} = I_{\tilde{\nu}}(\tau_{\tilde{\nu}}, \mu, \phi) - B_{\nu}(\tau_{\tilde{\nu}}). \tag{6.71}$$

For scattering atmospheres, the equations become more complicated in that particles may scatter light out of a beam and also scatter light into a beam that was initially traveling in other directions. The equation of transfer under these conditions becomes (Hanel *et al.*, 2003)

$$\mu \frac{dI_{\tilde{\nu}}(\tau_{\tilde{\nu}}, \mu, \phi)}{d\tau_{\tilde{\nu}}} = I_{\tilde{\nu}}(\tau_{\tilde{\nu}}, \mu, \phi) - \frac{\varpi(\tau_{\tilde{\nu}})}{4\pi} \int_0^{2\pi} \int_{-1}^{1} p(\tau_{\tilde{\nu}}; \mu, \phi; \mu', \phi') I_{\tilde{\nu}}(\tau_{\tilde{\nu}}, \mu', \phi')\, d\mu'\, d\phi' - (1 - \varpi(\tau_{\tilde{\nu}})) B_{\tilde{\nu}}(\tau_{\tilde{\nu}}) \tag{6.72}$$

where $p(\tau_{\tilde{\nu}}; \mu, \phi; \mu', \phi')$ is the phase function at the atmospheric level of optical depth $\tau_{\tilde{\nu}}$; and μ', ϕ' are the zenith and azimuth angles of radiation incident upon the scattering particles there. The single scattering albedo $\varpi(\tau_{\tilde{\nu}})$ at a particular level in the atmosphere is the ratio of $\sigma_{sca}/\sigma_{ext}$ for both aerosols *and* gas (i.e., it includes the scattering of aerosols and Rayleigh scattering of the gas together with absorption by both the aerosols and gas). This equation clearly reduces to the thermal emission form for nonscattering atmospheres, where $\varpi = 0$.

The scattering second term of Equation (6.72) contains the contribution of both the scattered diffuse field and the scattering of direct sunlight that has reached a particular optical depth $\tau_{\tilde{\nu}}$. It is very useful to be able to discriminate between the two. Suppose that the incident sunlight (considered to be a beam of collimated radiation) carries a spectral flux F_0 (W m^{-2} (cm^{-1})$^{-1}$) normal to its direction, and suppose that the cosine of the zenith angle of the Sun is μ_0. The magnitude of the flux directly vertically downwards is $\mu_0 F_0$ and Equation (6.72) may be re-expressed as

$$\mu \frac{d[e^{-\tau_{\tilde{\nu}}/\mu} I_{\tilde{\nu}}(\tau_{\tilde{\nu}}, \mu, \phi)]}{d\tau_{\tilde{\nu}}} = -\frac{\varpi(\tau_{\tilde{\nu}})}{4\pi} e^{-\tau_{\tilde{\nu}}/\mu} \int_0^{2\pi} \int_{-1}^{1} p(\tau_{\tilde{\nu}}; \mu, \phi; \mu', \phi') I_{\tilde{\nu}}(\tau_{\tilde{\nu}}, \mu', \phi')\, d\mu'\, d\phi' - \frac{\varpi(\tau_{\tilde{\nu}}) F_0(\tilde{\nu})}{4\pi} \exp\left[-\left(\frac{1}{\mu} + \frac{1}{\mu_0}\right)\tau_{\tilde{\nu}}\right] p(\tau_{\tilde{\nu}}; \mu, \phi; -\mu_0, \phi_0) - (1 - \varpi(\tau_{\tilde{\nu}})) e^{-\tau_{\tilde{\nu}}/\mu} B_{\tilde{\nu}}(\tau_{\tilde{\nu}}) \tag{6.73}$$

where the first term contains scattered diffuse light; the second term contains directly scattered sunlight; and the third term contains thermally emitted radiation. Equation (6.73) in its entirety may not be solved analytically, but must instead be solved numerically. There are a number of ways of doing this and one of the most common is the matrix–operator (or doubling–adding) method. This technique basically applies a Gaussian quadrature technique to integration over zenith angle and the Fourier method to integration over azimuth angle and is well described by Goody and Yung (1989), Hansen and Travis (1974), and Plass *et al.* (1973). Alternative techniques include discrete ordinates (e.g., Hanel *et al.*, 2003) and successive orders (Hansen and Travis, 1974). However, a particularly useful technique for cases where the scattering optical depth is small is the *single-scattering approximation*, which assumes that thermal emission is negligible and that the probability of a photon being scattered more than once is so small that only the second term on the right-hand side of Equation (6.73) need be considered. This leads to the directly integrable equation (where we have assumed that the optical depth of the atmosphere varies from zero at the top to infinity at the base)

$$I_{\tilde{\nu}}(\mu,\phi) = \frac{F_0(\tilde{\nu})}{4\pi}\int_0^\infty \varpi(\tau_{\tilde{\nu}})p(\tau_{\tilde{\nu}};\mu,\phi;-\mu_0,\phi_0)\exp\left[-\left(\frac{1}{\mu}+\frac{1}{\mu_0}\right)\tau_{\tilde{\nu}}\right]\frac{d\tau_{\tilde{\nu}}}{\mu}. \quad (6.74)$$

A further particularly simple approximation to scattering, which is applicable for atmospheres with thin cloud layers, is the so-called *reflecting layer approximation*. Here, gas absorption spectra are used to calculate the transmission of a path from the Sun to a particular level in the atmosphere and back to the observer. If a thin cloud exists at that pressure level, then the observed spectrum may be approximated by multiplying this transmission by the effective reflectivity of the cloud layer. Several clouds may be approximated by summing the "reflections" from a number of levels and the technique is closely related to the single-scattering approximation.

6.6.2 Spherical atmospheres and limb viewing: Monte Carlo simulations

The scattering equations just derived relate to plane-parallel atmospheres and thus are only applicable when the zenith angles are not too near to 90°. Nearer the limb, and for limb-viewing geometries, the equations are unusable and thus more complicated approaches must be used, of which the most general is the *Monte Carlo* method.

As the name suggests, the Monte Carlo technique basically "fires" a large number of model photons into an atmosphere and tracks where they go using scattering probability functions and a random number generator. The technique is computationally expensive and slow to converge. However, if enough photon paths are simulated, accuracy is as good as the more conventional techniques, and the technique has the advantage of being able to model any geometry.

From Beer's law of absorption we know that the probability that a photon will pass through a slab of optical thickness τ is given simply by $\exp(-\tau)$. Hence, by inversion we know that the optical thickness traveled by a random photon before absorption or scattering is given simply by

$$\tau = -\log(R) \tag{6.75}$$

where R is a random number between 0 and 1. From a given starting position and direction we may thus calculate the new position of the photon given this random optical thickness. At the new position, the probability of scattering is simply the single scattering albedo ϖ and thus the photon is scattered if $\varpi > R$, where R is a new random number between 0 and 1. If the photon is scattered, then the new photon direction may be calculated from the phase function, where the scattering angle θ_0 is given by

$$\int_{\theta=0}^{\theta_0} p(\theta) \sin\theta \, d\theta = 2R \tag{6.76}$$

where R is another random number between 0 and 1, and the azimuthal rotation of the new direction around the old direction is governed by a further random angle between 0° and 360°. The new optical thickness for the next photon path is then calculated and the process iterated until either the photon is absorbed or it leaves the atmosphere.

For cases where scattered sunlight dominates, the most efficient way of proceeding is to fire a sequence of photons at the planet from the direction of the Sun and track the proportion that are reflected by the atmosphere in different angles. For cases where thermal emission dominates, the most efficient way of proceeding is to fire a sequence of photons at the planet from the observer's position, calculate where in the atmosphere they are absorbed, and then average the Planck functions from these absorption regions. Monte Carlo thermal emission calculations converge much faster than reflected sunlight calculations.

6.7 GIANT PLANET SPECTRA

6.7.1 General features of giant planet spectra: UV to microwave

At UV wavelengths, the atmospheres of the giant planets are optically thick due to Rayleigh scattering. Hence, at these wavelengths most of the light we see is Rayleigh-scattered sunlight, modified by the absorption of high-altitude haze layers. Towards the poles, auroral glow is also seen, especially for Jupiter. Superimposed on this spectrum are the absorption features of several gases that suffer photolysis in the upper atmosphere (discussed in Chapter 4). As can be seen in Table 4.4, as the wavelength increases, Rayleigh scattering rapidly becomes less important and thus at visible wavelengths sunlight may penetrate to, and be reflected from, the deeper cloud layers at several bar pressures. Towards the red end of the visible spectrum, weak vibration–rotation bands of methane (and also ammonia) appear, which become increasingly strong in the near-infrared. In the center of these bands, the atmosphere is optically thick and thus any light that is detected must be reflected from the upper cloud and haze layers of the atmosphere. Between the bands the atmosphere is optically thin and thus sunlight may be reflected from both upper and deeper cloud layers. Hence, analysis of the near-IR reflection spectrum is very important in determining the vertical cloud structure of the giant planets. The solar spectrum diminishes at longer wavelengths, and thus in the mid-infrared the spectrum is dominated by thermal emission from the atmosphere itself, modulated by the presence of numerous vibration–rotation (Section 6.3.4) absorption bands of several molecules. In the far-infrared the thermal emission spectra of the giant planets becomes dominated by collision-induced H_2–H_2 and H_2–He absorption (Section 6.3.6) together with the rotational bands of several molecules (Section 6.3.3). At submillimeter and microwave wavelengths these sources of absorption become increasingly weak and thus thermal emission from the deep levels of the atmospheres may be detected at wavelengths other than near 1.3 cm, where ammonia has an inversion band (Section 6.3.5).

6.7.2 Near-IR and visible reflectance spectra

Estimates of the disk-averaged geometric albedos of the giant planets are shown in Figure 6.15. Data below 1 μm is taken directly from the ground-based observations of Karkoschka (1994, 1998a). Above 1 μm, data come from a variety of sources, which are listed in the figure. These spectra are formed by sunlight scattering off aerosol particles at different altitudes in the atmosphere. Sunlight scattering from deep clouds passes through a longer path of methane (and for Jupiter and Saturn, ammonia) before reaching the observer than sunlight scattering from haze layers in the upper troposphere. In the near-infrared the absorption of methane (and ammonia) is significant and thus only wavelengths close to the methane absorption minima of 1.05 μm, 1.3 μm, 1.6 μm, 2.0 μm, and 3.0 μm may be scattered from deeper clouds, leading to a series of narrow reflection peaks at these wavelengths. Sunlight scattering of higher clouds suffers less methane absorption and thus the reflection peaks are broader, while sunlight scattering off upper-tropospheric and lower-stratospheric hazes suffers very little absorption and thus has very broad reflection peaks. Hence, by analyzing the shape of the observed reflection spectra, we can deduce the vertical cloud structure of the giant planet atmospheres. In addition, the particle size of clouds may be estimated from analyzing how the reflectivity of cloud layers varies with wavelength since we saw in Section 6.5.2 that the extinction cross-section of particles tends to zero at long wavelengths at different rates, depending on the particle size.

The main absorption bands of methane (shown previously in Figure 6.6) are clearly seen in Figure 6.15 for all four giant planets. At wavelengths greater than 1 μm, the shape of the spectra tell us that the reflection spectra of Jupiter and Saturn are formed from reflections from a vertically extended cloud system, since a combination of reflections from deep, middle, and high cloud layers is required to simulate the observed spectra. In contrast, the reflection spectra of Uranus and Neptune have narrow peaks near the main methane passbands indicating that the dominant reflection comes from the main clouds at $\sim$3 bar, although some sunlight is also reflected from the tropospheric and stratospheric hazes, particularly for Neptune. Towards the visible, the absorption of methane rapidly decreases, and the reflectivity of the small

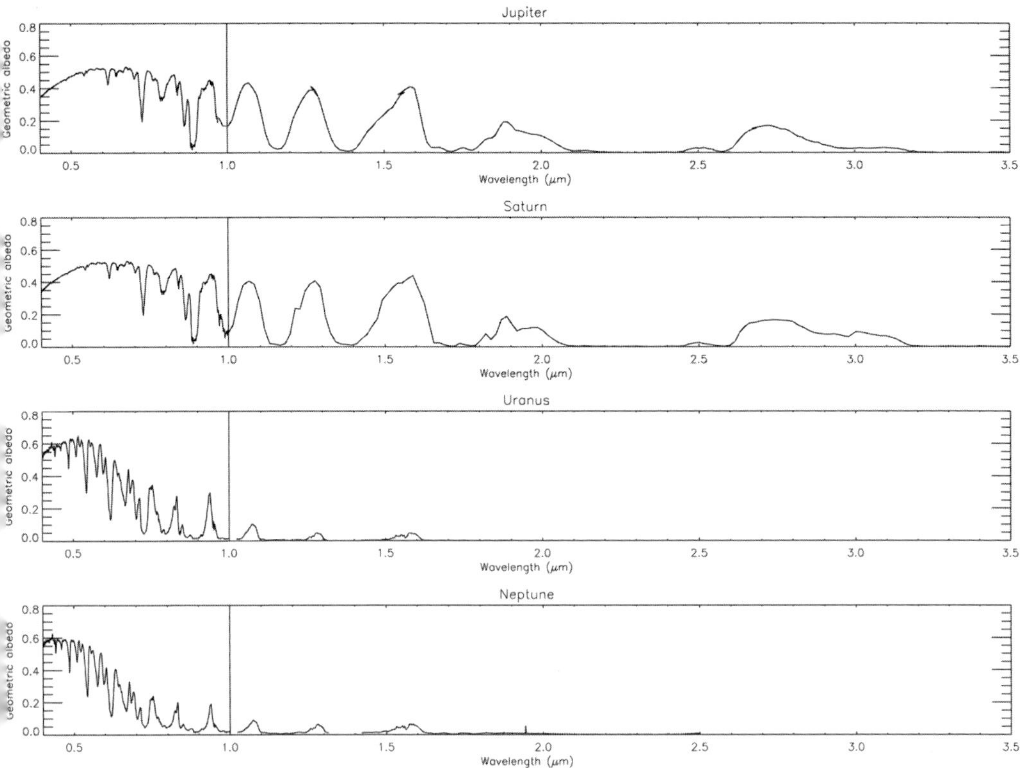

Figure 6.15. Measured and calculated geometric albedo spectra of the giant planets. Below 1 μm, the spectra are those measured from the ground-based observations of Karkoschka (1994, 1998). The spectra above 1 μm are from the following sources: Jupiter, *Galileo* NIMS real-time spectra (Irwin *et al.*, 1998); Saturn, *Cassini* VIMS typical spectrum (Roos-Serote, 2008); Uranus, ground-based UKIRT observation (Irwin *et al.*, 2007); and Neptune, ground-based UKIRT observation (Irwin *et al.*—unpublished).

particles found in the tropospheric hazes increases leading to the increase in albedo observed for all four planets. The visible spectra of both ice giants can be seen to be significantly weighted towards the blue end, especially Neptune, which gives these planets their characteristic blue colors. Although this coloration arises partly from the red-absorbing nature of methane, which can be clearly seen, it is found that in addition the main cloud deck at ~3 bar must also be significantly red-absorbing, as was mentioned in Chapter 4.

The variation of reflectivity of a planet across the visible disk may also tell us something about the scattering properties and vertical distribution of cloud particles. The least ambiguous definition of reflectivity is the *bidirectional reflectivity function*, or BDRF, defined as

$$R = \frac{\pi I}{\mu_0 F_0 / D^2} \tag{6.77}$$

where I is the measured reflected radiance; μ_0 is the cosine of the solar zenith angle; F_0 is the solar flux at the distance of the Earth; and D is the distance of the planet from the Sun in AU (Hanel *et al.*, 2003). This formulation correctly approximates the fact that the flux of sunlight arriving per unit area of a horizontal surface depends on the cosine of the zenith angle. For a *Lambertian* reflecting surface (where the BDRF is the same in all directions and equal to a constant R_L), Equation (6.77) may be rearranged as $I = R_L\mu_0 F_0/(\pi D^2)$ or $I = I_0\mu_0$, where I_0 is the maximum reflected radiance viewed when the Sun is directly overhead ($\mu_0 = 1$). Hence, for a single Lambertian cloud layer, the observed reflectivity is expected to be limb-darkened with reflectivity decreasing towards the limb of the planet as μ_0 tends to zero. A cloud of particles which is optically thick and thus multiply-scattering is found to approximate well to a Lambertian reflector regardless of the phase functions of individual scatterers. However, the extended vertical distribution of clouds in the giant planets together with significant gaseous atmospheric absorption means that scattering conditions approach the single scattering limit at some wavelengths and significant departures from this simple limb-darkening rule are then observed. A common approximation to the limb-darkening curve for non-Lambertian cases is the semi-empirical *Minnaert* limb-darkening equation (Minnaert, 1941)

$$I = I_0 \frac{(\mu_0\mu)^k}{\mu} \tag{6.78}$$

where μ is the cosine of the observer's zenith angle; and k is a constant between 0 and 1, which may be fitted to the experimental data. For a Lambert reflector $k = 1$. Once measured, the observed limb-darkening curves may be fitted with scattering models to determine the vertical distribution of particles and their scattering properties.

6.7.3 Thermal-IR spectra

Calculated thermal emission spectra of the giant planets are shown in Figure 6.16 for nadir-viewing geometry, and for the case of zero cloud opacity for Jupiter and Saturn, and for the case of a deep thick cloud for Uranus and Neptune with an optical depth of unity at ~3 bar for all wavelengths. These cloud structures were chosen to reflect the observation that for Jupiter and Saturn the clouds are mostly above or below the line-forming region, while for Uranus and Neptune near-IR observations detect a substantial cloud (presumably composed of H_2S) in the middle of the line-forming region at 2 bar to 3 bar. Atmospheric compositions were set to the current estimates of the gaseous composition outlined in Tables 4.6–4.9. For reference, the Planck function for a number of temperatures has also been plotted. Thermal emission diminishes rapidly with temperature and wavenumber and thus the emission of the ice giants is significantly smaller than for gas giants, as can be seen by the vertical scales of Figure 6.16 and by the steady diminishment of 5 μm and mid-IR radiance as we go from Jupiter to Neptune. Hence, it can be immediately seen that it is very much more difficult to measure the thermal emission spectra of the ice giants than the gas giants since the radiance levels are so much lower. An alternative way of

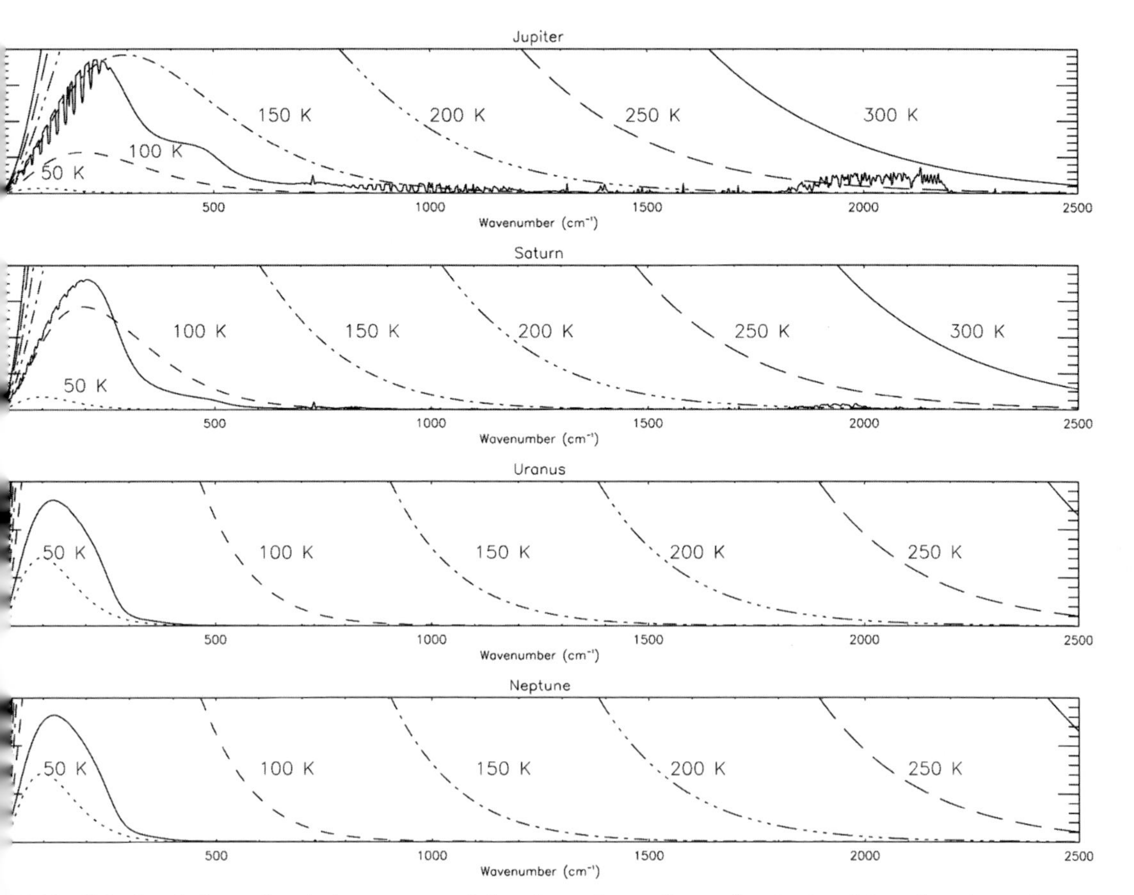

6.16. Calculated thermal emission spectra of the giant planets for nadir viewing. Note that the scales are nt since the integrated flux from the planets decreases as T^4, where T is mean emission temperature. The Planck ns at temperatures 50, 100, 150, 200, 250, and 300 K have also been plotted for reference. Radiance units are $^{-2}$ sr^{-1} (cm^{-1})$^{-1}$.

comparing the thermal spectra of the giant planets is to instead plot the log of the radiance (as shown in Figure 6.17), or to plot their *brightness temperature* spectra (as has been done in Figure 6.18). The brightness temperature is defined as the temperature of a black body that emits the same radiance as that observed at a given wavenumber.

Jupiter

To interpret the spectrum shown in Figures 6.16–6.18, it is useful to consider also the peak levels of the weighting functions for this planet, shown earlier in Figure 6.9. Starting in the far-infrared, the weighting function at $\sim$5 cm^{-1} peaks fairly deeply, but the Planck function for all temperatures tends to zero in this region leading to the

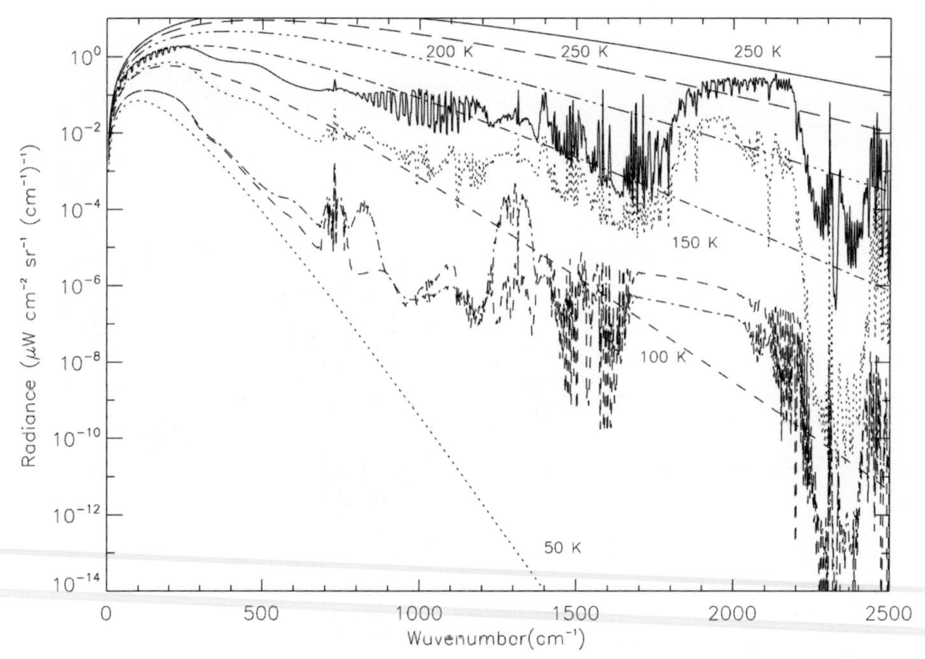

Figure 6.17. Calculated thermal emission spectra of the giant planets on a log scale. Jupiter: solid line; Saturn: dotted line; Uranus: dashed line; Neptune: dash-dotted line. The rapid decrease in brightness, due to the decrease in mean thermal temperature is clearly seen, as is the disappearance of spectral features due to ammonia and water vapor as we go outwards through the solar system. Absorption of methane is clearly visible for all planets as is the emission of hydrocarbons such as ethane and acetylene from the stratospheres. The Planck functions at temperatures 50, 100, 150, 200, 250, and 300 K have also been plotted for reference. The spectra of Jupiter and Saturn have been calculated for cloud-free conditions while deep "H_2S" clouds have been assumed for Uranus and Neptune with optical depth of unity at ~3 bar.

small radiance seen. As the wavenumber increases, opacity increases due to H_2–H_2 and H_2–He CIA, rotational absorption lines of mainly NH_3, and to a lesser extent PH_3 and CH_4. Although weighting functions move upwards to cooler levels, the Planck function increases rapidly and thus the spectrum peaks at $\sim 210\,\mathrm{cm}^{-1}$. Between $220\,\mathrm{cm}^{-1}$ and $600\,\mathrm{cm}^{-1}$ the spectrum is smooth and arises due almost entirely to H_2–H_2, H_2–He CIA. As the weighting function continues to drift slowly upwards, the decreasing temperature and the decay of the Planck function causes radiance to decrease. Between $600\,\mathrm{cm}^{-1}$ and $700\,\mathrm{cm}^{-1}$ the weighting function starts to drift downwards again until the $729\,\mathrm{cm}^{-1}$ vibration–rotation band of stratospheric acetylene appears, which in the central Q-branch introduces a second peak to the weighting function at $\sim 1 \times 10^{-5}$ bar (not shown in Figure 6.9). Since the stratosphere is warm at this altitude, the acetylene Q-branch introduces the characteristic "spike" seen in the spectrum at this wavelength. A perpendicular band of ethane next appears

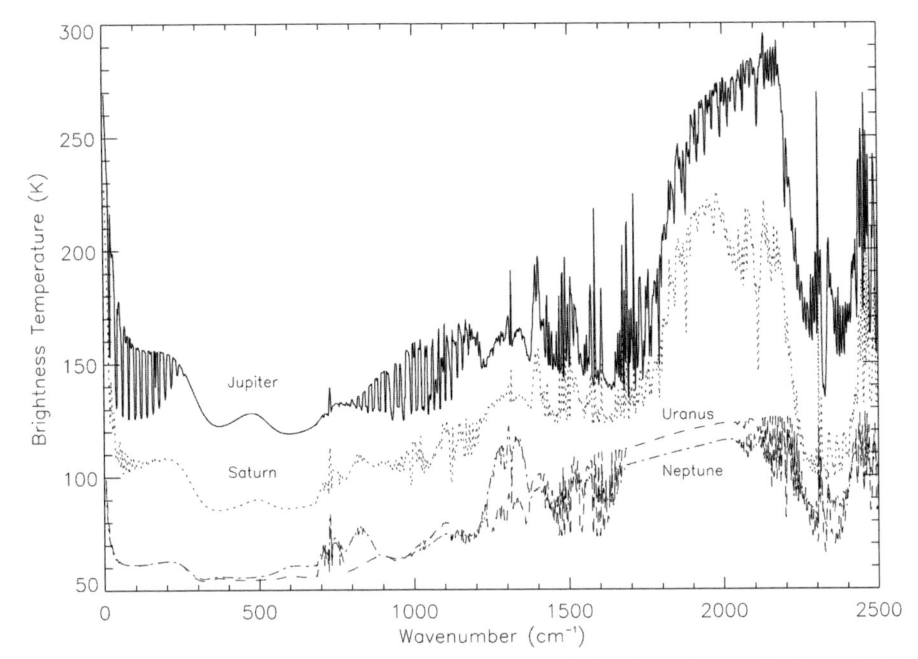

Figure 6.18. Calculated brightness temperature spectra of the giant planets. Jupiter: solid line; Saturn: dotted line; Uranus: dashed line; Neptune: dash-dotted line.

at $\sim$820 cm^{-1}, although this is more apparent at higher zenith angles. The spectrum between 800 cm^{-1} and 1,200 cm^{-1} is mostly dominated by vibration–rotation transitions of upper-tropospheric ammonia, and to a lesser extent phosphine and CH_3D. Between 1,200 cm^{-1} and 1,400 cm^{-1} there appears one of the main vibration–rotation bands of methane. The high absorption of methane pushes the weighting function high into the stratosphere and the characteristic P, Q, R-branch system is visible. The spectrum between 1,400 cm^{-1} and 1,800 cm^{-1} is composed of ammonia, CH_3D, and weaker bands of methane. The strength of these absorptions decreases rapidly after 1,800 cm^{-1}, and thus between 1,800 cm^{-1} and 2,100 cm^{-1} the weighting functions peak deep in the atmosphere, providing there are no clouds. The absorption features seen in this 5 μm window arise from deep H_2O, NH_3, CH_3D, CO, AsH_3, GeH_4, and PH_3 absorptions and thus allow for abundance determinations of these molecules in the 5 bar to 8 bar pressure region. Above 2,100 cm^{-1}, a strong vibration–rotation band of phosphine appears, which pushes the weighting function back up to approximately the 1 bar level. At higher wavenumbers, the spectrum becomes dominated by reflected sunlight on the day side. There is a small contribution of reflected sunlight in the 5 μm window, although this is usually negligible provided that the clouds are not too thick, as can be seen in Figure 6.19 where thermal emission from 1,500 cm^{-1} to 2,500 cm^{-1} has been over-plotted with the calculated reflected solar radiance from a cloud layer at 1 bar with albedo of 0.1.

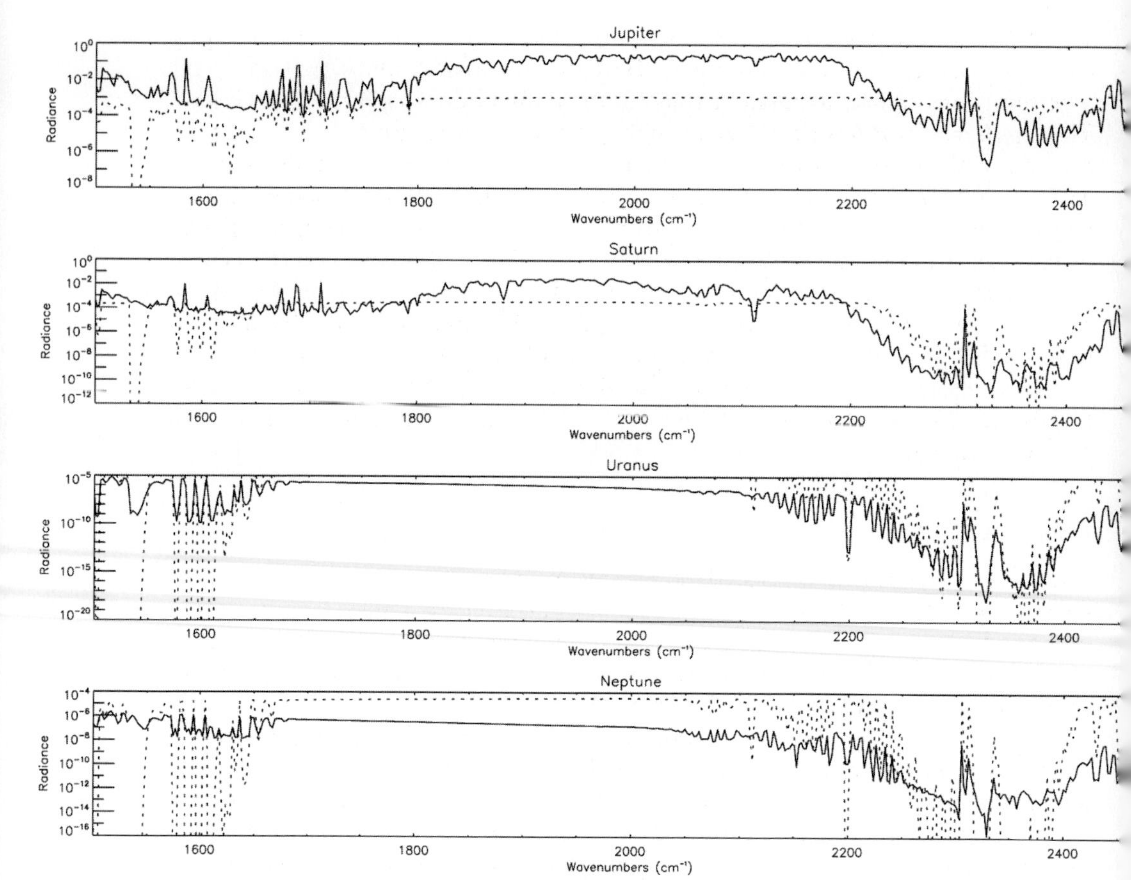

Figure 6.19. Overlap spectral regions between thermal emission (solid) and reflected sunlight (dotted) for the planets. Calculated reflection is from a Lambertian layer with albedo 0.1 placed at 1 bar for Jupiter and Saturn 3 bar for Uranus and Neptune. Radiance units are again $\mu W\,cm^{-2}\,sr^{-1}\,(cm^{-1})^{-1}$.

Saturn

The thermal emission spectrum of Saturn is in many respects rather similar to that of Jupiter, but of significantly lower overall brightness. Spectral features are formed from similar gas absorptions, with the main differences in the observed spectra coming from the lower mean atmospheric temperatures and the greater scale height of Saturn's atmosphere. Between $0\,cm^{-1}$ and $600\,cm^{-1}$, the overall shape of the spectrum is similar to that of Jupiter's with the exception that the low abundance of ammonia above 1 bar leads to greatly reduced rotational absorption lines at wavenumbers less than $200\,cm^{-1}$. Stratospheric emission from acetylene is again seen at $729\,cm^{-1}$, together with the clear appearance of the ethane perpendicular band at $820\,cm^{-1}$. The absorption bands of ammonia between $750\,cm^{-1}$ and $1{,}100\,cm^{-1}$ are much less strong, and instead the absorption features of phosphine are more prominent in this region. The strong vibration–rotation band of methane is again seen between 1,200 cm^{-1} and 1,400 cm^{-1} and the radiation in this region thus again comes mostly from the stratosphere. Between 1,400 cm^{-1} and 1,800 cm^{-1}, the spectrum looks similar to Jupiter's (but colder) until we arrive at the 5 μm window. The weighting functions here peak at slightly lower pressures than for Jupiter, both because of the extra opacity introduced by significantly supersolar abundances of gases such as phosphine and ammonia, and also because of Saturn's greater scale height. Combined with Saturn's lower-tropospheric temperatures, thermal emission in the 5 μm window is substantially smaller than that of Jupiter. In fact, thermal emission is so low that reflected sunlight is found to be a substantial component in the 5 μm spectrum at longer wavenumbers (Figure 6.19) and thus must be carefully modeled for day-side observations. At wavenumbers greater than 2,100 cm^{-1}, the strong absorption of PH_3 pushes the weighting function back up into the upper troposphere and the thermal emission drops to levels insignificant compared with reflected sunlight.

Uranus and Neptune

The thermal emission spectra of Uranus and Neptune were poorly known before recent observations with *Spitzer*, but are substantially similar. The spectra are formed in the same way as for Jupiter and Saturn, but since the atmospheres of these planets are so cold, the power of the spectra is extremely low, which makes them experimentally very difficult to measure. Between 200 cm^{-1} and 400 cm^{-1}, the spectrum was fairly well measured by the IRIS instrument on *Voyager 2*, and has the appearance shown in Figures 6.16 to 6.18. Ground-based microwave observations indicate very low abundances of ammonia in the observable troposphere and thus the rotational absorption lines of this gas are predicted to be completely lacking at all wavenumbers as shown. The acetylene spike at 729 cm^{-1} has been detected on both planets, but the signature of ethane is very weak in Uranus' atmosphere, both due to its low estimated stratospheric abundance of 1×10^{-8} and to Uranus' lower-stratospheric temperature. A similar effect can be seen in the center of the methane vibration–rotation band between 1,200 cm^{-1} and 1,400 cm^{-1}, where Neptune shows enhanced stratospheric emission compared with the neighboring tropospheric emissions, while Uranus shows little contrast. While there are expected to be no ammonia absorption features between 750 cm^{-1} and 1,200 cm^{-1} as shown, there may conceivably be phosphine absorption features observable. We have assumed a solar abundance in these calculations, since to date the abundance of phosphine in both planets' atmospheres has not been measured, although an upper limit of 2× the solar value has been derived for Uranus from VLT observations (Encrenaz *et al.*, 2004). Additional absorption features in this region are due to CH_3D and H_2–H_2, H_2–He CIA. In the range between 1,400 cm^{-1} and 1,800 cm^{-1}, absorption features of CH_3D and CH_4 dominate, but then the absorption becomes due effectively to H_2–H_2, H_2–He CIA alone since the abundance of the strong 5 μm absorbing gases—water vapor and ammonia—is estimated to be zero and the abundance of other 5 μm absorbing gases is unknown (although we have assumed solar abundances of GeH_4 and AsH_3, in addition to PH_3). Hence, the calculated 5 μm spectra appear rather smooth.

Unfortunately the real, observable 5 μm spectra of these planets are not at all well known due to the very low power of emitted radiance in this spectral region. However, the spectra from regions on the sunlit side of these planets are expected to be dominated by reflected sunlight, as can be seen in Figure 6.19 where the reflection from a Lambertian layer placed at 3 bar and with an albedo of 0.1 has also been plotted. An early attempt to measure Uranus' 5 μm spectrum, Orton and Kaminski (1989) found the 5 μm spectrum to be very different in character from that of Jupiter or Saturn, but the data did not have sufficient accuracy to allow unambiguous conclusions concerning composition and cloud structure. More recently, VLT observations of Uranus' 5 μm spectrum have allowed the detection of CO, and revised the upper limit on the abundance of phosphine (Encrenaz *et al.*, 2004).

6.7.4 Microwave spectra

Although the Planck function tends to zero at longer wavelengths, the thermal emission signal is measurable with ground-based microwave telescopes and has proved extremely useful since the opacity of any aerosols, and most of the atmospheric gases, becomes small, allowing the weighting function to probe down to almost 100 bar in some cases. For Jupiter the spectrum is complicated by synchrotron emission radiated by relativistic particles trapped in Jupiter's strong magnetic field, which dominates at decametric wavelengths. The wavelength where thermal emission from the atmosphere and synchrotron emission from the radiation belts are equal is approximately 7 cm (Berge and Gulkis, 1976). Fortunately the thermal and synchrotron components of Jupiter's microwave spectrum have different polarizations and other properties which make them separable. Synchrotron emission from the radiation belts of the other giant planets is negligible.

The microwave spectra of all four giant planets are well reviewed by de Pater and Lissauer (2001), de Pater and Massie (1985), and de Pater and Mitchell (1993), and the mean spectral observations are shown in Figure 6.20. The main gaseous absorber in this region is ammonia, which has an inversion band at 1.3 cm, and the absorption of ammonia is clearly visible in ground-based microwave spectra of Jupiter and Saturn. With the excellent spatial resolution achievable with current microwave observatories such as the Very Large Array (VLA), described in Section 7.6.2, the giant planets can be mapped from the Earth at microwave wavelengths. For Jupiter it has been possible to map the abundance of ammonia across the planet and it is found that belts are regions of depleted ammonia and zones regions of enhanced ammonia (de Pater, 1986; de Pater and Dickel, 1986) (Figure 6.21) as is expected from the generally accepted view that, above the water cloud, zones are regions of upwelling, moist air, and belts are regions of downwelling, desiccated air. Similar belt/zone contrasts are seen at these wavelengths on Saturn. Although the microwave spectra of Uranus and Neptune apparently lack a clear ammonia absorption feature, indicating depletion of this species at great depth, we saw in Section 5.7.1 that variations in the abundance of this species may account for the latitudinal variation of brightness temperature seen on Uranus at microwave wavelengths; imaging of Neptune's disk at

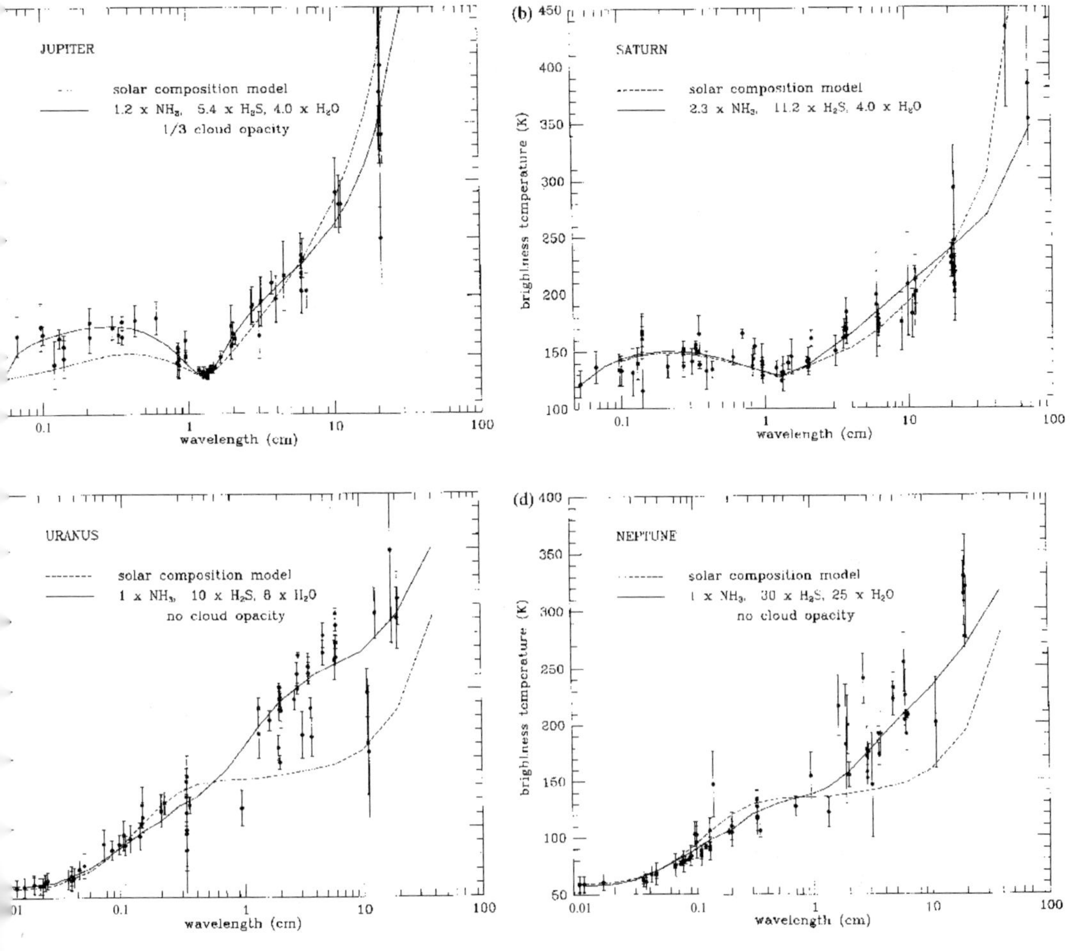

6.20. Microwave and radio emission spectra of the giant planets (from de Pater and Lissauer, 2001; and de nd Mitchell, 1993). Courtesy of Cambridge University Press.

these wavelengths has also been achieved. In addition, carbon monoxide and HCN have been detected at these wavelengths on Neptune.

6.8 APPENDIX

6.8.1 Planck function

The Planck function defines the radiance emitted by a surface of unit emissivity as a function of wavelength. There are two forms that are commonly used in IR spectroscopy depending on the unit of wavelength.

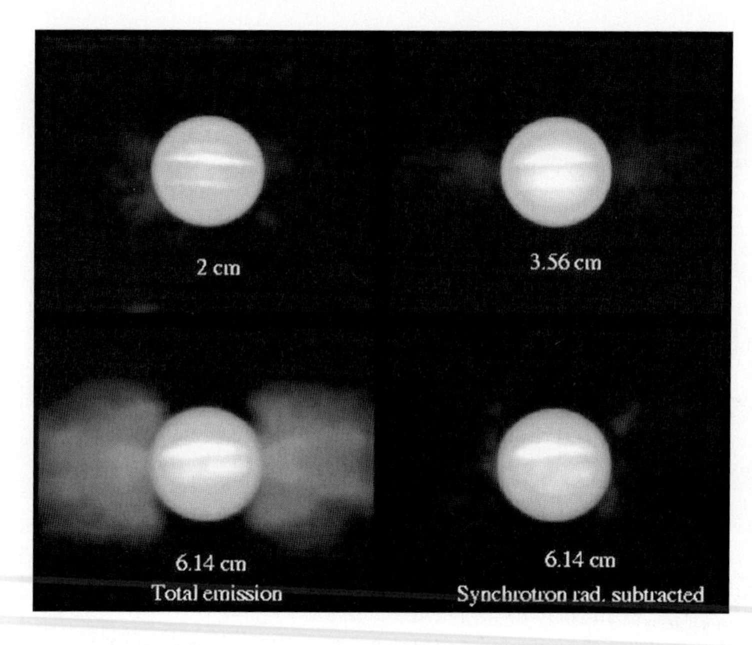

Figure 6.21. Appearance of Jupiter at 2.0, 3.56, and 6.14 cm as observed by the VLA (courtesy of Imke de Pater). The increasing contribution of synchrotron emission from the radiation belts with wavelength is clear.

In terms of wavenumbers (cm^{-1}), the Planck function is defined as

$$B(\tilde{\nu}, T)\, d\tilde{\nu} = \frac{1.1911 \times 10^{-12} \tilde{\nu}^3}{\exp(1.439\tilde{\nu}/T) - 1} d\tilde{\nu} \tag{6.79}$$

where the units of $B(\tilde{\nu}, T)$ are $\text{W cm}^{-2}\,\text{sr}^{-1}\,(\text{cm}^{-1})^{-1}$. In terms of wavelength (μm), the Planck function is defined as

$$B(\lambda_{\mu m}, T)\, d\lambda_{\mu m} = \frac{1.1911 \times 10^4}{\left(\exp\dfrac{14{,}390}{\lambda_{\mu m} T} - 1\right)\lambda_{\mu m}^5} d\lambda_{\mu m} \tag{6.80}$$

where the units are $\text{W cm}^{-2}\,\text{sr}^{-1}\,\mu\text{m}^{-1}$.

6.9 BIBLIOGRAPHY

Andrews, D.G. (2000) *An Introduction to Atmospheric Physics*. Cambridge University Press, Cambridge, U.K.

Goody, R.M. and Y.L. Yung (1989) *Atmospheric Radiation: Theoretical Basis* (Second Edition). Oxford University Press, Oxford, U.K.

Hanel, R.A., B.J. Conrath, D.E. Jennings, and R.E. Samuelson (2003) *Exploration of the Solar System by Infrared Remote Sensing* (Second Edition). Cambridge University Press, Cambridge, U.K.

Herzberg, G. (1945) *Molecular Spectra and Molecular Structure, II: Infrared and Raman Spectra of Polyatomic Molecules*. Van Nostrand Reinhold, New York.

Hollas, J.M. (1992) *Modern Spectroscopy* (Second Edition). John Wiley & Sons, New York.

Houghton, J.T. (1986) *The Physics of Atmospheres* (Second Edition). Cambridge University Press, Cambridge, U.K.

7

Sources of remotely sensed data on the giant planets

7.1 INTRODUCTION

In the previous chapter we saw how the ultraviolet (UV), visible, infrared (IR), and microwave spectra of the giant planets are formed, and how the absorption features of different gases (and theoretically aerosols) are visible in these spectra. Clearly much can be learned about the atmospheres from observing the spectra of these planets, and in this chapter we will review the measurements that have been made to date and how they may be used.

In this chapter we will review some of the technical details of measuring the UV–microwave spectra of the planets and in Section 7.2 we will briefly review how such radiation is detected and how spectra are measured. Prior to 1973, the only measurements of the giant planets that were available were telescope observations from the surface of the Earth in the visible, IR, and microwave wavelengths. Such observations have the obvious attraction that they are relatively easy to do, and have a number of other advantages, although there are drawbacks as we shall see in Section 7.3. We shall also see that the technology of detection and data processing has improved dramatically over the years and thus ground-based observations continue to be a rich source of information on the giant planets to this day. In Sections 7.4 and 7.5 we will look at some of the major ground and airborne visible/IR telescope facilities around the world that are engaged in outer planet observations and in Section 7.6 we will look at ground-based microwave observations. Many of the problems encountered by ground-based visible/IR telescopes are negated by placing the telescope in orbit around the Earth, and thus in Section 7.7 we will look at recent space telescopes such as the Hubble Space Telescope (HST), the Infrared Space Observatory (ISO), and the *Spitzer* Space Telescope.

In 1973, the first spacecraft mission arrived at Jupiter and since then a number of spacecraft have flown past the giant planets, and more recently have been placed in orbit to conduct extended campaigns. These missions have enormous advantages over ground-based and space-based telescopes, but are of course immensely expensive and difficult to achieve. In Section 7.8 we will review the flyby missions of the giant planets and then in Section 7.9 we will consider the orbiting missions of *Galileo* and *Cassini/Huygens*. Finally we will discuss retrieval methods in Section 7.10 where we will see how remotely sensed observations are actually used to infer atmospheric properties in planetary atmospheres.

7.2 MEASUREMENT OF VISIBLE, IR, AND MICROWAVE SPECTRA

Before going on to look at the current sources of remotely sensed spectral data on the giant planets and how these data are reduced to infer atmospheric properties, we will briefly look at how the spectra of these planets are actually measured from the visible through to microwave wavelengths.

7.2.1 Detection of IR radiation

For all remote-sensing instruments, the incident radiance must first be collected and focused onto detecting elements in order to record a signal. In any detection system there are a number of sources of noise and the design of remote-sensing instruments aims to minimize these in order to achieve a high signal-to-noise ratio. Suppose that radiance B ($\mathrm{W\,m^{-2}\,sr^{-1}(cm^{-1})^{-1}}$) is incident upon a remote-sensing system. The power $P(W)$ incident on the detector is given by

$$P = A\Omega \int F(\tilde{\nu})B(\tilde{\nu})\,d\tilde{\nu} + E \tag{7.1}$$

where $F(\tilde{\nu})$ is the spectral transmission of the optical system; and the $A\Omega$ product is given either by the area of the entrance aperture multiplied by the solid angle of the field of view (FOV) observed by the instrument, or equivalently by the area of the detector multiplied by the solid angle of the cone of radiation condensed onto the detector by the instrument optics. The quantity E in Equation (7.1) refers to thermal radiation self-emitted by the telescope and optics, which is incident on the detector. The signal detected by the instrument (e.g., volts or amps) is then given by

$$S = A\Omega \int F(\tilde{\nu})R(\tilde{\nu}, T)B(\tilde{\nu})\,d\tilde{\nu} + E' \tag{7.2}$$

where $R(\tilde{\nu}, T)$ is the *spectral responsivity* of the detector, and has units of V/W or A/W. In the detection and pre-amplification stages of instruments there are numerous sources of noise such as Johnson noise (or voltage noise, which appears across resistances), Shot noise (or current noise, arising from the fact that a "steady current" is actually composed of a stream of individual electrons), noise arising from the incident radiation itself, and radiation from the optical elements if we are considering the thermal-infrared (e.g., Hanel *et al.*, 2003). The Shot noise associated with a current I_0 is given by $I_S^2 = 2eI_0\Delta f$, where Δf is the bandwidth and e is the electron charge, and is clearly minimized by limiting the currents in the detection stages of amplification. Johnson noise, however, given by $V_J^2 = 4k_BTR\Delta f$, where R is the resistance and k_B is the Boltzmann constant, depends on temperature, as does the noise of radiation thermally emitted by the optics and filters. Hence, to maximize the signal-to-noise ratio of the detected radiance, especially when working in the thermal-infrared, the detectors, filters, and as much of the telescope optical system as is possible must be cooled to low temperatures. Adding all sources of noise together, a common figure of merit of the instrument is the noise equivalent power (NEP), which is defined as the power of incident radiation that, when viewed in a 1 Hz bandwidth, gives a signal equivalent to all the sources of noise. Another very useful figure of merit for comparing the sensitivities of detector is D^* which is defined as

$$D^* = \frac{(A\Delta f)^{1/2}}{\text{NEP}} \tag{7.3}$$

where A is the area of the detector. Highly sensitive detectors have a high D^*. Finally, for thermal-IR observations, all the sources of noise may be analyzed in terms of their noise equivalent radiance (NER), defined as the incident spectral radiance which gives a signal-to-noise ratio of unity.

There are two main ways of detecting radiation: (1) photon detectors, which detect individual photons; and (2) thermal detectors, or *bolometers*, which detect the temperature rise of elements exposed to radiation. Examples of photon detectors are: photovoltaic cells, where absorbed photons promote the production of an electron–hole pair in a p–n junction and thus produce a transient voltage; and photoconductive detectors, where absorbed photons again promote the production of an electron–hole pair in an element of semiconductor material, which temporarily alters its conductivity. Examples of thermal conductors include: thermopiles, which are basically a stacked array of bimetallic junctions that produce a voltage dependent on their temperature via the thermocouple effect; and pyroelectric detectors, which use a dielectric material with a temperature-sensitive dipole moment sandwiched between the plates of a capacitor. Absorption of thermal radiation modifies the permittivity and thus the capacitance. The choice of detector depends on many things: cost, required signal-to-noise ratio, and response time. The reader is referred to more specialized texts for further information (Hanel *et al.*, 2003; Houghton and Smith, 1966; Smith *et al.*, 1968).

7.2.2 Radiometers/Photometers

Many remote-sensing instruments simply record the incident radiation received within a bandwidth defined by a set of spectral filters, and most imaging cameras operate in this way. Photometers record accurately the flux level of visible or near-IR light within narrow spectral channels, while radiometers perform a similar function at thermal-IR wavelengths. Such an instrument design is cheap, reliable, and ideally suited to imaging although it does require *a priori* knowledge of the planetary spectrum in order to place the channel filters at suitable wavelengths.

Where the planetary spectrum is less well known, spectrometers must be used of which in the visible/IR there are two main types: grating spectrometers and interferometers. For really high-resolution visible/IR work, Fabry–Pérot interferometers may also be used, usually in conjunction with a grating spectrometer, which correctly limits the range of wavelengths that are passed through the Fabry–Pérot interferometer to eliminate aliasing.

7.2.3 Grating spectrometers

Grating spectrometers are often used for IR spectroscopy, particularly at near-IR wavelengths and have the advantage of relative simplicity. Light is collected from the planet via a telescope system and then the collimated light illuminates a reflecting diffraction grating, which disperses a spectrum onto the focal plane (Figure 7.1). The grating is "blazed" to maximize the throughput at the central wavelength of the region of interest. In its simplest form, there is a single IR detector at the focal plane, and the spectrum is measured by recording the detector signal as the grating is scanned through a small angle. Since IR detectors are usually sensitive to a wide range of wavelengths, an "order-sorting" filter must in practice also be added to limit the range of wavelengths that can be detected so that the spectrometer only operates in the spectral order desired.

While such a design is simple, only a small range of wavelengths are recorded at a time, and thus much of the radiation that is dispersed by the grating is wasted. Also it takes a certain length of time to scan the grating through the angular range required

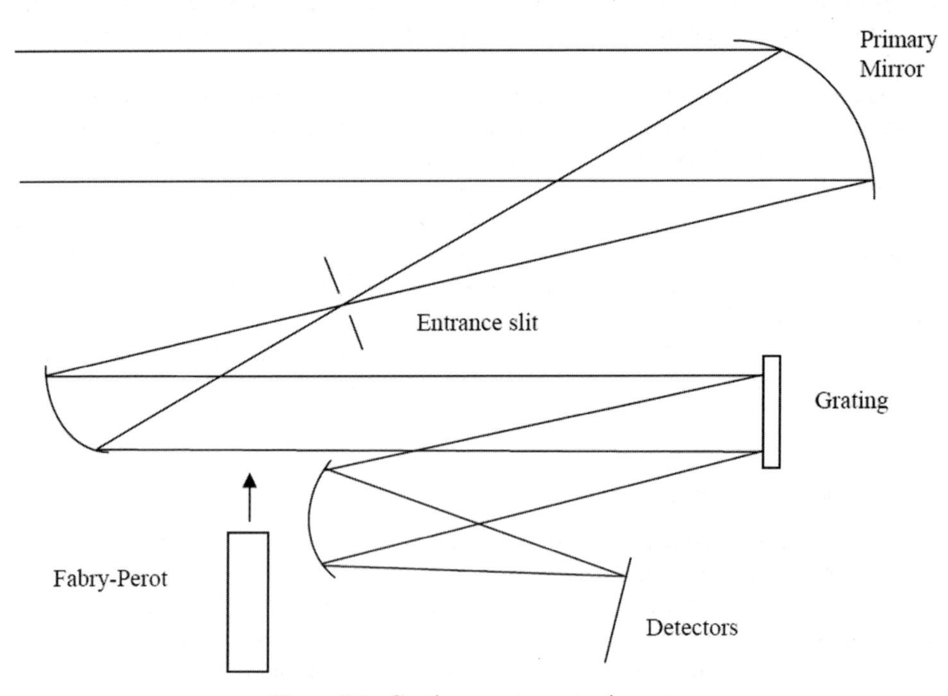

Figure 7.1. Grating spectrometer layout.

to build up a spectrum, and this can be particularly problematic from an observational point of view when the instrument is observing a planet whose spectrum alters significantly with position. During the time the grating is scanned, the spectrometer may be looking at a very different region at the end of the scan to that observed at the beginning, and thus the recorded spectrum may be very hard to interpret. An alternative approach, sometimes called a spectrograph, is to have a whole array of closely spaced, contiguous detectors in the focal plane, each recording different wavelengths, and thus leave the grating angle fixed. While such instruments are much harder and more expensive to build, a spectrum is recorded much more quickly, and if the region of the planet observed should vary, all parts of the recorded spectrum are equally affected. An intermediate approach is to use fewer detectors, spaced farther apart on the focal plane and then scan the grating over a short range, such that the final spectrum is built up from a number of subspectra recorded by each individual detector. This design is easier to build, records a spectrum in a reasonably short space of time, but can suffer from variable scene problems which reveal themselves as mismatches between the individual detector subspectra where they overlap.

Clearly, for planetary work it is of great interest to record the spectrum at a number of locations on the planet in order to build up a multispectral image. This can be achieved by scanning the instrument in both x and y directions, but in more recent years it has been possible to construct imaging spectrographs where a two-dimensional array of detectors is placed in the focal plane, and thus multiple spectra are simultaneously recorded from a small range of viewing angles perpendicular to the grating dispersion direction. The instrument then need only be scanned in one direction (the grating dispersion direction)—not two—in order to construct a multispectral image.

The spectral resolution of a grating spectrometer is fixed by the size of the entrance aperture to the collimator, the dispersion of the grating, and by the physical dimensions of the detectors. For higher resolution spectroscopy, some instrument designs allow for a Fabry-Pérot interferometer to enter the beam as indicated in Figure 7.1. This allows very high–resolution spectrometry to be conducted over a small spectral range with the resolving power ($\lambda/\Delta\lambda$) of the order of 10,000. Other designs of grating spectrometers use Echelle spectroscopy, where the spectrum is dispersed in two directions onto a two-dimensional pixel array. In this way many orders of interference are observed simultaneously, allowing an Echelle spectrometer to have very much higher resolution than a conventional grating spectrometer.

7.2.4 Michelson interferometers

While grating spectrometers perform spectroscopy by "division of wavefront", interferometers are also often used at mid-IR to far-IR wavelengths, which operate by division of amplitude. The simplest example is a Michelson interferometer, where light is split into two beams by a beam-splitting mirror and then recombined onto a detector (as in Figure 7.2). For monochromatic light of intensity I_0, the intensity at the detector varies with scanning mirror position x as (e.g., Hecht and

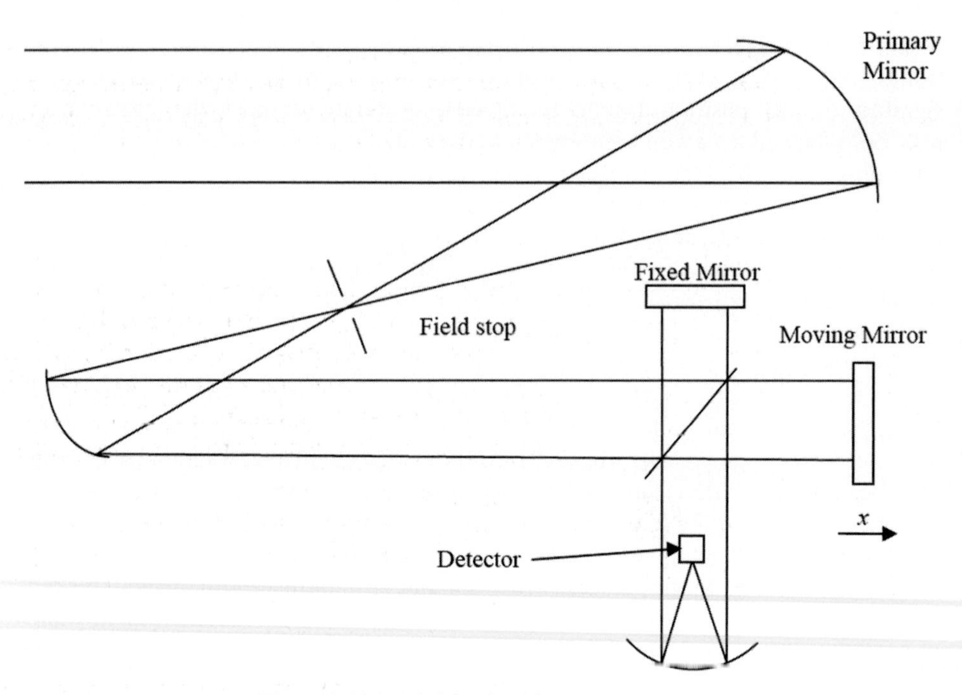

Figure 7.2. Michelson interferometer layout.

Zajac, 1974)

$$I(\Delta) = \frac{I_0}{2}(1 + \cos k\Delta) \tag{7.4}$$

where $k = 2\pi/\lambda$, and $\Delta = 2x$. For incident light with a spread of wavelengths and spectral density $I(k)\,dk$, Equation (7.4) must be modified to

$$I(\Delta, k)\,dk = \frac{I(k)}{2}(1 + \cos k\Delta)\,dk \tag{7.5}$$

and since IR detectors are typically sensitive to a wide range of wavelengths $\Delta\tilde{\nu}$, the detected intensity must be integrated over all wavelengths in this range:

$$I(\Delta) = \frac{1}{2}\int_{\Delta\tilde{\nu}} I(k)(1 + \cos k\Delta)\,dk = \frac{I_0}{2} + \frac{1}{2}\int_{\Delta\tilde{\nu}} I(k)\cos k\Delta\,dk. \tag{7.6}$$

Hence, in the limit that $\Delta\tilde{\nu}$ is large it can be seen that the intensity detected is equal to a mean value plus the Fourier cosine transform of the incident spectrum. The signal $I(\Delta)$, recorded as a function of mirror position, is called an *interferogram* and the incident spectrum may then theoretically be perfectly reconstructed by the inverse Fourier cosine transform as

$$I(k) = \frac{2}{\pi}\int_0^\infty [I(\Delta) - I_0/2]\cos k\Delta\,d\Delta. \tag{7.7}$$

Such instruments are thus often referred to as Fourier Transform Spectrometers

(FTS). Let us consider the case where the interferometer views monochromatic light of wavevector k_0. From Equation (7.4), the resulting interferogram is a pure cosine wave and substituting this into Equation (7.4) we find

$$I(k) = \frac{2}{\pi} \int_0^\infty (I_0/2) \cos k_0 \Delta \cos k\Delta \, d\Delta \equiv I_0 \delta(k - k_0) \tag{7.8}$$

where $\delta(k - k_0)$ is the Dirac delta function, and thus the reconstructed spectrum is found to be purely monochromatic as expected. However, in reality the pathlength $2x$ may only be scanned up to a maximum value of Δ_m, and this limits the spectral resolution of the interferometer. If we limit Δ to Δ_m in Equation (7.8), then the reconstructed spectrum is no longer monochromatic, but instead has finite width since

$$\begin{aligned} I(k) &= \frac{2}{\pi} \int_0^{\Delta_m} (I_0/2) \cos k_0 \Delta \cos k\Delta \, d\Delta \\ &= \frac{I_0}{\pi} \int_0^{\Delta_m} \tfrac{1}{2} [\cos(k_0 + k)\Delta + \cos(k_0 - k)\Delta] \, d\Delta \\ &= \frac{I_0}{2\pi} \left\{ \frac{\sin(k_0 + k)\Delta_m}{(k_0 + k)} + \frac{\sin(k_0 - k)\Delta_m}{(k_0 - k)} \right\}. \end{aligned} \tag{7.9}$$

Since k_0 is typically large (except for the very long wavelengths) this may usually be approximated as

$$I(k) \approx \frac{I_0}{2\pi} \frac{\sin(k_0 - k)\Delta_m}{(k_0 - k)} \tag{7.10}$$

and thus the reconstructed spectrum has a finite spread of wavelengths, although this becomes more monochromatic as Δ_m is increased since $I(k) \to \delta(k - k_0)$ as $\Delta_m \to \infty$, as expected. Hence, an interferogram which is truncated with a maximum path difference of Δ_m has an effective spectral resolution of $\Delta\tilde{\nu} = 1/\Delta_m$. Because of this feature, the spectral resolution of a Fourier Transform Spectrometer is easily adjustable by simply recording longer or shorter interferograms, and thus the same instrument may record spectra with multiple resolutions. In addition, the shape of the spectral instrument function may also be adjusted through the process of *apodization*. We can see in Equation (7.10) that the instrument function, when an interferogram is transformed directly with a hard cut-off at Δ_m, is essentially a sinc function. Suppose that there was a weak feature in the true spectrum very close to a strong feature. Because the sinc function has non-negligible ripples next to it, the weak feature might easily get lost in the ripples of the strong feature in the reconstructed spectrum, especially if noise was also present. To avoid this, the interferogram may first be multiplied by an apodizing function $A(\Delta)$, which instead of imposing a hard cut-off at Δ_m forces the interferogram to decay smoothly to zero at Δ_m. It may easily be shown that this changes the effective shape of the instrument function and if applied correctly may completely remove the ripples from strong features at the expense of slightly lowering the overall spectral resolution. This technique is thus called apodization which derives from Greek words meaning literally "removal of feet"!

Another factor to consider in real Fourier Transform Spectrometers is *aliasing*. In practice an interferogram is sampled at a finite resolution of the path difference $\Delta = 2x$. Higher frequencies in the observed spectrum can be seen from Equation (7.5) to appear as higher and higher frequency components in the interferogram. If the minimum sampling path difference is Δ_S, then the maximum frequency in the incident spectrum that may unambiguously be reconstructed, according to the Nyquist sampling theorem, is $\nu_0 = 1/(2\Delta_S)$ (James and Stern, 1969; Vanasse, 1983). Hence, if any higher frequencies are detectable by the system then they may artificially appear at lower and incorrect frequencies between 0 and ν_0. This phenomenon is known as aliasing and may be removed by ensuring that the frequency response of the actual detection system is limited to the frequency range defined by the sampling limit.

Together with allowing variable instrument functions, one of the major advantages of interferometers is that they simultaneously record data over a wide wavelength range and thus little radiation that is collected by the telescope system is wasted. This property also means that interferometers suffer less from FOV variations that may occur during an interferogram scan than grating spectrometers. One possible drawback of the classic Michelson interferometer design for space operation is that the flat mirrors need to be very precisely aligned in order that the central spot of the interference pattern falls on the detector. Such a precise alignment can easily be destroyed during the vibrations which accompany the launch of spacecraft, although the IRIS instruments on *Voyager* (and previous Earth and Mars missions), which had simple Michelson designs, were successful. More recently, the CIRS interferometer on the *Cassini* spacecraft (Section 7.9.2) uses corner reflectors, and roof reflectors, which are much less sensitive to misalignment.

7.2.5 Detection of microwave radiation

The microwave emission of the giant planets is extremely weak and difficult to detect, but technology is rapidly improving. Two main types of receivers are currently used: (1) bolometers; and (2) heterodyne receivers.

Bolometers

With bolometers the radiation is again collected in a tiny absorber, whose temperature changes are converted to electrical signals, which are then amplified and measured. Bolometers can detect broadband radiation (e.g., within the microwave atmospheric windows described in Section 7.3.1), with high sensitivity, but cannot give information on detailed spectral energy distribution within that band.

Heterodyne receivers

Heterodyne receivers of microwave radiation operate by first converting the microwave signals to a lower frequency by nonlinear mixing with a local oscillator signal. The converted lower frequency signal may then be amplified and measured with conventional radio frequency (RF) electronics. Such devices have very high spectral resolution (of the order of 10^6) and so can examine individual absorption lines.

The two critical components of such receivers are the local oscillator and the heterodyne mixer. The local oscillator defines the spectral frequency observed and should be very stable, have low noise, and ideally have low power. The heterodyne mixer allows the measured spectrum to be converted to lower frequencies and be analyzed with conventional RF electronics. The most sensitive receivers at present use the strong heterodyne mixing provided by superconductor–insulator–superconductor (or SIS) tunnel junctions. An SIS junction consists of two superconducting electrodes separated by a very thin insulating barrier. Electrons tunneling across this barrier give rise to a very nonlinear current–voltage characteristic, which is the key to heterodyne mixing. Such junctions need to operate at very cold temperatures in order to achieve superconductivity depending on the material used in their construction, typically the temperature of liquid helium (4.2 K). As an example the SIS junctions currently used by IRAM (Section 7.6.1) consist of a superposition of a thin layer (of the order of 4 μm) of aluminum oxide between two layers of the superconducting metal niobium. However, SIS junctions are still very much under development and thus a number of different superconducting alloys and insulating layer materials are currently under investigation all over the world. In addition to SIS junctions, other types of mixers are being studied, such as Schottky diodes, which can operate at higher temperatures and this is an area of rapid technological development.

Apart from low-noise characteristics, the main advantage of heterodyne receivers is their capability to provide high-resolution spectroscopy, which is very important for detecting the absorption lines of microwave absorbers such as CO and HCN, and determining their abundance by accurately measuring their line depth.

7.3 GROUND-BASED OBSERVATIONS OF THE GIANT PLANETS

There are a number of extremely large and sensitive telescopes across the world that may be used for planetary observations, each of which is equipped with a range of instruments for recording images or spectra of the giant planets. There are enormous advantages in ground-based observations including: (1) long-term monitoring of slow changes (e.g., such as decadal variation in the disk-averaged albedo of Neptune); (2) the flexibility of being able to record with variable spectral resolutions, in particular very high resolution and thus discriminate between individual gaseous absorption lines; and (3) the ability to observe at short notice should something unusual and unexpected occur on the planet in question. However, there are also disadvantages in that the available observing time for these telescopes is very limited, and any time that might be allocated for investigating a particular question may be foiled by weather conditions. In addition, there are also other significant problems of observing the giant planets from the surface of the Earth, including terrestrial atmospheric absorption, angular resolution, and brightness, which will now be discussed.

7.3.1 Terrestrial atmospheric absorption

From the UV through to microwave wavelengths the absorption of the Earth's atmosphere makes some regions of the giant planet spectra completely unobservable from the ground. This can be seen in Figure 7.3 where the vertical transmission to space from the ground has been calculated from UV to microwave wavelengths, both from sea level in a standard atmosphere and from an altitude of 4,000 m in a dry atmosphere. A number of strong absorptions throughout can be seen. At UV wavelengths below 0.3 μm, the Earth's atmosphere is effectively opaque due to photolysis of ozone (O_3) and molecular oxygen (O_2) is the stratosphere. In the visible/near-infrared most of the absorption below 3 μm is due to water vapor. At 4.3 μm a strong absorption band of CO_2 appears, but then between 6 μm and 9 μm (1,600 to

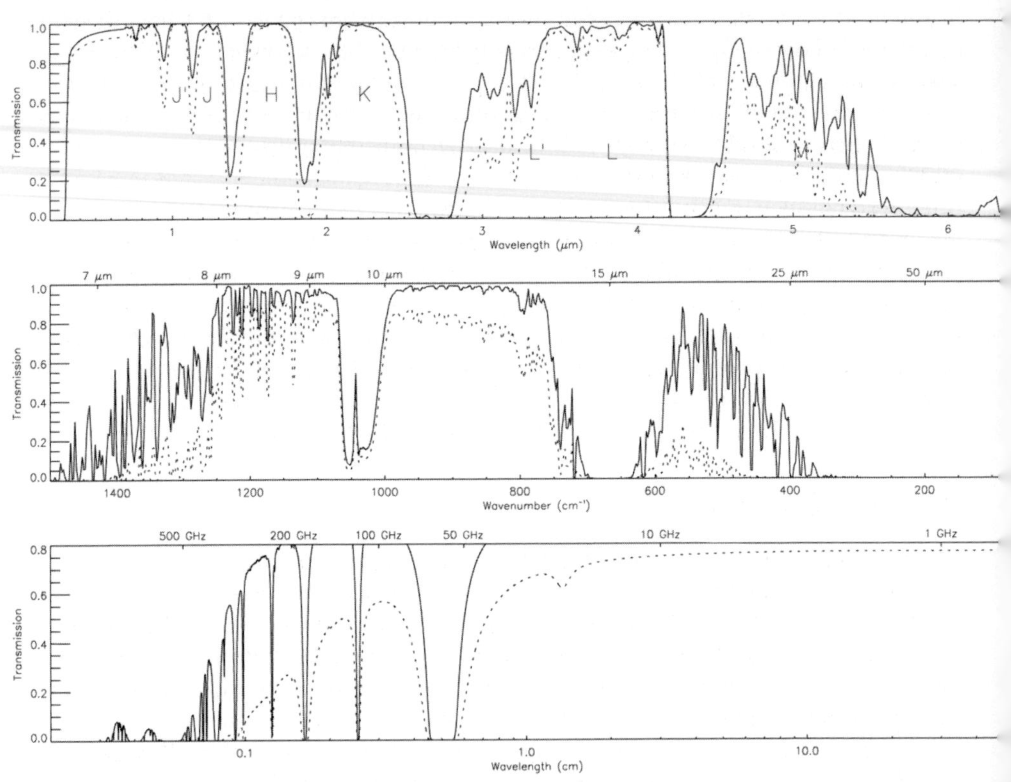

Figure 7.3. Transmission of Earth's atmosphere from ground to space (vertical path). The dotted lin[e is] calculated transmission from sea level in a nominal standard atmosphere while the solid line is from [an] altitude of 4,000 m in a midlatitude summer atmosphere. The transmission advantage of plac[ing] telescopes on mountain tops is clear. In addition, since most of the absorption features are due to wa[ter] vapor, telescopes are preferentially located in the driest regions of the world. In the near-IR range spectral regions where atmospheric absorption is minimum, known as spectral "windows", are commo[nly] called by the letter indicated. Hence, for example, the 2.1–2.4 μm window is commonly known as K-band and so on.

1,100 cm^{-1}) most of the absorption is again due to water vapor. At 9.6 μm (1,040 cm^{-1}) there appears a strong absorption band of ozone, while at 15 μm (667 cm^{-1}) the strong ν_2 absorption band of CO_2 appears. At even longer wavelengths, absorption is dominated by the rotational absorption lines of water vapor, which then tends to zero towards microwave and radio wavelengths. At millimeter wavelengths, in addition to water vapor, O_2 is also strongly absorbing at 2.5 mm and 5 mm due to magnetic dipole absorptions. At longer wavelengths, the atmosphere is effective transparent up to wavelengths of approximately 30 m, where it then becomes opaque again due to ionospheric effects.

Clearly the absorption features of water vapor are a major problem for terrestrial observatories, but fortunately most of the water vapor is held in the lower, warmer levels in the atmosphere. Thus, by placing the telescope at higher altitudes, and/or in desert regions, the absorption of water vapor can be greatly diminished as can be seen in Figure 7.3. In addition, since the pressure of the atmosphere, and thus column abundance of overlying air falls exponentially with height, the absorption of the other gases such as O_2 and CO_2 is also reduced, together with any absorption due to dust or haze. The regions of low atmospheric absorption between the main absorption bands are known as spectral "windows". In the near-IR range these spectral "windows" are often called by the designations shown in Figure 7.3. Hence, the 1.9–2.5 μm window is often referred to as the K-band, and so on. Clearly, even on dry mountain tops the absorption of the terrestrial atmosphere means that several very interesting regions of the giant planet spectra are unobservable. In addition, where the atmosphere is partially clear, the absorption depends on the abundance of highly variable atmospheric constituents such as water vapor and dust. Hence, whenever the spectrum of a planet is recorded, the spectrum of a nearby standard reference star must also be recorded so that the terrestrial absorption may be determined and corrected for. This correction leads inexorably to additional errors in the final recorded spectra.

7.3.2 Angular resolution

The second historical problem with ground-based observations of the giant planets is the achievable angular resolution. Figure 7.4 compares the physical sizes of all the giant planets and the Earth. However, for ground-based telescopes and Earth-orbiting space telescopes, the apparent angular size of the planets decreases greatly as we go from Jupiter to Neptune, and Figure 7.5 compares the apparent sizes of the planets as they appear at opposition where, for reference, the apparent diameter of Jupiter is $\sim 40''$ (i.e., 40 arcsec). Clearly the apparent size of Neptune is very small, making it very difficult to discern variable cloud features, although the disk-averaged brightness is easier to determine and has been monitored from the ground for decades (e.g., Lockwood and Jerzykiewicz, 2006; Lockwood and Thompson, 2002). While the angular resolution of space-based telescopes depends on aperture and optical quality alone, the angular resolution of ground-based telescopes is severely limited by the turbulence of the overlying atmosphere. The strength of this turbulence depends on local atmospheric conditions and is particularly noticeable in winter giving rise to the twinkling of the stars. Typically the "seeing" is limited to approximately $1''$ and in

Figure 7.4. Comparative sizes of the giant planets (Jupiter, Saturn, Uranus, and Neptune) and the Ear

Figure 7.5. Relative apparent sizes of the giant planets as seen at opposition from the Earth wit telescope of infinite resolution.

Figure 7.6, the images have been blurred to this approximate resolution. While considerable cloud detail can still be seen on Jupiter the effect on the other planets becomes increasingly severe as the apparent angular diameter decreases.

The blurring of ground-based telescope images caused by atmospheric turbulence arises due to refractive index variations in the column of air between the telescope and the object. Light from a distant source, such as a star, arrives at the top of the Earth's atmosphere effectively as a plane wave. However, after passing

Figure 7.6. Relative appearance of the giant planets as seen at opposition from the Earth with typi "seeing" of approximately 1 arcsec resolution.

through the Earth's atmosphere to reach the telescope, variations in temperature caused by turbulence introduce small variations in the refractive index of the air which introduce randomly changing phase variations that continuously distort the wavefront and make it impossible to form a diffraction-limited image. The typical correlation time of the distortions is of the order of a few milliseconds and the problem is most evident at visible wavelengths, becoming progressively smaller at longer wavelengths due to variations in the flatness of the wavefront becoming smaller and smaller compared with the wavelength of the light observed. This problem severely affected terrestrial astronomical observations for many years, but recently technology has developed to such an extent that new techniques have been developed that go a long way to negating this problem, as will now be described.

Adaptive optics

The most ambitious technique for correcting the problem of "seeing" is *adaptive optics*, which attempts to remove distortions from the wavefront before the image is formed. This is achieved by simultaneously observing either a bright star close to the target or, if no such star is available, observing a simulated star formed by light scattered back from sodium atoms in the upper atmosphere from a powerful laser situated at the telescope facility. The important thing is to observe the "star" and the target through as nearly the same column of air as possible. Light from the guide star is collimated by the telescope and the flatness of the wavefront sensed. Data from the wavefront sensor is then used, via a suitable control algorithm, to modify a corrector plate, which attempts to null atmospheric distortions before the image is formed. The wavefront sensor and corrector plate operate in a closed loop and when operating correctly can effectively fully flatten the wavefront from the guide star. An image simultaneously recorded of the nearby target will be similarly corrected and will thus have an angular resolution much closer to the diffraction limit of the telescope depending on the sensitivity and resolution of the wavefront sensor and corrector plate, the correlation time of the distortions, and the efficiency of the control algorithm. An example of the power of adaptive optics is demonstrated in Figure 7.7, which shows images of Uranus observed by the Keck Observatory with and without the adaptive optics system engaged. The difference is startling.

Complex adaptive optics systems use correcting plates that can fully flatten the wavefront, while the simplest use is a basic tip–tilt system: simply a flat mirror whose mean angle is continually adjusted to keep the centroid of the guide star fixed.

Speckle imaging

A simpler method of image correction is *speckle imaging*. Since the correlation time of variations is of the order of a few milliseconds, images recorded with shorter exposures will each have a constant, nonvarying distortion. If thousands of short-exposure images are recorded, they can be analyzed and then suitably averaged to reconstruct near diffraction–limited images. Obviously this technique will work best for bright images due to the low signal-to-noise ratio of short-exposure images, and

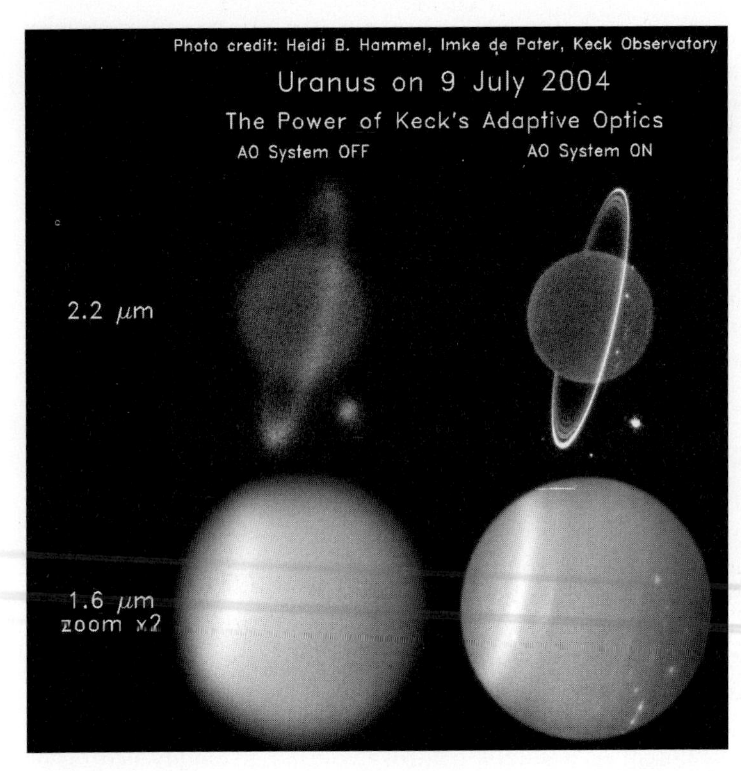

Figure 7.7. Images of Uranus observed with the Keck Observatory in ambient observing conditions (left-hand images) and with the adaptive optics system turned on (right-hand images). The improvement in spatial resolution given by the AO system is enormous. Courtesy of Keck Observatory, California Association for Research in Astronomy.

the technique has been successfully applied by amateur astronomers to Jupiter observations.

Deconvolution

The final method of improving angular resolution is *deconvolution*. In its simplest form, atmospheric turbulence blurs the image of the object by convolving the true image with an effective point spread function (PSF). If we have a blurred image and we know the PSF, then theoretically we can simply deconvolve the measured image to recover the original image. Unfortunately, in practice things are not that simple since both the PSF and the blurred image include random noise, and with the simplest deconvolution methods this noise can propagate through to yield enormous errors in the deconvolved image. This phenomenon is known as *ill-conditioning*, and is also encountered when trying to retrieve vertical atmospheric profiles from measured IR spectra (as we shall see in Section 7.10). Practical deconvolution routines must somehow constrain the deconvolved solution to prevent noise error building up and a commonly used technique is the Richardson–Lucy (RL) deconvolution algorithm (e.g., Sromovsky *et al.*, 2001a). If we represent the point spread function (PSF) as $P(i|j)$ (where the PSF represents the fraction of light from true pixel j which gets scattered into pixel i) then the noiseless blurred image $I(i)$ is formed from the

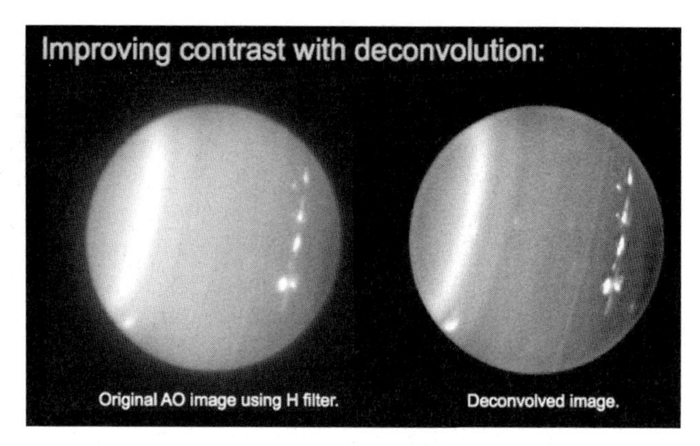

Figure 7.8. Two images of Uranus recorded with the Keck Observatory in AO mode, showing the additional improvement to resolution that can be achieved by using deconvolution. Courtesy of Keck Observatory, California Association for Research in Astronomy, and Larry Sromovsky.

unblurred image $O(j)$ as

$$I(i) = \sum_j P(i\,|\,j)O(j) \tag{7.11}$$

The RL algorithm then takes the nth estimate of the unblurred image and improves it by the iteration equation

$$O_{n+1}(j) = O_n(j)\frac{\sum_i P(i\,|\,j)\dfrac{D(i)}{I_n(i)}}{\sum_i P(i\,|\,j)} \tag{7.12}$$

where $D(i)$ is the observed image; and $I_n(i)$ is the nth fit to the blurred image constructed using $O_n(j)$ in Equation (7.11). Deconvolution algorithms may be used to improve the spatial resolution of images recorded by ground-based telescopes without adaptive optics, and may also be used to further process images recorded with adaptive optics, as is shown in Figure 7.8.

Interferometry

Interferometry is a technique routinely used in radio astronomy, and more recently microwave observations, which is currently under development at several optical/IR observatories also. Assuming that the effects of atmospheric turbulence are negated by the use of adaptive optics at visible wavelengths, or by observing at much longer wavelengths, the angular resolution of a single telescope is diffraction-limited by the diameter of the entrance aperture as

$$\theta_d = \frac{1.22\lambda}{D} \tag{7.13}$$

where D is the diameter of the entrance aperture; and λ is the wavelength. The 1.22 factor comes from the fact that the entrance aperture is assumed to be circular and θ_d is specifically the angle between the center and first minimum of the *Airy* function. Hence, as we go to longer wavelengths, the aperture size required to achieve a specific angular resolution increases linearly with wavelength, which means that it is simply impractical to build a single-dish radio telescope with the same angular resolution as optical telescopes. However, an alternative approach is to use several telescopes, spaced over large distances, and combine their signals together with the appropriate phase delay to simulate, in effect, a giant mirror of diameter equal to the maximum separation of individual telescopes. The details of the recombination of the signals are complex, but as a simple example, two telescopes of diameter D placed a distance L apart, would have an effective angular resolution in the direction parallel to the line connecting the telescopes of λ/L, and an angular resolution of λ/D perpendicular to this direction. To achieve high resolution at all angles, interferometers usually have several telescopes arranged in a "T" or "Y" shape and the telescopes may usually be placed at a variety of separations in order to increase sensitivity. The imaging properties of such arrangements are complicated, but effectively such interferometers have an angular resolution of λ/L, where L is the maximum baseline, and a field of view of λ/D.

7.3.3 Brightness

We saw in Section 6.7 that the thermally emitted radiance ($\mathrm{W\,m^{-2}\,sr^{-1}\,(cm^{-1})^{-1}}$) of the giant planets decreases rapidly as we go out through the solar system due to decreasing atmospheric temperatures. What was not explicitly stated, however, is that reflected solar radiance also drops rapidly as $1/D^2$, where D is the distance of the planet from the Sun, due to these planets' greater and greater distance from the Sun. This decrease in reflected and thermally emitted radiance affects all remote-sensing observations, not just ground-based ones, and makes remote observation increasingly difficult as we go from Jupiter to Neptune. However, some ground-based observations of the giant planets, such as microwave observations or some thermal-IR spectroscopic observations, which have limited angular resolution, are unable to resolve the disks of these planets, especially Uranus and Neptune. Where this is the case there is a further factor decreasing the measured disk-averaged irradiance of the giant planets due to the rapidly decreasing projected solid angle of these planets as we go out through the solar system. At opposition, when the Earth is closest to the planet, the solid angle is given by

$$\Delta\Omega = \frac{\pi R^2}{((D-1)D_{\mathrm{AU}})^2} \tag{7.14}$$

where D is the planet's distance to the Sun in AU; D_{AU} is 1 AU (in kilometers); and R is the planetary radius (also in kilometers). Hence, the observed irradiance is given by the calculated disk-averaged radiance multiplied by the above solid angle and so drops even more rapidly as we go from Jupiter to Neptune due both to the increase in D, and to the decrease in R. For example, at visible wavelengths the opposition magnitudes of Jupiter, Saturn, Uranus, and Neptune are, respectively, -2.7, 0.67, 5.52, and 7.84, where the magnitude m (designed to formalize the observing convention of a 100-fold decrease in irradiance when going from magnitude-1 to

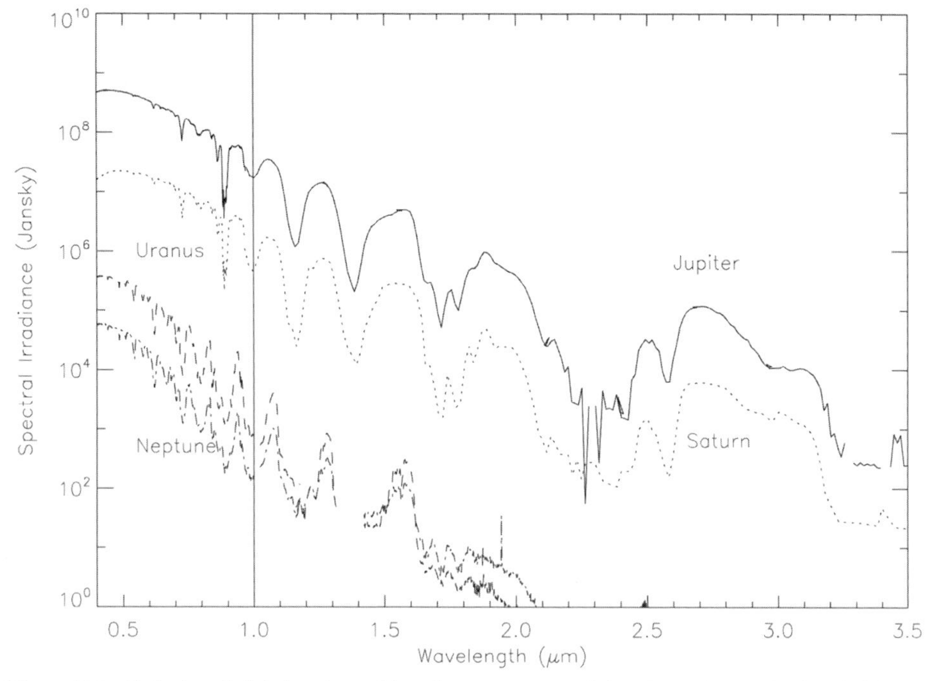

Figure 7.9. Calculated disk-integrated irradiance spectra of the giant planets in the visible and near-infrared as seen from the Earth at opposition, in Janskys ($1\,\mathrm{Jy} = 10^{-26}\,\mathrm{W\,m^{-2}\,Hz^{-1}}$). Jupiter: solid line; Saturn: dotted line; Uranus: dashed line; Neptune: dot-dashed line.

magnitude-6 stars) is defined as

$$m = 2.512 \log(B_0/B) \tag{7.15}$$

where B is the measured irradiance from the source; and B_0 is a reference irradiance approximately equal to $B_0 = 7 \times 10^{10}\,\mathrm{W\,m^{-2}\,nm^{-1}}$ at visible wavelengths. The calculated disk-averaged spectral irradiances at the Earth of the giant planets at opposition in the visible/near-IR, and mid-IR to far-IR spectral ranges are shown in Figures 7.9 and 7.10, where spectral irradiance has been expressed in terms of Janskys, a unit often used in ground-based IR spectroscopy and defined as $1\,\mathrm{Jy} = 10^{-26}\,\mathrm{W\,m^{-2}\,Hz^{-1}}$. The rapid decrease in irradiance with distance is clearly apparent in both spectral ranges.

7.4 GROUND-BASED VISIBLE/IR OBSERVATORIES

We saw earlier that the problems of atmospheric absorption, especially due to water vapor, may be limited by placing telescopes at high altitudes and thus limiting the mass of air above the telescope. In addition, it helps to place telescopes in regions where the air is statically stable and thus atmospheric turbulence is minimal. Very

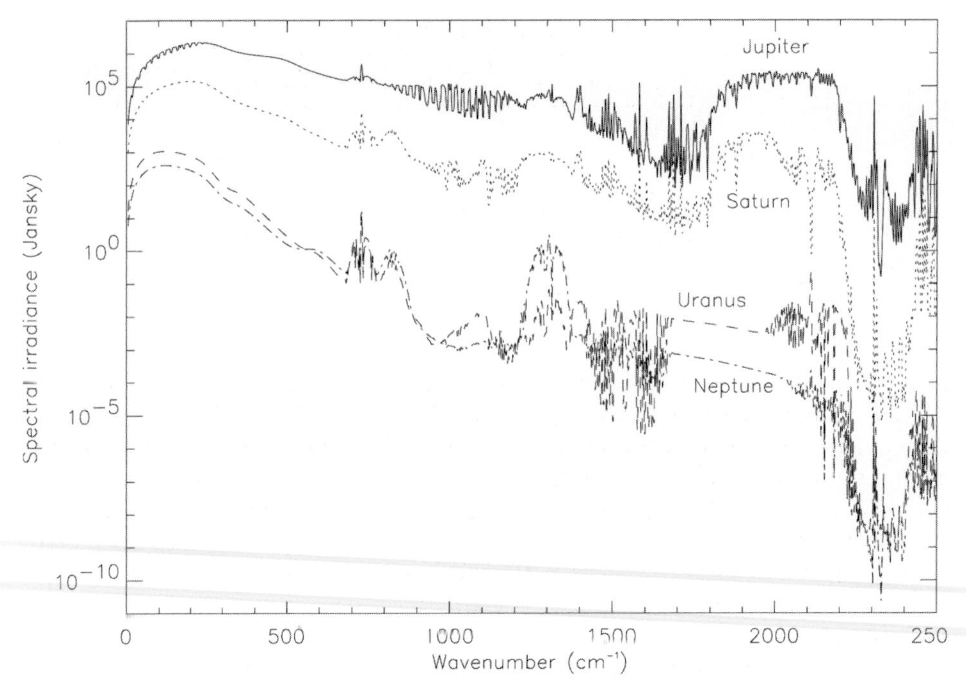

Figure 7.10. Calculated disk-integrated irradiance spectra of the giant planets in the mid-infrared to far-infrared as seen from the Earth at opposition (in Janskys). Jupiter: solid line; Saturn: dotted line; Uranus: dashed line;Neptune: dot-dashed line.

good regions for ground-based telescopes are thus near the tropics of Cancer and Capricorn since these are latitudes where air which has risen near the equator at the Inter-Tropical Convergence Zone (ITCZ) and traveled towards the poles in the terrestrial equatorial Hadley cell circulation then descends and heats adiabatically forming a particularly stable air column. The air is also very dry as most of its moisture precipitated in the ITCZ as rain, which has additional advantages for ground-based telescopes as we have just seen, and is also the reason the Earth's major deserts are found to lie at these latitudes. Such regions are also clearly relatively free of clouds, an obvious advantage for visible/IR astronomy! It is thus no accident that most of the major telescope sites on the Earth: Hawaii, the Canary Islands, Chile, Southwest U.S.A, etc. are near these desert latitudes. In this section we will review some of the major astronomical observatories that among many other things are currently undertaking observations of the giant planets.

7.4.1 European Southern Observatory (ESO); Very Large Telescope (VLT)

ESO is an intergovernmental organization of 15 member states that operates two astronomical observatories in the Atacama desert of Chile, one of the driest places on Earth. The La Silla observatory (29°15′S, 70°44′W), at an altitude of 2,400 m, is the

Table 7.1. La Silla telescopes.

ESO-operated telescopes	*National telescopes*
3.6 m	DENIS 1 m
2.2 m	Geneva 1.2 m
Danish 1.5 m	
New Technology Telescope 3.5 m	

oldest of the three ESO observatories and comprises a number of optical telescopes of diameter between 0.6 m and 3.6 m, which are listed in Table 7.1. The 12 m Atacama Pathfinder Experiment (APEX) is based at Llano de Chajnantor (23°00′S, 67°46′W) at an altitude of 5,104 m in the Atacama desert and is the largest submillimeter telescope in the southern hemisphere. Finally, the Paranal Observatory (24°40′S, 70°25′W) at an altitude of 2,645 m is home to ESO's Very Large Telescope (VLT) facility and was chosen for its excellent atmospheric conditions and remoteness, ensuring that its operation is not disturbed by the effects of human settlement such as dust and light from roads and mines.

The VLT is the world's largest and most advanced astronomical observatory (Figure 7.11). It comprises four 8.2 m reflecting unit telescopes, which may operate individually or in combined mode providing the total light-collecting power of a single 16 m telescope. In addition, the telescopes can be used in an interferometric mode together with three movable 1.8 m auxiliary telescopes to form the VLT Interferometer (VLTI). The four large telescopes, initially designated UT1–UT4, have been renamed in the indigenous Mapuche language as *Antu* (the Sun), *Kueyen* (the Moon), *Melipal* (the Southern Cross), and *Yepun* (Venus). A schematic of the observatory showing the telescopes and instruments that will eventually be available is shown in Figure 7.12. To save weight, the primary mirrors of the unit telescopes are rather thin and thus flexible. Hence, their shape is dynamically controlled using *active optics* to apply correcting forces to the primary mirror and move the secondary mirror in order to cancel out the errors. The active optics system was originally developed for the ESO 3.5 m New Technology Telescope (NTT) at La Silla. The VLT telescopes use the Ritchey-Chrétien optical design and each telescope may operate in the Cassegrain, Nasmyth, or Coudé focus. The image quality of the VLT telescopes is very impressive with a record angular resolution of 0.18″. More usually, the seeing is in the 0.5″ to 1.0″ range, but the VLT may use its adaptive optics to achieve near diffraction–limited observations of angular resolution 0.05″. Should no guide star be available then UT4 (*Yepun*) incorporates a laser system to generate a laser guide star. UT1 (*Antu*) achieved first light on May 25, 1998, and the last completed telescope, UT4 (*Yepun*), achieved first light on September 3, 2000. The VLT provides excellent data on the giant planets due to its high angular resolution, and is equipped with a number of instruments for imaging and spectroscopy from the UV to approximately 25 μm.

Figure 7.11. The European Southern Observatory Very Large Telescope (VLT) at La Paranal, Chile. Courtesy of ESO.

7.4.2 The Mauna Kea observatories

The Mauna Kea volcano on Big Island, Hawaii is an excellent place for astronomical observations due to its high altitude of 4,200 m and its dry, stable atmospheric conditions. Hence, the site is currently home to the world's largest and most powerful telescopes (Figure 7.13). The largest of these are the twin Keck telescopes (Figure 7.14), which have primary mirrors that are 10 m in diameter and are composed of 36 individual actively controlled hexagonal elements that operate together as a single, high-precision mirror. The Keck Observatory is jointly operated by NASA and the California Institute of Technology (Caltech). The Keck I telescope began operations in May 1993, while Keck II began observing in October 1996.

Other telescopes which are situated on the Mauna Kea site are listed in Table 7.2 and of particular note are the Japanese Subaru 8.2 m telescope, the Gemini North 8.1 m telescope, the NASA Infrared Telescope Facility (IRTF), the Canada–France–

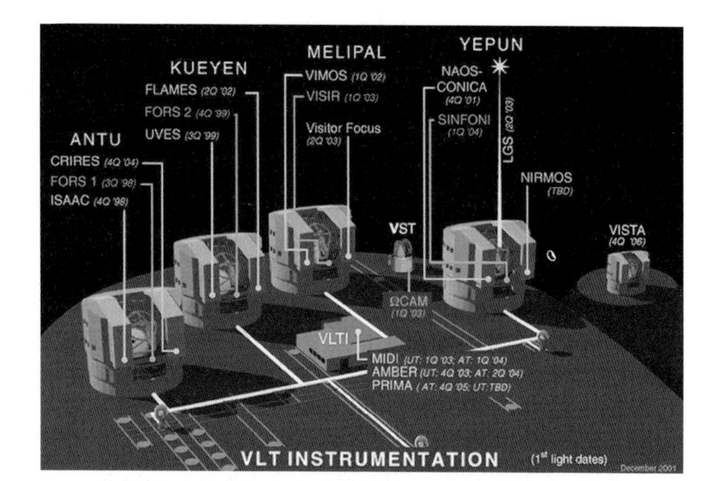

Figure 7.12. Schematic design of VLT site. Courtesy of ESO.

Figure 7.13. Mauna Kea site in Hawaii. The Keck telescopes are the twin domes seen at center right. The 8 m Subaru Telescope (with a cylindrical dome) is in front of the Kecks and behind it are (left to right): the NASA IRTF, the Canada–France–Hawaii Telescope, the Gemini North 8 m telescope (in this picture still under construction), the University of Hawaii 2.2 m telescope, and the 3.8 m United Kingdom Infrared Telescope (UKIRT). In the lower foreground are (left to right): the Smithsonian–Taiwan submillimeter array, the James Clerk Maxwell 15 m submillimeter telescope, and the Caltech 10 m submillimeter telescope. Courtesy of Keck Observatory, California Institute of Technology, and Richard Wainscoat/ILA.

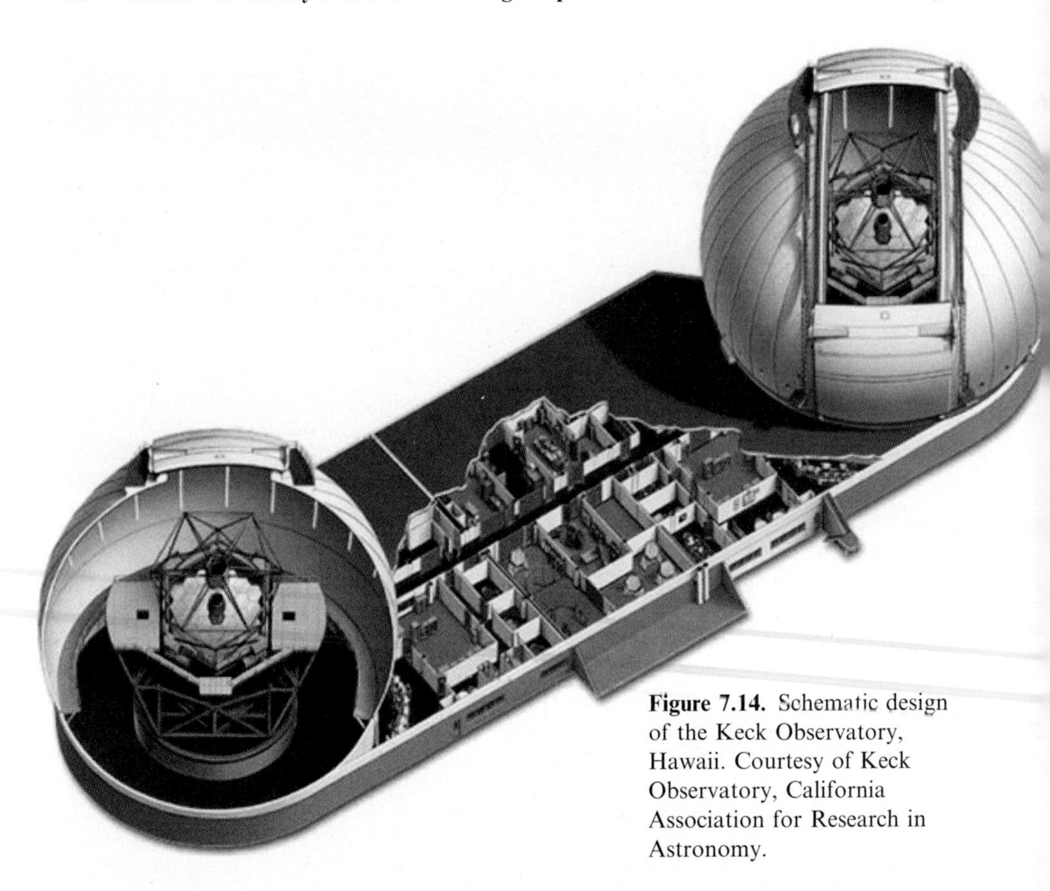

Figure 7.14. Schematic design of the Keck Observatory, Hawaii. Courtesy of Keck Observatory, California Association for Research in Astronomy.

Hawaii Telescope (CFHT), and the United Kingdom Infrared Telescope (UKIRT). The Subaru 8.2 m telescope incorporates the largest single-piece, or monolithic, mirror ever built and achieved "first light" in January 1999. The Gemini project is an international collaboration between seven countries (U.S.A., U.K., Canada, Chile, Australia, Argentina, and Brazil) to place two identical 8.1 m telescopes in both northern and southern hemispheres to allow unobstructed observation of both northern and southern skies simultaneously. While Gemini North is located on Mauna Kea, its identical brother, Gemini South, is located at Cerro Pachón in Chile (2,737 m). The Gemini Telescope mirrors are single-piece thin mirrors whose shape is continuously adjusted by an active optics system using 120 actuators at the back of the mirror to maintain the mirror shape. Gemini North was dedicated on June 25, 1999, and Gemini-South was dedicated on January 18, 2002. The NASA IRTF is a 3 m IR telescope operated and managed for NASA by the University of Hawaii (which reserves 50% of its observing time for studying solar system objects), while CFHT is a 3.6 m telescope and UKIRT is a 3.8 m IR telescope, which have both been used for a number of planetary observations. Recent images of Jupiter and Uranus observed by the author at UKIRT are shown in Figures 7.15 and 7.16, see color section for both, respectively.

Table 7.2. Mauna Kea telescopes.

Telescope	*Mirror diameter* (m)	*Operator*
ubaru	8.2	National Astronomical Observatory of Japan (NAOJ)
NASA Infrared Telescope Facility (IRTF)	3	NASA and University of Hawaii
Canada–France–Hawaii Telescope (CFHT)	3.6	Canada and France
Gemini North	8.1	International Gemini Consortium
University of Hawaii	2.2	University of Hawaii
United Kingdom Infrared Telescope UKIRT)	3.8	Joint Astronomy Centre (JAC) U.K., Holland, Canada
ames Clark Maxwell submillimeter Telescope (JCMT)	15	Joint Astronomy Centre (JAC) U.K., Holland, Canada
Caltech submillimeter Telescope	10	California Institute of Technology
Smithsonian–Taiwan Submillimeter Array	6×8	Smithsonian Institute and Taiwan

7.4.3 Other major observatories

There are a number of other major observatories across the world, many of which are involved in giant planet observations. In this section we will briefly list some of the largest and most famous of these observatories and outline the telescopes available.

(1) *Calar Alto*. This observatory, sited at an altitude of 2,168 m in Andalusia, Spain, is operated by the Max-Planck-Institut für Astronomie, Heidelberg, Germany, and has been used extensively for planetary observations. Three main telescopes are provided with apertures of 1.23 m, 2.2 m, and 3.5 m, together with a 1.5 m telescope operated by the Observatory of Madrid.
(2) *AAO*. The Anglo-Australian Observatory is situated in northern New South Wales, Australia and operates the 3.9 m Anglo-Australian telescope and the 1.2 m U.K. Schmidt telescope. The AAO is not currently involved in giant planet observations, but is involved in the search for extrasolar planets, by the "Doppler wobble" technique described in Chapter 8.
(3) *Pic-du-Midi*. The Pic-du-Midi Observatory is situated at an altitude of 2,872 m in the French Pyrenees. The observatory operates a 1 m and 2 m telescope and has undertaken numerous investigations of the giant planets.
(4) *La Palma*. The La Palma Observatory, run by the Isaac Newton Group of Telescopes (ING) operates the 4.2 m William Herschel Telescope, the 2.5 m

Isaac Newton Telescope, and the 1.0 m Jacobus Kapteyn Telescope on behalf of British, Dutch, and Spanish research agencies. The La Palma Observatory does not currently undertake giant planet observations.

(5) *Kitt Peak National Observatory*. The Kitt Peak National Observatory is located high above the Sonoran Desert in Arizona and is home to 22 optical and two radio telescopes representing eight astronomical research institutions. The largest of these telescopes are the 4 m Mayall, the 3.5 m WIYN, the 2.1 m and the 0.9 m. The Kitt Peak National Observatory is also involved in the search for extrasolar planets.

(6) *Palomar*. The Palomar Observatory in California is owned and operated by the California Institute of Technology. Its principal instruments are the 200-inch (5 m) Hale Telescope, the 48-inch (1.2 m) Oschin Telescope, the 18-inch (0.45 m) Schmidt telescope, and the 60-inch (1.5 m) reflecting telescope. For many years the Hale telescope was the largest in the world.

(7) *Las Campanas*. The Las Campanas Observatory, at an altitude of 2,438 m in the Chilean Andes, is operated by the Observatories of the Carnegie Institution of Washington (OCIW). The site includes the two 6.5 m Magellan telescopes which began operations in September 2000 and September 2002, respectively.

7.5 AIRBORNE VISIBLE/IR OBSERVATIONS

We saw earlier that IR astronomy from ground-based telescopes is severely hampered by the absorption of gases in the Earth's atmosphere, especially water vapor. Most of this water vapor lies in the lower, warmer parts of the atmosphere and thus, as mentioned earlier, most ground-based IR facilities are placed at high altitudes, in climatically dry parts of the world. However, the number of locations on the Earth that satisfy these requirements is limited, and they are often also remote, making the construction and operation of these telescopes rather difficult. An alternative approach to Earth-based IR astronomy is to mount an infrared telescope on to an aircraft and observe in-flight at high altitude almost anywhere in the world. At the high altitudes attainable by jet aircraft, the obscuration by overlying water vapor is reduced by a factor of 1,000 compared with observations at sea level. In addition, airborne observations are unaffected by cloud obscuration, and such observatories may be used to observe both the southern and northern skies. A further advantage is that an airborne observatory is ideal for observing stellar occultation events since it can fly to the optimum position in the world to observe them.

7.5.1 Kuiper Airborne Observatory

The first airborne observations were made in the 1960s, but airborne IR observations really came of age with the commission of NASA's Kuiper Airborne Observatory (KAO) named after the American planetary scientist Gerard P. Kuiper (1905–1973). The Kuiper aircraft (Figure 7.17) was a Lockheed C-141A jet transport plane with a range of 6,000 nautical miles and was capable of conducting research operations to

Figure 7.17. The Kuiper Airborne Astronomy (KAO) aircraft. Courtesy of NASA.

45,000 feet (14 km). The aircraft was modified to carry a 0.91 m aperture Cassegrain IR telescope operating in the 1 μm to 500 μm spectral range and which observed through a hole cut in the side of the aircraft at an elevation of between 35° and 75° (Figure 7.18). The KAO flew out of NASA Ames Research Center, at Moffett Field, California, and began operating in 1971, finally finishing service in October 1995.

The telescope was mounted such that it moved independently from the aircraft and an automatic system kept the telescope pointed at the selected target even when the aircraft moved in turbulence. The telescope mirrors were cooled with liquid nitrogen, and other cryogenic liquids were used to reduce their background IR emission.

7.6 GROUND-BASED MICROWAVE OBSERVATORIES

The submillimeter to microwave part of the giant planets' thermal emission spectra is a very interesting one since it allows for the probing of deep-pressure levels of these planets with weighting functions extending down to almost 100 bar for Uranus and Neptune. The spectral range contains absorption features of ammonia, CO, HCN and other constituents, and thus the deep abundance of these molecules may be

Figure 7.18. The KAO telescope looking through the aperture in the aircraft's side. Courtesy of NASA.

determined. The transmission of the Earth's atmosphere at microwave wavelengths was shown in Figure 7.3, and ground-based microwave observations are thus restricted to the spectral windows between the main absorption bands. Since the main absorber is once again water vapor, microwave observatories are, like visible/IR observatories, preferentially located at high altitude in dry regions of the world. Observing the giant planets at microwave wavelengths poses considerable problems, not least of which is the very low power of microwave emission radiated by the giant planets, which means that antennas must be very large and detectors must be very sensitive. A second problem, for Jupiter observations, is that synchrotron emission is also observed from the radiation belts, which needs to be subtracted for wavelengths longer than about 4 cm. The final problem is that at these long wavelengths, it is technically very difficult to make antennas large enough to resolve structure in the atmospheres of these planets. For example, an antenna able to resolve down to 1 arcsec at the wavelength of the inversion absorption band of ammonia at 1.3 cm needs to be 3.2 km across! For this very reason, microwave and radio observatories have led the way in developing interferometric arrays to increase their angular resolution. There are currently several large millimeter arrays in the world that have been used for giant planet observations, which will now be reviewed.

7.6.1 The Institut de RadioAstronomie Millimétrique (IRAM)

IRAM is an international institute for research in millimeter astronomy founded in 1979 by the Centre National de la Recherche Scientifique (CNRS), France, and the Max Planck Gesellschaft (MPG), Germany. The organization was joined by the Instituto Geográfico Nacional (IGN), Spain, in September 1990. IRAM operates two major facilities: a 30 m radio telescope on Pico Veleta in the Sierra Nevada (southern Spain), and an array of six 15 m radio telescopes on the Plateau de Bure in the French Alps. IRAM telescopes observe in microwave "windows" at 0.85 mm, 1.3 mm, 2 mm, and 3 mm.

The Pico Veleta observatory is located at an altitude of 2,920 m on the second highest mountain of the Iberian peninsula. Its high altitude, southern location, and dry climate are extremely favorable for millimetric observations due to the low column abundances of water vapor which, on cold, dry winter days, can drop to as low as 1 precipitable-mm. The 30 m Cassegrain telescope at Pico Veleta is currently the world's largest telescope operating at wavelengths between 0.8 mm and 3.5 mm (350 to 80 GHz), with a collection area of 700 m^2, and received its first millimetric "light" in May 1984. The angular resolution of the Pico Veleta observatory is purely diffraction-limited and thus depends inversely on wavelength, with an angular resolution of 10″ at a wavelength of 1.3 mm. Radio telescopes have historically operated with a single receiver (or pixel), and thus an image was built up by scanning the entire telescope across the sky. More recently, multiple detectors are located in imaging arrays at the focal plane of these telescopes to provide instant imaging of objects.

The Plateau de Bure observatory is located at an altitude of 2,552 m in the French Alps. Atmospheric conditions are good for millimetric observations and the column amount of water vapor can drop below 2 precipitable-mm on dry winter days. Work on the construction of the site began in 1985, and the first interferometric fringes were obtained in 1988. The site consists of six 15 m antennas operating between 2.6 mm and 3.7 mm (81 to 115 GHz) and also between 1.2 mm and 1.4 mm (205 to 245 GHz), which can move on rail tracks up to a maximum separation of 408 m in the east–west direction and 232 m in the north–south direction (Figure 7.19). Each dish has a collecting area of 175 m^2 and thus the combined collecting area of the six telescopes is 1,050 m^2, which make this one of the most sensitive, and highly resolving interferometers in the world. As mentioned earlier, for an interferometer the field of view is defined by the ratio of wavelength to diameter for each antenna, while the resolution is determined by the ratio of the wavelength to the maximum separation between antennas. Hence, for example, at 1.3 mm the field of view of the interferometer is 20″ while the best angular resolution is 0.5″.

7.6.2 Very Large Array (VLA)

The VLA is one of the world's greatest astronomical radio observatories. It consists of twenty-seven 25 m radio antennas in a Y-shaped configuration on the plains of San Agustín in New Mexico at an altitude of 2,124 m. Each arm is 21 km in length (Figure 7.20) and the data from the antennas may be combined interferometrically to give the resolution equivalent to an antenna 36 km across in the facility's highest resolution configuration. Each antenna has a collecting area of 491 m^2, giving a total facility collection area of 13,250 m^2, equivalent to a single 130 m diameter dish. Construction of the VLA began in 1973, and the facility was formally dedicated in 1980.

The radio dishes of the VLA may be moved on a rail system to take up one of four positions with different maximum separations: A (36 km), B (10 km), C (3.6 km),

Figure 7.19. The IRAM millimetre array at the Plateau de Bure Observatory, France. Courtesy of IRAM.

Figure 7.20. The Very Large Array (VLA) in New Mexico. Courtesy of NRAO/AUI.

Table 7.3. Receivers available at the VLA.

	4-band	*P-band*	*L-band*	*C-band*	*X-band*	*U-band*	*K-band*	*Q-band*
Frequency (GHz)	0.073–0.0745	0.30–0.34	1.34–1.73	4.5–5.0	8.0–8.8	14.4–15.4	22–24	40–50
Wavelength (cm)	400	90	20	6	3.6	2	1.3	0.7
Primary beam (arcmin)	600	150	30	9	5.4	3	2	1
Highest resolution (arcsec)	24.0	6.0	1.4	0.4	0.24	0.14	0.08	0.05

and D (1 km). The configuration is typically switched every four months or so. The reason for changing the separations is that although configuration A has the highest resolution, it also has the lowest sensitivity to faint objects. Conversely, configuration D has the highest sensitivity, but the lowest angular resolution. Hence, the optimal configuration depends on the use to which the VLA is to be put. The VLA can observe in various bands between 75 MHz (400 cm) and 43 GHz (0.7 cm), summarized in Table 7.3, and the beam size and angular resolution in configuration A vary inversely with wavelength between 10° and $1'$, and $24''$ and $0.05''$, respectively.

7.6.3 Very Large Baseline Array (VLBA)

The VLBA is a series of ten 25 m diameter radio antennas located at sites across the continental United States, and on Mauna Kea, Hawaii, and St. Croix in the U.S. Virgin Islands. Work on its construction began in 1985 and the last VLBA station on Mauna Kea, Hawaii, was completed in 1993. Because of its extremely long baseline and large collecting area, the VLBA has a maximum angular resolution of less than one-thousandth of an arcsecond (at $\lambda = 7$ mm), and covers a wavelength range of 3 mm to 90 cm in ten bands. Clearly, to achieve its high resolution the data from the antennas must be combined interferometrically, but since the sites are so far apart, this is not easy to do in real time. Instead, each site records its data onto magnetic tape and the data "time-tagged" with a reference signal generated by a hydrogen maser at each site. The tapes are then sent to the VLBA station in Socorro, New Mexico, where the tapes from each station are read and combined with the appropriate time delays consistent with the station's different positions across the globe to form a single interferometric device.

7.6.4 Combined Array for Research in Millimeter-wave Astronomy (CARMA)

CARMA is the merger of two university-based millimeter arrays: the Owens Valley Radio Observatory (OVRO) millimeter array and the Berkeley Illinois Maryland Association (BIMA) millimeter array.

OVRO was run by the California Institute of Technology and consisted of six 10.4 m radio telescopes, where the individual telescopes could be moved to various observing stations along a "T"-shaped rail track, which was 440 m long in the north–south direction and 400 m long in the east–west direction. BIMA was run by a consortium consisting of the University of California Berkeley, the University of Illinois Urbana, and the University of Maryland and had an array of ten 6.1 m dishes, which could be moved to one of three configurations along a "T"-shaped concrete track 305 m long in the east–west direction, and 183 m long in the north–south direction.

The OVRO and BIMA telescopes were combined and moved to their new home at Cedar Flat, California (altitude: 2,195 m) in 2004 and the new combined array began operating in 2006. The total collecting area of CARMA is 772 m^2 and operates at 115 GHz, 230 GHz, and 345 GHz. CARMA has a number of configurations (A–E), with baselines ranging from 30 m to 2 km, giving it a maximum angular resolution of 0.1 arcsec at 230 GHz. CARMA is able to observe millimetric radio emission from a number of objects including nearby starburst galaxies, blue dwarf galaxies, nearby molecular clouds forming clusters of stars, newly born stars emerging from their present clouds, comets, and the cosmic radiation left over from the Big Bang.

7.6.5 Nobeyama Millimeter Array (NMA)

The NMA in Japan consists of six transportable 10 m antennas, equipped with cryogenically cooled SIS receivers, covering three wavelength bands: 3 mm (85–116 GHz), 2 mm (126–152 GHz), and 1 mm (213–237 GHz), respectively (Figure 7.21). A 45 m dish is also at the site. There are 30 antenna stations located along two tracks, one extending 560 m in the east–west direction, and another 520 m long inclined at an angle of 33° from the north–south direction. The antennas may be arranged in a number of configurations.

7.7 SPACE-BASED TELESCOPES

Space-based visible/IR telescopes offer considerable advantages over ground-based and airborne telescopes in that the effects of the terrestrial atmosphere are almost completely eliminated (although even the HST at an altitude of 600 km experiences some sensitivity to terrestrial UV airglow). Hence, the complete, unobscured spectra of the giant planets may be observed and angular resolution is, if the telescope and detection system are correctly constructed, diffraction-limited. Of course the angular diameters of the planets are unchanged and so very fine resolution, highly sensitive instruments are still required to image the farthest giant planet, Neptune. Although a number of space observatories have now been launched, only five have been used for planetary observations: the Hubble Space Telescope (HST), the Infrared Space Observatory (ISO), the Submillimeter Wave Astronomy Satellite (SWAS), the *Spitzer* Telescope, and the AKARI space telescope, which will now be described.

Figure 7.21. The Nobeyama Millimeter Array (NMA) in Japan. Courtesy of NMA.

7.7.1 HST

The HST is a collaborative ESA/NASA mission that was launched on April 25, 1990 by the space shuttle *Discovery* (STS-31) into a low-Earth 600 km altitude orbit. The telescope is 13.2 m long, 4.2 m in overall diameter, and weighs 11,110 kg (Figure 7.22). With a primary mirror of diameter 2.5 m, the theoretical angular resolution of HST is approximately $0.1''$ in the visible and near-infrared, which is ten times better than can be achieved with most ground-based observations without adaptive optics. Unfortunately, soon after launch, the primary mirror was discovered to have been ground incorrectly and thus the telescope initially suffered from spherical aberration, which significantly impaired its performance. Some of these defects were dealt with by corrective optics installed by the December 1993 shuttle servicing mission (STS-61, *Endeavour*), and subsequent servicing missions have now completely replaced the original instruments launched on HST such that they all now correct for the aberration of the primary. The instruments currently onboard HST, at the time of writing, prior to Servicing Mission 4, will now be reviewed.

Wide-Field/Planetary Camera 2 (WFPC2)

WFPC2 was a replacement for the original Wide-Field/Planetary Camera (WFPC1) launched on HST and was installed by the December 1993 shuttle servicing mission to

Figure 7.22. The Hubble Space Telescope in orbit about the Earth. Courtesy of NASA/ESA.

negate the aberration in the primary mirror. WFPC2 records high-resolution images of astronomical objects over a relatively wide FOV and a broad range of wavelengths (115 to 1,100 nm), which are defined by a large selection of filters. WFPC2 is composed of four cameras, each of which has an 800×800 element silicon CCD array. Three of the cameras make up the wide-field camera (WFC) and operate at f/12.9 giving a pixel size of $0.1''$ and a FOV of $150 \times 150''$. The pointing of these cameras is arranged to form a projected "L" shape in the sky. The fourth camera, known as the planetary camera (PC), operates at f/28.3 and thus covers a $35 \times 35''$ area with a pixel size of $0.046''$. This camera points in the gap left by the WFC.

Space Telescope Imaging Spectrograph (STIS)

STIS covers a similar spectral range to WFPC2 (115–1,000 nm) and uses three detector arrays. The spectral regions 115 nm–170 nm and 165 nm–310 nm each use a Multi-Anode Microchannel Array (MAMA), while a CCD array covers the 305 nm to 1,000 nm range. All three arrays have 1,024 × 1,024 elements and the field of view is 25 × 25″ for each MAMA (0.024″/pixel), and 50 × 50″ for the CCD (0.05″/pixel). Although STIS covers a smaller region of the sky than WFPC2, it has higher angular resolution and other advantages at UV wavelengths.

Near-IR Camera and Multi-Object Spectrometer (NICMOS)

NICMOS covers the wavelength range of 0.8 μm and 2.5 μm, and its highly sensitive HgCdTe detector arrays must be cooled to liquid-nitrogen temperatures for optimum sensitivity. Initially the cooling was provided by an onboard exhaustible liquid-nitrogen supply and it operated as such between February 1997 and November 1998 yielding exciting new results on the giant planets, among many other things. However, the cryogen was eventually exhausted and thus the detectors warmed up, reducing the usefulness of the instrument. NICMOS was subsequently revived by the shuttle Servicing Mission 3B (STS-109) in March 2002, which installed an active cooling system for the instrument to return the performance of NICMOS to its designed level.

NICMOS is composed of three cameras: NIC1, NIC2, and NIC3, which form three adjacent FOVs at different angular resolutions. Each camera has a 256 × 256 detector array and 19 different combinations of filters, gratings, and prisms. NIC1 covers a 11 × 11″ region of the sky at a resolution of 0.043″, NIC2 covers a 19.2 × 9.2″ region at a resolution of 0.075″, and NIC3 covers a 51.2 × 51.2″ region at an angular resolution of 0.2″. Most of the filters of NIC1 cover the shortwavelength end of the NICMOS range, while NIC2 has more filters covering the long-wavelength end. NIC3 has yet another selection of filters across the range, and can also operate as a spectrometer with a resolving power of 200 and three spectral ranges of 0.8–1.2 μm, 1.1–1.9 μm, and 1.4–2.5 μm, respectively. In addition, NICMOS can also measure the mean polarization at 0°, 120°, 240° between 0.8-1.3 μm for NIC1, and 1.9–2.1 μm for NIC2.

Advanced Camera for Surveys (ACS)

ACS, installed in March 2002 during Servicing Mission 3B, provides HST with: (1) a deep, wide-field survey capability from the visible to near-infrared; (2) high-resolution imaging from the near-ultraviolet to the near-infrared; and (3) solar blind far-UV imaging. The primary ACS design goal is to achieve a factor of 10 improvement in the discovery efficiency of new objects, compared with WFPC2, where discovery efficiency is defined as the product of imaging area and instrument throughput. ACS is composed of three cameras: (1) a wide-field camera (WFC), which covers the spectral range 350–1100 nm and a 202 × 202″ region of the sky at an angular resolution of 0.049″; (2) a high-resolution camera (HRC), which covers the spectral range of 200–1,100 nm and a 29.1 × 26.1″ region of the sky at an angular resolution of 0.028 × 0.025″; and (3) a solar blind camera (SBC), which covers the spectral range of 115–170 nm, and a 34.6 × 30.8″ region of the sky at an angular resolution of 0.033 × 0.030″.

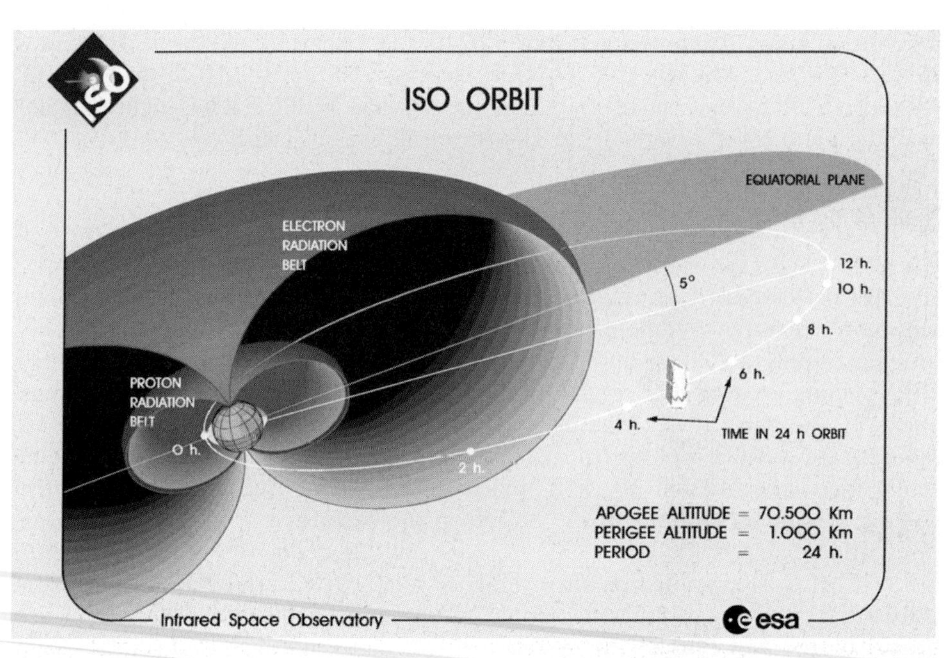

Figure 7.23. Orbit of the Infrared Space Observatory (ISO) about the Earth. Courtesy of ESA.

7.7.2 ISO

The ESA ISO was launched on November 17, 1995 by an *Ariane* 44P launcher from Kourou, French Guiana into a highly elliptical orbit with a perigee altitude of 1,000 km, an apogee altitude of 70,500 km, and a period of approximately 24 hours (Figure 7.23). The design of the orbit allowed for very long integration times to record extremely faint objects and also ensured that ISO minimized its time within the Earth's radiation belts, which interfere with the operation of its detectors. The orbit chosen allowed 17 hours of continuous telescope operation per orbit.

Although the physical size of ISO is large ($5.3 \times 3.6 \times 2.8$ m), the telescope itself is relatively small with a primary mirror diameter of only 0.6 m (Figure 7.24). Most of the volume was filled up with liquid helium, which cooled the entire optical system to a temperature of 4 K, reducing the noise of the detectors to their minimum theoretical values and allowing the measurement of images and spectra of extremely cold objects over the spectral range of 2.3 μm to 240 μm (41–4,300 cm^{-1}). ISO was originally planned to be operational for 20 months, but eventually the working life was stretched to more than 28 months; ISO operated until May 1998. ISO had four main instruments which will now be reviewed.

Infrared camera: ISOCAM

The ISO camera, ISOCAM, provided imaging in the spectral range 2.5 μm to 17 μm. The instrument was split into a short wavelength (SW) channel covering 2.5 μm to

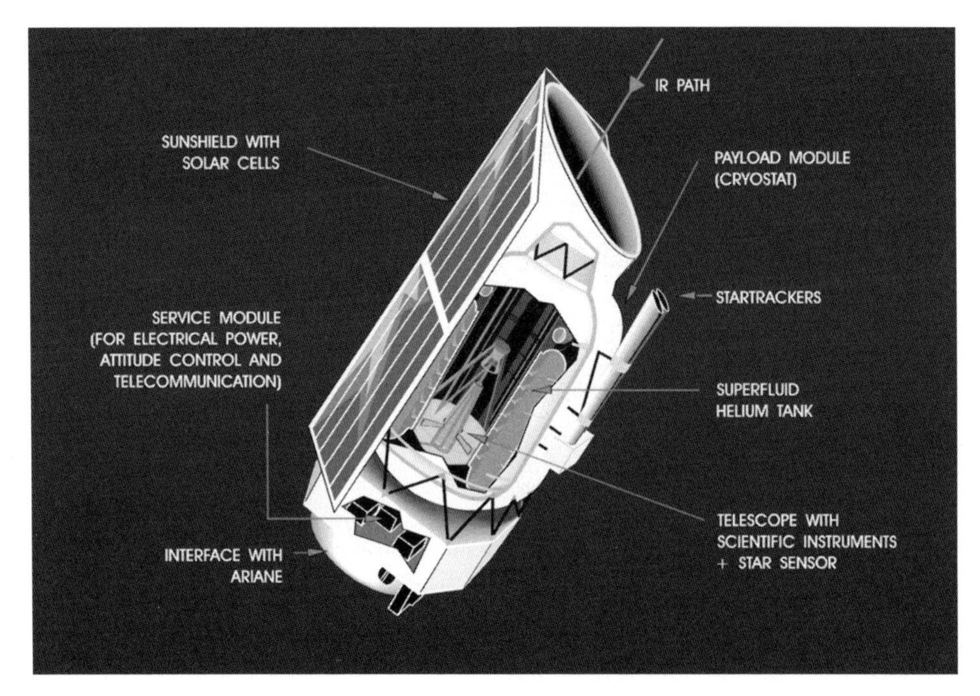

Figure 7.24. Schematic design of ISO. Courtesy of ESA.

5.2 μm, and a long wavelength (LW) channel covering 4 μm to 17 μm. Each channel contained a 32×32 pixel IR detector array, fixed filters, continuous variable filters (CVF) with a resolving power of $\Delta\lambda/\lambda = 40$, and a set of magnification lenses giving angular resolutions of 1.5″, 3″, 6″, and 12″, respectively. These effective pixel sizes, however, neglect diffraction effects, which can become very significant at longer wavelengths for a telescope with this small mirror size. The calculated diffraction-limited resolution is 2″ at 2.5 μm, and 13.6″ at 17 μm.

Photo-polarimeter: ISOPHOT

The ISO imaging photo-polarimeter, ISOPHOT, was composed of three subsystems optimized for specific photometric modes. These were: (1) PHT-P, a multi-band, multi-aperture photometer with three single detectors covering the wavelength range 3 μm to 120 μm from the near-infrared to the far-infrared; (2) PHT-C, two photometric far-IR cameras for the wavelength range 50 μm to 240 μm; and (3) PHT-S, two grating spectrophotometers, operated simultaneously, for the wavelength ranges of about 2.5 μm to 5 μm and 6 μm to 12 μm. In addition, ISOPHOT was equipped with two sets of three polarizers, one set for the PHT-P detector group and one for the PHT-C detectors, covering the whole wavelength region from 3 μm to 240 μm, although during the mission, polarization observations were actually only ever done at 25 μm and 170 μm. The instrument design incorporated a number of design features to maximize the performance of the instrument including the fact that a range of apertures (5–180″) could be selected to match the point spread function at each selected filter wavelength and thus optimize source-to-background contrast.

Short-Wavelength Spectrometer: SWS

The ISO Short-Wavelength Spectrometer (SWS) provided medium and high spectral resolution in the wavelength region 2.4 μm to 45.2 μm. It consisted of two largely independent grating spectrometers operating in the SW range of 2.4 μm to 12 μm, and the LW range of 12 μm to 45 μm, and had a spectral resolving power of $R = \lambda/\Delta\lambda \approx 1{,}000$–$2{,}000$. By inserting Fabry–Pérot (FP) interferometers into the beam, one for the range 11–26 μm and the other for the region 26–45 μm, the spectral resolving power could be increased to $R \approx 30{,}000$.

The SW and LW parts of the spectrometer each had two sets, or blocks, of 12 detectors (photoconductive and photodiode), and by the switching in of different order-sorting filters each block recorded the spectrum in a "band" of wavelengths covering approximately half of the SW and LW spectral ranges, respectively, when the gratings were scanned. Hence, in grating mode the SW and LW spectra were each made up of 24 individual subspectra from each of the 24 detectors, which were then overlapped. In addition, the FP interferometers each had 2 double detectors (only one of each pair being used to gather valid data) giving a total of 52 detectors in all.

The aperture size of the spectrometer in grating mode was $14 \times 20''$ for $\lambda < 12$ μm (detector blocks 1 and 2), $14 \times 27''$ or $20 \times 27''$ (depending on the order-sorting filter) for $\lambda > 12$ in detector block 3, and $20 \times 33''$ in detector block 4. In FP mode, the aperture size was $10 \times 39''$ for $\lambda < 26$ μm, and $17 \times 40''$ for $\lambda > 26$ μm.

ISO/SWS recorded many spectra of the giant planets, and averaged spectra are shown in Figure 7.25. For Uranus and Neptune, whose angular diameters are much less than the FOV, these are pure disk-averaged spectra. However, the spectra for Saturn and Jupiter are not quite disk-averaged since Saturn's apparent diameter was comparable with the FOV, and Jupiter completely filled it. Although there are some gaps in the spectra as we go from Jupiter to Neptune, the spectra that were measured can be seen to be in good agreement with the synthetic spectra shown in Figure 7.10 and in Chapter 6.

Long-Wavelength Spectrometer: LWS

The ISO Long-Wavelength Spectrometer (LWS) covered the long wavelength spectral range between 43 μm and 197 μm with a single grating operating in either first or second order, and a block of ten order-sorting filters and photoconductive detectors, each covering approximately 1/10th of the total spectral range when the grating was scanned. Hence, like SWS, the spectrum recorded by LWS was composed of ten subspectra from the ten individual detectors, which were then overlapped. The resolving power varied between $\Delta\lambda/\lambda = 150$–$200$ in the individual subspectra, but again, like SWS, this could be greatly increased to $\Delta\lambda/\lambda = 6{,}800$–$9{,}700$ by the addition of one of two FP interferometers placed in the beam.

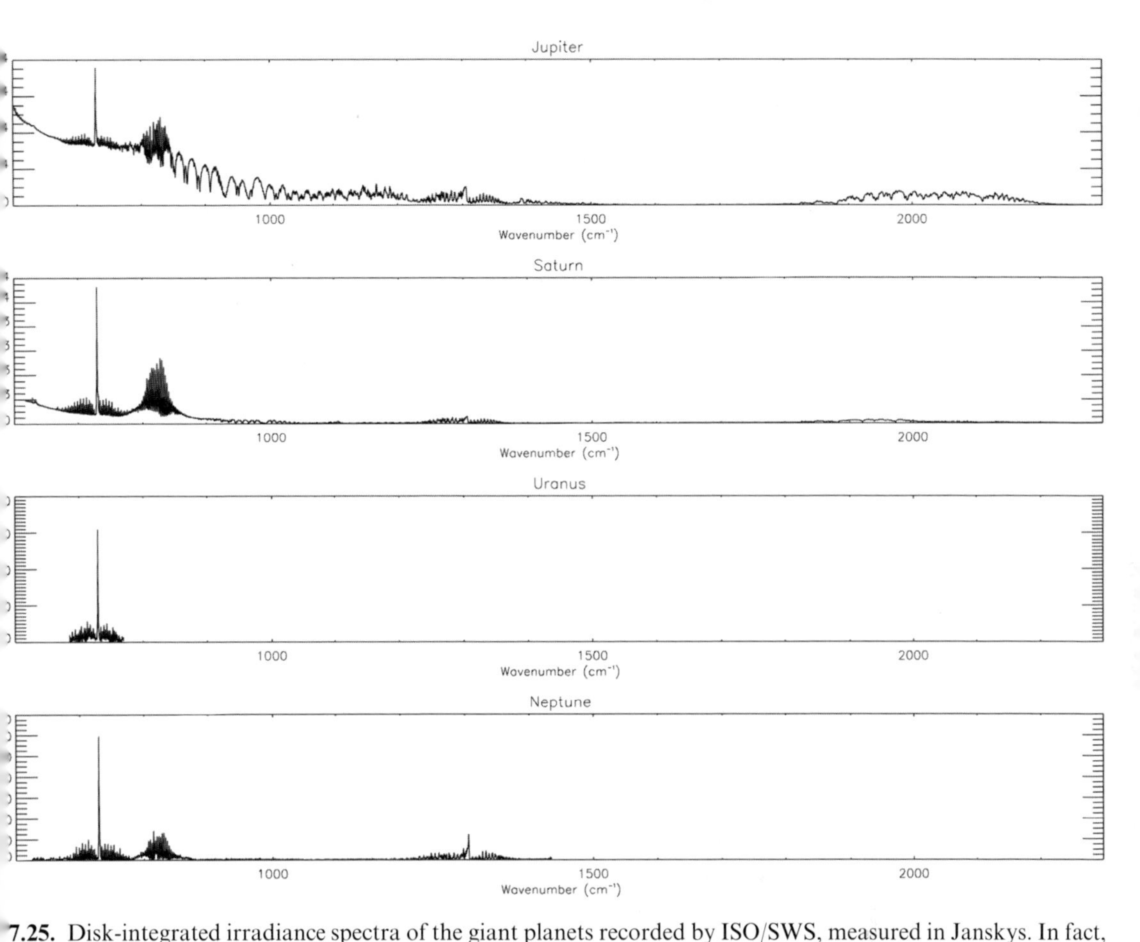

7.25. Disk-integrated irradiance spectra of the giant planets recorded by ISO/SWS, measured in Janskys. In fact, ctra for Jupiter and Saturn are not quite disk-averaged, as is described in the text. Courtesy of T. Encrenaz and T. t.

The aperture size of the instrument was set to be approximately equal to the diffraction-limited angular resolution of the ISO telescope at 200 μm, the longest operating wavelength of LWS, and during in-flight operation the FOV was observationally determined to be 80″.

Observed disk-averaged LWS spectra of Jupiter and Saturn are shown in Figure 7.26. It proved in practice somewhat difficult to radiometrically calibrate this instrument, and thus the detector subspectra here do not overlap well. However, individual subspectra are correct relative to themselves and thus the depths of the ammonia absorption lines are reliable. The available LWS and SWS spectra of the giant planets have been combined in Figure 7.27 to give near-complete disk-averaged spectra. The gap between 16 μm and 45 μm (222–625 cm^{-1}), which should nominally have been recorded by SWS, was not reliably measured. The ISO instruments were designed

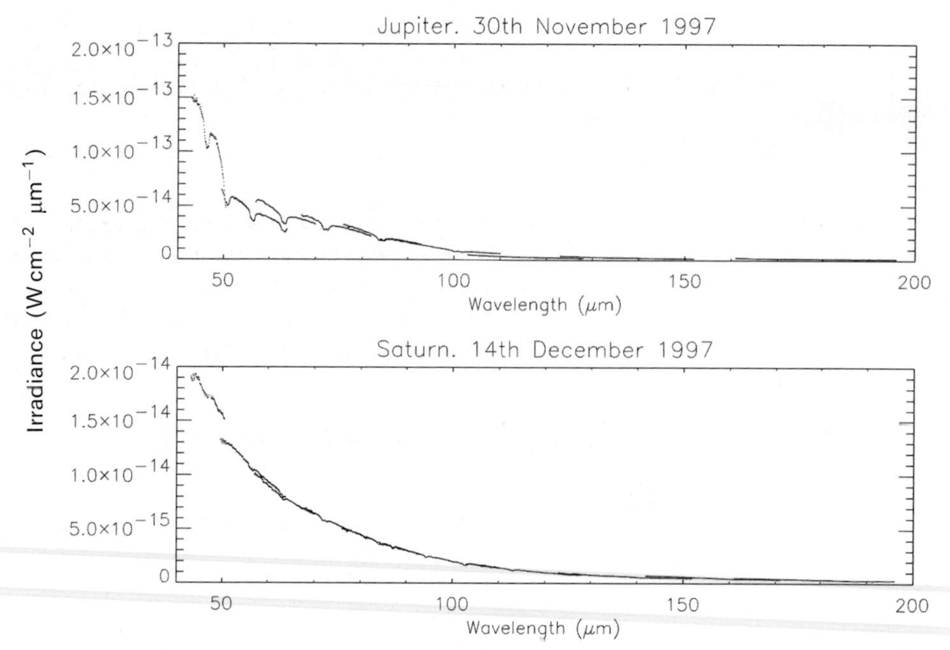

Figure 7.26. Disk-integrated irradiance spectra of Jupiter and Saturn recorded by ISO/LWS, in units of W cm^{-2} μm^{-1}. Courtesy of G. Davies, JAC, Hawaii.

primarily to observe very cold, distant astronomical objects, and it was found that Jupiter SWS observations suffered saturation effects in this spectral range.

7.7.3 Submillimeter Wave Astronomy Satellite (SWAS)

The NASA SWAS was launched in 1998 on a two-year mission to observe mainly interstellar dust clouds at submillimeter wavelengths from 487 GHz to 557 GHz (540–620 μm, 16.2–18.6 cm^{-1}). The wavelength region was chosen to search for emission lines of water vapor and SWAS detected water in almost every dust cloud that it observed, a key indicator in the cooling process of these clouds that eventually leads to their collapse to form stars. In addition to these observations, SWAS also observed water vapor emission lines from Jupiter (Lellouch *et al.*, 2002).

SWAS was launched into a 600 km altitude near-polar orbit; its primary mirror was 0.68 × 0.58 m in size and its total mass was 288 kg. Hence, SWAS was a reasonably small spacecraft, but provided a valuable precursor to future far-IR/submillimeter missions such as the *Herschel Space Observatory*, described in Chapter 8.

7.7.4 *Spitzer*

The NASA *Spitzer* Space Telescope (formerly known as the Space Infrared Telescope Facility, SIRTF), launched in August 2003 from Cape Canaveral, is a spaceborne,

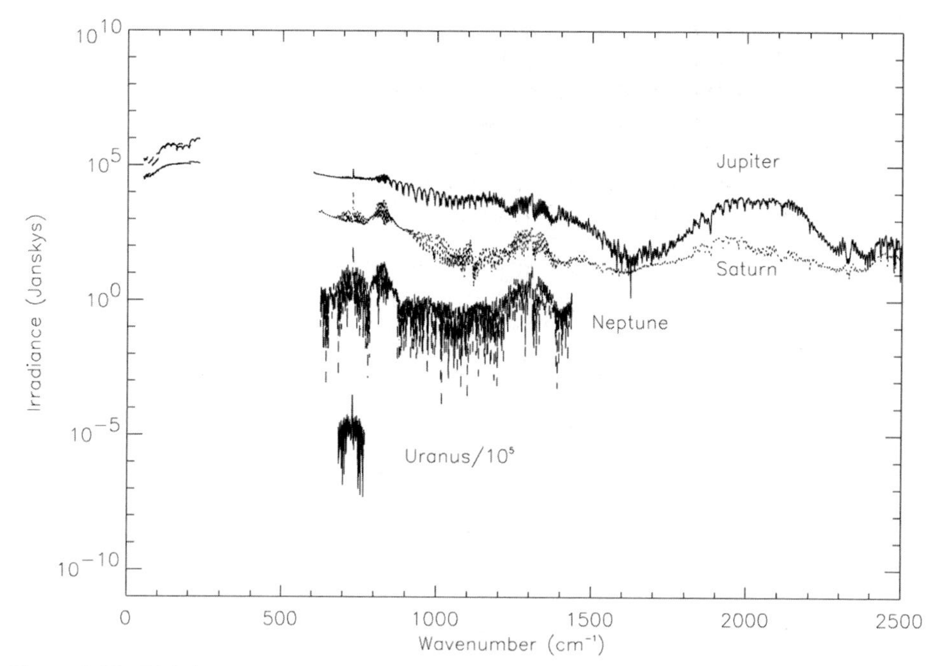

Figure 7.27. Disk-integrated irradiance spectra of the giant planets recorded by both ISO/SWS and ISO/LWS, plotted together on a log-scale. Jupiter: solid line; Saturn: dotted line; Uranus: dashed line; Neptune: dot-dashed line. The SWS Uranus spectrum has been divided by 10^5 in order to distinguish it from the SWS Neptune spectrum. The spectra can be seen to be in good agreement with the synthetic spectra shown previously in Figure 7.8.

cryogenically cooled telescope that forms the final element in NASA's Great Observatory Program. The four major scientific objectives of *Spitzer* are: (1) to study the early universe; (2) to search for and study brown dwarfs and superplanets; (3) to study ultraluminous galaxies and active galactic nuclei; and (4) to discover and study protoplanetary and planetary debris disks.

The telescope incorporates a 0.85 m primary mirror cooled to 5.5 K by liquid helium and at launch weighed 950 kg (Figure 7.28). The telescope was placed in an Earth-trailing, heliocentric orbit, where the background radiation levels are low and where it may observe chosen targets for long integration times, uninterrupted by orbital considerations, as was the case for the Infrared Space Observatory (ISO) (Section 7.7.2). The low thermal radiation environment means that *Spitzer*'s optics naturally cool to low temperatures without the need for a very large supply of cryogen, and the cryogens that are carried are only used to cool the mirror and detectors to their final operating temperatures. Hence, the lifetime of *Spitzer* will be much longer than it was for ISO, and *Spitzer* is expected to remain operational until April 2009.

Spitzer has three instruments and performs imaging, photometry, and spectrometry from 3 μm to 180 μm, as will now be described.

Figure 7.28. An artist's impression of the *Spitzer* Space Telescope, launched in 2003. Courtesy of NASA.

InfraRed Array Camera (IRAC)

IRAC is a four-channel camera that provides simultaneous $5.17 \times 5.17''$ images at 3.6 μm, 4.5 μm, 5.8 μm, and 8 μm. Two adjacent fields of view are imaged in pairs (3.6 and 5.8 μm; 4.5 and 8.0 μm) using a dichroic beamsplitter. All four cameras use 256×256 pixel arrays, giving a pixel size of $1.2''$. The two short wavelength channels use InSb detectors and the two long wavelength channels use SiAs detectors.

IRAC is mainly an astronomical survey instrument due to its high sensitivity, large field of view, and simultaneous four-color imaging and thus does not provide useful data on the giant planets.

InfraRed Spectrograph (IRS)

IRS provides *Spitzer* with low-resolution and moderate-resolution spectroscopy from 5.3 μm to 40 μm. The IRS is composed of four separate modules, each of which is built around a 128×128 pixel-detecting array and each module has its own entrance slit in the focal plane. The IRS instrument has no moving parts and is designed to achieve high sensitivity through the use of cooled detectors.

Low-resolution ($R = \lambda/\Delta\lambda = 60$–$124$), long-slit spectra can be obtained from 5.3 μm to 14 μm with a FOV of $3.6 \times 54.6''$, and from 14 μm to 40 μm with a FOV of $9.7 \times 151.3''$. The pixel sizes of these two modes are $1.8''$ and $4.8''$, respectively. In this mode, both spectral and one-dimensional spatial information are acquired simultaneously on the same detector array. Two small, imaging subarrays (peak-up arrays) in the shortwavelength, low-resolution module (SL) also allow objects to be placed accurately into any of the IRS entrance slits. Moderate-resolution spectra ($\lambda/\Delta\lambda = 600$) may also be obtained in Echelle mode from 10 μm to 19.5 μm with a FOV of $5.3 \times 11.8''$ and from 19 μm to 37 μm with a FOV of $11.1 \times 22.4''$. The pixel sizes for these two high-resolution modes are $2.4''$ and $4.8''$, respectively.

Although spectral resolution is less than that of the ISO SWS, spatial resolution and sensitivity are significantly better and thus this instrument has proved useful in measuring the abundance and mapping the spatial distribution of gases in the giant planet atmospheres, especially those of Uranus and Neptune. Initial observations of Uranus by *Spitzer* IRS (Burgdorf *et al.*, 2006) were reported in Chapter 4. However, on a practical note, *Spitzer* is primarily designed to look at extremely cold targets, which explains its very low operating temperature. Hence, to *Spitzer* the larger giant planets, in particular Jupiter, appear very warm, and thus any observations have to be strictly limited to avoid excessive heating of the telescope and thus loss of precious cryogen.

Multiband Imaging Photometer for Spitzer (MIPS)

MIPS provides imaging and photometry in broad spectral bands centered nominally at 24 μm, 70 μm, and 160 μm, together with low-resolution spectroscopy between 52 μm and 99 μm. The nominal 24 μm channel uses a 128×128 pixel SiAs array giving a pixel size of 2.45″, while the nominal 70 μm channel uses a $32 \times 32''$ GeGa array with a pixel size of either 4.9″ or 9.9″. The nominal 160 μm channel uses a 2×20 pixel stressed GeGa array giving a pixel size of 16″. All three arrays view different areas of the sky simultaneously, and thus multiband imaging is provided through telescope pointing. This instrument is of limited use for the study of the giant planets of our own solar system.

7.7.5 AKARI

The AKARI spacecraft telescope (formerly known as ASTRO-F), was launched by the Japanese Aerospace Exploration Agency (JAXA) on February 21, 2006 by an M-V rocket from the Uchinoura Space Center into a Sun-synchronous low-Earth orbit of altitude 695 km. The AKARI spacecraft is three axis–stabilized and keeps its telescope pointed away from the Earth and thus shaded from Earth's thermal emission, by the bus module of the spacecraft.

AKARI uses a 0.67 m Richey-Chrétien telescope and weighed 955 kg at launch. Its primary mission was to survey the entire sky from the near-infrared through to the far-infrared.

Observations at mid-IR to far-IR wavelengths ended in August 2007, when its supply of liquid helium was used up, but observations at near-IR wavelengths continue. AKARI is included here as the grism mode of its infrared camera (IRC) instrument has been used to investigate the D/H ratio on Neptune. These observations are still being analyzed.

7.8 FLYBY SPACECRAFT

While ground-based and Earth-orbiting telescopes have recorded a wealth of information concerning the giant planets, the vast majority of what we know about these worlds comes from spacecraft observations of flyby, and more recently, orbiting spacecraft. The advantages of spacecraft observations for understanding the atmospheres of the giant planets are enormous. By recording the strength of the signal from the spacecraft as it travels behind the planet (or emerges from behind) the number density of molecules in the atmosphere may be almost directly determined and, using simple assumptions, the temperature–pressure profiles may be extracted with a high degree of precision. Such *radio occultation* measurements have been made for all the giant planets and provide the bedrock of the models of atmospheric structure. Another advantage is that the spacecraft can get very close to the planets and thus record images at much higher spatial resolution than can usually be achieved from the ground, or indeed from Earth orbit, especially for the more distant worlds of Neptune and Uranus. Reflectance and thermal emission spectra may be recorded either at high spatial resolution, allowing studies of the spatial variation of atmospheric constituents or, by averaging a large number of spectra or averaging over large solid angles, with high sensitivity at the expense of spatial resolution. In addition, these thermal emission spectra can be recorded at a wide range of emission angles, thus allowing for much better determination of the vertical profiles by increasing the vertical spread of the weighting functions. Similarly, reflectance spectra can be measured with a wide range of incident solar, and reflected zenith angles, which not only allows for much better vertical discrimination of cloud structure, but also allows better estimates of the aerosol properties since the reflected intensity and polarization may be sampled over a wide range of phase angles (the angle between the incident and reflected beam). In contrast, ground-based and Earth-orbiting telescope observations are limited to phase angles close to zero. A final advantage is that if the FOVs of the instruments are small enough, limb-sounding may be performed, which offers significant advantages in terms of vertical resolution and enhanced sensitivity to trace species, as we saw in Chapter 6.

A number of spacecraft have now visited the giant planets, each armed with a wide selection of remote-sensing instruments which will now be reviewed. The angular resolution of these instruments is typically not as fine as that of terrestrial instruments, but since they are so much closer to their targets, both their spatial resolution and sensitivity are usually significantly better. In order to compare between the spatial resolutions quoted in the following sections and the terrestrial observations described previously, Table 7.4 converts the angular resolutions possible by terrestrial and space-based observatories to maximum possible spatial resolution (at opposition) for the four giant planets. Table 7.5 converts a range of possible angular resolutions for flyby and orbiting spacecraft observations to spatial resolution, dependent on the distance from the target.

7.8.1 Pioneer

The first spacecraft to visit some of the giant planets were the *Pioneer 10* and *11* spacecraft managed by the NASA Ames Research Center. The spacecraft were designed as low-cost, simple missions to demonstrate the viability of sending spacecraft through the Asteroid Belt to these planets, before more sophisticated and

Table 7.4. Spatial resolutions of terrestrial and Earth-orbiting telescopes at the giant planets.

gular resolution	*Jupiter*		*Saturn*		*Uranus*		*Neptune*	
	km	f_D	km	f_D	km	f_D	km	f_D
$10''$	30,460	0.213	61,650	0.511	132,000	2.582	211,057	4.261
$1'$	3,046	0.021	6,165	0.051	13,200	0.258	21,105	0.426
$0.5''$	1,523	0.011	3,082	0.026	6,600	0.129	10,552	0.213
$0.1''$	305	0.002	616	0.005	1,320	0.026	2,110	0.043

f_D is the fraction of the disk covered by spatial resolution.

Table 7.5. Conversion of angular resolution to spatial resolution (in kilometers) as a function of distance.

Angular resolution	*Distance* (km)			
	25,000	75,000	750,000	1,500,000
10 μrad	0.25	0.75	7.5	15
0.5 mrad	12.5	37.5	375	750
1 mrad	25	75	750	1,500
10 mrad	250	750	7,500	15,000

expensive spacecraft were flown (Figure 7.29). The spacecraft weigh 270 kg each and the diameter of the main radio dish is 2.7 m. The spacecraft are spin-stabilized, spinning at 5 rpm, and are powered by four radioisotope thermoelectric generators (RTGs), each of which provided 40 W power at launch.

Pioneer 10 was launched on March 2, 1972 and flew past Jupiter on December 3, 1973 at an altitude of only 1.82 R_J (130,354 km). It was then directed to the outer reaches of the solar system and is now heading in the direction of the star Aldebaran, as can be seen in Figure 7.30. *Pioneer 10* is still in radio contact and it may have been deflected in 1999 by an encounter with a Kuiper Belt Object. In addition, detailed tracking of *Pioneer 10* spacecraft (and other spacecraft now at great distance from the Sun) has revealed a tiny unexplained acceleration towards the Sun (Anderson *et al.*, 1998). The nature of this mysterious anomaly has so far defied analysis and may even indicate a small discrepancy in gravitational theory itself! As the mission has proceeded, the power generated by the RTGs has steadily fallen, partially because of the 92-year half-life of the Plutonium-238 isotope used, but mainly due to degradation of the thermocouple junctions that convert the heat to electricity. *Pioneer 10* is expected

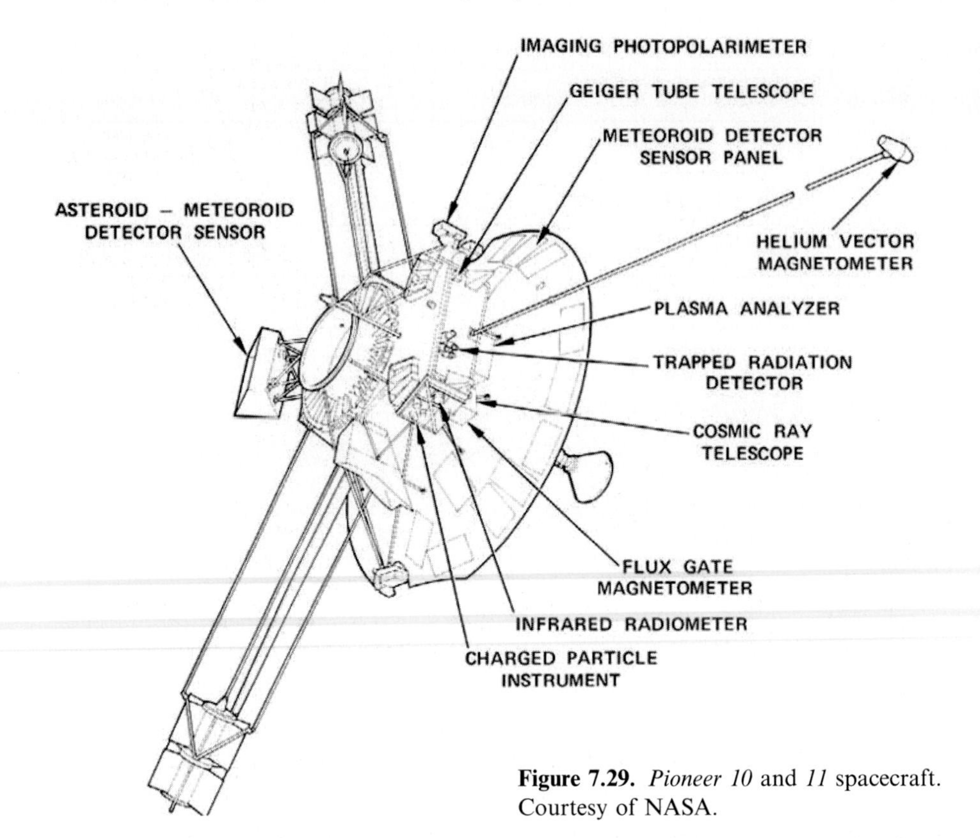

Figure 7.29. *Pioneer 10* and *11* spacecraft. Courtesy of NASA.

to run out of power, and thus lose contact with the Earth, sometime in the next 20 years.

Pioneer 11 was identical to *Pioneer 10* except that a flux-gate magnetometer was also added. The spacecraft was launched on April 5, 1973 and flew past Jupiter on December 2, 1974 at an even lower altitude of 0.6 R_J (43,000 km). *Pioneer 11* arrived from south of Jupiter's equator and left from above allowing imaging of Jupiter's North Polar region. Its trajectory then took it across the solar system to fly past Saturn on September 1, 1979 at an altitude of 0.21 R_S (13,000 km). Since then it has been heading out of the solar system and contact with the spacecraft was lost in November 1995. It is not known if *Pioneer 11* is still transmitting.

Pioneer 10 and *11* both carried three remote-sensing instruments useful for atmospheric studies: a UV photometer, an imaging photopolarimeter, and an IR radiometer.

UV photometer

This was a very simple, two-channel photometer which recorded the UV reflectivity of the giant planets at 1216 Å (hydrogen-α) and 584 Å (helium). The instrument made

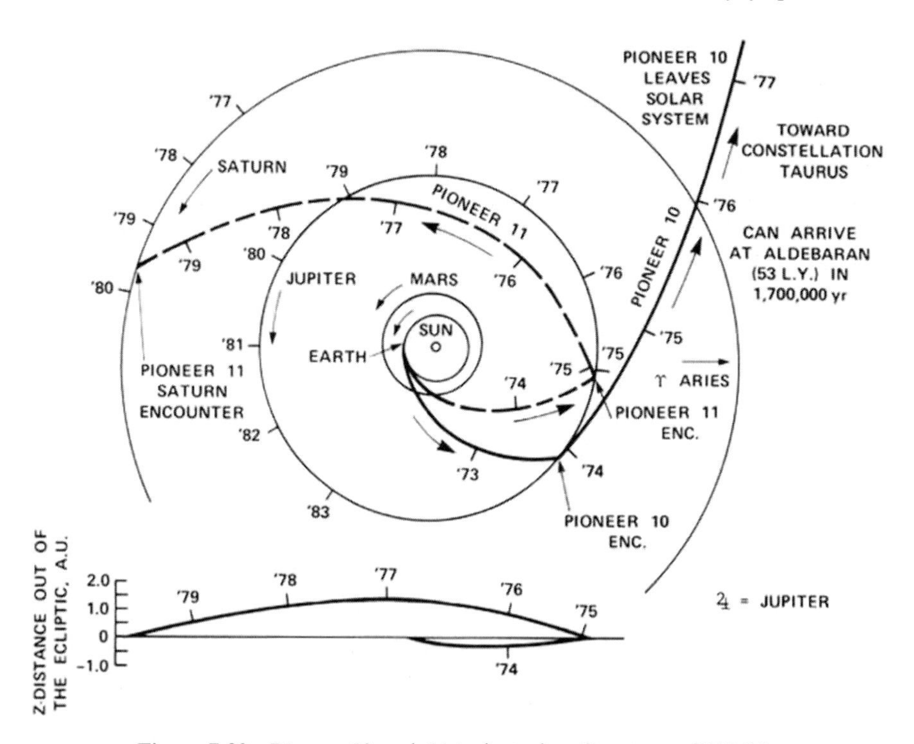

Figure 7.30. *Pioneer 10* and *11* trajectories. Courtesy of NASA.

the first detection of helium on Jupiter and enabled the first estimation of the H_2/He ratio.

Imaging photopolarimeter

The imaging photopolarimeter had a single 0.5 mrad FOV and used the spinning motion of the spacecraft to build up images. Each rotation of the spacecraft provided one line of the image, and a pointing mirror was then adjusted before the next line was recorded and so on. The pointing mirror allowed the instrument to view at angles between 27° and 170° of the rotation axis. The imaging photopolarimeter could record images in two spectral channels: red (595–720 nm) and blue (390–500 nm), and could also determine the polarization of the light.

IR radiometer

The IR radiometer had a FOV of 17.4 × 5 mrad and had two channels covering the spectral ranges (14–25 μm) and (30–56 μm), respectively. Since both spectral regions are dominated by the collision-induced absorption of hydrogen and helium, temperature sounding was achieved by viewing a location on the planet at multiple emission angles, allowing a range of weighting functions peaking at different altitudes.

7.8.2 Voyager

Once the *Pioneer 10* and *11* spacecraft had demonstrated that spacecraft missions to the giant planets were possible and of high scientific interest, more sophisticated spacecraft were built. The *Voyager* spacecraft were managed by the NASA Jet Propulsion Laboratory (JPL) and were considerably larger and more complex than *Pioneers 10* and *11* (Figure 7.31). The spacecraft weigh 825 kg each and the main

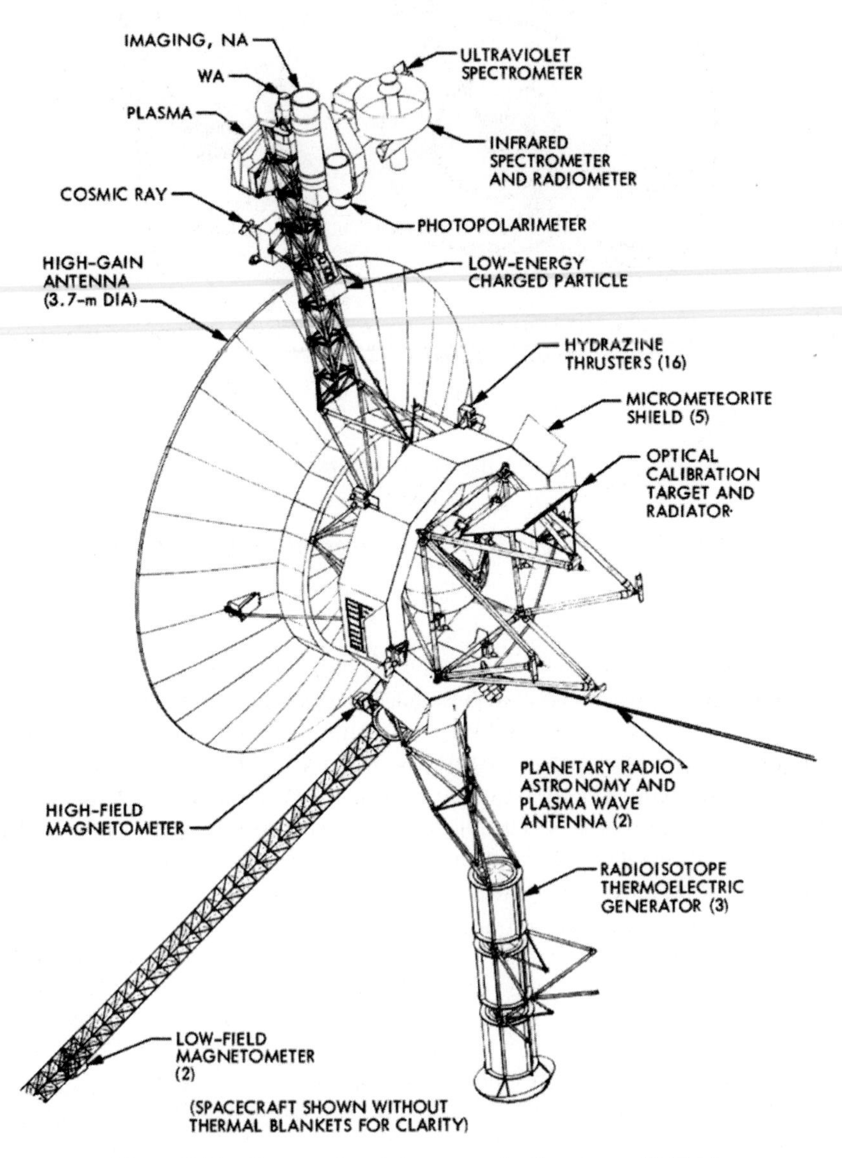

Figure 7.31. *Voyager 1* and *2* spacecraft. Courtesy of NASA.

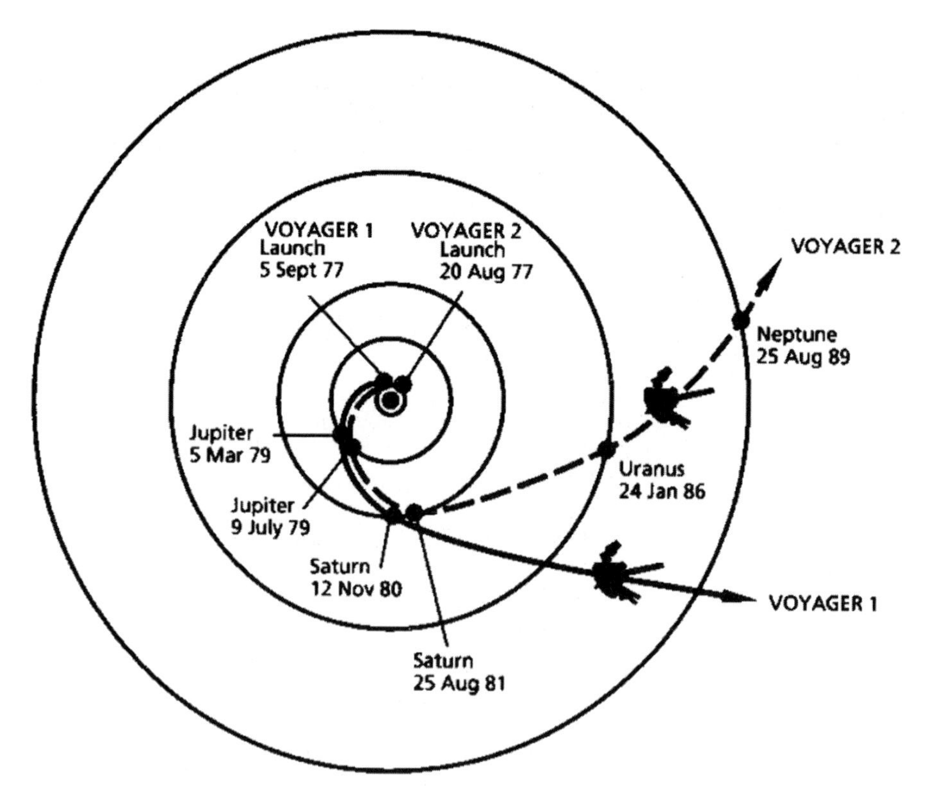

Figure 7.32. *Voyager 1* and *2* trajectories. Courtesy of NASA.

communications dish is 3.7 m in diameter. The spacecraft are three axis–stabilized and are powered by three RTGs, which together generated 470 W at launch.

Voyager 1 was launched from Cape Canaveral on September 5, 1977 and flew past Jupiter on March 5, 1979 at an altitude of 2.9 R_J (206,700 km). The spacecraft then flew by Saturn on November 12, 1980 at an altitude of 1.1 R_S (64,200 km) and the trajectory also provided *Voyager 1* with a close encounter with Titan. Subsequently, *Voyager 1* left the ecliptic plane and is now heading out to interstellar space at an angle of 35° to it (Figure 7.32). On February 17, 1998, *Voyager 1* passed *Pioneer 10* to become the most distant man-made object in space (Figure 7.33).

Voyager 2 was actually launched before *Voyager 1* on August 20, 1977 from Cape Canaveral and subsequently made the "Grand Tour" of all four giant planets. *Voyager 2* flew past Jupiter on July 9, 1979 at an altitude of 8 R_J (570,000 km) and then flew on to Saturn, which it passed on August 25, 1981 at an altitude of 0.7 R_S (41,000 km). The flyby of Uranus took place on January 24, 1986 at an altitude of 3.2 R_U (81,500 km) and *Voyager 2* made a very close flyby of Neptune on August 25, 1989 at an altitude of just 0.2 R_N (5,000 km). The resulting trajectory is taking *Voyager 2* south out of the ecliptic plane at an angle of 48°. Both spacecraft have

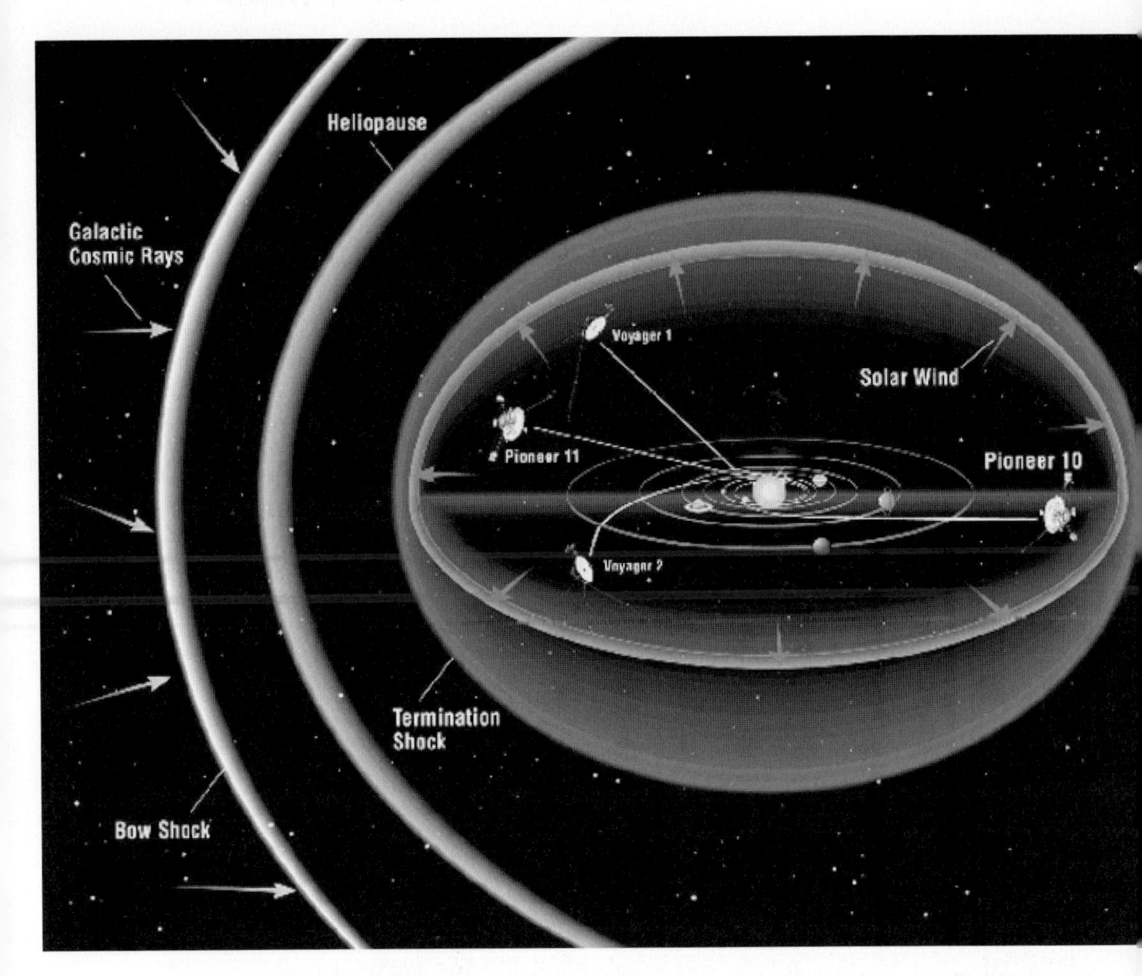

Figure 7.33. Current position of *Pioneer* and *Voyager* spacecraft. Courtesy of NASA.

sufficient power (currently ~280 W/spacecraft), and attitude control propellant to keep them operational until probably 2020.

The *Voyager* spacecraft carried four instruments of relevance to atmospheric remote sensing.

Imaging Science Subsystem (ISS)

The *Voyager* ISS consisted of a high-resolution narrow-angle (NA) video camera and a more sensitive, lower resolution wide-angle (WA) camera. Each camera had an 800×800 element detecting array and the FOV was 7.4×74 mrad for the narrow-angle camera, and 55.31×55.31 mrad for the wide-angle camera, giving pixel sizes of 10 μrad and 70 μrad, respectively. Each camera was equipped with a set of eight filters covering various wavelengths across the visible and UV.

Ultraviolet spectrometer (UVS)

The *Voyager* UVS was a reflection diffraction grating spectrometer, which dispersed the spectrum from 50 to 170 nm onto an array of 128 adjacent detectors. Hence, the spectral resolution of the instrument was 1 nm. The instrument had two fields of view: (1) 1.7×15 mrad boresighted with the camera (ISS); and (2) 4.4×15 mrad offset from the boresight by 20°.

Photopolarimeter Subsystem (PPS)

The *Voyager* PPS was a photoelectric photometer and used a 15 cm telescope and a set of eight filters between 235 nm and 750 nm, eight polarizers, and four field stop apertures, each located on a separate wheel to allow any combination of filter, polarizer, and field stop. The allowed FOVs had diameters of 2.1 mrad, 5.8 mrad, 17 mrad, and 61 mrad, respectively.

Infrared Spectrometer and Radiometer (IRIS)

The *Voyager* IRIS comprised two instruments sharing a single large-aperture telescope system and is shown in Figure 7.34. The large primary mirror of diameter 0.5 m was needed to record the extremely low thermal emission and reflected flux from the more distant giant planets. Shortwave radiation (0.3–2 μm) was monitored

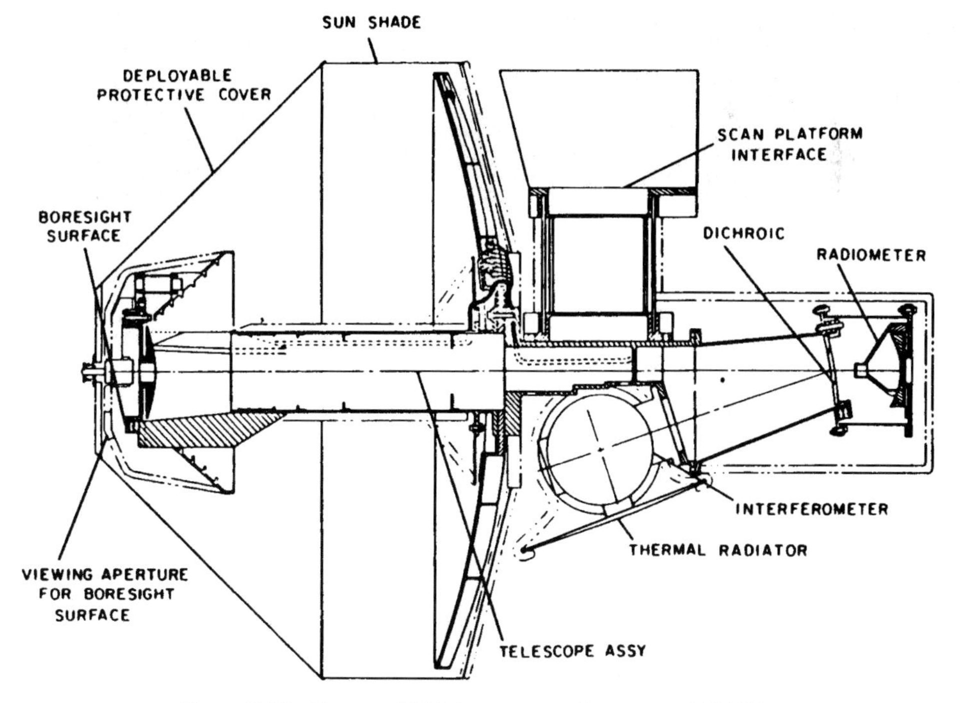

Figure 7.34. *Voyager* IRIS instrument. Courtesy of NASA.

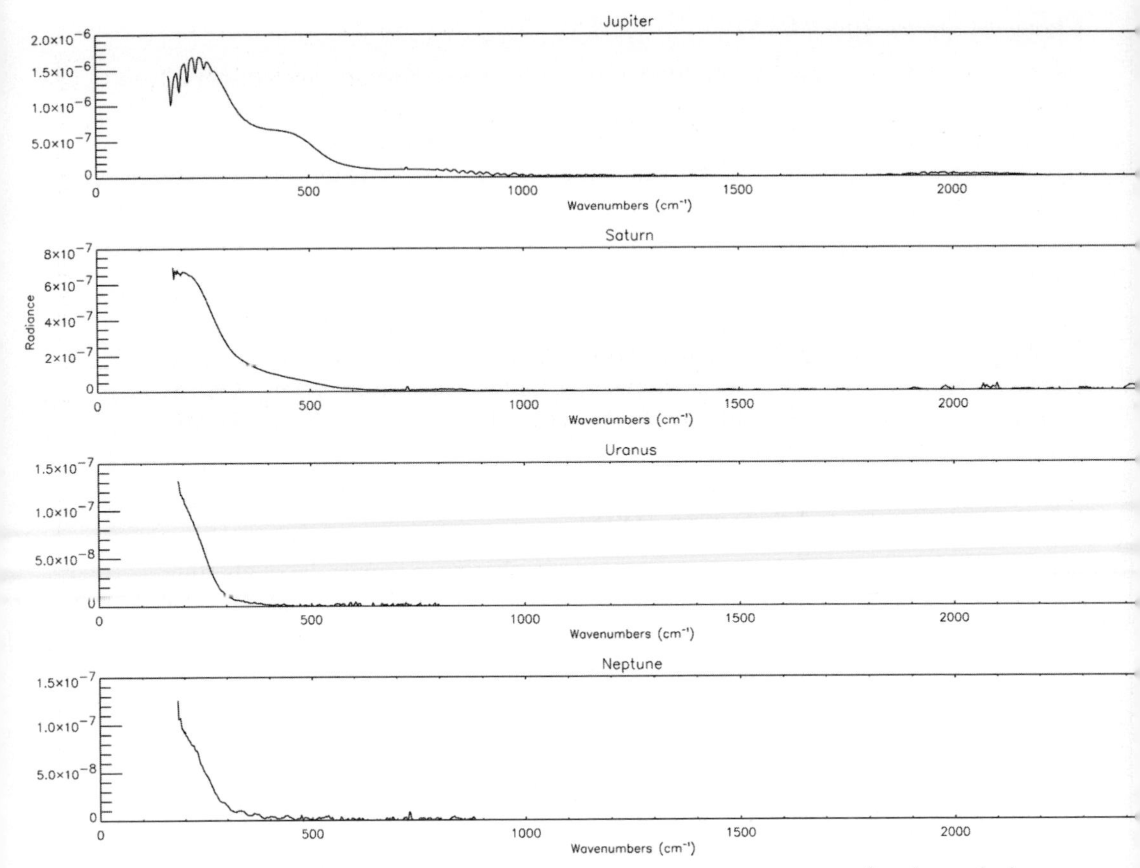

Figure 7.35. *Voyager* IRIS radiance spectra of the giant planets (in units of $W\,cm^{-2}\,sr^{-1}\,(cm^{-1})^{-1}$). Averages spectra with emission angle less than 20°.

by a simple radiometer utilizing a thermopile detector, while the long wavelength radiation passed to a Michelson interferometer, which recorded the IR spectrum from $180\,cm^{-1}$ to $2{,}500\,cm^{-1}$ at a spectral resolution of up to $4.3\,cm^{-1}$. Both halves of the instruments shared the same 4.4 mrad diameter circular FOV, which was boresighted with the ISS camera system. The mean nadir spectra of the giant planets recorded by IRIS are shown in Figures 7.35 and 7.36 and have yielded a great deal of information on the temperature and composition of these planets' atmospheres.

7.8.3 Ulysses

The *Ulysses* spacecraft was built by ESA and was not intended for planetary observations. However, it did record data covering fields and particles about Jupiter in the early 1990s and so is included here for completeness. *Ulysses* was launched on October 6, 1990 by the space shuttle *Discovery* (STS-41) and its mission was to

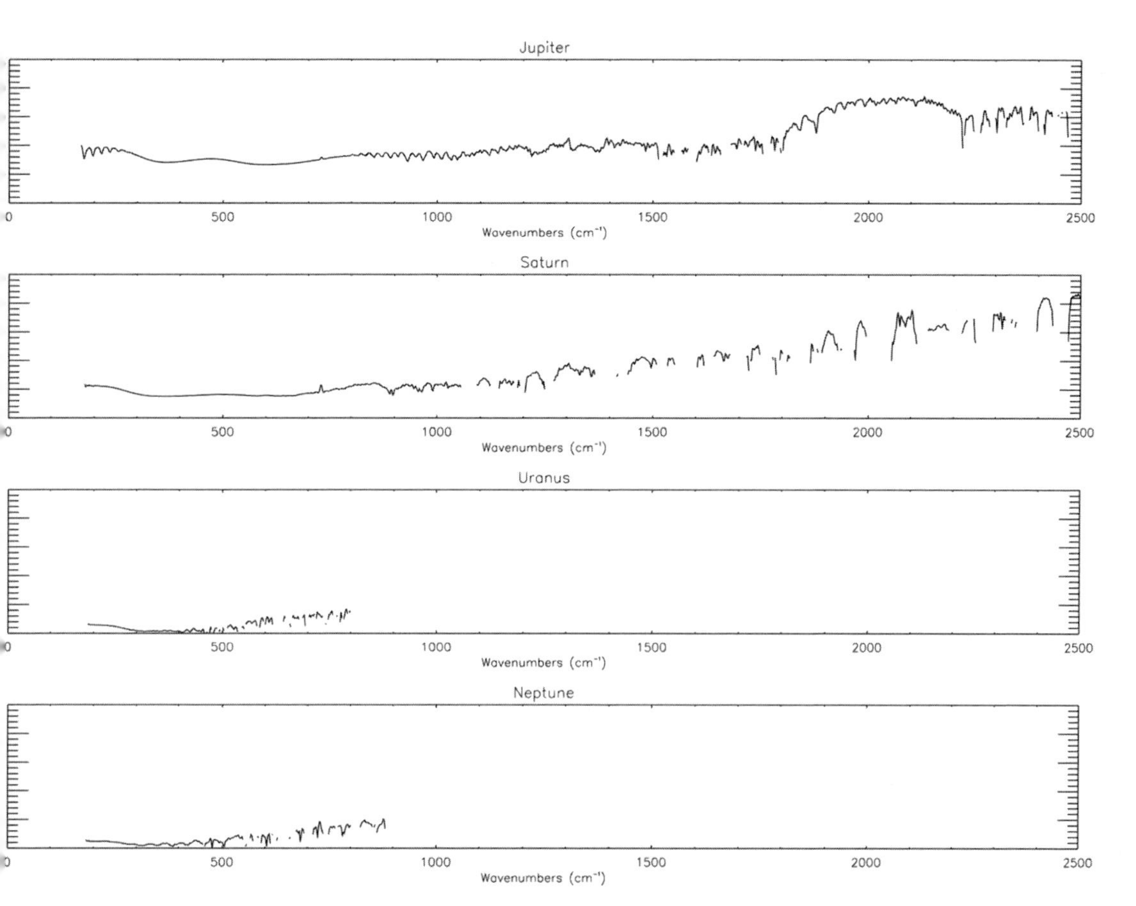

Figure 7.36. *Voyager* IRIS average spectra of the giant planets, expressed as brightness temperatures.

observe the magnetic field and plasma environment in the solar wind from above the poles of the Sun. To achieve this viewing position, *Ulysses* flew first to Jupiter and made a gravity assist flyby maneuver on February 8, 1992, which placed it into a high-inclination, highly eccentric orbit about the Sun. It passed over the Sun's South Pole in 1994 and then its North Pole in 1995. The most recent passes over the Sun's South and North Poles took place in 2000/2001 and again in 2007/2008. However, although *Ulysses'* mission was extended until March 2009 the power output from the spacecraft's RTG steadily declined and in July 2008 mission operations formally ceased due to the freezing of the spacecraft's attitude control fuel, hydrazine.

7.8.4 *New Horizons*

New Horizons is a NASA mission to Pluto and the Kuiper Belt, which was launched on January 19, 2006 from Cape Canaveral on an Atlas-Centaur launch vehicle. *New*

Horizons is due to arrive at Pluto–Charon in 2015 and will make many encounters with Kuiper Belt Objects between 2016 and 2020.

New Horizons was launched into a Jupiter gravity assist trajectory to Pluto and made a close flyby of Jupiter on February 28, 2007, which it passed at a distance of 2.3 million km (32.2 R_J), which is approximately one-fourth of the closest approach distance during the *Cassini/Huygens* flyby. The spacecraft is three axis–stabilized, relatively small with a total mass, including fuel, of 460 kg (net weight without fuel is 385 kg). Power comes from a single RTG generating 228 W, and the main communication dish has a diameter of 2.5 m.

The spacecraft carries a number of instruments for remote sensing, relevant to its observations of Jupiter. RALPH is a visible and IR imager/spectrometer, ALICE is an UV imaging spectrometer, REX is *New Horizons*' Radio Science Experiment, while LORRI (Long Range Reconnaissance Imager) is the spacecraft's camera system. All instruments have very simple designs ensuring low mass and low power.

During *New Horizon*'s flyby of Jupiter, many of the spacecraft instruments were turned on and its main findings were discussed in Chapters 4 and 5.

7.9 ORBITING SPACECRAFT

7.9.1 *Galileo*

The *Galileo* mission to Jupiter was the first space mission designed to place a spacecraft into orbit about a giant planet and also the first space mission to deploy an entry probe directly into the atmosphere of such a planet. The *Galileo* spacecraft, managed by JPL, was a huge structure with a mass of 2,223 kg and was over 6 m tall (Figure 7.37). The spacecraft used a novel "dual-spin" design with a top section, incorporating the communications systems, booms, and other systems, spinning at 3 rpm and a lower three axis–stabilized section, upon which the remote-sensing instruments were placed on a pointable platform allowing them to stop and look in almost any direction. The *Galileo* entry probe was stowed at the bottom of the spacecraft until it was deployed on approach to Jupiter. The spacecraft was named after Galileo Galilei (1564–1642), who first observed Jupiter's major moons: Io, Europa, Ganymede, and Callisto, now known as the Galilean satellites, in January 1610.

The spacecraft was launched on October 18, 1989 from Cape Canaveral onboard the space shuttle *Atlantis* and embarked upon an extended Venus–Earth–Earth Gravity Assist (VEEGA) trajectory to Jupiter, as shown in Figure 7.38, before going into orbit about that planet on December 7, 1995. Six months prior to its arrival, the probe was deployed, which then proceeded purely under its own momentum and the pull of Jupiter's gravity, before entering the atmosphere just prior to the entry of the main part of the spacecraft into Jupiter orbit.

The orbital design of *Galileo*'s tour primary is shown in Figure 7.39 and the distance at each perijove was typically 15 R_J. The high eccentricity of the orbits allowed *Galileo* to not only pass close to Jupiter on each revolution, but also to pass

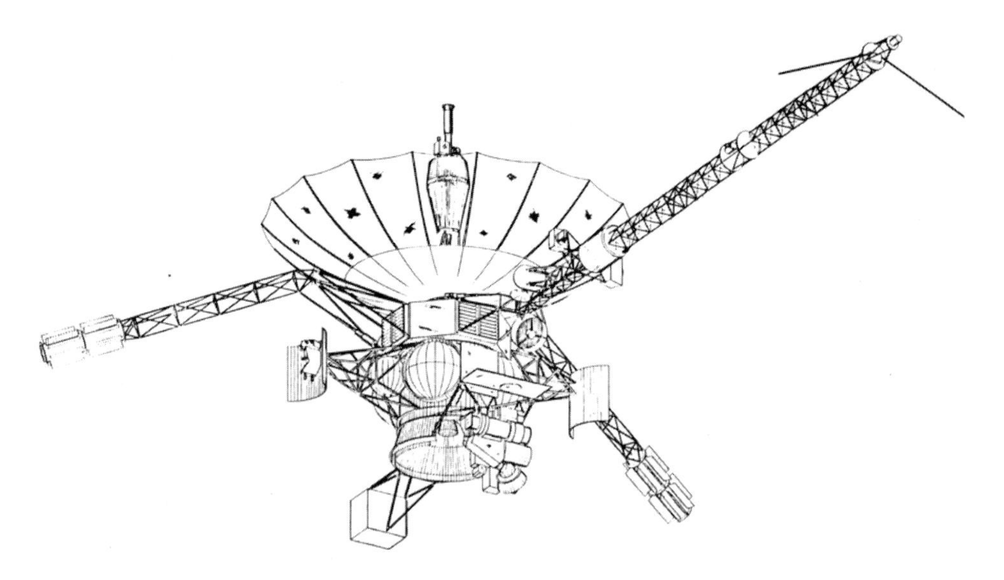

Figure 7.37. *Galileo* spacecraft. Courtesy of NASA.

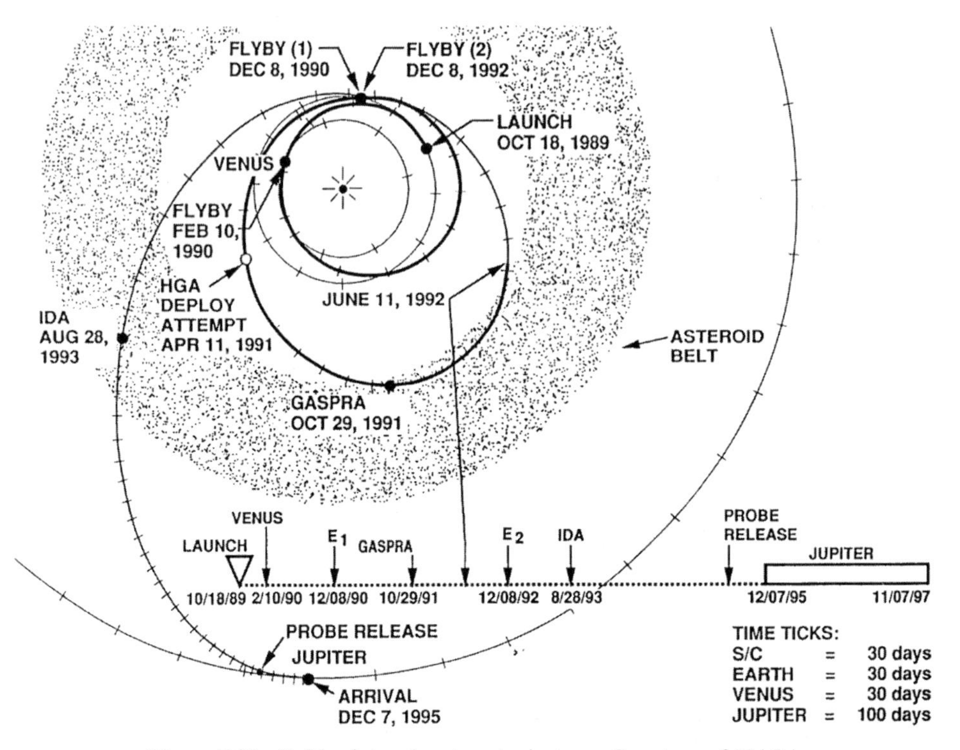

Figure 7.38. *Galileo* interplanetary trajectory. Courtesy of NASA.

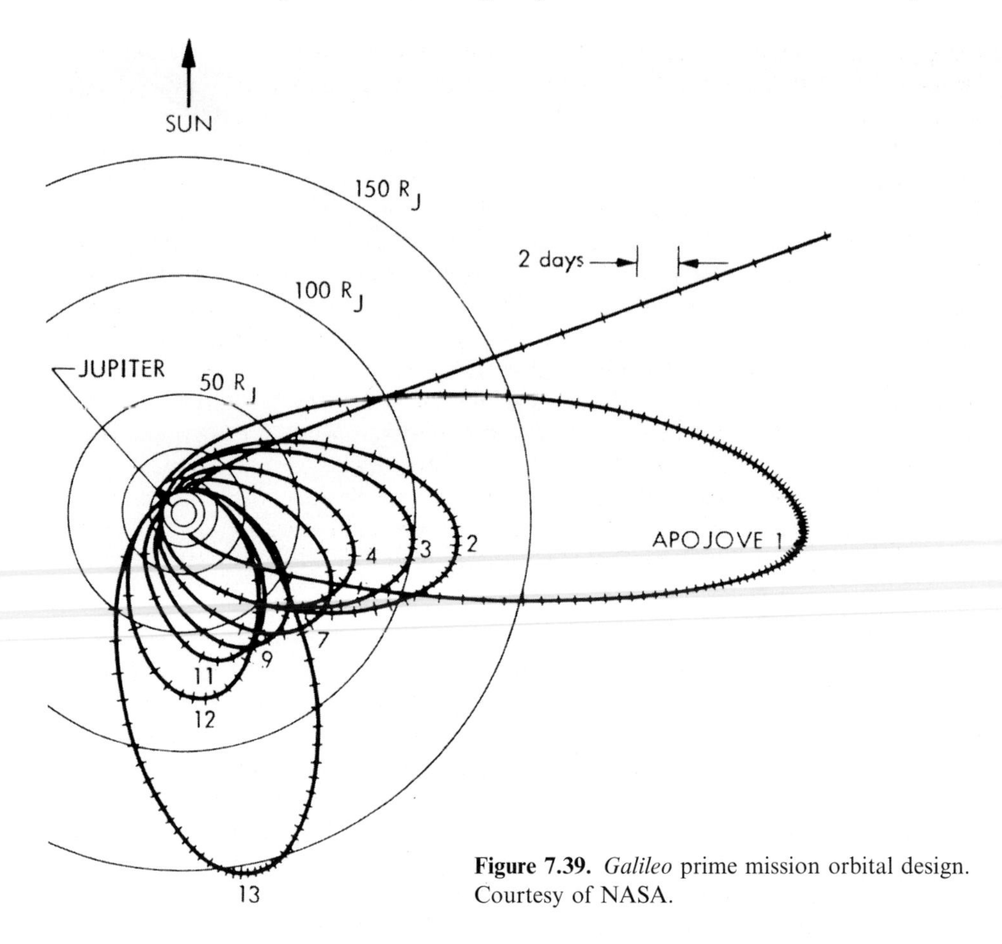

Figure 7.39. *Galileo* prime mission orbital design. Courtesy of NASA.

close by one of the four Galilean satellites either on the inward or outward legs. The orbital design also minimized the time that *Galileo* spent close to Jupiter, and thus within its highly energetic and damaging radiation belts. The orbital design, however, had a further, unforeseen, and wholly serendipitous advantage. The main communications high-gain antenna (HGA) of *Galileo* had a novel, lightweight, deployable design, rather like an umbrella. Unfortunately, when this was commanded to open soon after launch, it became stuck and could not subsequently be moved. This meant that *Galileo* had to communicate with Earth via its much smaller low-gain antenna, which at Jupiter's distance from Earth limited the communication speed to initially only 10 bits/s! It was feared that this would greatly reduce the amount of data that could be returned and thus the scientific value of the mission. However, since each orbit lasted approximately two months, and encounters with Jupiter and the satellites took only a few days, the spacecraft was comparatively idle for most of the time. Hence, data were recorded onto an onboard tape recorder during each encounter and then subsequently "trickled" back to Earth in the relatively inactive parts of the orbit at the low data rate. This solution, together with the use of sophisticated data compression techniques and improvements in the Deep Space Network receiving stations meant that the *Galileo* mission achieved scientifically almost everything it set out to achieve.

The *Galileo* orbiter operated for many years longer than originally planned. However, the spacecraft and its instruments slowly became more and more damaged by particle impacts from passing though Jupiter's radiation belts and the spacecraft slowly ran out of attitude control fuel. Hence, it was decided to terminate the mission by crashing the spacecraft into Jupiter's atmosphere where it burned up and vaporized in September 2003. The advantage of destroying *Galileo* in this way rather than leaving it in orbit about Jupiter was to avoid collision with the Galilean satellites and thus contamination of those worlds with any terrestrial organic matter that might have been on the spacecraft. This is of particular importance with respect to Europa, which is believed to have a substantial salty ocean beneath its icy crust and could thus, as some researchers suggest, have possibly evolved its own forms of life!

The *Galileo* orbiter had four remote-sensing instruments suitable for atmospheric study placed on its remote-sensing platform and a further instrument mounted on the spinning section, which will now be reviewed, together with the probe mission.

Solid State Imaging (SSI)

The SSI used a 17.6 cm Cassegrain telescope to form an image on a 800 × 800 pixel solid-state CCD array. The harsh nature of Jupiter's radiation belt meant that this CCD array had to be shielded by a 1 cm thick shroud of tantalum. The device contained an eight-position filter wheel, which had narrowband filters covering various wavelengths between 400 nm and 1,100 nm. The central filter wavelengths were: 611 nm, 404 nm, 559 nm, 671 nm, 734 nm, 756 nm, 887 nm, and 986 nm and the overall field of view of the camera was 8.1 × 8.1 mrad, which converts to 10.16 × 10.16 μrad/pixel. Both the optical system and the filter wheel were inherited from the *Voyager* ISS narrow-angle camera. At closest approach the spatial resolution was 11 km.

Ultraviolet Spectrometer (UVS) and Extreme Ultraviolet Spectrometer (EUV)

Galileo carried two instruments for measuring UV radiation, the UVS and the EUV. The UVS was mounted on the scan platform and could thus be freely pointed. The EUV, however, was on the spun-section and thus observed only a narrow ribbon of space perpendicular to the spin axis.

The UVS was a grating spectrometer utilizing a 25 cm diameter Cassegrain telescope. There were three photomultiplier detectors, which by scanning the grating covered the spectral ranges 113 nm to 192 nm, 282 nm to 432 nm, and 162 nm to 323 nm, respectively, at a spectral resolution of 0.7 nm below 190 nm and 1.3 nm above. The FOV was 17.4 × 1.7 mrad (1 × 0.1°) for the first two detectors, and 7 × 1.7 mrad (0.4 × 0.1°) for the third detector.

The EUV also had a 25 cm aperture telescope system, but had a fixed grating and array of 128 contiguous detecting elements. Hence, the design was very similar to the *Voyager* UVS. The spectral range was 54 nm to 128 nm and thus each detector had a theoretical resolution of 0.59 nm. However, other instrumental effects meant that the resolution was 3.5 nm for extended sources and 1.5 nm for point sources. The instrument FOV was 3 × 15 mrad (0.17 × 0.87°).

Photopolarimeter-Radiometer (PPR)

The *Galileo* PPR used a 10 cm diameter Cassegrain telescope to observe a 2.5 mrad diameter circular FOV in a number of photometric, polarimetric, and radiometric channels. Polarimetry was done in three spectral channels at 410 nm, 678 nm, and 945 nm, while photometry was done in seven narrowband channels at 619 nm, 633 nm, 648 nm, 789 nm, 829 nm, 840 nm, and 892 nm, which covered various important methane and ammonia absorption features. PPR also had seven radiometry bands. One of these used no filters and observed all the radiation, both solar and thermal, while another let only solar radiation through (i.e., wavelengths less than 4 μm). The difference between the solar-plus-thermal and the solar-only channels gave the total thermal radiation emitted. Five further broadband channels were included at 17 μm, 21 μm, 27 μm, 36 μm, and >42 μm. Unfortunately, for a substantial part of the mission the PPR filter wheel became stuck, which somewhat limited the data return from this instrument.

Near-Infrared Mapping Spectrometer (NIMS)

The *Galileo* NIMS was a single-grating spectrometer utilizing a 22.8 cm mirror telescope system (Figure 7.40). Light diffracted from the grating was focused onto an array of 15 InSb detectors cooled to 64 K by passive radiative cooler, each cover-

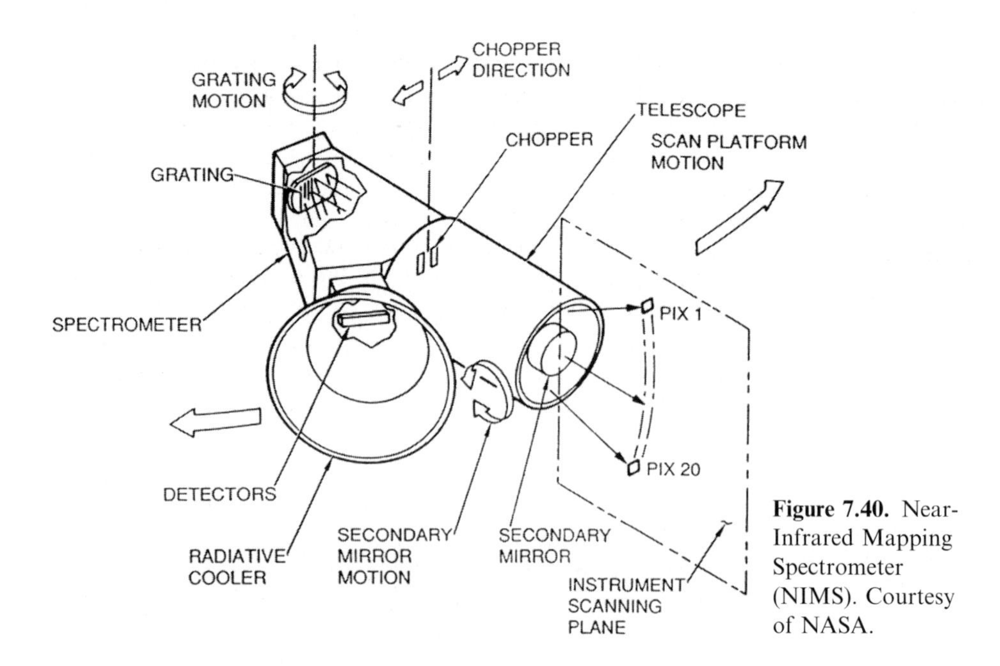

Figure 7.40. Near-Infrared Mapping Spectrometer (NIMS). Courtesy of NASA.

ing just over 1/15th of the spectral range 1 μm to 5.2 μm (in first order), and two Si detectors covering the submicron range 0.7 μm to 1 μm (in second order). The spectral resolution was 0.0125 μm below 1 μm, and 0.025 μm above. Angular resolution was 0.5×0.5 mrad, corresponding to approximately 500×500 km at perijove.

A spectrum was constructed by scanning the grating over a small angular range such that 17 subspectra were recorded by the individual detectors, which were then overlapped and combined. In addition to the grating scan, the secondary mirror of the telescope could also be scanned over 20 contiguous positions in the cross-dispersion direction. Hence, by scanning the grating for each mirror position, a single line of a spectral image could be constructed with 20 individual spectra of 408 wavelengths (24 grating steps $\times$ 17 detectors). To record an image, the whole instrument was simultaneously scanned by the remote-sensing platform in the grating dispersion direction, as is shown in Figure 7.40, building up an image with 20 rows (secondary mirror scan) $\times$ a variable number of columns (remote-sensing platform scan) $\times$ up to 408 wavelengths (grating scan). The format of the NIMS data were thus referred to as "cubes" composed of a number of two-dimensional images recorded at multiple wavelengths. In practice, the instrument could be operated with a variable number of grating steps and to reduce data volume, not all detectors were read during a particular observation. The instrument could thus very flexibly trade off between spatial and spectral coverage. This was particularly fortunate given *Galileo*'s communications difficulties and meant that NIMS could adapt to best utilize whatever observation time/data storage was available.

Galileo probe

The 340 kg *Galileo* entry probe entered the atmosphere of Jupiter on December 7, 1995 at a speed of 170,000 km hr^{-1} and a shallow entry angle as shown in Figure 7.41. The probe was "aero-captured" by Jupiter's atmosphere (experiencing a maximum deceleration of 230g), and once it had slowed sufficiently, its heat shield was jettisoned, a parachute deployed, and the probe then descended slowly down through the atmosphere recording information with several instruments on the way. Instruments included a particle nephelometer (Ragent *et al.*, 1998), a mass-spectrometer (Niemann *et al.*, 1998), a net flux radiometer (Sromovsky *et al.*, 1998), and a host of thermometers and accelerometers to record vertical structure (Seiff *et al.*, 1998). In addition to the *in situ* observations, the probe signal was also tracked from the orbiter and the Doppler-shifting of the signal used to deduce horizontal wind speeds down to depths of nearly 20 bar, while the strength of the signal was used to determine the deep NH_3 abundance.

The probe collected data for 58 minutes as it descended through Jupiter's atmosphere, with the transmission terminating at a pressure level of ~20 bar (Young, 2003). The findings of the probe have already been discussed in Chapters 4 and 5 and the probe data provide the only "ground truth" we have for conditions in the Jovian atmosphere. However, as discussed previously, it was somewhat unfortunate that the probe entered the atmosphere in a rather unrepresentative 5 μm hotspot region and it

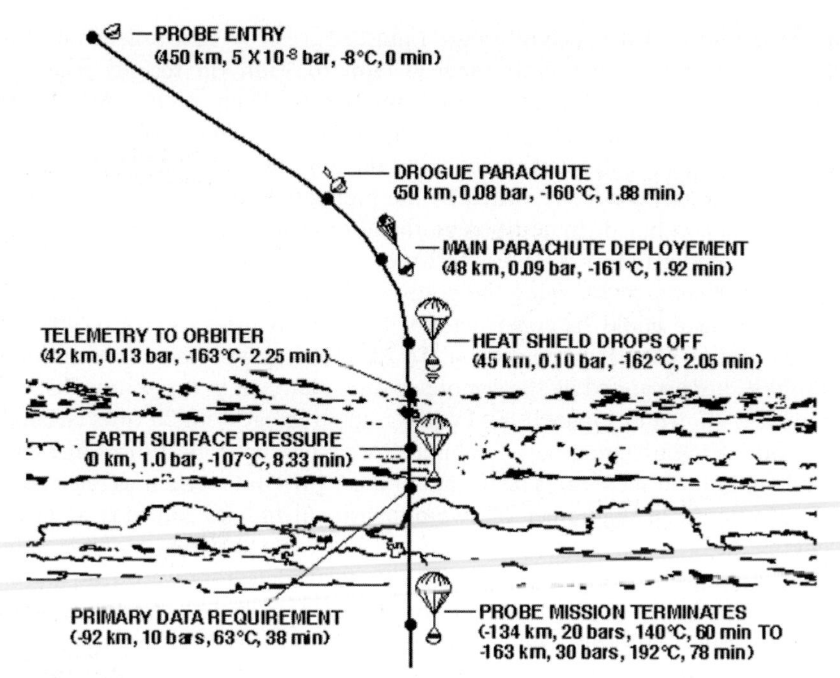

Figure 7.41. *Galileo* probe descent trajectory. Courtesy of NASA.

would be highly desirable to fly further multiple probe missions to explore other more typical regions.

7.9.2 *Cassini/Huygens*

Cassini/Huygens is the first space mission dedicated to the study of the Saturnian system. *Cassini* is a NASA-led spacecraft designed to go into orbit about Saturn and observe the planet and its moons with a range of remote-sensing instruments. The *Cassini* orbiter is named after Jean-Dominique Cassini (1625–1712) who made many early observations of Saturn and discovered the major gap in its ring system now known as the Cassini division. *Cassini* also carried an ESA-led entry probe, *Huygens*, which parachuted through, and directly sampled, the atmosphere of Titan on January 14, 2005. *Huygens* was named after Christiaan Huygens (1629–1695) who discovered Titan in 1655.

The *Cassini* spacecraft was launched from Cape Canaveral on October 15, 1997 by a Titan IV launcher into a similar multiple gravity–assist trajectory (Venus–Venus–Earth–Jupiter Gravity Assist or VVEJGA) taken by the *Galileo* spacecraft (Figure 7.42). The spacecraft flew past Jupiter on December 30, 2000 at a distance of 136 R_J (9,700,000 km), and went into orbit about Saturn in July 2004. During *Cassini*'s flyby of Jupiter, its remote-sensing instruments were activated to record a huge amount of information regarding the Jupiter system and Jupiter's atmosphere.

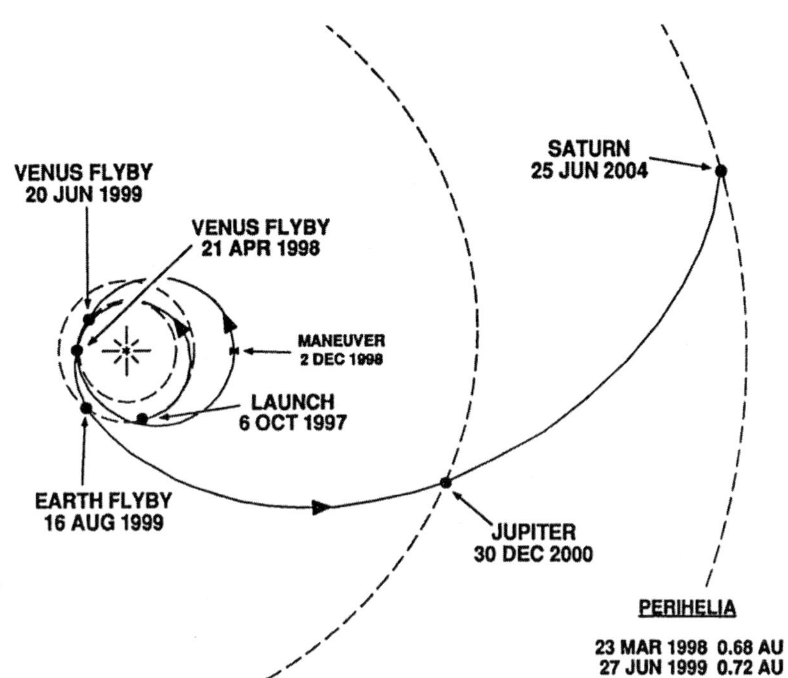

Figure 7.42. *Cassini* interplanetary trajectory. Courtesy of NASA.

During this "Millennium Mission", data were also recorded simultaneously by the *Galileo* spacecraft, already in orbit, and the combined data from two different positions has proved to be very useful for studies of the magnetic field and the particle environment.

The *Cassini* spacecraft is a truly huge three axis–stabilized spacecraft. It stands over 6.7 m tall (Figure 7.43), weighs 2,175 kg, and has a high-gain antenna with a diameter of 4 m. The spacecraft is powered by three RTGs developing a total of 630 W. Unlike *Voyager* and *Galileo*, for financial reasons there is no pointable remote-sensing platform. Instead, the remote-sensing instruments are hard-bolted onto the side of the spacecraft and pointing is achieved by using the attitude control system of the spacecraft. Hence, during observations the HGA may not point at Earth and thus data recorded are temporarily stored on a solid-state recorder until such time as wideband communications with the spacecraft are re-established and the data can be relayed to Earth.

Just like the *Galileo* mission to Jupiter, *Cassini* entered into a complex series of petal-shaped orbits when it arrived at Saturn on July 1, 2004 (Figure 7.44), which allows sampling of a wide region of the magnetosphere and also allows close flybys of the satellites and Saturn itself during each orbit. The prime mission of *Cassini* ended in 2008, but has already been extended until 2010–2012 in order to allow its instruments to observe the Northern Spring Equinox and the return of sunshine to Saturn's and Titan's North Poles and may be extended further if sufficient attitude control propellant remains.

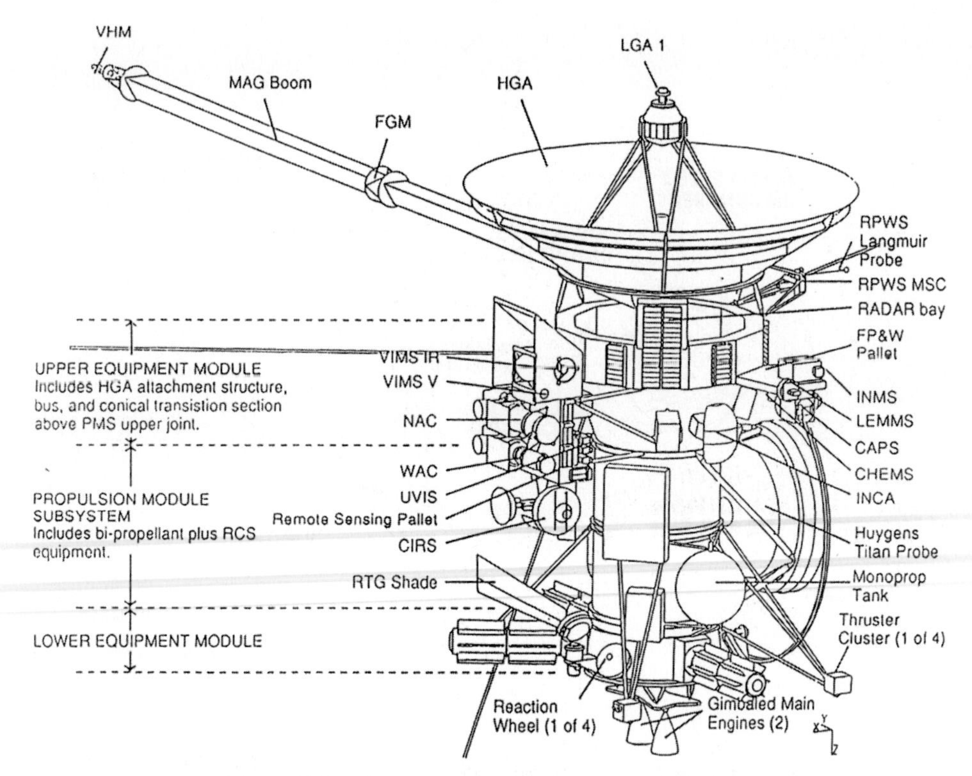

Figure 7.43. *Cassini* spacecraft. Courtesy of NASA.

Cassini is equipped with six fields and particles instruments, and six remote-sensing instruments, which are outlined in Tables 7.6 and 7.7. However, the instruments on *Cassini* most suited for atmospheric studies are ISS, UVIS, VIMS, and CIRS.

Imaging Science Subsystem (ISS)

The *Cassini* ISS includes a narrow-angle camera (NAC) and a wide-angle camera (WAC), each focusing images onto a $1,024 \times 1,024$ CCD array. The WAC field of view is 61×61 mrad (60×60 µrad per pixel) and the NAC field of view is 6.1×6.1 mrad (6×6 µrad per pixel). The WAC has 18 filters between 380 nm and 1,100 nm, and the NAC has 24 filters between 200 nm and 1,100 nm. The CCD arrays are cooled to 180 K by a passive radiative cooler to improve the signal-to-noise ratio.

ISS may record numerous high spatial resolution images in a number of spectral channels ranging from 200 nm to 1,100 nm with either WAC or NAC. The atmospheric science goals of the ISS instrument at Saturn and Titan are: (1) to map the 3-dimensional structure and motions within the atmospheres; (2) to study the composition, distribution, and physical properties of clouds and aerosols; (3) to investigate scattering, absorption and solar heating within the atmospheres; and (4) to

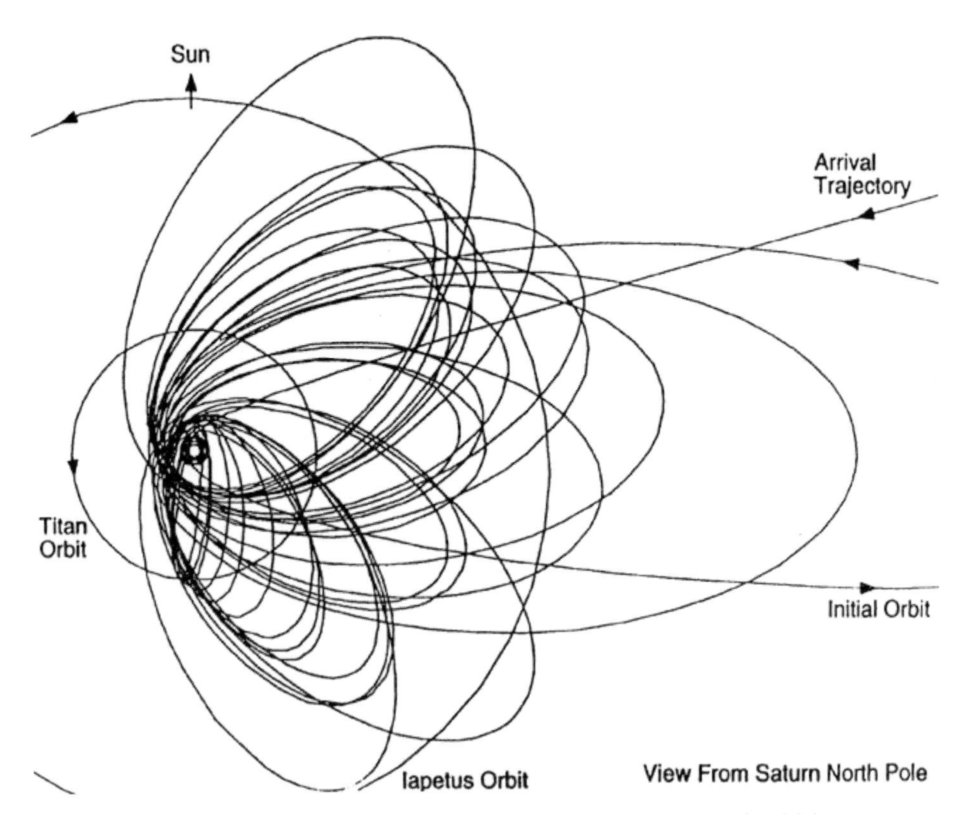

Figure 7.44. *Cassini* prime mission orbital tour. Courtesy of NASA.

search for evidence of lightning, aurorae and airglow. Just as the *Galileo* SSI instrument has yielded important new observations of Jupiter's horizontal and vertical cloud structure, so *Cassini* ISS instrument has substantially improved our understanding of the cloud morphology of Saturn. By recording images at wavelengths of low, medium, and high gaseous absorption near the methane absorption bands in the near-infrared, considerable vertical discrimination has been achieved, which is further enhanced by observing the same location at multiple illumination and viewing zenith angles. Our understanding of global circulation has been considerably improved since wide-area and indeed global images have been used to generate movies showing how the clouds in Saturn's atmosphere move at different altitudes and thus reveal the wind circulation. This was also achieved successfully for Jupiter during the December 2000 Jupiter flyby and revealed organized zonal flow on that planet extending right the way to the poles.

Ultraviolet Imaging Spectrograph (UVIS)

The *Cassini* UVIS has two main channels: the extreme ultraviolet channel (EUV) and the far ultraviolet channel (FUV), together with a high-speed photometer channel and a hydrogen–deuterium absorption cell channel.

Table 7.6. *Cassini* orbiter field-and-particle instruments.

Instrument	*Acronym*	*Purpose*
Cassini Plasma Spectrometer	CAPS	*In situ* study of plasma within and near Saturn's magnetic field
Cosmic Dust Analyzer	CDA	*In situ* study of ice and dust grains in the Saturn system
Dual Technique Magnetometer	MAG	Study of Saturn's magnetic field and interactions with the solar wind
Ion and Neutral Mass Spectrometer	INMS	*In situ* study of compositions of neutral and charged particles within the magnetosphere
Magnetospheric Imaging Instrument	MIMI	Global magnetospheric imaging and *in situ* measurements of Saturn's magnetosphere and solar wind interactions
Radio and Plasma Wave Science	RPWS	Measurement of electric and magnetic fields, and electron density and temperature in the interplanetary medium and within Saturn's magnetosphere

The EUV and FUV channels are of similar construction and use a grating to form a spectrum on to a $60 \times 1{,}024$ array of detecting elements with the 1,024 element dimension in the spectral direction. The EUV records from 55.8 nm to 118 nm, and the FUV from 110 nm to 190 nm. In addition, the FOV has three settings for each spectrograph: $(1, 2, 6) \times 64$ mrad for the EUV and $(0.75, 1.5, 6) \times 64$ mrad for the FUV. UVIS is different from previous UV spectrometers in that rather than recording a spectrum from a single location, it records a series of spectra from 60 pixels across a field of view of length 64 mrad. Hence, by scanning the instrument in the dispersion direction, multispectral images may be rapidly constructed. The width of individual pixels' FOV is adjustable, but is of the order of 1 mrad.

The photometer channel is a wideband channel recording between 115 nm and 185 nm with a FOV of 6×6 mrad and an integration time of 2 ms. It is designed to perform stellar occultation measurements of Saturn's atmosphere and ring system. The hydrogen–deuterium absorption channel may also view the Saturnian system through onboard hydrogen, deuterium, and oxygen gas absorption cells to measure the abundances of hydrogen and deuterium.

The atmospheric science goals of the instrument are: (1) to map the vertical and horizontal composition of Saturn's and Titan's upper atmospheres and thus inves-

Table 7.7. *Cassini* orbiter remote-sensing instruments.

Instrument	*Acronym*	*Purpose*
Cassini Radar	RADAR	Radar imaging, altimetry, and passive radiometry of Titan's surface
Composite Infrared Spectrometer	CIRS	Infrared studies of temperature and composition of surfaces, atmospheres and rings within the Saturn system
Imaging Science Subsystem	ISS	Multispectral imaging of Saturn, Titan, rings and icy satellites
Radio Science Subsystem	RSS	Uses communication signals in X-, S-, and Ka-bands to study atmospheric and ring structure, gravity fields and gravitational waves (S-band: 2–4 GHz; X-band: 7–8 GHz; Ka-band: 23–32 GHz)
Ultraviolet Imaging Spectrograph	UVIS	Ultraviolet spectra and low resolution imaging of atmospheres and rings for structure, chemistry and composition
Visible and Infrared Mapping Spectrometer	VIMS	Visible and IR spectral mapping to study composition and structure of surfaces, atmospheres and rings

tigate photochemical reactions; (2) to map the distribution and properties of aerosols; and (3) to study the nature of circulation in the upper atmospheres of these worlds. We saw in Chapters 4 and 5 that UV images are sensitive to the abundance of stratospheric hazes. It is hoped that UVIS will be able to take so many images that it will be able to generate movies showing how such material is moved in the upper atmospheres of Saturn and Titan, rather like expected ISS movies, but at lower wavelengths. The observed UV spectra may reveal absorption signatures of several upper-tropospheric/lower-stratospheric gases such as ammonia. In addition, UVIS measures the fluctuations of starlight and sunlight as the Sun and stars move behind the atmospheres of Titan and Saturn and it is hoped it will be able to make new estimates of the D/H ratio in these atmospheres. Such observations are currently being analyzed.

Visible and Infrared Mapping Spectrometer (VIMS)

VIMS is a development of the NIMS instrument flown on *Galileo* and is designed to measure reflected and emitted radiation from 0.35 μm to 5.1 μm to determine composition, temperature, and structure. It actually consists of two instruments: a visible channel (VIMS-V), recording the spectrum in 96 channels between 0.35 μm and 1.07 μm, and an IR channel (VIMS-IR), recording the spectrum in 256 channels between 0.85 μm and 5.1 μm. The FOV of both channels is 32 × 32 mrad and both focal planes are cooled by a passive radiative cooler to 190 K for VIMS-V and as low as 56 K for VIMS-IR.

In the visible section (VIMS-V), light collected by a 4.5 cm Shafer telescope is dispersed by a holographic grating onto a 256 × 512 silicon CCD array. The data are averaged into 96 spectral channels and 64 cross-dispersion pixels giving a pixel size of 0.5 mrad. Imaging is then achieved by a single axis–scanning mirror.

The IR section (VIMS-IR) consists of a 23 cm Cassegrain telescope and a linear array of 256 cooled InSb detectors. During the time VIMS-V makes an exposure, VIMS-IR must record 64 individual spectra by stepping its two-axis scan mirror in the cross-dispersion direction which, together with the fact that the reflected radiance of Jupiter decreases with wavelength, is why the entrance aperture of VIMS-IR has to be so much larger. Imaging is then achieved by stepping the scan mirror in the dispersion direction in the same way as VIMS-V.

The atmospheric science goals of VIMS are: (1) to map the temporal behavior of winds, eddies, and other features on Saturn and Titan at multiple wavelengths (and thus altitudes) and to perform long-term studies of cloud movement and morphology in Saturn's atmosphere; (2) to study the composition of the atmospheres and distribution of different cloud species; (3) to determine the temperature, internal structure, and rotation of Saturn's deep atmosphere; and (4) to search for lightning and, for the case of Titan, for active volcanism. As has been reported in Chapters 4 and 5, many of these goals have already been achieved.

Composite Infrared Spectrometer (CIRS)

CIRS is a development of the IRIS spectrometers flown on *Voyagers 1* and *2* (Calcutt *et al.*, 1992). Light is gathered by a 50 cm telescope and fed to two interferometers, one working in the mid-infrared from 600 cm^{-1} to 1,400 cm^{-1} and one operating in the far-infrared from 10 cm^{-1} to 600 cm^{-1} (Figure 7.45). Both sections share the same mirror drive assembly and the maximum path difference that can be introduced is 2 cm. Hence, the maximum spectral resolution is 0.5 cm^{-1}.

The mid-IR section is itself split into two parts. The spectrum from 600 cm^{-1} to 1,100 cm^{-1} is recorded by the FP3 array of ten mercury–cadmium–telluride (HgCdTe) photoconductive detectors of 0.273 × 0.273 mrad FOV arranged as shown in Figure 7.46. Similarly, the spectrum from 1,100 cm^{-1} to 1,400 cm^{-1} is recorded by the FP4 array of ten HgCdTe photovoltaic detectors with the same FOV. The mirror system of the mid-IR interferometer utilizes corner cube reflectors, to negate any misalignments and the detectors are cooled to 80 K by a passive radiative cooler, which gives them high sensitivity.

The far-IR section, FP1, uses a polarizing interferometer and a pair of bolometer detectors. The polarizing beamsplitter and polarizing plates are formed from finely

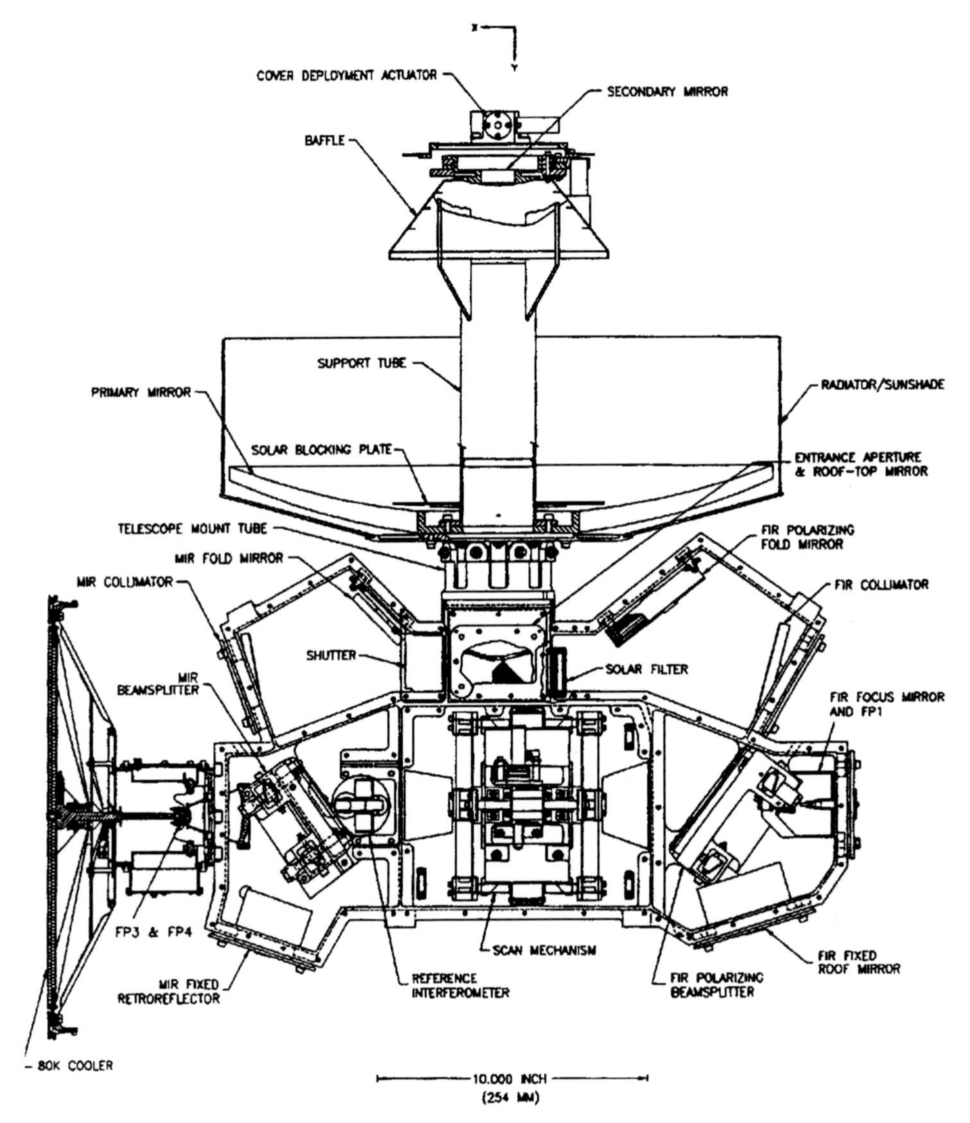

Figure 7.45. *Cassini* CIRS instrument. The mid-IR section is to the left and is cooled by the passive radiative cooler. The far-IR polarizing interferometer is to the right. Courtesy of NASA.

etched metal grids, and the use of polarization grids allows for the cancellation of the offset term in the interferogram (Equation 7.6) by subtracting the signals recorded by the detector pair. This makes the spectrometer far less susceptible to instrumental drifts and increases measurement precision. The far-IR FOV is 3.9 mrad in diameter and roof top reflectors are used to guard once more against any possible misalignments in the beams.

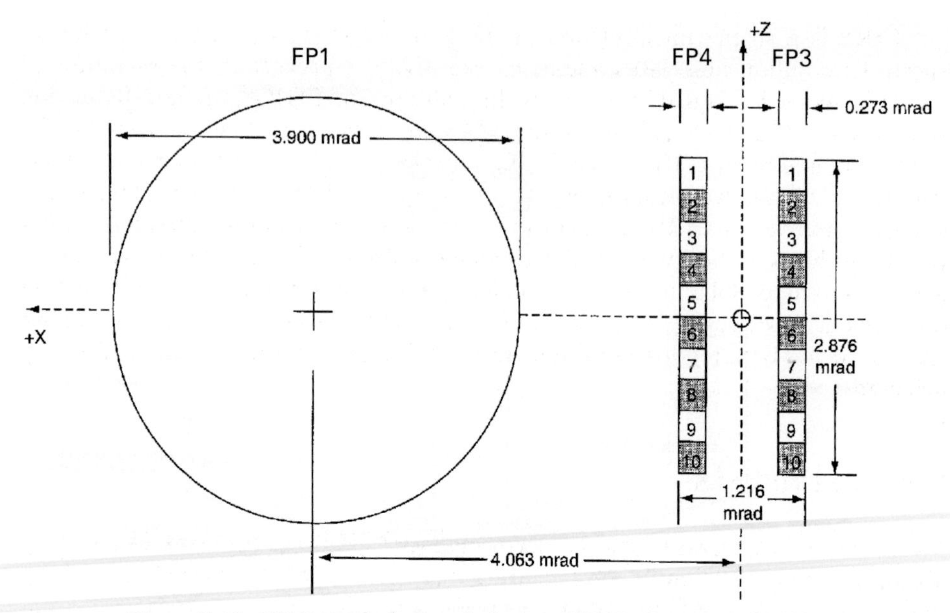

Figure 7.46. CIRS focal plane pointing and FOV. Courtesy of NASA.

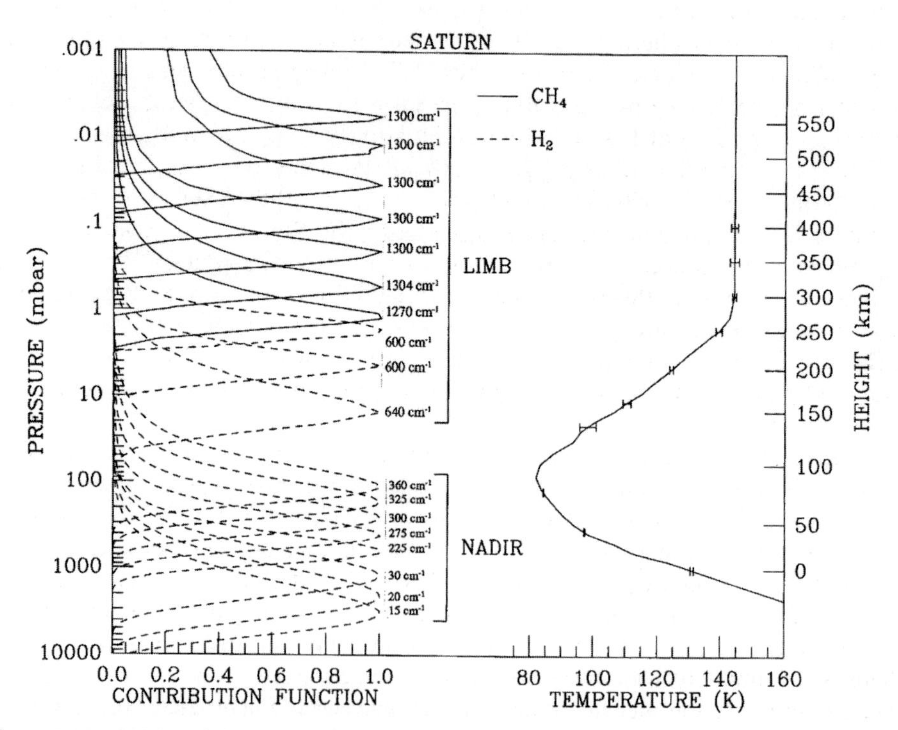

Figure 7.47. CIRS weighting functions for both limb-viewing and nadir-viewing at Saturn (from Calcutt *et al.*, 1992). Courtesy of the British Interplanetary Society.

CIRS is a significant improvement on *Voyager* IRIS, with up to 10× higher spectral resolution and vastly increased sensitivity, especially in the mid-infrared owing to its cooled HgCdTe detectors. In addition, the FOV of the mid-IR section is very small (0.273 × 0.273 mrad) allowing limb-viewing at these wavelengths for both Titan and Saturn. The transmission weighting functions for Saturn limb observations at various wavelengths are shown in Figure 7.47. The main atmospheric science objectives of CIRS are: (1) to map the global temperature structure within the atmospheres of these worlds; (2) to map the global composition of various known gases; (3) to map global information on hazes and clouds; (4) to collect information on energetic processes within these atmospheres; and (5) to search for new molecular species. As has been reported in Chapters 4 and 5, these goals have been handsomely achieved.

7.10 RETRIEVALS

We have seen in Chapter 6 how the electromagnetic spectra of the planets have many absorption and reflection features that are unique to particular constituents. Using measured and estimated absorption coefficients and assumed atmospheric profiles of temperature, cloud, and composition, radiative transfer models may be constructed which can simulate the spectrum of a planet as seen from the Earth or a passing spacecraft. These synthetic spectra may then be compared with measured spectra and any differences interpreted in terms of how the assumed profiles need to be modified in order to achieve the best possible fit between the two. This is the essence of the *retrieval theory*, and at first sight appears relatively straightforward. For example, consider a region of the thermal-infrared where atmospheric absorption is due only to a well-mixed gas. The 600 cm^{-1} to 700 cm^{-1} spectrum of the giant planets is a good example of such a region since absorption here is due almost entirely to H_2–H_2 and H_2–He collision-induced absorption. Suppose our simulated spectrum was too bright at some wavelength in this range, then we would correctly deduce that our assumed, or *a priori*, temperature profile was too warm at roughly the altitude where the weighting function (Section 6.4) peaked. Hence, we would need to slightly "cool" our model profile at this level in order to improve the fit between synthetic and measured spectra.

Although this seems straightforward it is in fact very difficult for a number of reasons. First of all, it can be seen from the radiative transfer equation (Section 6.4.1 and Equation 6.40) that radiance at any particular wavelength is actually the weighted average of thermal emission from a continuous range of altitudes governed by the transmission weighting function. The width of the transmission weighting function for nadir-viewing is approximately one scale height and so information on the vertical temperature structure is considerably vertically smoothed. Hence, there is an infinitely wide range of possible temperature profiles whose synthetic spectra would fit the measured spectrum equally well! Fortunately, using spectral data from a range of wavelengths, or a range of emission angles, such that the peaks of the weighting functions cover a certain vertical span, reduces this ambiguity somewhat, but even then it must be remembered that the spectra are measured at a finite number of wavelengths (or emission angles), whereas an atmospheric profile is a continuous function. Hence, the retrieval problem is in effect one of attempting to calculate an infinite set of parameters using a finite set of measurements, which is known as an *ill-posed* problem. While there are techniques for solving such ill-posed problems, retrievals also suffer from a further complication in that they are *ill-conditioned.* This means that, without care, experimental noise in the measured spectrum may be greatly amplified by the fitting process leading to wholly unreliable vertical oscillations in the fitted atmospheric profiles.

Since there is literally an infinite number of possible atmospheric profiles whose simulated spectra will fit a measured spectrum to within the measurement error, how then may we hope to extract any meaningful information from the measurement of planets' electromagnetic spectra? Fortunately nature comes to our rescue since, although atmospheric properties such as temperature are continuous functions of height, we know that in practice they are also generally *smoothly varying* functions. Hence, by imposing sufficient vertical smoothing, meaningful vertical profiles may be determined from the measurements, as we shall see in Sections 7.10.1 to 7.10.4.

7.10.1 Exact, least-squares, and Backus–Gilbert solutions

Suppose that we represent a set of measurements (channel radiances, or a whole spectrum measured at a set of discrete wavelengths) by the vector $\mathbf{y}_m$ of m elements, known as the *measurement vector*. We may also represent atmospheric conditions with the *state vector* $\mathbf{x}_n$ of n elements, which may contain the temperature, composition, cloud abundances, at appropriate levels in the atmosphere, etc. We may represent the radiative transfer model by a forward model $\mathbf{F}(\mathbf{x}_n)$, from which we may calculate the synthetic spectrum $\mathbf{y}_n$

$$\mathbf{y}_n = \mathbf{F}(\mathbf{x}_n). \tag{7.16}$$

Retrievals of properties such as temperature may be linearized by expanding the state vector $\mathbf{x}$ about an initial first-guess, or *a priori*, solution $\mathbf{x}_0$ and hence

$$\mathbf{y}_n = \mathbf{y}_0 + \Delta\mathbf{y} = \mathbf{F}(\mathbf{x}_0) + \mathbf{K}(\mathbf{x}_n - \mathbf{x}_0) \tag{7.17}$$

where $\mathbf{K}$ is a matrix containing the rate of range of each element of $\mathbf{y}_c$ with respect to each element of $\mathbf{x}$; and $\mathbf{y}_0$ is the spectrum calculated with the *a priori* state vector.

If we choose the number of elements n of the state vector to be equal to the number of measured spectral radiances m, then the $\mathbf{K}$-matrix is square and may be inverted. Hence, substituting the measured spectrum $\mathbf{y}_m$ for the calculated spectrum $\mathbf{y}_n$ in Equation (7.17), and solving for $\mathbf{x}_n$ we find

$$\mathbf{x}_n = \mathbf{x}_0 + \mathbf{K}^{-1}(\mathbf{y}_m - \mathbf{y}_0). \tag{7.18}$$

This so-called "exact" solution provides a perfect fit to the measured spectrum, but does so at a very heavy price. Such solutions are extremely ill-conditioned and small errors in the measured spectrum may lead to huge errors in $\mathbf{x}_n$ and hence it is difficult to assess the reliability of the derived solution.

A particularly widespread approach when interpreting giant planet spectra is thus to parameterize the atmospheric temperature and composition profiles with far fewer parameters than the number of spectral points. For example, the abundance of hydrocarbons in the stratosphere is so little known that often all that can be meaningfully retrieved from the data is the approximate mean mole fraction. Assuming a temperature profile derived either from previous radio occultation measurements, or from radiances measured in the nearby 1,300 cm^{-1} methane band (methane is well-mixed in the atmospheres of the giant planets and its mole fraction is known), synthetic spectra may be generated for a range of mole fractions likely to be consistent with photochemical models and compared with the measured spectrum to find the best fit. Another example is ammonia retrievals for Jupiter and Saturn, where the ammonia profile may be represented by a mean value up to the condensation level, whereupon it is assumed to follow either a relative humidity curve, or alternatively decreases at a certain fractional scale height. Representing the atmospheric profile with fewer points than the number of measurements and solving for $\mathbf{x}_n$ results in a "least-squares" solution, which minimizes the difference between the measured and calculated spectra by minimizing the "cost function" ϕ

$$\phi = (\mathbf{y}_m - \mathbf{y}_n)^{\mathrm{T}}(\mathbf{y}_m - \mathbf{y}_n) \tag{7.19}$$

where ϕ is simply the sum of the squares of the differences between the measured and calculated spectra.

In cases where more vertical resolution is required, or where the spectra are particularly noisy, it is found that we need to apply some *a priori* assumptions to prevent meaningless solutions. There are a number of approaches to the problem, which are discussed at length by Hanel *et al.* (2003), Houghton *et al.* (1984), and Rodgers (2000). However, all methods basically arrive at the same conclusion: namely, that the precision of a retrieved profile depends on the vertical averaging applied or assumed. Hence, remotely sensed spectra may be inverted to yield very accurate smoothed perturbations to the assumed profile, but increasingly less accurate retrievals as less and less vertical smoothing is applied. There is thus essentially a trade-off between error and vertical resolution, which is formalized by the Backus–Gilbert approach (1970) and further discussed by Hanel *et al.* (2003) and Rodgers (2000). There are a number of formalisms to solving the inversion problem, of which the two most commonly used for planetary retrievals are the constrained linear inversion technique (Conrath *et al.*, 1998; Hanel *et al.*, 2003) and a modified form of optimal estimation (Rodgers, 2000), which is described here.

7.10.2 Linear optimal estimation

Optimal estimation was developed for use with terrestrial retrievals where satellite observations are used to improve upon the measurements already provided by other sources. Although the necessary *a priori* information is not available for planetary observations, the technique is relatively easy to understand and if used correctly the solutions from this model are effectively identical to a number of other retrieval methods that have been developed specifically for planetary retrievals.

Suppose that we again represent a set of measurements by the measurement vector $\mathbf{y}_m$ and atmospheric conditions by the state vector $\mathbf{x}_n$. For terrestrial applications, the properties of the atmosphere are not a complete mystery and from models and previous measurements, we have some *a priori* knowledge of the state of the atmosphere to within some initial error. We may then use this profile and the assumed constraints on it as the first guess and find the solution to the state vector $\mathbf{x}_n$ which minimizes the modified cost function

$$\phi = (\mathbf{y}_m - \mathbf{F}(\mathbf{x}_n))^{\mathrm{T}}\mathbf{S}_{\varepsilon}^{-1}(\mathbf{y}_m - \mathbf{F}(\mathbf{x}_n)) + (\mathbf{x}_n - \mathbf{x}_0)^{\mathrm{T}}\mathbf{S}_x^{-1}(\mathbf{x}_n - \mathbf{x}_0) \tag{7.20}$$

where $\mathbf{y}_m$ is the measured spectrum;
$\mathbf{F}(\mathbf{x}_n)$ is the spectrum calculated with the forward model;
$\mathbf{S}_{\varepsilon}$ is the measurement covariance matrix, which contains both estimated random and systematic measurement errors, as well as the estimated accuracy of the radiative transfer model used;
$\mathbf{x}_n$ is the model state vector;
$\mathbf{x}_0$ is the *a priori* state vector;
$\mathbf{S}_x$ is the *a priori* covariance matrix, which contains assumed errors on the *a priori* state vector, together with vertical correlations.

The cost function is simply a combination of how closely the synthetic spectrum matches the measured spectrum and how far the solution deviates from the assumed *a priori* state vector. The optimal solution to the state vector is thus found that maximizes the closeness of fit to the measured spectrum without deviating too greatly from the *a priori* state vector, and the degree to which the final solution will depart from the *a priori* solution will depend upon the relative size of the errors contained within the *a priori*, and measurement covariance matrices. For linear models such as temperature retrievals, for which the synthetic spectrum is calculated as $\mathbf{y}_n - \mathbf{y}_0 = \mathbf{K}(\mathbf{x}_n - \mathbf{x}_0)$, the optimal solution is found to be (Rodgers, 2000)

$$\hat{\mathbf{x}} = \mathbf{x}_0 + \mathbf{S}_x\mathbf{K}^{\mathrm{T}}(\mathbf{K}\mathbf{S}_x\mathbf{K}^{\mathrm{T}} + \mathbf{S}_{\varepsilon})^{-1}(\mathbf{y}_m - \mathbf{y}_0 - \mathbf{K}(\mathbf{x}_0 - \mathbf{x}_n)). \tag{7.21}$$

The error covariance matrix of the optimal solution to the state vector is then found to be

$$\hat{\mathbf{S}} = \mathbf{S}_x - \mathbf{S}_x\mathbf{K}^{\mathrm{T}}(\mathbf{K}\mathbf{S}_x\mathbf{K}^{\mathrm{T}} + \mathbf{S}_{\varepsilon})^{-1}\mathbf{K}\mathbf{S}_x. \tag{7.22}$$

Since the optimal estimation technique was developed for inverting Earth observation data, where *a priori* knowledge of the expected atmospheric profiles is mostly good, how can such a model be used for inverting spectra from other planetary atmospheres, where we have very little *a priori* knowledge? While we can apply some *a priori* information, such as the known saturated vapor pressure profiles of certain gaseous constituents, the *a priori* profiles and covariance matrices for planetary retrievals are really little more than first guesses. However, it is possible to formalize the technique for use with planetary retrievals, as is discussed by Irwin *et al.* (2008). First of all the measurement covariance matrix, $\mathbf{S}_{\varepsilon}$, may be calculated accurately from the known noise performance of the instrument and/or noting the standard variation of measured radiances. Errors in the forward model may also be estimated and added to $\mathbf{S}_{\varepsilon}$. Once $\mathbf{S}_{\varepsilon}$ is set, the retrieval model may be tuned by considering Equation (7.21). If the errors in the *a priori* covariance matrix are very large, then $\mathbf{KS}_x\mathbf{K}^T$ dominates $\mathbf{S}_\varepsilon$ in the first bracket, and an unwanted "exact" solution is obtained. Conversely, if the errors in the *a priori* covariance matrix are very small, then $\mathbf{S}_\varepsilon$ dominates $\mathbf{KS}_x\mathbf{K}^T$ in the first bracket and a solution is achieved that deviates very little from the *a priori* estimate. Hence, it can be seen that the relative size of the errors in $\mathbf{S}_\varepsilon$ and $\mathbf{S}_x$ may be used to tune the retrieval, with the best trade-off achieved when errors in $\mathbf{S}_\varepsilon$ and $\mathbf{S}_x$ have approximately equal weight. How then may we construct a suitable *a priori* covariance matrix that incorporates sufficient vertical smoothing? The normal procedure is to make reasonable guesses of the expected errors at each altitude of the profile to be retrieved and set the diagonal elements of the *a priori* covariance matrix to the square of these errors. Off-diagonal elements may then be set assuming a certain degree of cross-correlation according to schemes such as (Rodgers, 2000):

$$S_{ij} = (S_{ii}S_{jj})^{1/2} \exp\left(-\frac{|\ln p_i/p_j|}{c}\right) \tag{7.23}$$

where p_i and p_j are the pressures at the ith and jth atmospheric levels; and c is a "correlation length", here equivalent to the number of scale heights over which we can assume the profile to be reasonably correlated. A value of $c = 1.5$ is commonly used (Irwin *et al.*, 2008). The retrieval must then be tuned to ensure that the *a priori* profile provides sufficient smoothing. Tests are performed with a range of *a priori* errors, usually by simply scaling the *a priori* covariance matrix. If the errors are too large, the solution is unconstrained and the retrieval does everything it can to minimize the difference between measured and modeled spectra leading to ill-conditioning and large vertical oscillations in the retrieved vertical profile. If, instead, the *a priori* errors are set to be very small, then the solution is over-constrained and the retrieved profile differs little from the *a priori* profile. By tuning the *a priori* errors we search for intermediate conditions where $\mathbf{KS}_x\mathbf{K}^T$ and $\mathbf{S}_\varepsilon$ contribute similarly and thus where the solution is constrained quasi-equally by the data and by the *a priori* profile. This leads to solutions that match the measured spectra well, but which still have sufficient smoothing, supplied by the *a priori* covariance matrix $\mathbf{S}_x$, to be well conditioned.

7.10.3 Nonlinear optimal estimation

The linearity of temperature retrievals comes from the fact that the elements of the **K**-matrix (the rate of change of spectral radiances with temperature) are not strongly dependent on the temperature profile. Hence, once the **K**-matrix is calculated, the best-fit temperature profile is found in a single step. This situation does not apply for composition retrievals, since small changes in the composition profile strongly affect atmospheric transmission, and thus the **K**-matrix. Hence, composition retrievals are highly nonlinear and computationally expensive since they must be performed iteratively, and the **K**-matrix recalculated at every step in the worst case.

Extending the principles of optimal estimation to such nonlinear cases the difference between the spectrum computed from the nth trial measurement vector $\mathbf{x}_n$ and that measured is used to calculate a new estimate of the trial vector $\mathbf{x}_{n+1}$ through the equation

$$\mathbf{x}_{n+1} = \mathbf{x}_0 + \mathbf{S}_x\mathbf{K}_n^{\mathrm{T}}(\mathbf{K}_n\mathbf{S}_x\mathbf{K}_n^{\mathrm{T}} + \mathbf{S}_\varepsilon)^{-1}(\mathbf{y}_m - \mathbf{y}_n - \mathbf{K}_n(\mathbf{x}_0 - \mathbf{x}_n)) \tag{7.24}$$

where $\mathbf{K}_n$ is the matrix of functional derivatives (for the nth iteration), or the Jacobian (i.e., the rate of change of radiance with state vector elements for all the wavelengths in the spectrum). It should be noted that the *a priori* vector, $\mathbf{x}_0$, is used at each iteration to ensure that *a priori* constraints remain applied throughout the retrieval. In practice, $\mathbf{K}_n$ can vary greatly between iterations and the simple iteration scheme of Equation (7.25) can become unstable. In the NEMESIS optimal estimation retrieval model (Irwin *et al.*, 2008) a modified iteration scheme, based on the Marquardt–Levenberg principal (e.g., Press *et al.*, 1992) is used where the actual modified state vector used in the next iteration, $\mathbf{x}'_{n+1}$, is calculated from $\mathbf{x}_{n+1}$ and $\mathbf{x}_n$ as

$$\mathbf{x}'_{n+1} = \mathbf{x}_n + \frac{\mathbf{x}_{n+1} - \mathbf{x}_n}{1 + \lambda}. \tag{7.25}$$

The parameter λ is initially set to 1.0. If the spectrum calculated from $\mathbf{x}'_{n+1}$ is found to reduce the cost function ϕ, then $\mathbf{x}_n$ is set to $\mathbf{x}'_{n+1}$, λ is multiplied by a factor of 0.3, and the next iteration started. If, however, the spectrum calculated from $\mathbf{x}'_{n+1}$ is found to increase the cost function ϕ, then $\mathbf{x}_n$ is left unchanged, the parameter λ increased by a factor of 10, and a new vector $\mathbf{x}'_{n+1}$ calculated. The choice of multiplication parameters (0.3, 10) is somewhat arbitrary, although it is important to ensure that they are not reciprocals of one another, as this can lead to endless loops. To ensure smooth convergence, the lower number (0.3) is chosen to be greater than the reciprocal of the larger number so that λ does not decrease too quickly. As the retrieval approaches its final solution, $\lambda \to 0$ and the model tends to the optimal estimate, whose error can be estimated from Equation (7.22).

For cases where the $\mathbf{K}_n$ matrix does not change very rapidly with the state vector, inversion is approximately linear and convergence is achieved in 2–3 steps. However, for volume mixing ratio retrievals, where $\mathbf{K}_n$ can vary greatly between iterations, convergence can be slower, requiring perhaps 10–20 steps.

It is interesting to compare and contrast the optimal estimation approach as applied to planetary applications with the constrained inversion technique (Conrath *et al.*, 1998; Hanel *et al.*, 2003) used by many research groups. In this approach, the nonlinear iterative solution may be written, using our optimal estimation formalism, as

$$\mathbf{x}_{n+1} = \mathbf{x}_0 + \hat{\mathbf{S}}_x\mathbf{K}_n^{\mathrm{T}}(\mathbf{K}_n\hat{\mathbf{S}}_x\mathbf{K}_n^{\mathrm{T}} + \gamma\mathbf{S}_\varepsilon)^{-1}(\mathbf{y}_m - \mathbf{y}_n - \mathbf{K}_n(\mathbf{x}_0 - \mathbf{x}_n)) \tag{7.26}$$

where $\hat{\mathbf{S}}_x$ is now the *a priori correlation* matrix (which provides vertical smoothing); $\mathbf{S}_\varepsilon$ and $\mathbf{K}_n$ are as before; and γ is an adjustable parameter used to fine-tune the balance between measurement and *a priori* constraint. When retrieving a single profile, the *a priori* correlation matrix $\hat{\mathbf{S}}_x$ is equal to our *a priori* covariance matrix $\mathbf{S}_x$, but with each row divided by the value of the diagonal component. Thus, comparing Equation

(7.26) with Equation (7.24) we can see that the two methods are essentially identical when it comes to retrieving continuous vertical profiles.

7.10.4 Joint retrievals

Consider a region of the thermal-infrared where the weighting functions peak in the upper troposphere and where, say, ammonia is strongly absorbing. Suppose that the simulated spectrum is less bright than the measured spectrum. This would suggest that the abundance of ammonia in our first simulated spectrum is too high near the peak of the weighting function and should thus be reduced. However, it could also mean that the assumed atmospheric temperatures are too low and must be increased. Or it could mean that the abundance of some other constituent such as aerosols needs to be modified. It could also conceivably mean that all three need to be modified in some way!

In such cases, a spectral region could first be selected that is sensitive only to well-mixed absorbers and a linear temperature retrieval conducted to give the "true" temperature profile. This temperature profile could then be used in an ammonia composition nonlinear retrieval, using another spectral region dependent on both temperature and ammonia. However, in some cases it is found to be more effective to consider both spectral regions together and retrieve both ammonia and temperature simultaneously in a nonlinear joint retrieval. Such an approach is useful in cases where the number of spectral points is limited or where there is significant noise.

7.11 BIBLIOGRAPHY

Hanel, R.A., B.J. Conrath, D.E. Jennings, and R.E. Samuelson (2003) *Exploration of the Solar System by Infrared Remote Sensing* (Second Edition). Cambridge University Press, Cambridge, U.K.

Hecht, E. and A. Zajac (1974) *Optics*. Addison-Wesley, Reading, MA.

Houghton, J.T., F.W. Taylor, and C.D. Rodgers (1984) *Remote Sounding of Atmospheres*. Cambridge University Press, Cambridge, U.K.

Houghton, J.T. and S.D. Smith (1966) *Infrared Physics*. Oxford University Press, Oxford, U.K.

James, J.F. and R.S. Stern (1969) *The Design of Optical Spectrometers*. Chapman & Hall, London.

Rodgers, C.D. (2000) *Inverse Methods for Atmospheric Sounding: Theory and Practice*. World Scientific, Singapore.

Vanasse, G.A. (Ed.) (1983) *Spectrometric Techniques*, Vol. III. Academic Press, San Diego, CA.

Much of the information presented in this chapter has been extracted from various very useful public websites which are now listed.

Useful public NASA websites for mission information include

NASA Jet Propulsion Laboratory	*http://www.jpl.nasa.gov/*
NASA Ames Research Center	*http://www.arc.nasa.gov/*
NASA Goddard Space Flight Center	*http://www.jpl.nasa.gov/*
General NASA missions	*http://history.nasa.gov/*
General ESA missions	*http://sci.esa.int/home/ourmissions/index.cfm*

Ground-based visible/IR telescope sites include

European Southern Observatory	*http://www.eso.org/*
Palomar Observatory	*http://www.astro.caltech.edu/palomarpublic/overview.html*
NASA IRTF	*http://irtfweb.ifa.hawaii.edu/*
Anglo-Australian Observatory	*http://www.aao.gov.au/*
Joint Astronomy Centre (JCMT, UKIRT)	*http://www.jach.hawaii.edu/*
Kitt Peak National Observatory	*http://www.noao.edu/kpno/*
Pic-du-Midi Observatory	*http://www.omp.obs-mip.fr/*
Keck Telescopes	*http://www.keckobservatory.org/*
Gemini	*http://www.gemini.edu/*

Ground-based microwave telescope sites include

CARMA	*http://www.mmarray.org/*
IRAM	*http://iram.fr/*
Nobeyama Millimeter Array	*http://www.nro.nao.ac.jp/~nma/index-e.html*
VLA	*http://www.vla.nrao.edu/*
VLBA	http://www.vlba.nrao.edu/

Airborne observatory

Kuiper Airborne Observatory	*http://vathena.arc.nasa.gov/curric/space/lfs/kao.html*

Earth-orbiting space telescope sites include

HST	*http://www.stsci.edu/hst/*
ISO	*http://www.iso.vilspa.esa.es/*
SWAS	*http://sunland.gsfc.nasa.gov/smex/swas/*
Spitzer	*http://www.spitzer.caltech.edu/*
AKARI	*http://www.ir.isas.jaxa.jp/AKARI/Outreach/index_e.html*

Flyby and orbiting mission sites include

Cassini-Huygens	*http://saturn.jpl.nasa.gov/index.cfm*
Galileo	*http://galileo.jpl.nasa.gov/*
Millennium Mission	*http://www.jpl.nasa.gov/jupiterflyby/*
Pioneer	*http://www.nasa.gov/mission_pages/pioneer/*
Voyager	*http://voyager.jpl.nasa.gov/*
Ulysses	*http://ulysses.jpl.nasa.gov/*

8

Future of giant planet observations

8.1 INTRODUCTION

In this book, we have discussed how the giant planets and their atmospheres formed 4.6 billion years ago and have reviewed what is known about their atmospheres from existing measurements, models, and theory. The story of the way in which our understanding of these giant worlds has advanced is essentially one of how improvements in technology have allowed more and more detailed observations of these worlds, which in turn have allowed more advanced and accurate analysis. The rate of improvement in these data has rapidly increased with time, and thus we have learned more about these worlds in the last 25 years than in all of the previous centuries put together! Technology continues to develop at an extraordinary rate, and thus many exciting new measurements are expected in the next few years which will further improve our understanding of these giant worlds, and how they formed from circumplanetary disks. Some of the key questions concerning the atmospheres of the giant planets that still remain are

- What are the upper visible clouds of Jupiter and Saturn actually composed of when the spectral signature of ammonia ice seems to be generally so elusive?
- What actually is the composition of the main cloud deck on Uranus and Neptune?
- What is the nature of the chromophores on the different giant planets?
- To what level are different heavy elements enriched on the giant planets and how did the observed differences come about?
- What is the enrichment of heavy elements other than carbon on Uranus and Neptune?
- Were the heavy elements introduced to the giant planet atmospheres through amorphous or crystalline water ice?

- Why is ammonia so scarce down to several tens of bars on Uranus and Neptune and why does it appear depleted in the upper atmosphere of Jupiter?
- Is there really N_2 in the atmosphere of Neptune and what are the abundances of other disequilibrium species on the ice giants?
- Why does Uranus have such a small source of internal energy?
- Why do Jupiter and Saturn have strong prograde equatorial jets, while Uranus and Neptune have retrograde jets?
- Why are almost all of the large ovals on the giant planets anticyclonic?
- Is the structure of our solar system actually typical compared with other star systems, or is it in some way unusual?

The list could go on and on since, although we have already learned a huge amount, we have only just begun to really understand the atmospheres of these planets. In this chapter we will review the likely projects in the next ten years, which should have major impacts on our understanding of these worlds. Major developments are expected in the fields of space-based telescopes, ground-based telescopes, and spacecraft missions. Further in the future, missions are currently being planned to search for planets about other stars, which should allow us to understand better the structure of our own solar system and how it most likely formed and evolved.

8.2 GROUND-BASED VISIBLE/INFRARED (IR) OBSERVATIONS

In this section we will review the major developments that are expected in ground-based visible/IR observations during the next ten years. Unfortunately it would be impossible to fully cover the plans and improvements of all the world's observatories here, and so instead just the largest telescopes will be considered for which giant planet observations are very likely.

8.2.1 Very Large Telescope Interferometer (VLTI)

The ESO VLT in Chile was described in Section 7.4.1. In addition to the main 8.2 m utility telescopes, the last of which is almost complete, the site also includes the Italian VLT 2.6 m Survey Telescope (VLST) and the British 4 m Visible and Infrared Telescope for Astronomy (VISTA), which were both installed in 2007/2008. One of the most exciting developments at the VLT, however, is the construction of the VLT Interferometer (VLTI). The VLTI will, for the first time, provide a long-baseline interferometric capability to the visible/infrared, routinely used in microwave observations, and thus greatly increase the angular resolution of ground-based observations at these wavelengths. Among other things, the VLTI provides an important precursor to possible future terrestrial planet-finding missions such as NASA's TPF and ESA's *Darwin* (Section 8.7). The VLTI system will be able to combine light from any of the four main unit telescopes (UTs), together with light gathered by any of three 1.8 m auxiliary telescopes (ATs), which may be moved on a rail system to be placed at any of 30 stations, from which light may be directed to the

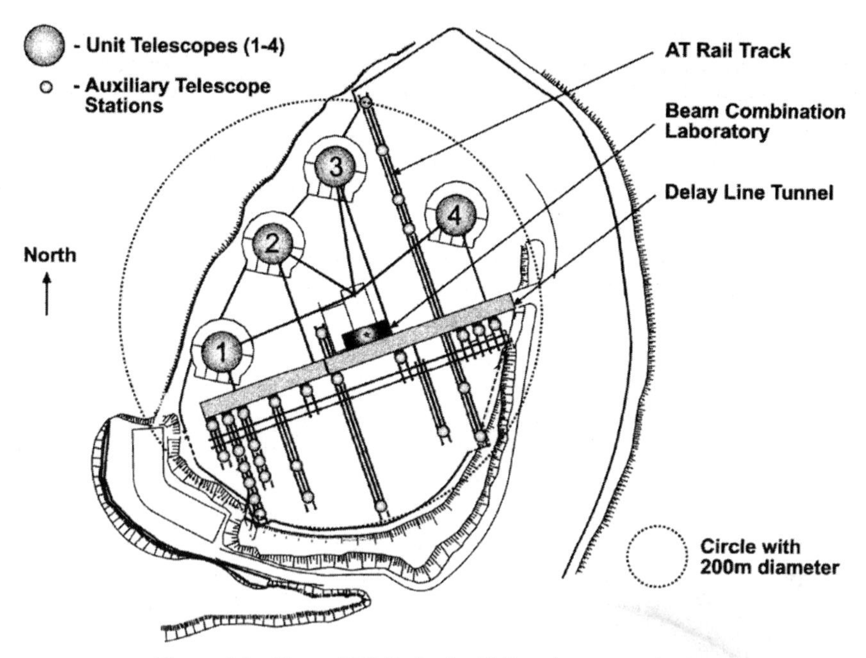

Figure 8.1. Plan of VLT site in Chile. Courtesy of ESO.

Beam Combination Laboratory (Figure 8.1). The maximum baseline achievable with the UTs is 130 m, while the ATs may be placed as far apart as 200 m, giving angular resolutions theoretically as high as 0.001″ at visible wavelengths. While the UTs will only occasionally be used for interferometry, the ATs are dedicated to the VLTI alone. Before the beams from two telescopes can be combined, the light from one must first be passed through optical delay lines, as shown in Figure 8.2, in order to compensate for the different distances between the observing stations and the Beam Combination Laboratory. The accuracy of pathlength compensation must be to within 0.05 μm over a distance of 120 m and the VLTI delay line system uses a retro-reflector fixed on a carriage running on two stainless steel rails. The carriage is driven by a 60 m linear motor and a piezo-transducer element, and a laser metrology system monitors the instantaneous distances between the mirrors. The Beam Combination Laboratory houses a number of instruments for performing interferometry across the wide wavelength range, and the proposed GENIE experiment (den Hartog *et al.*, 2006; Wallner *et al.*, 2006) will test the principle of "nulling" interferometry, proposed for use by both the TPF and *Darwin* missions to cancel the light of stars allowing direct imaging of orbiting extrasolar planets. Nulling interferometry will be discussed further in Section 8.7.3.

8.2.2 Keck Interferometer

Work is currently underway to combine the light from the Keck telescopes to form another large stellar interferometer, in just the same way as is being done at VLT.

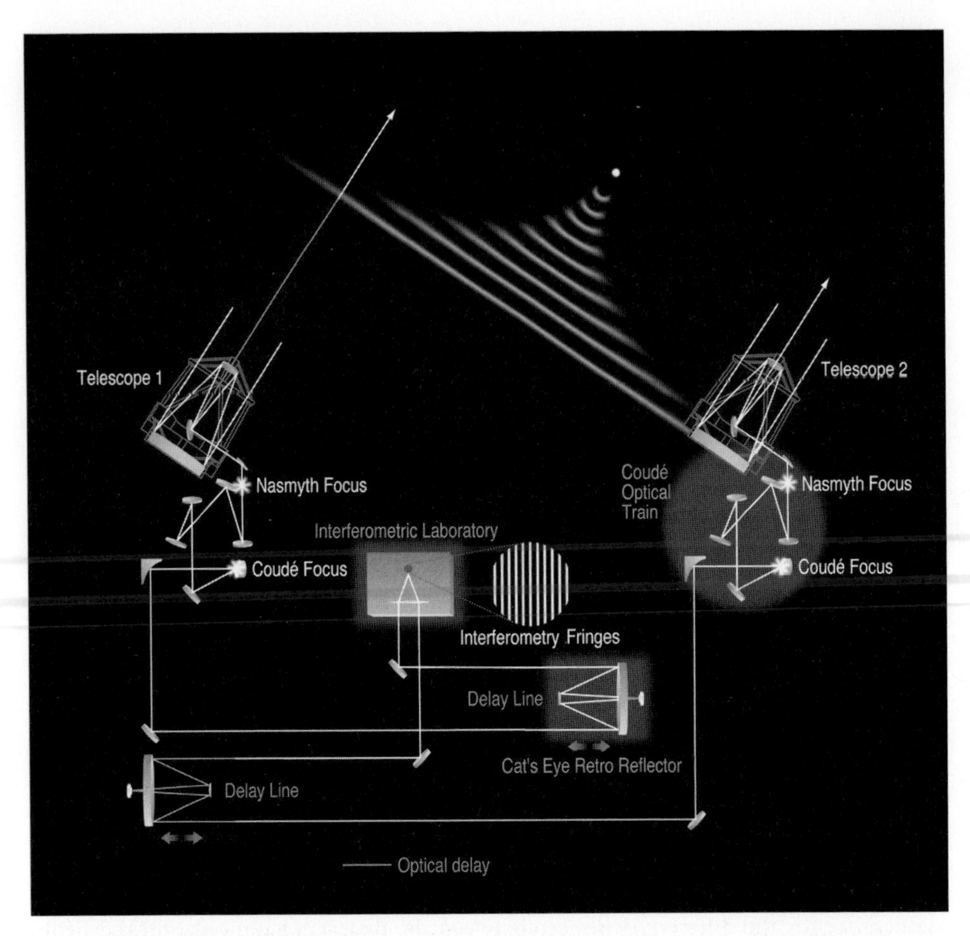

Figure 8.2. Schematic design of the VLT Interferometer. Courtesy of ESO.

Similarly, there are also plans to install several 2 m outrigger telescopes, and the Keck adaptive optics/interferometry project is a significant cornerstone of NASA's Origins Program, which seeks to identify and characterize possible Earth-like planets around neighboring stars. In 2005 the Keck Interferometer team announced that it had achieved a 100:1 nulling of the light from Vega and two other stars, and work is ongoing to increase nulling to the point where light from circumstellar disks and perhaps planets becomes visible.

8.2.3 Large Binocular Telescope (LBT)

The LBT, currently being completed on Mt. Graham in Arizona, consists of two 8.4 m primary mirrors mounted in a single structure, 14.4 m apart (Figure 8.3). The first primary achieved "first light" in October 2005, while the second achieved "first light" in January 2008. The telescopes can be used separately or, by sending the light

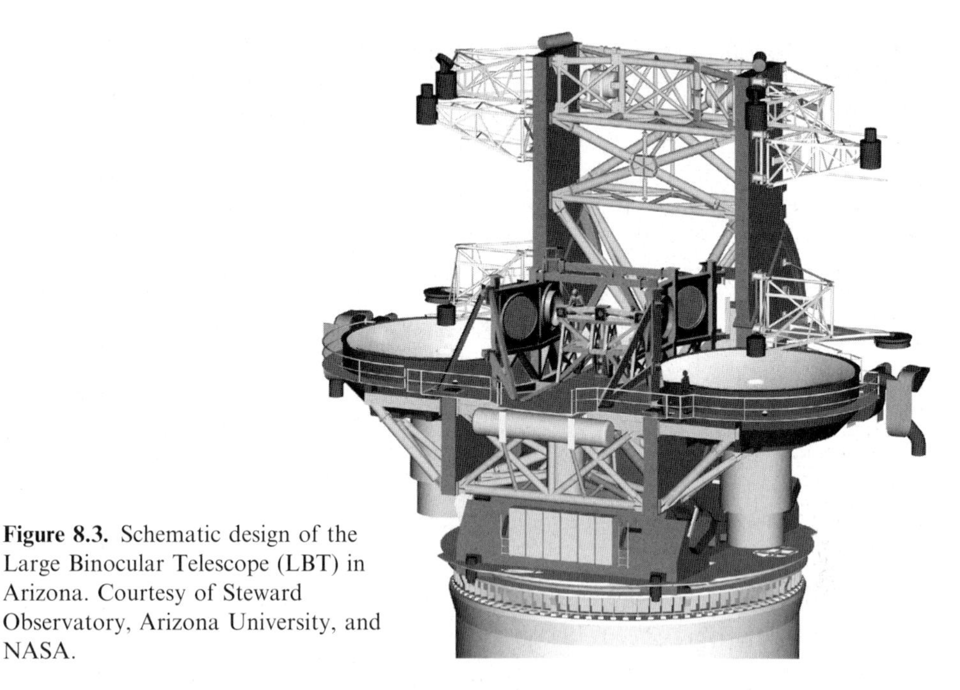

Figure 8.3. Schematic design of the Large Binocular Telescope (LBT) in Arizona. Courtesy of Steward Observatory, Arizona University, and NASA.

to a single camera between the telescopes, be used as an interferometer with a maximum baseline of 22.8 m, giving LBT extremely high angular resolution. It is intended to use the LBT Interferometer (LBTI) to search for extrasolar planets through the use of "nulling" interferometry, described in Section 8.7.3, to cancel out the light from stars in stellar systems to leave just the light reflected, or emitted, from nearby planets, and thus allow their direct imaging. In addition to looking for extrasolar planets, LBTI will be able to directly image faint zodiacal dust disks (indicative of planetesimals) around other stars. A survey of a large sample of stars close to our Sun will reveal how common the makeup of our own solar system is and identify which stars have planetary systems potentially suitable for life-bearing, terrestrial planets.

8.2.4 Extremely large telescopes (ELTs)

The largest optical telescopes in the world are currently the 10 m Keck telescopes on Mauna Kea, Hawaii. Astronomers always want telescopes with ever larger apertures in order to image fainter and fainter objects, but the size of new telescope plans currently being considered is astonishing in their ambition! The next generations of telescopes, called "extremely large telescopes", are likely to have apertures of at least 30 m. Apart from their enormous light-gathering power, with adaptive optics such telescopes would have fine enough angular resolution to greatly improve observations of solar system objects such as the giant planets and would even be able to directly image extrasolar planets and, by measuring their visible and near-IR spectra, place constraints on their atmospheric composition.

Such enormous telescopes are not the stuff of science fiction and detailed plans and proposals are already being considered. Almost all use the idea of segmented mirrors, along the design of the Keck telescopes, although it may now be technically feasible to make a monolithic mirror as large as 50 m in diameter! The multi-mirror approach will probably use a spherical primary since all the individual elements would then be ground to the same, simple shape. The disadvantage is, of course, spherical aberration, which would need to be corrected for by additional optics, slightly reducing the overall throughput through reflective losses. Making a parabolic mirror with individual elements is technically possible, but much more difficult and expensive since the shape of the mirrors will differ depending on their final position in the assembled primary.

There are currently a number of ELT designs under serious consideration.

European Extremely Large Telescope (E-ELT)

The European Southern Observatory is currently considering plans for a 30 m to 50 m diameter ELT. An earlier study into the so-called "Overwhelmingly Large Telescope (OWL)", which was a concept to construct a ground-based 100 m class optical and near-IR adaptive optics telescope, has for the time being been shelved due to the extremely high projected costs.

The current E-ELT design is for an alt-azimuthal telescope with a segmented, aspherical 42 m primary mirror and then one of two secondary mirror systems: (1) a Gregorian system with a 4.8 m concave secondary and also perhaps a 5 m flat tertiary mirror to bring the light to the Nasmyth focus, or (2) a new optical design with an active convex 6 m secondary followed by a concave, aspheric tertiary and two flat mirrors to relay the beam to the Nasmyth focus. The use of a segmented mirror with individual mirrors no larger than 1.45 m reduces the risks and projected costs, and it is clear that such a system will require active optics to maintain optical quality, which has already been successfully demonstrated by the ESO New Technology Telescope (NTT) and VLT. Compensation for atmospheric turbulence will be taken care of by multi-conjugate adaptive optics modules located between the telescope focus and the science instruments. The telescope structure is designed to be all-steel and would weigh something in the region of 5,500 tonnes. The telescope would probably operate in the open air, since an enclosure with a 42 m wide slit would not provide much protection from the wind anyway, and hence in operation the enormous enclosure would probably roll right away. Modeling the possible vibrations induced by surface winds, which would severely affect optical quality, is one of the main activities of the current study.

Thirty Meter Telescope (TMT)

In a project formerly known as the California Extremely Large Telescope (CELT) (Irion, 2002) the TMT Consortium—made up of the California Institute of Technology, the University of California, and the Association of Canadian Universities for

Research in Astronomy (ACURA)—are designing a 30 m class telescope, which they hope to begin constructing in 2009 and achieve "first light" by 2016. The current design is for an alt-azimuth mounted Ritchey–Chrétien telescope with a 492-segment, 30 m primary mirror, an active secondary mirror, and an articulated tertiary mirror and will use adaptive optics. To maximize performance, such a telescope needs to be built in a dry region at high altitude and a number of possible sites are currently being considered.

Giant Magellan Telescope (GMT)

The Giant Magellan Telescope (GMT) Organization consists of the Carnegie Institution of Washington, Harvard University, the Smithsonian Astrophysical Laboratory, Texas A&M University, the University of Texas at Austin, the Australian National University, the Univerisity of Arizona, and Astronomy Australia Ltd. The consortium plan to build an ELT consisting of seven 8.4 m mirrors at a site in Chile, and hope to complete in around 2017.

8.3 AIRBORNE VISIBLE/IR OBSERVATIONS

The major development in airborne observations in the next ten years will be the commission of the Stratospheric Observatory for Infrared Astronomy (SOFIA) airborne observatory.

8.3.1 SOFIA

SOFIA is a collaboration between NASA and the German Space Agency DLR to provide a new airborne observatory with much greater sensitivity than its predecessor, the Kuiper Airborne Observatory (KAO) described in Chapter 7.

The SOFIA aircraft is a Boeing 747SP aircraft, named the *Clipper Lindbergh*, which has been modified to accommodate a 2.5 m reflecting telescope (Figure 8.4). Work began in 1996 and SOFIA made its first test flight in 2007. When it begins operations in 2009 SOFIA will be the largest airborne observatory in the world, making observations that are impossible for even the largest and highest of ground-based telescopes. Like the KAO, SOFIA's science operations will be managed by the NASA Ames Research Center, but the aircraft itself will be based at NASA's Dryden Flight Research Center, Edwards, California. The aircraft will cruise at an altitude of between 12,400 m and 13,700 m (41,000–45,000 feet) for periods of 8 hours at a time and will make approximately 160 flights per year for a total operational lifetime of 20 years.

During operations the SOFIA telescope will be at the ambient temperature of about −35°C (240 K) and will be equipped with a number of cryogenically cooled instruments to investigate IR radiation over a very wide range from 0.3 μm to 655 μm. The telescope will view out of one side of the aircraft over a range of elevation angles from +20° to +60°. The telescope will be isolated against aircraft

Figure 8.4. The SOFIA airborne observatory aircraft. Courtesy of NASA/DLR.

vibrations by air bladders (or air springs) and will be further stabilized by gyroscopes, which will counter sudden gusts from outside the 4 m wide opening in the aircraft into the telescope cavity. In flight, air flows past this door at 800 km/h and so observations at visible wavelengths will be somewhat blurred by the accompanying turbulence. However, observations are affected less and less as one goes to longer wavelengths and seeing is expected to be diffraction-limited at wavelengths greater than 15 μm.

8.4 GROUND-BASED MICROWAVE OBSERVATIONS

The major development in ground-based microwave observations in the next ten years will be the commission of the Atacama Large Millimeter Array (ALMA) in Chile.

8.4.1 Atacama Large Millimeter Array Project (ALMA)

The ALMA project is a joint European/American/Canadian/Japanese effort to build a huge millimeter array with a collecting area of approximately 7,000 m^2. This will be roughly ten times the collecting area of the largest currently operating millimeter array in the world, the Institut de RadioAstronomie Millimétrique (IRAM) array of six 15 m dishes at the Plateau de Bure Observatory in France.

ALMA will be able to image in all atmospheric windows between 0.3 mm and 9.6 mm (31–950 GHz) and will have a very large baseline extending up to

15 km–18 km, providing an angular resolution of 0.005 arcsec at the highest frequencies observed. To achieve these requirements ALMA will consist of an array of fifty 12 m dishes, which can be moved to form different array configurations (e.g., Schilling, 1999c). In addition, there will also be a more compact array (the Atacama Compact Array or ACA) consisting of twelve 7 m dishes and four 12 m dishes, which will be used in fixed configuration to image large-scale structures that are not well sampled by the main ALMA array.

To capitalize fully on such a massive investment, ALMA needs to be placed at as high an altitude as possible and in one of the driest regions of the world in order to minimize obscuration by terrestrial water vapor. The proposed site for the instrument is thus in Cerro Chajnantor, Chile which is a vast plain at an altitude of 5,000 m in the Chilean Andes, perfectly suited for the purpose. Unlike most interferometers, the dishes of the interferometer will not be limited to a fixed set of configurations, but can be moved to a continuous range of baselines between 200 m and 18 km. While the 18 km baseline gives the best angular resolution, smaller baselines will be better suited for the detection of weak signals. The Atacama Compact Array (ACA) will have essentially one configuration. Because the site is so high, altitude sickness is likely to be a real problem and thus there will probably be no scientists at the site. Instead, day-to-day maintenance will be conducted by local Chilean workers, acclimatized to the altitude.

Construction of ALMA at the observatory site started in 2005. When it is completed, hopefully by 2012, ALMA will provide unprecedented access to the millimeter spectra of astronomical bodies and in particular it will investigate high-redshifted galaxies and planetary formation in circumstellar disks. Its high sensitivity and angular resolution will also allow it to investigate, as never before, the microwave spectra of the giant planets, allowing it to probe the temperature and composition of these planets well below the visible cloud decks. It should also be able to image extrasolar planets directly, allowing for investigation of their composition and temperature (Schilling, 1999a).

8.5 SPACE TELESCOPE OBSERVATIONS

8.5.1 *Herschel*

The *Herschel* Space Observatory, formerly known as FIRST (Far-infrared Space Telescope), will be ESA's fourth cornerstone mission and will conduct far-IR and submillimeter imaging photometry and spectroscopy from 60 μm to 670 μm. The telescope is due for an *Ariane 5* launch in early 2009 and it will be placed into orbit about the Second Lagrangian (L2) point of the Earth–Sun system (1.5 million km from Earth in the direction directly away from the Sun), where it will operate for a period of about 3 years. The L2 orbital position has been chosen because of its low and stable background radiation field, and orbital stability.

The *Herschel* spacecraft is approximately 7.5 m high and 4 × 4 m wide, with a launch mass of around 3.3 tonnes (Figure 8.5). It carries a 3.5 m radiatively cooled

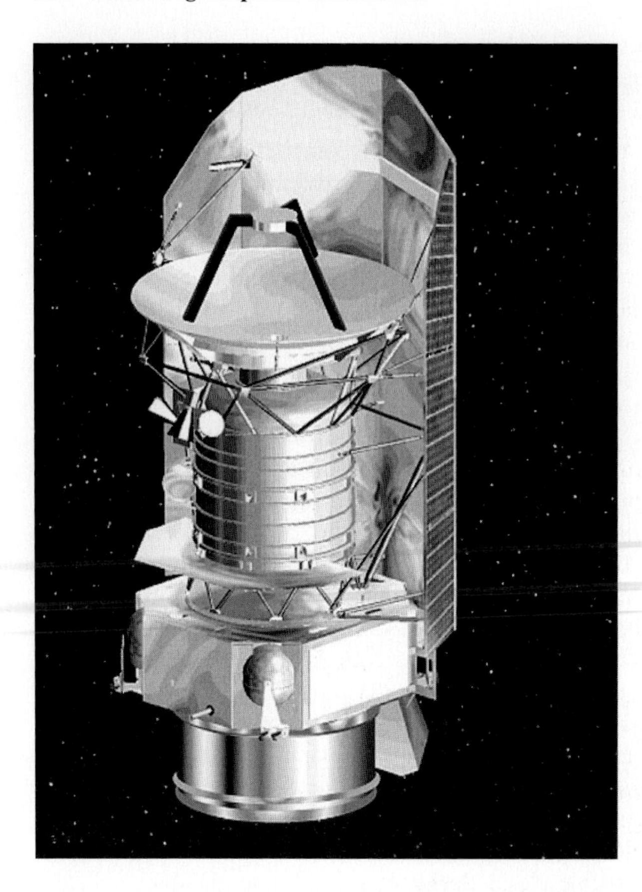

Figure 8.5. Schematic appearance of the *Herschel* Space Telescope, due for launch in 2009. Courtesy of ESA.

Ritchey–Chrétien telescope (the largest ever built for space operation) and three scientific instruments: PACS, SPIRE, and HIFI, housed in a superfluid helium cryostat and thus cooled to approximately 2 K. *Herschel*'s long wavelength range will primarily allow it to investigate the formation of stars and galaxies, since it is only at these long wavelengths that radiation may penetrate the circumstellar and circum-galactic dust clouds in detectable quantities. In addition, the wavelength region contains features useful for the study of interstellar medium (ISM), astrochemistry, and the atmospheres of the giant planets. *Herschel* should prove particularly useful in providing new data on the D/H ratio of Uranus, Neptune, and the abundance of water vapor in the giant planets.

The three planned scientific instruments of *Herschel* will now be reviewed.

Photodetector Array Camera and Spectrometer (PACS)

PACS is an imaging photometer and spectrometer which covers the wavelength range 60 μm to 210 μm. For photometry, radiation is imaged in both a "blue" and a "red" channel, each covering a FOV of $105 \times 210''$. The blue channel is composed of a 32×64 bolometer pixel array covering the wavelength ranges of either 60–90 μm or

90–130 μm. The longer wavelength red channel is composed of a 16 × 32 bolometer detector array and covers the 130 μm–210 μm wavelength range. In spectroscopy mode, PACS will image a FOV of about 50 × 50″, resolved into 5 × 5 pixels. The light is then dispersed by a grating onto one of two 16 × 25 GeGa arrays covering the ranges 40–120 μm and 100–210 μm, respectively. The resolving power of the instrument will be approximately 1,500 and the detectors will be cooled to 300 mK by an internal ^{3}He sorption cooler to improve its sensitivity.

Spectral and Photometric Imaging Receiver (SPIRE)

SPIRE comprises an imaging photometer and a symmetrical Mach–Zehnder imaging spectrometer. The instrument has five bolometer detector arrays, which are cooled to 300 mK by an internal ^{3}He sorption cooler.

In its broadband photometry mode ($R \sim 3$) SPIRE will simultaneously image a 240 × 480″ field of the sky in three bands centered at 250 μm, 350 μm, and 500 μm with angular resolutions of 71″, 24″, and 35″, respectively. This mode will not be useful for giant planet observations. However, in its spectrometer mode SPIRE will operate as a Mach–Zehnder configuration Fourier Transform Spectrometer (FTS) with a wavelength range of 200 μm to 670 μm, and a circular FOV of diameter 156″. The spectral resolution is tunable between 0.04 cm^{-1} and 1 cm^{-1}. The FTS has two detector arrays and may operate in either the 200–300 μm or 300–670 μm ranges. SPIRE should provide very useful observations on the far-IR disk-averaged spectra of the giant planets.

Although SPIRE has been optimized to address the questions of how stars and galaxies form it should also provide important measurements not only for the giant planets, but also for comets and galactic ISM.

Heterodyne Instrument for the Far Infrared (HIFI)

HIFI is a very high–resolution heterodyne spectrometer offering resolving powers in the range 1,000–1,000,000 combined with low noise detection using superconductor–insulator–superconductor (SIS) and hot electron bolometer (HEB) mixers. Although not an imaging instrument, HIFI will provide continuous coverage in five bands over the range of 480 GHz to 1,250 GHz (240–625 μm) in five bands and two additional bands between 1,410 GHz and 1,910 GHz (157–212 μm).

8.5.2 James Webb Space Telescope (JWST)

The NASA/ESA JWST, named after NASA's second administrator, is the successor to the Hubble Space Telescope (HST). It is scheduled for launch in 2013 onboard an *Ariane 5*, and, like *Herschel*, will take up position at the Second Lagrange Point (L2). Unlike HST its great distance from the Earth will mean that it will not be serviceable by shuttle missions, and hence its instrument complement will remain fixed after launch. The mission is planned to last for at least 5 years, extendable to 10 years.

The designed primary mirror size for JWST is 6.5 m in diameter, and so this mirror will have to be folded and stowed before and during launch. The mirror

Figure 8.6. An artist's impression of the James Webb Space Telescope (JWST), due for launch in 2013. Courtesy of NASA and ESA.

"petals" will only be commanded to open once the spacecraft reaches the L2 position. The primary mirror will probably comprise three to eight hinged segments that are individually monolithic or segmented in the manner of the Keck telescopes (Section 7.4.2). The whole telescope will be shielded from sunlight by a large sunshade (currently planned to be 22×10 m) composed of many layers of lightweight reflecting material, sufficient to sustain a 300 K temperature drop from front to back (Figure 8.6). With a back sunshade temperature of $\sim$90 K, the primary mirror, the optical truss, and the instrument payload can radiate their heat to space and reach cryogenic temperatures of 30 K to 50 K. Combined with the L2 orbit, the large sunshade provides a stable, cold environment with a minimum of background radiation. The mass of the telescope at launch is currently estimated to be $\sim$5,400 kg.

The 6.5 m diameter mirror will provide a diffraction-limited angular resolution of $0.2''$ at 5 μm, and the telescope is planned to provide observations between 0.6 μm and 28 μm. It is planned that the JWST will carry a near-IR camera, a multi-object spectrometer, and a mid-IR camera/spectrometer, all mounted within an Integrated Science Instrument Module (ISIM), which will include a thermal system capable of passively cooling the near-IR detectors to ~27 K. JWST should help to establish how galaxies first emerged out of the darkness that initially followed the Big Bang since the light from the youngest galaxies is seen in the infrared—not the visible—due to the universe's expansion. JWST will also investigate the formation of planets in disks around young stars, study supermassive black holes in other galaxies, and should also provide high-quality measurements of the giant planets. The current designs for JWST's main instruments will now be reviewed.

Mid-Infrared Instrument (MIRI)

MIRI will provide imaging and spectroscopy in the wavelength range 5 μm–27 μm at an angular resolution of 0.2″. The imaging part of the instrument is planned to have a FOV of 90 × 90 and approximately 12 spectral channels across the 5 μm–28 μm wavelength range. The spectroscopic part of MIRI is planned to provide medium to high resolution ($R \sim 3{,}000$) spectra from 5 μm to 10 μm, with the resolving power decreasing to no lower than 1,400 μm by 20 μm, with a pixel size of 0.2″. The design of this instrument is still under development. MIRI will be invaluable in understanding the chemical evolution of the ISM and accretion process in the early stages of star formation when most of a protostar's emission is in the mid-infrared to far-infrared. The wavelength range also covers many important absorption features in the giant planets' thermal emission spectra, which given the high angular resolution and high sensitivity of JWST should prove to be extremely illuminating.

Near Infrared Camera (NIRCam)

NIRCam will provide imaging in the wavelength range 0.6 μm to 5 μm and addresses the core astronomical science goals of JWST including: (1) detection of the early phases of star and galaxy formation; (2) morphology and color of galaxies at very high redshift; (3) detection and study of distant supernovae; (4) mapping dark matter via gravitational lensing; and (5) the study of stellar populations in nearby galaxies.

The current design of NIRCam consists of four modules: two broadband and intermediate-band imaging modules and two tunable filter-imaging modules, each with a FOV of 140 × 140″. The imaging modules will have a short wavelength and a long wavelength channel, taking images simultaneously with light split by a dichroic at about 2.35 μm. The short wavelength channels will be sampled by a 4,096 × 4,096 detecting array (giving a pixel size of 0.03″), and the long wavelength channels by a 2,048 × 2,048 detector array (giving a pixel size of 0.06″). The tunable filter-imaging modules will have a resolving power of about 100 with one filter module optimized for wavelengths from about 1.2 μm to 2.5 μm, and the other from 2.5 μm to 4.5 μm. It is planned to incorporate coronagraphs into all modules, which allow for the observation of faint targets lying very close to bright targets by blocking the light from the central object (see Section 8.7.3). The detectors will most likely be either InSb or HgCdTe.

The position of the segmented optical elements of JWST must be measured and re-aligned on a regular basis to reach the required optical performance, and NIRCam will provide this wavefront-sensing function. To maintain the optimum configuration will probably require dedicated observations of bright stars by NIRCam on a weekly or perhaps more frequent basis.

Near Infrared Spectrograph (NIRSpec)

NIRSpec, to be provided by ESA, will provide spectroscopy in the wavelength range 0.6 μm to 5 μm with a spectral resolving power of between ~100 and 1,000. NIRSpec is planned to be able to obtain simultaneous spectra of more than 100 objects in a 9 arcmin^2 FOV. To improve sensitivity, the pixel size will be larger than for NIRCam and the detectors will again be either InSb or HgCdTe, although the design is yet to be finalized.

While NIRSpec will again be designed to primarily address astronomical studies such as galaxy formation, clustering, chemical abundances, star formation, etc., its great sensitivity and wavelength range would make it a useful tool for studies of the giant planets, particularly Uranus and Neptune, whose near-IR reflection spectra are not well-known and whose 5 μm spectra may provide clues to these planets' deep composition.

8.6 SPACECRAFT MISSIONS TO THE GIANT PLANETS

Over the past few decades NASA has led the way in the exploration of the giant planets with a series of ever larger and more complicated spacecraft ranging from the simple *Pioneer 10* and *11* spacecraft to the highly advanced *Cassini/Huygens* mission introduced in the last chapter. More recent spacecraft have involved a significant contribution from ESA and other individual countries, which is especially true of the *Cassini/Huygens* mission: the entire Titan *Huygens* probe was ESA-funded. Such international collaboration is likely to underpin future missions to these worlds. In addition, NASA has recently steered away from giant multifunctional spacecraft to more numerous, cheaper missions aimed at particular scientific questions. This "faster, cheaper, better" philosophy underlies the Discovery Program of NASA, and a number of giant planet missions have so far been proposed, although none has yet been selected. Hence, the only assured giant planet mission in the next ten years is the continuing *Cassini/Huygens* mission to Saturn, previously described in Chapter 7, and the NASA *Juno* mission described next, although it is very likely that NASA and ESA will collaborate to send a spacecraft to either the Jupiter or Saturn systems in the 2015–2025 timeframe.

8.6.1 *Juno*

Juno is a NASA New Frontiers mission to place a spacecraft in an elliptical polar orbit about Jupiter to study Jupiter's interior structure. *Juno* is due for launch on an Atlas V launch vehicle in 2011 and after an Earth flyby in 2013 should go into orbit about Jupiter in 2016.

The final polar orbit of *Juno* will have a period of about 11 days and will be highly elliptical with a perijove at 1.6 R_J above the North Pole and apojove at $\sim$39 R_J. The nominal mission is planned to last for 32 orbits, or roughly 1 year, and *Juno* will communicate its data to Earth with a 2.5 m high gain antenna. Accurate observations of *Juno*'s trajectory in this polar orbit will allow Jupiter's gravitational field to be determined to a very high degree of precision and constrain its higher order J-coefficients (Section 2.7.1). The *Juno* mission should thus be able to establish whether or not Jupiter has a core. The polar orbit also limits the damage done by Jupiter's powerful radiation belts and keeps the spacecraft continuously in sunlight throughout its mission. Through this, and also through designing the spacecraft to have very low power consumption, *Juno* will the first outer planets' mission not to be powered by radioisotope thermal generators (RTGs). Instead, it will be powered by three 2×9 m solar panels as can be seen in Figure 8.7. The spacecraft will be spin-stabilized, rather than three axis–stabilized to dispense with the need of power-hungry reaction wheels. The electronics are protected from radiation by placement in a shielded instrument electronics vault.

The polar regions are fascinating areas of the giant planet atmospheres, but the polar regions of Jupiter are relatively hard to see from the Earth due to Jupiter's low

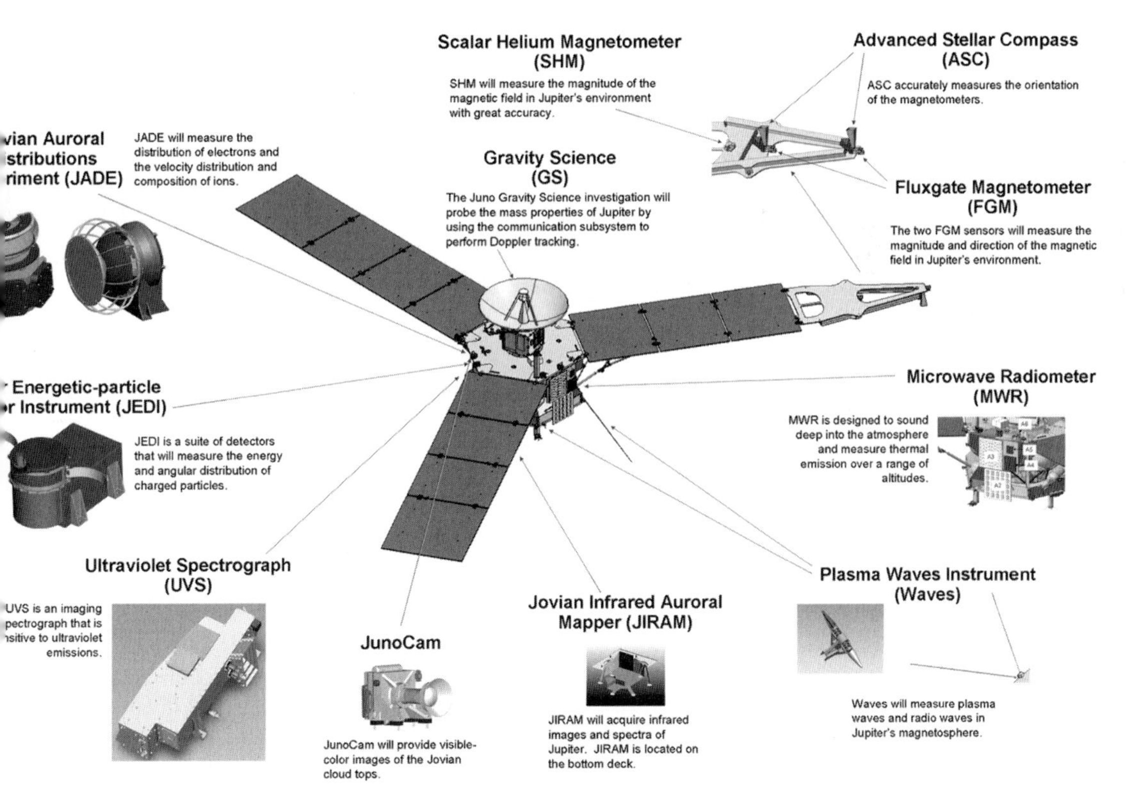

Figure 8.7. Payload system of the *Juno* spacecraft. Courtesy of NASA.

obliquity and were not well observed by previous space missions such as *Galileo* and *Cassini*, which remained close to the equatorial plane. *Juno*'s polar orbit allows its suite of remote-sensing instruments to look straight down at Jupiter's North Pole from a very close distance allowing it to map the polar region with unprecedented detail.

Juno will carry a number of remote-sensing instruments relevant to exploring Jupiter's atmosphere. For the most part these will concentrate on trying to determine conditions in Jupiter's deep atmosphere to complement the mission's main goal of determining the interior structure of the planet.

Microwave Radiometer (MWR)

The Microwave Radiometer will probe the deep atmosphere of Jupiter down to around 200 bar using six separate radiometers sounding from 1.3 cm to 50 cm that sweep across the planet as the spacecraft spins. These observations and their limb-darkening can then be used to determine the temperature profile of the deep atmosphere and also constrain the deep abundance of ammonia and water. The latitudinal dependence of the retrieved deep-temperature profiles will allow the circulation of Jupiter's deep atmosphere to be inferred to a much greater depth than was obtained by the *Galileo* probe.

Jupiter Infrared Auroral Mapper (JIRAM)

JIRAM is *Juno*'s near-IR mapping spectrometer and is the successor to the *Galileo* NIMS and *Cassini* VIMS instruments. The instrument will observe in the 2 μm–5 μm range and is being contributed to the mission by the Italian Space Agency. JIRAM's auroral specialization refers to one of its goals of observing H_3^+ emission at 3.4 μm, but it has many other goals including mapping the deep abundances of ammonia and phosphine at 5 μm.

JunoCam (JCM)

JunoCam is *Juno*'s camera, and is mainly included for education and public outreach. It has an $18 \times 3.4^\circ$ FOV and will use three filters mounted directly on the detector to obtain the first close-up color images of Jupiter's poles.

8.6.2 *Rosetta*

The ESA *Rosetta* mission is designed to examine a cometary nucleus from close quarters as the comet approaches the Sun and then develops its tail. While clearly not a giant planet mission, understanding the composition and structure of these remnants of solar system formation is important for understanding the origin and evolution of the solar system, and the composition of all the planets.

Rosetta was launched on March 2, 2004 from French Guiana on an *Ariane 5G* launcher and weighed 2,970 kg (1,300 kg plus 1,670 kg of propellant) at launch. After an extended gravitationally assisted tour, *Rosetta* will begin its rendezvous maneuver

Figure 8.8. Artist's impression of the *Rosetta* spacecraft approaching its target comet. Courtesy of ESA.

with Comet 67P/Churyumov–Gerasimenko in May 2014 at a heliocentric distance of $\sim$4 AU (during the comet's approach to the Sun) and will achieve its final close orbit with the comet by November 2014. At closest approach, the orbiter will be only 1 km above the nucleus. Active observations will be taken up until the nominal end of mission, which coincides with the perihelion passage of the comet in December 2015. The mission is named after the Rosetta Stone found by French soldiers in Egypt in 1799 and now in the British Museum in London. Inscribed on the stone is a declaration in three different languages—Greek, Egyptian Demotic, and Hieroglyphics—which were analyzed by both the English physicist Thomas Young, and the French egyptologist Jean François Champollion who in 1822 finally succeeded in decrypting the hieroglyphic system (Adkins and Adkins, 2000). The main structure of the *Rosetta* orbiter is a $2.8 \times 2.1 \times 2.0$ m cube shown in Figure 8.8 and is powered by two solar panels each of which when fully extended has an area of 32 m^2 and total span of 32 m.

The *Rosetta* orbiter will be the first spacecraft ever to go into orbit around a comet nucleus and will be the first to observe the changes that occur in comet activity as the comet travels towards the inner solar system. The orbiter's scientific payload includes 11 experiments (listed in Table 8.1) and a small 90 kg lander, which will be the first spacecraft ever to make a soft landing on the surface of a comet nucleus and

Table 8.1. *Rosetta* instruments.

Instrument	*Acronym*	*Purpose*
Remote-sensing instruments		
Optical, Spectroscopic, and Infrared Remote Imaging System	OSIRIS	A wide-angle and narrow-angle camera to image the comet's nucleus
Ultraviolet Imaging Spectrometer	ALICE	Analyzes gases in the coma and tail and measures the comet's production rates of water, CO, and CO_2
Visible and Infrared Thermal Imaging Spectrometer	VIRTIS	Maps and studies the nature of the solids and the temperature on the surface of the nucleus. Also identifies comet gases, and characterizes the physical conditions of the coma
Microwave Instrument for the *Rosetta* Orbiter	MIRO	Used to determine the abundances of major gases, the surface outgassing rate, and the nucleus subsurface temperature
Composition analysis		
Rosetta Orbiter Spectrometer for Ion and Neutral Analysis	ROSINA	Sensors will directly determine the composition of the comet's atmosphere and ionosphere
Cometary Secondary Ion Mass Analyser	COSIMA	Analysis of the nature of dust grains emitted by the comet
Micro-Imaging Dust Analysis System	MIDAS	Provides information on particle population, size, volume, and shape of dust around the comet
Nucleus large-scale structure		
Comet Nucleus Sounding Experiment by Radiowave Transmission	CONSERT	Probes the comet's interior by measuring radio waves that are reflected and scattered by the nucleus. Consists of a transmitter/receiver on the Orbiter and of a transmitter/receiver on the Lander
Grain Impact Analyzer and Dust Accumulator	GIADA	Measures the number, mass, momentum, and velocity distribution of dust grains coming from the nucleus and from other directions

Table 8.1 (*cont.*)

Instrument	*Acronym*	*Purpose*
Comet plasma environment and solar wind interaction		
Rosetta Plasma Consortium	RPC	Incorporates five sensors that measure the physical properties of the nucleus, examine the structure of the inner coma, monitor cometary activity, and study the comet's interaction with the solar wind
Radio Science Investigation	RSI	Shifts in the spacecraft's radio signals are used to measure the mass, density, and gravity of the nucleus; define the comet's orbit; and study the inner coma

will focus on the *in situ* study of the composition and structure of the nucleus material. The prime objective of the *Rosetta* mission is to study the origin of comets and the relationship between cometary and interstellar material. To address these goals *Rosetta* will make the following measurements of Comet 67P/Churyumov–Gerasimenko: (1) determine the dynamic properties, surface morphology, and composition of the nucleus; (2) determine the chemical, mineralogical, and isotopic compositions of volatiles and refractories in the nucleus and coma; and (3) study the development of cometary activity and the processes in the surface layer of the nucleus and the inner coma (dust/gas interaction).

8.6.3 Future outer-planet missions

At the time of writing, both ESA and NASA are considering an outer-planet mission to be launched in the 2015–2025 timeframe to either the Jovian or Saturnian systems. In addition to further studying the atmospheres of these planets, the missions will also explore in more detail the moons of the giant planets, in particular Europa and Ganymede for the Jupiter system mission, and Titan and Enceladus for the Saturn system mission. The decision of which mission study to explore further will be made towards the end of 2008.

8.7 EXTRASOLAR PLANET SPACE MISSIONS

In Chapter 2 we reviewed the current theories of formation of the solar system and put forward explanations for the current position, composition, and mass of the planets. If such theories are sound then they should apply equally well to the planetary systems of other stars, but until relatively recently no "extrasolar" planets had ever been observed and thus there was no way of testing these ideas. Then, in 1995, the first extrasolar planet was observed about the solar-type star 51 Pegasi by Mayor and Queloz (1995) and then about 70 Virginis by Marcy and Butler (1996). The technique used involved measuring the frequency of light emitted by the star and observing how this was modulated by the Doppler effect caused by the star's motion about the stellar system barycenter. The planets inferred by these two early studies were both large (with a mass of the order of Jupiter's) and were both very close to their stars (within 0.5 AU), leading them to become known as "Hot Jupiters" and to the conclusion that these planetary systems are obviously nothing like our own solar system! Since then, many more extrasolar planets have been detected, not just with this radial velocity technique, but also by transits (where the reduction in light from the star as the planet passes in front of it is observed) and gravitational lensing, and the number of known exoplanets currently stands at more than 300. Most of the exoplanets discovered so far lie within a few astronomical units of their star, but the maximum detectable orbital radius is steadily increasing as the techniques become more precise and extend over longer timeframes. Although the radial velocity technique is biased to find heavy, short-period planets, most exoplanets discovered so far have a mass $<10\ M_J$. However, the close "Hot Jupiters" are by no means uncommon and preliminary studies suggest that approximately 6% of solar-type stars have close Jupiter-sized companions (Marcy and Butler, 1998) within 2 AU (Irwin, 2008).

There is a clear need to establish just how typical our own solar system is, and on a more philosophical level to determine whether there are other planets in the galaxy which are more Earth-like and fall within the so-called "habitable zone", where liquid water may exist on the surface. Such planets would be suitable for the evolution of life and it would be of profound significance indeed if life could be detected to have evolved elsewhere in the universe independently of the life that has evolved here on the Earth. The search for terrestrial extrasolar planets is one of the most exciting fields of research in planetary physics and at this time there are a number of missions on the drawing board which hope to address these questions. Of course a system capable of detecting terrestrial planets is also likely to detect giant planets as well, and thus one of the spin-offs from this program will be determination of the true "average" planetary system and variants, which will provide much stronger constraints on stellar and planetary formation theories than can be gleaned from consideration of our solar system alone. The major new space missions planned in the next 10–20 years to search for extrasolar terrestrial planets will now be reviewed.

8.7.1 *Kepler*

The *Kepler* mission, led by NASA Ames, is a spaceborne telescope designed to search for planets about other stars by observing the small dip in light received from the star caused by a planet's transit in front of it. The mission is named after the German astronomer Johannes Kepler (1571–1630), who deduced his three famous orbital laws in 1609 (First and Second Law) and 1619 (Third Law). *Kepler* will continuously and simultaneously monitor the brightnesses of approximately 100,000 A–K dwarf

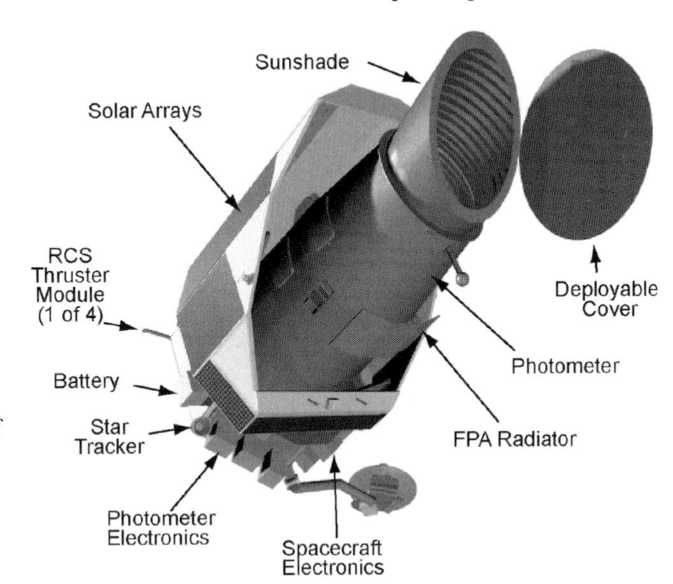

Figure 8.9. A schematic of the *Kepler* telescope, due for launch in 2009 to search for extrasolar planetary transits. Courtesy of NASA.

(main-sequence) stars brighter than magnitude 14 in the region of the Cygnus constellation along the Milky Way. The star field in this region is very dense and far enough from the ecliptic plane so as not to be obscured by the Sun at any time of the year and also to avoid confusion resulting from possible occultations by asteroids and Kuiper Belt Objects.

The *Kepler* spacecraft is basically a Schmidt telescope with a 1.4 m diameter primary mirror and a 0.95 m aperture (Figures 8.9, 8.10). The current planned spacecraft mass is ~900 kg, and the spacecraft will be placed in an Earth-trailing, heliocentric orbit where it will continuously monitor the Cygnus star field for a period of 4 years. It is currently planned to launch *Kepler* in 2009 using a *Delta II* launcher. The *Kepler* telescope will operate as a differential photometer utilizing forty-two 50×25 mm CCD arrays, each containing $2{,}200 \times 1{,}024$ pixels, and the FOV of the combined instrument will be roughly 12° in diameter. The CCDs will not record images as such, but instead will monitor and return data from those CCD pixels where there are stars with a visible magnitude greater than 14. The image system is intentionally defocused to an angular resolution of $10''$ to improve photometric precision and the signals from the CCD arrays will be integrated for 15 minutes. The orbital position of the telescope chosen is required to ensure low background loading and to allow unbroken observations which are essential for the mission.

Although the probability of a planet passing in front of an individual star during *Kepler*'s mission is small, it is certainly not insignificant and *Kepler* will monitor so many stars that current formation and accretion models predict the following results:

1. Observation of about 50 transits of Earth-mass planets at a distance of ~1 AU.
2. Observation of thousands of transits of Earth-mass planets for orbits significantly less than 1 AU.

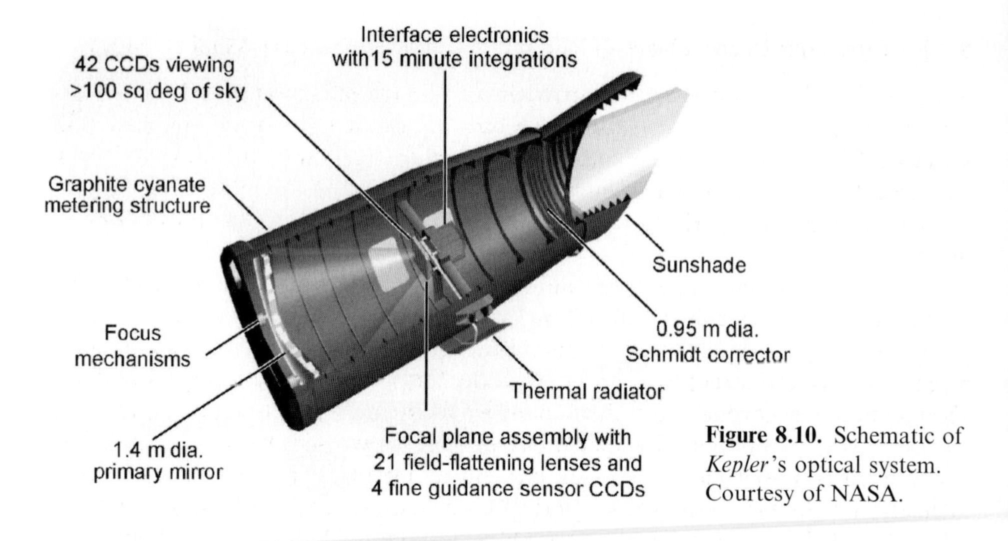

Figure 8.10. Schematic of *Kepler*'s optical system. Courtesy of NASA.

3. Observation of reflected light from ~900 giant planets with periods <1 week.
4. Observation of about 130 transits of inner-orbit giants, with albedo determination for ~100 of these.
5. Determination of the density of ~35 inner-orbit giant planets.
6. Observation of the transit of ~30 outer-orbit giants.

While the *Kepler* mission is designed primarily to search for terrestrial planets in the "habitable zone" about other stars, the detection method is less biased than the Doppler shift method, and so *Kepler* should be able to build up a much more representative statistical picture of the distribution of planet sizes and orbital distances in our galaxy. Such data will provide a strong test of current formation theories, and indeed on the representativeness of our own solar system.

8.7.2 Convection, Rotation and Transits (COROT) mission

The COROT mission is a French-led project (CNES) to place a small space telescope into orbit dedicated to the study of astroseismology and extrasolar planet transit detection, in the same manner as for *Kepler*.

COROT was launched on December 27, 2006 and is equipped with a 0.27 m diameter telescope and four CCD detector arrays cooled to −40°C; the combined instrument has a total field of view of 2.8 × 2.8°. The spacecraft is 4.2 m long, weighed 650 kg at launch, and was placed into a circular polar orbit at an altitude of 827 km. The nominal mission lifetime is at least $2\frac{1}{2}$ years. While COROT is not as sensitive as *Kepler*, there is still a strong possibility of it detecting several extrasolar planets during its mission lifetime, and indeed COROT announced the detection of its first extrasolar planet on May 3, 2007.

8.7.3 Terrestrial Planet Finder (TPF) and Space Interferometry Mission (SIM)

The NASA TPF is a proposed mission to directly image extrasolar planets and, in one design, record their thermal emission spectra to search for biogenic indicators. If selected, TPF would take the form of either a coronagraph operating at visible wavelengths or a large-baseline interferometer operating in the infrared and would be placed either in an Earth-trailing heliocentric orbit, or at the Earth–Sun L2 position.

Preliminary designs for a visible light coronograph (TPF-C) use a single telescope with an effective diameter of 6.5 m–8 m, operating at room temperature. A coronograph works by blocking the direct and diffracted light of a bright object so that faint nearby objects and structures can be seen. Such instruments were originally developed to study the corona about the Sun, hence the name. The diffraction pattern of a telescope with a simple circular aperture is the well-known Airy pattern, consisting of a bright central spot surrounded by concentric, circular rings of ever decreasing intensity. To observe a faint planet close to a star requires that the first several bright rings of starlight are suppressed without blocking out the light of a planet. By using masks to simulate a telescope with a different effective entrance aperture shape, the diffraction pattern may be modified so that the starlight is much dimmer closer to the center at some rotation angles, and brighter in others. Hence, by then rotating the telescope about the line of sight to the star, the degree of obscuration of a planet orbiting that star will be periodically modulated and thus the planet may be detected. To achieve the sensitivity needed will require the use of active optics in order to correct for wavefront distortions arising from optical imperfections.

Preliminary IR TPF (TPF-I) concepts would use multiple, smaller telescopes (3–4 m mirrors) configured as an interferometer and spread out either over a large boom (of less than 40 m in length) or alternatively operated on separate spacecraft over distances of a few hundred meters (Figure 8.11). At thermal-IR wavelengths an Earth-like planet at a distance of 1 AU from a Sun-type star would appear a million times fainter than the star, and hence discriminating its thermal emission from that of the star is by no means an easy task. However, at visible wavelengths the light reflected by such a planet would appear a billion times fainter than the star, and thus it is theoretically easier to detect a terrestrial planet in the thermal-infrared. In order to discriminate between the radiation from the star and the planet it is proposed to use the technique of nulling interferometry outlined in Figure 8.12. The interferometer system is centered on the star and the beams from two telescopes separated by distance D are combined with a carefully controlled $\lambda/2$ path difference, maintained by an active delay line system, such that the starlight interferes destructively and is almost completely canceled at the detector. The light from a nearby planet however, offset by an angle θ, introduces an extra path difference of $D \sin\theta \approx D\theta$. If $D\theta = \lambda/2$, then the planet's light will interfere constructively and so can be detected. Several telescopes (3–5) would need to be used at a number of separations and rotations in order to scan the different separation and rotation angles of planets in the system and these telescopes would need to be precisely aligned with respect to each other. In addition the telescopes would need to be extremely cold in order to

Figure 8.11. Artist's impression of one scenario (Lockheed Study) of an IR nulling interferometer concept proposed for the NASA Terrestrial Planet Finder (TPF). Courtesy of NASA.

detect the faint, cold planetary emission, although the requirements on the telescopes' optical quality are less than for the visible coronograph system, and the mirror sizes are less. A major advantage of such a system would be that once planets have been detected, the thermal-IR radiation emitted by them could be integrated for a very long time in order to build up a thermal emission spectrum of the planet, from which the atmospheric composition may be determined together with the temperature of the atmosphere and surface. The IR designs that are planned are tailored to detect warm terrestrial planets and will thus not be able to detect the emission of cold, distant giant planets, which emit more in the far-infrared. However, the really exciting thing about TPF (and *Darwin* outlined in Section 8.7.5) is that it will be optimized to detect the major absorption bands of CH_4, O_3, CO_2, and H_2O from 6 μm to 15 μm. The atmosphere of the Earth is peculiar in that it has a high abundance of molecular oxygen O_2, a highly reactive gas not found freely in such high concentrations, as far as we know, anywhere else in the universe. The oxygen is photolyzed to create a significant abundance of ozone (O_3) in the stratosphere, which has a clear absorption band at 9.6 μm, easily detectable from space. What makes our atmosphere even more peculiar is that there are also significant levels of methane, a highly combustible gas existing alongside the molecular oxygen. The reason for Earth's peculiar atmospheric

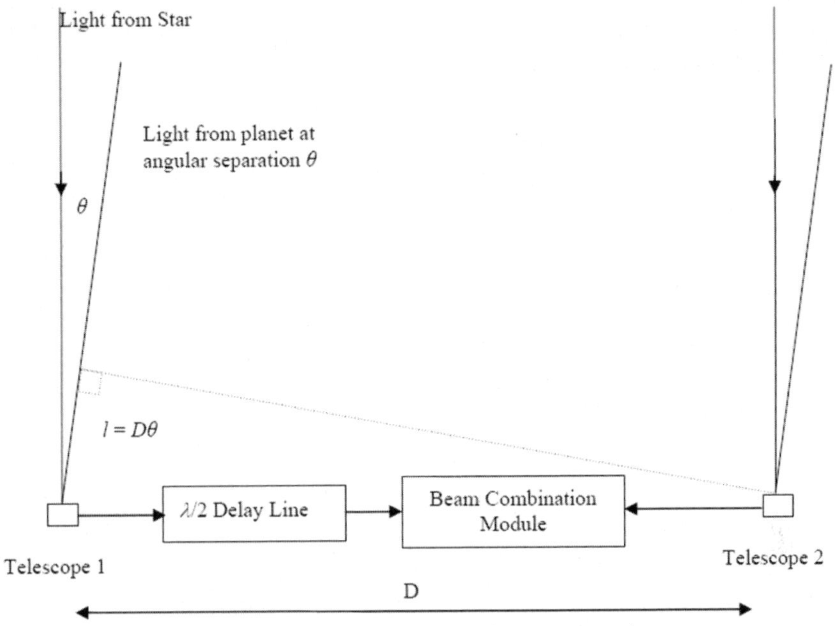

Figure 8.12. How nulling interferometry works.

composition is life. Plants generate O_2 through photosynthesis, and animals generate CH_4 as a by-product of digestion. As far as is known, there is no other natural way of generating such an atmosphere, and thus if a similar atmosphere could be detected about an extrasolar planet, it would provide strong evidence that life had independently evolved there. Such a conclusion would be profound indeed.

A precursor to the TPF is the Space Interferometry Mission (SIM), which will use optical interferometry to determine the positions and distances of stars several hundred times more accurately than has previously been possible. The predicted accuracy will allow SIM to determine the distances to stars throughout the galaxy and to probe nearby stars for Earth-sized planets by observing the "wobble" in the star's position as its planets orbit around it. SIM will use three co-linear interferometers mounted on a 10 m long boom, and will be placed in an Earth-trailing heliocentric orbit. The interferometer mirrors will be 0.3 m in diameter; SIM is designed to last for 5 years.

Unfortunately, the TPF mission is currently in an indefinitely deferred status and the launch of its precursor mission, SIM, has now been delayed until 2015–2016 at the earliest.

8.7.4 *Darwin*

Darwin is a proposed ESA mission which has many similarities with the IR option of NASA's TPF mission. Recent *Darwin* designs have a flotilla of six free-flying telescopes, each with a mirror diameter in excess of 1.5 m to observe the thermal-IR spectra of terrestrial extrasolar planets through the technique of nulling interferometry. It is proposed to place the flotilla of spacecraft at the Earth–Sun L2 point, where the telescopes will cool to ~40 K and the detecting elements themselves to as low as 8 K (Figure 8.13).

To demonstrate the feasibility of the nulling interferometric approach on the ground, ESA and the European Southern Observatory (ESO) are developing a nulling interferometer to be used at the VLTI facility in Chile, outlined earlier in Section 8.2.1.

Since *Darwin* shares so many similarities with TPF it is quite possible that NASA and ESA may join forces in the next few years to develop a joint mission to search for terrestrial planets about other stars.

8.8 CONCLUSION

From the projects covered in this chapter it can be seen that the study of giant planet atmospheres will continue to be provided for many years with a large amount of ever increasingly accurate measurements with which theories may be tested. The improvements in our understanding of these worlds that are expected, or hoped for, will now be briefly summarized.

Jupiter

The *Galileo* and *Cassini* spacecraft provided an enormous quantity of data on the largest of the giant planets, and these data are still being analyzed. Indeed, as knowledge of the spectroscopy of Jovian gases and candidate aerosols improves, both through better modeling and new laboratory studies, these data, and the data from the previous *Voyager* missions, are likely to be reanalyzed for many more years to come. Particular advances in spectroscopy that are expected are improvements in the knowledge of cold-temperature ammonia and methane gas absorption in the near-infrared, and new measurements of the real and imaginary refractive indices of Jovian candidate aerosols such as NH_3, NH_4SH and $(NH_4)_2S$. Clearly, retrievals of planetary spectra are of limited value when the absorption spectra of candidate absorbers are themselves somewhat unclear, and such studies will greatly improve the accuracy of existing Jovian analyses. Together with the analysis of existing data, new spacecraft observations are planned and in particular the *Juno* spacecraft, arriving in 2016, will map Jupiter's internal structure, hopefully determining if it has an icy/rocky core, and will also make detailed measurements of Jupiter's North Pole to see if Jupiter, like the other giant planets (except Uranus) has a warm cyclonic vortex there.

In parallel with the continued analysis of spacecraft data, the quality of ground-based data continues to improve. One particular advantage of ground-based observations of significance to Jovian studies is the ability to observe at very high spectral resolution and thus observe individual absorption lines. For example, such

Figure 8.13. Artist's impression of ESA's proposed *Darwin* nulling interferometer mission (Alcatel study). Courtesy of NASA

studies have led to a significantly more accurate determination of the $(D/H)_{CH_4}$ ratio by observing close, near-equal strength CH_4 and CH_3D lines (Bézard *et al.*, 2003).

Finally, the ability to model the dynamics of the giant planets continues to improve. Numerical models such as EPIC (Dowling *et al.*, 1998) have been developed which provide new tools for analyzing existing dynamical observations. General circulation models (GCMs), originally developed for terrestrial weather forecasting, are currently being modified for giant planet studies which should help in the interpretation of observations by spacecraft and by ground-based observatories.

Saturn

The *Cassini* mission has greatly improved our understanding of the Saturnian system and has returned many stunning images. The *Cassini* spacecraft will continue to orbit Saturn and return data until at least 2010. Hence, *Cassini* will be observing Saturn during its Northern Spring Equinox in 2009 and there is great anticipation in how the loss of sunlight from the South Pole might affect the South Polar Vortex (SPV) and how the return of sunlight to the North Polar region might affect the North Polar Hexagon (NPH) and warm North Polar Vortex. In particular, once sunlight returns to the North Pole it will be possible to map the NPH in visible light at the cloud top level and compare its structure with that seen by *Voyager 2*, and compare this with the structures determined at the 2 bar to 3 bar level by *Cassini* VIMS, and above the clouds up to the tropopause by *Cassini* CIRS.

In addition to continuing Cassini observations, ground-based observations continue to improve in terms of both spatial and spectral (e.g., Greathouse *et al.* 2005, 2006) resolution, and the use of revised laboratory reference data should lead to a steadily improving understanding together with ever more advanced dynamical analysis, as described in the previous section.

Uranus

While Uranus had a very bland appearance during the *Voyager 2* encounter in 1986, we have seen that the activity of its atmosphere has increased greatly in the run-up to its Northern Spring Equinox in 2007. Observing how Uranus' appearance changes in the years after this equinox should reveal more about how the circulation of its atmosphere is driven. Advances in our understanding of Uranus' 5 μm spectrum are also expected with planned ground-based observations, and improvements in the determination of Uranus D/H ratio are expected from recent *Spitzer* IRS observations. Finally, the far-IR capabilities of *Herschel* (Section 8.5.1), planned for launch in 2009, should also hopefully address the question of D/H enrichment in Uranus' atmosphere, and other issues.

Neptune

Neptune subtends the smallest disk size of the giant planets as seen from the Earth and thus presents considerable observational difficulties. However, Neptune has the most vigorous circulation of any of the giant planets and improvements in both ground-based and space-based astronomy should improve spatial resolution and spectral coverage allowing for more detailed understanding of the observed discrete "convective clouds", dark spots, and the South Polar Wave (SPW). The current season on Neptune is just after summer solstice, and the planet will reach its autumnal equinox in around the year 2040. Hence, in the next few years, more and more of the northern hemisphere will become observable, which has never before been explored by modern detection systems. In the meantime it is also planned to address the D/H enrichment question with data from the AKARI space telescope and, after 2009, with the *Herschel* Space Observatory.

Concluding remarks

The field of giant planet study is truly multi-disciplined, drawing as it does on atmospheric physics, retrieval theory, IR spectroscopy, formation theory, and many other branches of physics. Understanding these worlds is of interest in its own right, and also in understanding how our own world came to be. While the Earth, and other terrestrial planets suffered a period of heavy bombardment by planetesimals early in its history, the inner Solar System has since been kept mostly clear of debris by the strong gravitational pull of Jupiter, which deflected such debris far out of the solar system and continues to protect the inner solar system from comets to this day. It is possible that life, and thus humans, may not have had a chance to evolve without the benign influence of these giant, distant worlds. The study of extrasolar systems has revealed that a significant proportion of them have giant planets orbiting close to the star, which would probably have the effect of jettisoning terrestrial planets into interstellar space, or into the star itself. Hence, it is of great interest to understand just how our solar system evolved into the configuration in which it is now found, and also to discern the likely formation distances of the planets prior to any migration. As for the giant planets themselves, their composition teaches us a great deal about the composition of material from which the solar system formed, and observations of the dynamics of the giant planet atmospheres provide intriguing tests of dynamical theories used in the world's weather-forecasting and climate prediction centers. Hence, while giant planet studies may seem to be pure "blue skies" research, the implication of these studies has significance for both our understanding of how the Earth came to be, and also for our understanding of its atmospheric structure and evolution.

8.9 BIBLIOGRAPHY

Much of the information in this chapter has been extracted from a number of public websites which are well worth investigating further.

Useful public NASA websites for mission information include

NASA Jet Propulsion Laboratory *http://www.jpl.nasa.gov/*
NASA Ames Research Center *http://www.arc.nasa.gov/*

NASA Goddard Space Flight Center *http://www.jpl.nasa.gov/*

The current round of Discovery missions may be reviewed at

http://discovery.nasa.gov/

ESA's main public page is at

http://www.esa.int/

ESA's planetary science missions *http://sci.esa.int/*

Other useful webpages include

JWST	*http://www.jwst.nasa.gov/, http://sci.esa.int/home/jwst/index.cfm*
Kepler	*http://www.kepler.arc.nasa.gov/*
TPF	*http://planetquest.jpl.nasa.gov/TPF/tpf_index.html*
Darwin	*http://sci.esa.int/home/darwin/index.cfm*
Herschel	*http://sci.esa.int/home/herschel/index.cfm*
Cassini/Huygens	*http://saturn.jpl.nasa.gov/index.cfm*
Galileo	*http://galileo.jpl.nasa.gov/*
Juno	*http://juno.wisc.edu/index.html*
Space Interferometry Mission (SIM)	*http://planetquest.jpl.nasa.gov/SIM/sim_index.html*
ALMA	*http://www.mma.nrao.edu/, http://www.eso.org/projects/alma/, http://www.arcetri.astro.it/science/ALMA*
SOFIA	*http://www.sofia.usra.edu/*
ESO (including VLT, VLTI, E-ELT)	*http://www.eso.org/*
TMT	*http://www.tmt.org/*
GMT	*http://www.gmto.org/*
E-ELT	*http://www.eso.org/public/astronomy/projects/e-elt.html*
Extrasolar planets	*http://www.obspm.fr/encycl/encycl.html*

References

Abbas, M.M., A. LeClair, T. Owen, B.J. Conrath, F.M. Flasar, V.G. Kunde, C.A. Nixon, R.K. Achterberg, G. Bjoraker, D.J. Jennings *et al.* (2004) The nitrogen isotopic ratio in Jupiter's atmosphere from observations by the Composite Infrared Spectrometer on the Cassini spacecraft. *Astrophys. J.*, **602**, 1063–1074.

Achterberg, R.K. and F.M. Flasar (1996) Planetary-scale thermal waves in Saturn's upper troposphere. *Icarus*, **119**, 350–369.

Achterberg, R.K. and A.P. Ingersoll (1994) Numerical-simulation of baroclinic Jovian vortices. *J. Atmos. Sci.*, **51**, 541–562.

Achterberg, R.K., B.J. Conrath, and P.J. Gierasch (2006) Cassini CIRS retrievals of ammonia in Jupiter's upper troposphere. *Icarus*, **182**, 169–180

Adkins, L. and R. Adkins (2000) *The Keys of Egypt: The Race to Read the Hieroglyphs*. Harper-Collins, London.

Aguiar, A.C.B., P.L. Read, Y.H. Yamazaki, and R.D. Wordsworth (2008) A laboratory model of Saturn's north polar hexagon. In preparation.

Alexander, A. (1965) *The Planet Uranus*. Faber & Faber, London.

Alibert, Y., O. Mousis, and W. Benz (2005a) On the volatile enrichments and composition of Jupiter. *Astrophys. J.*, **622**, L45–L48.

Alibert, Y., C. Mordasini, W. Benz, and C. Winisdoerffer (2005b) Models of giant planet formation with migration and disc evolution. *Astron. Astrophys.*, **434**, 343–353.

Alibert, Y., C. Mordasini, O. Mousis, and W. Benz (2005c) Formation of giant planets: An attempt in matching observational constraints. *Space Sci. Rev.*, **116**, 77–95.

Allen, C.W. (1976) *Astrophysical Quantities* (Third Edition). Athlone Press, University of London, London.

Allison, M. (1990) Planetary waves in Jupiter's equatorial atmosphere. *Icarus*, **83**, 282–307.

Allison, M., D. Godfrey, and R.F. Beebe (1990) A wave dynamical interpretation of Saturn's polar hexagon. *Science*, **247**, 1061–1063.

Allison, M., R.F. Beebe, B.J. Conrath, D.P. Hinson, and A.P. Ingersoll (1991) Uranus atmospheric dynamics and circulation. In J. Bergstrahl, E.D. Miner, and M.S. Matthews (Eds.), *Uranus*. University of Arizona Press, Tucson, AZ.

Anders, E. and N. Grevesse (1989) Abundances of the elements: Meteoritic and solar. *Geochimica et Cosmochimica Acta*, **53**, 197–214.

Anderson, J.D. and G. Schubert (2007) Saturn's gravitational field, internal rotation, and interior structure. *Science*, **317**, 1384–1387.

Anderson, J.D., P.A. Laing, E.L. Lau, A.S. Liu, M.M. Nieto, and S.G. Turyshev (1998) Indication, from Pioneer 10/11, Galileo, and Ulysses data, of an apparent anomalous, weak, long-range acceleration. *Phys. Rev. Lett.*, **81**, 2858–2861.

André, Ph. and T. Montmerle (1994) From T Tauri stars to protostars: Circumstellar material and young stellar objects in the ρ Ophiuchi cloud. *Astrophys. J.*, **420**, 837–862.

Andrews, D.G. (2000) *An Introduction to Atmospheric Physics*. Cambridge University Press, Cambridge, U.K.

Andrews, D.G., J.R. Holton, and C.B. Leovy (1987) *Middle Atmosphere Dynamics*. Academic Press, London.

Aurnou, J.M. and P.L. Olson (2001) Strong zonal winds from thermal convection in a rotating spherical shell. *Geophys. Res. Lett.*, **28**, 2557–2559.

Asplund, M., N. Grevesse, and A.J. Sauval (2005) The solar chemical composition. *Nuclear Physics A*, **777**, 1–4.

Atkinson, D.H., J.B. Pollack, and A. Seiff (1998) The Galileo Probe Doppler Wind Experiment: Measurement of the deep zonal winds on Jupiter. *J. Geophys. Res.*, **103**, 22911–22928.

Atreya, S.K. (1986) *Atmospheres and Ionospheres of the Outer Planets and Their Satellites*. Springer-Verlag, Berlin.

Atreya, S.K. and A.S. Wong (2005) Couple clouds and chemistry of the giant planets: A case for multiprobes. *Space Sci. Rev.*, **116**, 121–136.

Atreya, S.K., M.H. Wong, T.C. Owen, P.R. Mahaffy, H.B. Niemann, I. de Pater, P. Drossart, and Th. Encrenaz (1999) A comparison of the atmospheres of Jupiter and Saturn: Deep atmospheric composition, cloud structure, vertical mixing, and origin. *Planet. Space Sci.*, **47**, 1243–1262.

Atreya, S.K., P.R. Mahaffy, H.B. Niemann, M.H. Wong, and T.C. Owen (2003) Composition and origin of the atmosphere of Jupiter: An update and implications for the extrasolar planets. *Planet. Space Sci.*, **51**, 105.

Atreya, S.K., A.S. Wong, K.H. Baines, M.H. Wong, and T.C. Owen (2005) Jupiter's ammonia clouds: Localized or ubiquitous? *Planet. Space Sci.*, **53**, 498–507.

Atreya, S.K., K.H. Baines, and P.A. Egeler (2006) An ocean of water-ammonia on Neptune and Uranus: Clues from tropospheric cloud structure. *Bull. of the Am. Astron. Soc.*, **38**, 489

Aubert, J., S. Jung, and H.L. Swinney (2002) Observations of zonal flow created by potential vorticity in a rotating fluid. *Geophys. Res. Lett.*, **29**, 1876.

Backus, G. and F. Gilbert (1970) Uniqueness in the inversion of inaccurate gross earth data. *Phil. Trans. Royal. Soc. London, A*, **266**, 123–192.

Baines, K.H. (1997a) *Uranus: Atmosphere. Encyclopaedia of the Planetary Sciences* (edited by J.H. Shirley and R.W. Fairbridge). Chapman & Hall, London.

Baines, K.H. (1997b) *Neptune: Atmosphere. Encyclopaedia of the Planetary Sciences*. Chapman & Hall, London.

Baines, K.H. and J.T. Bergstralh (1986) The structure of the Uranian atmosphere: Constraints from the geometric albedo spectrum and H_2 and CH_4 line profiles. *Icarus*, **65**, 406–441.

Baines, K.H. and H. Hammel (1994) Clouds, hazes, and stratospheric methane abundance in Neptune. *Icarus*, **109**, 20–39.

Baines, K.H. and W.H. Smith (1990) The atmospheric structure and dynamical properties of Neptune derived from ground-based and IUE spectrophotometry. *Icarus*, **85**, 65–108.

Baines, K.H., H.B. Hammel, K.A. Rages, P.N. Romani, and R.E. Samuelson (1995a) Clouds and hazes in the atmosphere of Neptune. In D.P. Cruikshank, M.S. Matthews, and A.M. Schumann (Eds.), *Neptune and Triton*, pp. 489–546. University of Arizona Press, Tucson, AZ.

Baines, K.H., M.E. Mickelson, L.E. Larson, and D.W. Ferguson (1995b) The abundances of methane and ortho/para hydrogen on Uranus and Neptune: Implications of new laboratory 4–0 H_2 quadrupole line parameters. *Icarus*, **144**, 328–340.

Baines, K.H., R.W. Carlson, and L.W. Kamp (2002) Fresh ammonia ice clouds in Jupiter: Spectroscopic identification, spatial distribution, and dynamical implications. *Icarus*, **159**, 79–94.

Baines, K.H., P. Drossart, T.W. Momary, V. Formisano, C. Griffith, G. Bellucci, J.-P. Bibring, R.H. Brown, B.J. Buratti, F. Capaccioni *et al.* (2005) The atmospheres of Saturn and Titan in the near-infrared: First results of Cassini/VIMS. *Earth, Moon, and Planets*, **96**, 119–147.

Baines, K.H., A.A. Simon-Miller, G.S. Orton, H.A. Weaver, A. Lunsford, T.W. Momary, J. Spencer, A.F. Cheng, D.C. Reuter, D.E. Jennings *et al.* (2007a) Polar lightning and decadal-scale cloud variability on Jupiter. *Science*, **318**, 226–228.

Baines, K, T. Momary, M. Roos-Serote, S. Atreya, R. Brown, B. Buratti, R. Clark, and P. Nicholson (2007b) Saturn's Polar Hexagon at depth: New images of stationary planetary waves in the North Polar Region by Cassini/VIMS. *Geophys. Res. Abst.*, **9**, 02109

Balsiger, H., K. Altwegg, and J. Geiss (1995) D/H and $^{18}O/^{16}O$ ratio in hydronium ion and in neutral water from in situ ion measurements in comet P/Halley. *J. Geophys. Res.*, **100**, 5827–5834.

Banfield, D., P.J. Gierasch, M. Bell, E. Ustinov, A.P. Ingersoll, A.R. Vasavada, R. West, and M.J.S. Belton (1998) Jupiter's cloud structure from Galileo imaging data. *Icarus*, **135**, 230–250.

Baraffe, I. (2005) Structure and evolution of giant planets. *Space Sci. Rev.*, **116**, 67–76.

Bar-Nun, A., I. Kleinfeld, and E. Kochavi (1988) Trapping of gas mixtures by amorphous water ice. *Physical Rev. B*, **38**, 7749–7754.

Beebe, R.F. (1997) *Saturn: Atmosphere. Encyclopaedia of the Planetary Sciences* (edited by J.H. Shirley and R.W. Fairbridge). Chapman & Hall, London.

Beebe, R.F., A.P. Ingersoll, G.E. Hunt, J.L. Mitchell, and J.-P. Muller (1980) Measurements of wind vectors, eddy momentum transports, and energy conversions in Jupiter's atmosphere from Voyager 1 images. *Geophys. Res. Lett.*, **7**, 1–4.

Belton, M.J.S., J.W. Head, A.P. Ingersoll, R. Greeley, A.S. McEwen, K.P. Klaasen, D. Senske, R. Pappalardo, G. Collins, A.R. Vasavada *et al.* (1996) Galileo's first images of Jupiter and the Galilean satellites. *Science*, **274**, 377–385.

Berge, G.L. and S. Gulkis (1976) Earth-based radio observations of Jupiter: Millimeter to meter wavelengths. In T. Gehrels (Ed.), *Jupiter*. University of Arizona Press, Tucson, AZ.

Bézard, B. and P. Romani (1991) Hydrocarbons in Neptune's stratosphere from Voyager infrared observations. *J. Geophys. Res.*, **96**, 18961–18975.

Bézard, B., D. Gautier, and B. Conrath (1984) A seasonal model of the saturnian upper troposphere comparison with Voyager infrared measurements. *Icarus*, **60**, 274–288.

Bézard, B., C. Griffith, J. Lacy, and T. Owen (1995) Non-detection of hydrogen cyanide on Jupiter. *Icarus*, **118**, 384–391.

Bézard, B., C.A. Griffith, D.M. Kelly, J.H. Lacy, T. Greathouse, and G. Orton (1997) Thermal infrared imaging spectroscopy of Shoemaker–Levy 9 impact sites: Temperature and HCN retrievals. *Icarus*, **125**, 94–120.

Bézard, B., H. Feuchtgruber, J.I. Moses, and T. Encrenaz (1998) Detection of methyl radicals (CH_3) on Saturn. *Astron. Astrophys.*, **334**, L41–L44.

Bézard, B., P.N. Romani, H. Feuchtgruber, and T. Encrenaz (1999) Detection of the methyl radical on Neptune. *Astrophys. J.*, **515**, 868–872.

Bézard, B., E. Lellouch, D. Strobel, J.-P. Maillard, and P. Drossart (2002) Carbon monoxide on Jupiter: Evidence for both internal and external sources. *Icarus*, **159**, 95–111.

Bézard, B., T. Greathouse, J. Lacy, M.J. Richter, and C. Griffith (2003) The D/H ratio in Jupiter and Saturn from high-resolution spectral observations near 8.6 μm (American Astronomical Society, DPS meeting #35, #50.01). *Bull. Am. Astron. Soc.*, **35**, 1017.

Birnbaum, G., A. Borysow, and G.S. Orton (1996) Collision-induced absorption of H_2–H_2 and H_2–He in the rotational and fundamental bands for planetary applications. *Icarus*, **123**, 4–22.

Bleaney, B.I. and B. Bleaney (1976) *Electricity and Magnetism* (Third Edition). Oxford University Press, Oxford, U.K.

Bockelée-Morvan, D., D. Gautier, D.C. Lis, K. Young, J. Keene, T. Phillips, T. Owen, J. Crovisier, P.F. Goldsmith, E.A. Bergin *et al.* (1998) Deuterated water in Comet C.1996 B2 (Hyukutake) and its implications for the origin of comets. *Icarus*, **133**, 147–162.

Bockelée-Morvan, D., J. Boissier, J. Crovisier, F. Henry, and H.A. Weaver (2005) The carbon monoxide extended source in Comet Hale–Bopp revisited. *Bull. Am. Astron. Soc.*, **37**, 633.

Bosak, T. and A.P. Ingersoll (2002) Shear instabilities as a probe of Jupiter's atmosphere. *Icarus*, **158**, 401–409. doi: 10.1006/icar.2002.6886.

Boss, A.P. (1997) Giant planet formation by gravitational instability. *Science*, **276**, 1836.

Boss, A.P. (1998) Evolution of the solar nebula, IV: Giant gaseous protoplanet formation. *Astrophys. J.*, **503**, 923–937.

Boss, A.P. (2002) Evolution of the solar nebula, V: Disk instabilities with varied thermodynamics. *Astrophys. J.*, **576**, 462.

Boss, A.P. (2004) Convective cooling of protoplanetary disks and rapid giant planet formation. *Astrophys. J.*, **610**, 456.

Briceño, C., A.K. Vivas, N. Calvet, L. Hartmann, R. Pacheco, D. Herrera, L. Romero, P. Berlind, G. Sánchez, J.A. Snyder *et al.* (2001) The CIDA-QUEST large-scale survey of Orion OB1: Evidence for rapid disk dissipation in a dispersed stellar population. *Science*, **291**, 93–97.

Briggs, F.H. and P.D. Sackett (1989) Radio observations of Saturn as a probe of its atmosphere and cloud structure. *Icarus*, **80**, 77–103.

Brittain, S.D. and T.W. Rettig (2002) CO and H_3^+ in the protoplanetary disk around the star HD 141569. *Nature*, **418**, 57–59.

Brooke, T., R.F. Knacke, T. Encrenaz, P. Drossart, D. Crisp, and H. Feuchtgruber (1998) Models of the ISO 3 micron reflection spectrum of Jupiter. *Icarus*, **136**, 1–13.

Bryden, G., M. Rozyczka, D.N.C. Lin, and P. Bodemheimer (2000a) On the interaction between protoplanets and protostellar disks. *Astrophys. J.*, **540**, 1091–1101.

Bryden, G., D.N.C. Lin, and S. Ida (2000b) Protoplanetary formation, I: Neptune. *Astrophys. J.*, **544**, 482–495.

Burgdorf, M., G.S. Orton, G.R. Davis, S.D. Sidher, H. Feuchtgruber, M.J. Griffin, and B.M. Swinyard (2003) Neptune's far-infrared spectrum from the ISO long-wavelength and short-wavelength spectrometers. *Icarus*, **164**, 244–253.

Burgdorf, M.J., G.S. Orton, T. Encrenaz, G.R. Davis, E. Lellouch, S.D. Sidher, and B.M. Swinyard (2004) Far-infrared spectroscopy of the giant planets: Measurements of ammonia and phosphine at Jupiter and Saturn and the continuum of Neptune. *Adv. Space Res.*, **34**, 2247–2250.

Burgdorf, M.J., G.S. Orton, J. van Cleve, V. Meadows, and J. Houck (2006) Detection of new hydrocarbons in Uranus' atmosphere by infrared spectroscopy. *Icarus*, **184**, 634–637.

Burles, S. and D. Tytler (1988) *Proceedings of the Second Oak Ridge Symposium on Atomic and Nuclear Astrophysics* (edited by P. Mezzacappa, p. 113). Institute of Physics, Bristol, U.K.

Busse, F.H. (1976) A simple model of convection in the Jovian atmosphere. *Icarus*, **29**, 255–260.

Busse, F.H. (1994) Convection driven zonal flows and vortices in the major planets. *Chaos*, **4**, 123–134.

Busse, F.H. (2002) Convective flows in rapidly rotating spheres and their dynamo action. *Phys. Fluids*, **14**, 1301–1314.

Calcutt, S., F. Taylor, P. Ade, V. Kunde, and D. Jennings (1992) The composite infrared spectrometer. *J. British Interplan. Soc.*, **45**, 811–816.

Caldwell, J., X.-M. Hua, B. Turgeon, J.A. Westphal, and C.D. Barnet (1993) The drift of Saturn's north polar SPOT observed by the Hubble Space Telescope. *Science*, **260**, 326–329.

Cameron, A.G.W. (1982) Elemental and nuclidic abundances in the solar system. In C.A. Barnes, D.D Clayton and D.N. Schramm (Eds.), *Essays in Nuclear Astrophysics*, pp. 23–43). Cambridge University Press, Cambridge, U.K.

Campbell, J.K. and J.D. Anderson (1989) Gravity field of the Saturnian system from Pioneer and Voyager tracking data. *Astron. J.*, **97**, 1485–1495.

Campbell, J.K. and S.P. Synott (1985) Gravity field of the Jovian system from Pioneer and Voyager tracking data. *Astron. J.*, **90**, 364–372.

Carlson, B., A. Lacis, and W. Rossow (1993) Tropospheric gas composition and cloud structure of the Jovian north equatorial belt. *J. Geophys. Res.*, **98**, 5251–5290.

Cauble, R., P.M. Celliers, G.W. Collins, L.B. da Silva, D.M. Gold, M.E. Foord, K.S. Budil, R.L. Wallace, and A. Ng (2000) Equation of state and material property measurements of hydrogen isotopes at the high-pressure, high-temperature, insulator–metal transition. *Astrophys. J. Suppl.*, **127**, 267–273.

Chamberlain, J.W. and D.M. Hunten (1987) *Theory of Planetary Atmospheres: An Introduction to Their Physics and Chemistry* (Second Edition). Academic Press, San Diego, CA.

Charney, J.G. (1971) Geostrophic turbulence. *J. Atmos. Sci.*, **28**, 1087–1095.

Christensen, U.R. (2001) Zonal flow driven by deep convection in the major planets. *Geophys. Res. Lett.*, **28**, 2553–2556.

Christensen, U.R. (2002) Zonal flow driven by strongly supercritical convection in rotating spherical shells. *J. Fluid Mech.*, **470**, 115–133.

Cochran, A.L. (2002) A search for N_2^+ in spectra of Comet C/2002 C1 (Iyeka–Zhang). *Astrophys. J. Lett.*, **576**, L165.

Cochran, A.L., W.D. Cochran, and E.S. Barker (2000) N_2^+ and CO^+ in Comets 122P/1995 S1 (deVico) and C/1995 O1 (Hale–Bopp). *Icarus*, **146**, 583.

Combes, M., C. de Bergh, J. Lecacheux, and J.P. Maillard (1975) Identification of C-13 methane in the atmosphere of Saturn. *Astron. Astrophys.*, **40**, 81–84.

Conrath, B.J. and D. Gautier (2000) Saturn helium abundance: A reanalysis of Voyager measurements. *Icarus*, **144**, 124–134.

Conrath, B.J. and J.A. Pirraglia (1983) Thermal structure of Saturn from Voyager infrared measurements: Implications for atmospheric dynamics. *Icarus*, **53**, 286–292.

Conrath, B.J., D. Gautier, R. Hanel, G. Lindal, and A. Marten (1987) The helium abundance of Uranus from Voyager measurements. *J. Geophys. Res.*, **92**, 15003–15010.

Conrath, B.J., F.M. Flasar, R. Hanel, V. Kunde, W. Maguire, J. Pearl, J. Pirraglia, R. Samuelson, P. Gierasch, A. Weir *et al.* (1989) Infrared observations of the Neptunian system. *Science*, **246**, 1454.

Conrath, B.J., P.J. Gierasch, and S.S. Leroy (1990) Temperature and circulation in the stratosphere of the outer planets. *Icarus*, **83**, 255–281.

Conrath, B.J., D. Gautier, G.F. Lindal, R.E. Samuelson, and W.A. Shaffer (1991) The helium abundance of Neptune from Voyager measurements. *J. Geophys. Res.*, **96**, 18907–18919.

Conrath, B.J., D. Gautier, T.C. Owen, and R.E. Samuelson (1993) Constraints on N_2 in Neptune's atmosphere from Voyager measurements. *Icarus*, **101**, 168–172.

Conrath, B.J., P.J. Gierasch, and E.A. Ustinov (1998) Thermal structure and para hydrogen fraction on the outer planets from Voyager IRIS measurements. *Icarus*, **135**, 501–517.

Courtin, R., D. Gautier, A. Marten, and V. Kunde (1983) The $^{12}C/^{13}C$ ratio in Jupiter from the Voyager infrared investigation. *Icarus*, **53**, 121–132.

Courtin, R., D. Gautier, A. Marten, B. Bézard, and R. Hanel (1984) The composition of Saturn's atmosphere at northern temperate latitudes from Voyager IRIS spectra: NH_3, PH_3, C_2H_2, C_2H_6, CH_3D, CH_4 and the Saturnian D/H isotopic ratio. *Astrophys. J.*, **287**, 899–916.

Courtin, R., D. Gautier, and D. Strobel (1996) The CO abundance on Neptune from HST observations. *Icarus*, **123**, 37–55.

Cruikshank, D.P., H. Imanaka, and C.M. Dalle Ore (2005) Tholins as coloring agents on outer Solar System bodies. *Adv. Space Res.*, **36**, 178–183.

Dahmen, G., T.L. Wilson, and F. Matteucci (1995) The nitrogen isotope abundance in the galaxy, 1: The galactic disk gradient. *Astron. and Astrophys.*, **295**, 194–198.

Davis, G.R., D.A. Naylor, M.J. Griffin, T.A. Clark, and W.S. Holland (1997) Broadband submillimeter spectroscopy of HCN, NH_3 and PH_3 in the troposphere of Jupiter. *Icarus*, **130**, 387–403.

de Bergh, C., B.L. Lutz, T. Owen, J. Brault, and J. Chauville (1986) Monodeuterated methane in the outer solar system, II: Its detection on Uranus at 1.6 microns. *Astrophys. J.*, **311**, 501–510.

de Bergh, C., B.L. Lutz, T. Owen, and J.-P. Maillard (1990) Monodeuterated methane in the outer solar system, IV: Its detection and abundance on Neptune. *Astrophys. J.*, **355**, 661–666.

de Graauw, Th., H. Feuchtgruber, B. Bézard, P. Drossart, Th. Encrenaz, D.A. Beintama, M. Griffin, A. Heras, M. Kessler, K. Leech *et al.* (1997) First results of ISO-SWS observations of Saturn: Detection of CO_2, CH_3C_2H, C_4H_2 and tropospheric H_2O. *Astron. Astrophys.*, **321**, L13–L16.

Del Genio, A.D., J.M. Barbara, J. Ferrier, A.P. Ingersoll, R.A.. West, A.R. Vasavada, J. Spitale, and C.C. Porco (2007) Saturn eddy momentum fluxes and convection: First estimates from Cassini images. *Icarus*, **189**, 479–492.

Deming, D., M.J. Mumma, F. Espenak, D.E. Jennings, T. Kostiuk, G. Wiedemann, R. Loewenstein, and J. Priscitelli (1989) A search for *p*-mode oscillations on Jupiter: Serendipitous observations of non-acoustic thermal wave structure. *Astrophys. J.*, **343**, 456–467.

Deming, D., D. Reuter, D. Jennings, G. Bjoraker, G. McCabe, K. Fast, and G. Wiedemann (1997) Observations and analysis of longitudinal thermal waves on Jupiter. *Icarus*, **126**, 301–312.

den Hartog, R., O. Absil, P. Gondoin, A. Stankov, J.C. Augerau, V. Coudé du Forresto, D. Mourard, M. Fridlund, P. Gitton, F. Puech *et al.* (2005) The prospects of detecting exoplanets with the Ground-based European Nulling Interferometer Experiment (GENIE). *Direct Imaging of Exoplanets: Science and Techniques Proceedings*, IAU Colloquium No. 200 (edited by C. Aime and F. Vakili,), pp. 233–239, doi: 10.1017/S1743921306009379.

de Pater, I. (1986) Jupiter's zone–belt structure at radio wavelengths. *Icarus*, **68**, 344–365.

de Pater, I. and J.R. Dickel (1986) Jupiter's zone–belt structure at radio wavelengths, I: Observations. *Astrophys. J.*, **308**, 459–471.

de Pater, I. and J.J. Lissauer (2001) *Planetary Sciences.* Cambridge University Press, Cambridge, U.K.

de Pater, I. and S. Massie (1985) Models of the millimeter–centimeter spectra of the giant planets. *Icarus*, **62**, 143–171.

de Pater, I. and D.L. Mitchell (1993) Radio observations of the planets: The importance of laboratory measurements. *J. Geophys. Res.*, **98**, 5471–5490.

de Pater, I., D. Dunn, P.N. Romani, and K. Zahnle (2001) Reconciling Galileo probe data and ground-based radio observations of ammonia on Jupiter. *Icarus*, **149**, 66–78.

de Pater, I., P.N. Romani, and S.K. Atreya (1989) Uranus' deep atmosphere revealed. *Icarus*, **82**, 288–313.

de Pater, I., P.N. Romani, and S.K. Atreya (1991) Possible microwave absorption by H_2S gas in Uranus' and Neptune's atmospheres. *Icarus*, **91**, 220–233.

Deloule, E., J.-C. Doukhan, and F. Robert (1998) Interstellar hydroxyls in meteorite chondrules: Implications for the origin of water in the inner solar system. *Geochim. Cosmochim. Acta*, **62**, 3367–3378.

Dowling, T.E. (1995) Dynamics of Jovian atmospheres. *Annu. Rev. Fluid Mech.*, **27**, 293–334.

Dowling, T.E. (1997) *Jupiter: Atmosphere. Encyclopaedia of the Planetary Sciences* (edited by J.H. Shirley and R.W. Fairbridge). Chapman & Hall, London.

Dowling, T.E. and A.P. Ingersoll (1988) Potential vorticity and layer thickness variations in the flow around Jupiter's Great Red Spot and White Oval. *J. Atmos. Sci.*, **45**, 1380–1396.

Dowling, T.E. and A.P. Ingersoll (1989) Jupiter's Great Red Spot as a shallow water system. *J. Atmos. Sci.*, **46**, 3256–3278.

Dowling, T.E., A.S. Fischer, P.J. Gierasch, J. Harrington, R.L. LeBeau, and C.M. Santori (1998) The Explicit Planetary Isentropic-Coordinate (EPIC) atmospheric model. *Icarus*, **132**, 221–238.

Dritschel D.G., de la Torre Juarez, M., and M.H.P. Ambaum (1999) The three-dimensional vortical nature of atmospheric and oceanic turbulent flows. *Phys. Fluids*, **11**, 1512–1520.

Drouart, A., B. Dubrulle, D. Gautier, and F. Robert (1999) Structure and transport in the solar nebula from constraints on deuterium enrichment and giant planets formation. *Icarus*, **140**, 129–155.

Dyudina, U. and A.P. Ingersoll (2002) Monte Carlo radiative transfer modelling of lightning observed in Galileo images of Jupiter. *Icarus*, **160**, 336–349.

Dyudina, U.A., A.D. del Genio, A.P. Ingersoll, C.C. Porco, R.A. West, A.R. Vasavada, and J.M. Barbara (2004) Lightning on Jupiter observed in the Hα line by the Cassini imaging science subsystem. *Icarus*, **72**, 24–36.

Dyudina, U.A., A.P. Ingersoll, S.P. Ewald, C.C. Porco, G. Fischer, W. Kurth, M. Desch, A. Del Genio, J. Barbara, and J. Ferrier (2007) Lightning storms on Saturn observed by Cassini ISS and RPWS during 2004–2006. *Icarus*, **190**, 545–555.

Dyudina, U.A., A.P. Ingersoll, S.P. Ewald, A.R. Vasavada, R.A. West, A.D. Del Genio, J.M. Barbara, C.C. Porco, R.K. Achterberg, F.M. Flasar *et al.* (2008) Dynamics of Saturn's South Polar Vortex. *Science*, **319**, 1801.

Eberhardt, P., M. Reber, D. Krankowsky, and R.R. Hodges (1995) The D/H and $^{18}O/^{16}O$ ratios in water from Comet P/Halley. *Astron. Astrophys.*, **302**, 301–316.

Encrenaz, Th. (1999) The planet Jupiter. *Astron. Astrophys. Rev.*, **9**, 171–219.

Encrenaz, Th., E. Serabyn, and E.W. Weisstein (1996) Millimeter spectroscopy of Uranus and Neptune: Constraints on CO and PH_3 tropospheric abundances. *Icarus*, **124**, 616–624.

Encrenaz, Th., H. Feuchtgruber, S.K. Atreya, B. Bézard, E. Lellouch, J. Bishop, S. Edgington, Th. de Graauw, M. Griffin, and M.F. Kessler (1998) ISO observations of Uranus: The stratospheric distribution of C_2H_2 and the eddy diffusion coefficient. *Astron. Astrophys.*, **333**, L43–L46.

Encrenaz, T., E. Lellouch, P. Drossart, H. Feuchtgruber, G.S. Orton, and S.K. Atreya (2004) First detection of CO in Uranus. *Astron. Astrophys.*, **413**, L5–L9.

Evonuk, M. and G.A. Glatzmaier (2004) 2D studies of various approximations used for modeling convection in giant planets. *Geophys. Astrophys. Fluid Dyn.*, **98**, 241–255.

Fernandez, J.A. and W.-H. Ip (1984) Some dynamical aspects of the accretion of Uranus and Neptune: The exchange of orbital angular momentum with planetesimals. *Icarus*, **58**, 109–120.

Fegley, B. Jr. and K. Lodders (1994) Chemical models of the deep atmospheres of Jupiter and Saturn. *Icarus*, **110**, 117–154.

Feuchtgruber, H., E. Lellouch, T. de Graauw, B. Bezard, T. Encrenaz, and M. Griffin (1997) External supply of oxygen to the atmospheres of the giant planets. *Nature*, **389**, 159.

Feuchtgruber, H., E. Lellouch, B. Bézard, Th. Encrenaz, Th. de Graauw, and G.R. Davis (1999) Detection of HD in the atmospheres of Uranus and Neptune: A new determination of the D/H ratio. *Astron. Astrophys.*, **341**, L17–L21.

Fink, U. and H.P. Larson (1979) The infrared spectra of Uranus, Neptune, and Titan from 0.8 to 2.5 microns. *Astrophys. J.*, **233**, 1021–1040.

Finn, C.B.P. (1993) *Thermal Physics* (Second Edition). Chapman & Hall, London.

Fischer, G., M.D. Desch, P. Zarka, M.L. Kaiser, D.A. Gurnett, W.S. Kurth, W. Macher, H.O. Rucker, A. Lecacheux, W.M. Farrell *et al.* (2006) Saturn lightning recorded by Cassini/RPWS in 2004. *Icarus*, **183**, 135–152.

Fischer, G., W.S. Kurth, U.A. Dyudina, M.L. Kaiser, P. Zarka, A. Lecacheux, A.P. Ingersoll, and D.A. Gurnett (2007) Analysis of a giant lightning storm on Saturn. *Icarus*, **190**, 528–544.

Flasar, F.M. and P.J. Gierasch (1986) Mesoscale waves as a probe of Jupiter's deep atmosphere. *J. Atmos. Sci.*, **43**, 2638–2707.

Flasar, F.M., B.J. Conrath, P.J. Gierasch, and J.A. Pirraglia (1987) Voyager infrared observations of Uranus' atmosphere: Thermal structure and dynamics. *J. Geophys. Res.*, **92**, 15011–15018.

Flasar, F.M., V.G. Kunde, M.M. Abbas, R.K. Achterberg, P. Ade, A. Barucci, B. Bézard, G.L. Bjoraker, J. C. Brasunas, S. Calcutt *et al.* (2004a) Exploring the Saturn system in the thermal infrared: The Composite Infrared Spectrometer. *Space Sci. Rev.*, **115**, 169–297.

Flasar, F.M., V.G. Kunde, R.K. Achterberg, B.J. Conrath, A.A. Simon-Miller, C.A. Nixon, P.J. Gierasch, P.N. Romani, B. Bézard, P. Irwin *et al.* (2004b) An intense stratospheric jet on Jupiter. *Nature*, **427**, 132–135.

Flasar, F.M., R.K. Achterberg, B.J. Conrath, J.C. Pearl, G.L. Bjoraker, D.E. Jennings, P.N. Romani, A.A. Simon-Miller, V.G. Kunde, C.A. Nixon *et al.* (2005) Temperatures, winds, and composition in the Saturnian system. *Science*, **307**, 1247–1251.

Fletcher, L.N., P.G.J. Irwin, N.A. Teanby, G.S. Orton, P.D. Parrish, S.B. Calcutt, N. Bowles, R. de Kok, C. Howett, and F.W. Taylor (2007a). The meridional phosphine distribution in Saturn's upper troposphere from Cassini/CIRS observations. *Icarus*, **188**, 72–88.

Fletcher, L.N., P.G.J. Irwin, N.A. Teanby, G.S. Orton, P.D. Parrish, R. de Kok, C. Howett, S.B. Calcutt, N. Bowles, and F.W. Taylor (2007b) Characterising Saturn's vertical temperature structure from Cassini/CIRS. *Icarus*, **189**, 457–478.

Fletcher, L.N., P.G.J. Irwin, G.S. Orton, N.A. Teanby, R.K. Achterberg, G.L. Bjoraker, P.L. Read, A.A. Simon-Miller, C. Howett, R. de Kok *et al.* (2008a) Temperature and composition of Saturn's polar hot spots and Hexagon. *Science*, **319**, 79–81, doi: 10.1126/science.1149514.

Fletcher, L.N., G.S. Orton, N.A. Teanby, P.G.J. Irwin, and G.L. Bjoraker (2008b) Methane and its isotopologues on Saturn from Cassini/CIRS observations. *Icarus* (in preparation).

Flowers, B.H. and E. Mendoza (1970) *Properties of Matter*. John Wiley & Sons, Chichester, U.K.

Folkner, W.M., R. Woo, and D. Nandi (2008) Ammonia abundance in Jupiter's atmosphere derived from the attenuation of the Galileo probe's radio signal. *J. Geophys. Res.*, **103**, 22847–22856.

Fouchet, T., E. Lellouch, B. Bézard, T. Encrenaz, P. Drossart, H. Feuchtgruber, and T. de Graaw (2000) ISO-SWS observations of Jupiter: Measurement of the ammonia tropospheric profile and the $^{15}N/^{14}N$ isotopic ratio. *Icarus*, **143**, 223–243.

Fouchet, T., E. Lellouch, and H. Feuchtgruber (2003) The hydrogen ortho-to-para ratio in the stratospheres of the giant planets. *Icarus*, **161**, 127–143.

Fouchet, T., G. Orton, P.G.J. Irwin, S.B. Calcutt, and C.A. Nixon (2004a) Upper limits on hydrogen halides in Jupiter from Cassini/CIRS observations. *Icarus*, **170**, 237–241.

Fouchet, T., P.G.J. Irwin, P. Parrish, S.B. Calcutt, F.W. Taylor, C.A. Nixon, and T. Owen (2004b) Search for spatial variation in the Jovian $^{15}N/^{14}N$ ratio from Cassini/CIRS observations. *Icarus*, **172**, 50–58.

Fraser, H.J., M.R.S. McCoustra, and D.A. Williams (2002) The molecular universe. *Astronomy and Geophysics*, **43**, 210–218.

French, R.G., J.L. Elliot, L.M. French, J.A. Kangas, K.J. Meech, M.E. Ressler, M.W. Buie, J.A. Frogel, J.B. Holberg, J.J. Fuensalida *et al.* (1988) Uranian ring orbits from earth-based and Voyager occultation observations. *Icarus*, **73**, 349–378.

Friedson, A.J. (1999) New observations and modelling of a QBO-like oscillation in Jupiter's stratosphere. *Icarus*, **137**, 34–55.

Friedson, A.J., R.A. West, A.K. Hronek, N.A. Larsen, and N. Dalal (1999) Transport and mixing in Jupiter's stratosphere inferred from Comet S-L9 dust migration. *Icarus*, **138**, 141–156.

Fry, P.M. and L.A. Sromovsky (2004) Keck 2 AO observations of Neptune in 2003 and 2004. *Bull. Am. Astron. Soc.*, **36**, 1103.

Gautier, D., F. Hersant, O. Mousis, and J.I. Lunine (2001a) Enrichments in volatiles in Jupiter: A new interpretation of the Galileo measurements. *Astrophys. J.*, **550**, L227–L230.

Gautier, D., F. Hersant, O. Mousis, and J.I. Lunine (2001b) Erratum: Enrichments in volatiles in Jupiter. *Astrophys. J.*, **559**, L183.

Geiss, J. and G. Gloeckler (1998) Abundances of deuterium and helium-3 in the protosolar cloud. *Space Sci. Rev.*, **84**, 239–250.

Giampieri, G., M.K. Dougherty, E.J. Smith, and C.T. Russell (2006) A regular period for Saturn's magnetic field that may track its internal rotation. *Nature*, **441**, 62–64.

Gibbard, S.G., I. de Pater, H.G. Roe, S. Martin, B.A. Macintosh, and C.E. Max (2003) The altitude of Neptune cloud features from high spatial-resolution near-infrared spectra. *Icarus*, **166**, 359–374.

Gierasch, P.J. and B.J. Conrath (1993) Dynamics of the atmospheres of the outer planets: Post-Voyager measurement objectives. *J. Geophys. Res.*, **98**, 5459–5469.

Gierasch, P.J. and R.M. Goody (1969) Radiative time constants in the atmosphere of Jupiter. *J. Atmos. Sci.*, **26**, 979–980.

Gierasch, P.J., B.J. Conrath, and J.A. Magalhães (1986) Zonal mean properties of Jupiter's upper troposphere from Voyager infrared observations. *Icarus*, **67**, 456–483.

Gierasch, P.J., A.P. Ingersoll, D. Banfield, S.P. Ewald, P. Helfenstein, A. Simon-Miller, A. Vasavada, H.H. Breneman, D.A. Senske and the Galileo Imaging Team (2000) Observation of moist convection in Jupiter's atmosphere. *Nature*, **403**, 628–630.

Gierasch, P.J., B.J. Conrath, and P.L. Read (2004) Nonconservation of Ertel potential vorticity in hydrogen atmospheres. *J. Atmos. Sci.*, **61**, 1953–1965.

Gill, A.E. (1982) *Atmosphere–Ocean Dynamics*. Academic Press, San Diego, CA.

Go, C., I. de Pater, J. Rogers, G. Orton, P. Marcus, M.H. Wong, P.A. Yanamandra-Fischer, and J. Joels (2007) The global upheaval of Jupiter (American Astronomical Society, DPS meeting #39, #19.09).

Godfrey, D.A. (1988) A hexagonal feature around Saturn's north pole. *Icarus*, **76**, 335–356.

Godfrey, D.A. and V. Moore (1986) The Saturnian ribbon feature: A baroclinically unstable model. *Icarus*, **68**, 313–343.

Goody, R.M. and Y.L. Yung (1989) *Atmospheric Radiation: Theoretical Basis* (Second Edition). Oxford University Press, Oxford, U.K.

Goody, R.M., R. West, L. Chen, and D. Crisp (1989) The correlated-*k* method for radiation calculations in nonhomogeneous atmospheres. *J. Quant. Spect. Rad. Trans.*, **42**, 539–550.

Greathouse, T.K., J.H. Lacy, B. Bézard, J.I. Moses, C.A. Griffith, and M.J. Richter (2005) Meridional variations of temperature, C_2H_2 and C_2H_6 abundances in Saturn's stratosphere at southern summer solstice. *Icarus*, **177**, 18–31.

Greathouse, T.K., J.H. Lacy, B. Bézard, J. Moses, M.J. Richter, and C. Knez (2006) The first detection of propane on Saturn. *Icarus*, **181**, 266–271

Grevesse, N. and A.J. Sauval (1998) Standard solar composition. *Space Sci. Rev.*, **85**, 161–174.

Grevesse, N., M. Asplund, and A.J. Sauval (2007) The solar chemical composition. *Space Sci. Rev.*, **130**, 105–114.

Griffith, C.A., B. Bézard, T. Greathouse, E. Lellouch, J. Lacy, D. Kelly, and M.J. Richter (2004) Meridional transport of HCN from SL9 impacts on Jupiter. *Icarus*, **170**, 58–69.

Grote, E., F.H. Busse, and A. Tilgner (2000) Regular and chaotic spherical dynamos. *Phys. Earth Planet. Int.*, **117**, 259–272.

Guillot, T. (1999a) A comparison of the interiors of Jupiter and Saturn. *Planet. Spac. Sci.*, **47**, 1183–1200.

Guillot, T. (1999b) Interiors of giant planets inside and outside the Solar System. *Science*, **286**, 72–77.

Guillot, T., D.J. Stevenson, W.B. Hubbard, and D. Saumon (2004) The interior of Jupiter. In F. Bagenal, W. McKinnon, and T. Dowling (Eds.), *Jupiter: The Planet, Satellites and Magnetosphere*, pp. 35–57. Cambridge University Press, Cambridge, U.K.

Guilloteau, S., A. Dutrey, A. Marten, and D. Gautier (1993) CO in the troposphere of Neptune: Detection of the $J = 1$–0 line in absorption. *Astron. Astrophys.*, **279**, 661–667.

Guilloteau, T., A. Dutrey, and F. Gueth (1997) Disks and outflows as seen from the IRAM Interferometer. In B. Reithpurth and C. Bertout (Eds.), *Herbig–Haro Flows and the Birth of Low Mass Stars*, pp. 365–380. International Astronomical Union Symposium No. 182, Kluwer Academic, Dordrecht, the Netherlands.

Gurnett, D.A., W.S. Kurth, G.B. Hospodarsky, A.M. Persoon, T.F. Averkamp, B. Cecconi, A. Lecacheux, P. Zarka, P. Canu, N. Cornilleau-Wehrlin *et al.* (2005) Radio and plasma wave observations at Saturn from Cassini's approach and first orbit. *Science*, **307**, 1255–1259.

Haisch, K.E., E.A. Lada, and C.J. Lada (2001) Disk frequencies and lifetimes in young clusters. *Astrophys. J. Lett.*, **553**, 153.

Hammel, H.B. and G.W. Lockwood (2007) Long-term atmospheric variability on Uranus and Neptune. *Icarus*, **186**, 291–301.

Hammel, H.B., S.L. Lawson, J. Harrington, G.W. Lockwood, D.T. Thompson, and C. Swift (1992) An atmospheric outburst on Neptune from 1986 through 1989. *Icarus*, **99**, 363–367.

Hammel, H.B., G.W. Lockwood, J.R. Mills, and C.D. Barnet (1995) Hubble Space Telescope imaging of Neptune's cloud structure in 1994. *Science*, **268**, 1740–1742.

Hammel, H.B., K. Rages, G.W. Lockwood, E. Karkoschka, and I. de Pater (2001) New measurements of the winds on Uranus. *Icarus*, **153**, 229–235.

Hammel, H.B., I. de Pater, S. Gibbard, G.W. Lockwood, and K. Rages (2005a) Uranus in 2003: Zonal winds, banded structure, and discrete features. *Icarus*, **175**, 534–545.

Hammel, H.B., I. de Pater, S. Gibbard, G.W. Lockwood, and K. Rages (2005b) New cloud activity on Uranus in 2004: First detection of a southern feature at 2.2 μm. *Icarus*, **175**, 284–288.

Hammel, H.B., D.K. Lynch, R.W. Russell, M.L. Sitko, L.S. Bernstein, and T. Hewagama (2006) Mid-infrared ethane emission on Neptune and Uranus. *Astrophys. J.*, **644**, 1326–1333.

Hammel, H.B., M.L. Sitko, D.K. Lynch, G.S. Orton, R.W. Russell, T.R. Geballe, and I. de Pater (2007) Distribution of ethane and methane emission on Neptune. *Astron. J.*, **134**, 637–641.

Hanel, R.A., B. Conrath, F.M. Flasar, V. Kunde, P. Lowman, W. Maguire, J. Pearl, J. Pirraglia, R. Samuelson, D. Gautier *et al.* (1979a) Infrared observations of the Jovian system from Voyager 1. *Science*, **204**, 972–976

Hanel., R.A., B. Conrath, F.M. Flasar, L. Herath, V. Kunde, P. Lowman, W. Maguire, J. Pearl, J. Pirraglia, and L. Horn (1979b) Infrared observations of the Jovian System from Voyager 2. *Science*, **206**, 952–956.

Hanel., R.A., B. Conrath, F.M. Flasar, V. Kunde, W. Maguire, J.C. Pearl, J. Pirraglia, R. Samuelson, L. Herath, M. Allison *et al.* (1981) Infrared observations of the Saturnian system from Voyager 1. *Science*, **212**, 192–200.

Hanel., R.A., B. Conrath, F.M. Flasar, V. Kunde, W. Maguire, J.C. Pearl, J. Pirraglia, R. Samuelson, D.P. Cruikshank, D. Gautier *et al.* (1982) Infrared observations of the Saturnian system from Voyager 2. *Science*, **215**, 544–548.

Hanel, R.A., B. Conrath, F.M. Flasar, V. Kunde, W. Maguire, J.C. Pearl, J. Pirraglia, R. Samuelson, L. Horn, and P. Schulte (1986) Infrared observations of the Uranian system. *Science*, **233**, 70–74.

Hanel, R.A., B.J. Conrath, D.E. Jennings, and R.E. Samuelson (2003) *Exploration of the Solar System by Infrared Remote Sensing* (Second Edition). Cambridge University Press, Cambridge, U.K.

Hansen, J.E. and L.D. Travis (1974) Light scattering in planetary atmospheres. *Space Sci. Rev.*, **16**, 527–610.

Harrington, J., T.E. Dowling and R.L. Baron (1996) Jupiter's tropospheric thermal emission, II: Power spectrum analysis and wave search. *Icarus*, **124**, 32–44.

Hart, J.E., J. Toomre, A.E. Deane, N.E. Hurlbert, G.A. Glatzmaier, G.H. Fichtl, F. Leslie, W.W. Fowlis, and P.A. Gilman (1986) Laboratory experiments on planetary and stellar convection performed on Spacelab 3. *Science*, **234**, 61–64.

Hartmann, L., N. Calvet, E. Gullbring, and P. D'Alessio (1998) Accretion and the evolution of T Tauri disks. *Astrophys. J.*, **495**, 385–400.

Hashizume, K., M. Chaussidon, B. Marty, and F. Robert (2000) Solar wind record on the Moon: Deciphering presolar from planetary nitrogen. *Science*, **290**, 1142–1145.

Hecht, E. (1998) *Optics* (Third Edition). Addison-Wesley, Reading, MA.

Hecht, E. and A. Zajac (1974) *Optics*. Addison-Wesley, Reading, MA.

Heimpel, M. and J. Aurnou (2007) Turbulent convection in rapidly rotating spherical shells: A model for equatorial and high latitude jets on Jupiter and Saturn. *Icarus*, **187**, 540–557, doi: 10.1016/j.icarus.2006.10.023.

Heimpel, M., J. Aurnou, and J. Wicht (2005) Simulation of equatorial and high-latitude jets on Jupiter in a deep convection model. *Nature*, **438**, 193–196, doi: 10.1038/nature04208.

Hersant, F., D. Gautier, and J.-M. Huré (2001) A two-dimensional model of the primordial nebula constrained by D/H measurements in the solar system: Implications for the formation of the giant planets. *Astrophys. J.*, **554**, 391–407.

Hersant, F., D. Gautier, and J. Lunine (2004) Enrichment in volatiles in the giant planets of the Solar System. *Planet. Space Sci.*, **52**, 623–641.

Herzberg, G. (1945) *Molecular Spectra and Molecular Structure, II: Infrared and Raman Spectra of Polyatomic Molecules*. Van Nostrand Reinhold, New York.

Hesman, B.E., G.R. Davis, H.E. Matthews, and G.S. Orton (2007) The abundance profile of CO in Neptune's atmosphere. *Icarus*, **186**, 342–353.

Hofstadter, M.D. and B.J. Butler (2003) Seasonal change in the deep atmosphere of Uranus. *Icarus*, **165**, 168–180.

Holton, J.R. (1992) *An Introduction to Dynamical Meteorology* (Third Edition). Academic Press, London.

Hooke, R. (1665) A spot in one of the belts of Jupiter. *Philos. Trans. R. Soc. London*, **1**, 3.

Houghton, J.T. (1986) *The Physics of Atmospheres* (Second Edition). Cambridge University Press, Cambridge, U.K.

Houghton, J.T. and S.D. Smith (1966) *Infrared Physics*. Oxford University Press, Oxford, U.K.

Howett, C.J.A. P.G.J. Irwin, N.A. Teanby, A. Simon-Miller, S.B. Calcutt, L.N. Fletcher, and R. de Kok (2007) Meridional variations in stratospheric acetylene and ethane in the Southern Hemisphere of the Saturnian atmosphere as determined from Cassini/CIRS measurements. *Icarus*, **190**, 556–572.

Huang, H.-P. and W.A. Robinson (1998) Two-dimensional turbulence and persistent zonal jets in a global barotropic model. *J. Atmos. Sci.*, **55**, 611–632

Hubbard, W.B. (1984) Interior structure of Uranus. *Uranus and Neptune, NASA-JPL CP*, **2330**, 291–325.

Hubbard, W.B. (1997a) *Jupiter: Interior Structure. Encyclopaedia of the Planetary Sciences* (edited by J.H. Shirley and R.W. Fairbridge). Chapman & Hall, London.

Hubbard, W.B. (1997b) *Saturn: Interior Structure. Encyclopaedia of the Planetary Sciences* (edited by J.H. Shirley and R.W. Fairbridge). Chapman & Hall, London.

Hubbard, W.B. (1997c) *Uranus: Interior Structure. Encyclopaedia of the Planetary Sciences* (edited by J.H. Shirley and R.W. Fairbridge). Chapman & Hall, London.

Hubbard, W.B. (1997d) *Neptune: Interior Structure. Encyclopaedia of the Planetary Sciences* (edited by J.H. Shirley and R.W. Fairbridge). Chapman & Hall, London.

Hubbard, W.B. (1999) Gravitational signature of Jupiter's deep zonal flows. *Icarus*, **137**, 357–359

Hubickyj, O., P. Bodenheimer, and J.J. Lissauer (2005) Accretion of the gaseous envelope of Jupiter around a 5–10 Earth-mass core. *Icarus*, **179**, 415–431.

Hueso, R. and T. Guillot (2003) Evolution of the protosolar nebula and formation of the giant planets. *Space Sci. Rev.*, **106**, 105–120.

Hueso, R. and A. Sánchez-Lavega (2004) A three-dimensional model of moist convection for the giant planets, II: Saturn's water and ammonia moist convective storms. *Icarus*, **172**, 255–271.

Huré, J.-M., D. Richard, and J.-P.Zahn (2001) Accretion discs models with the β-viscosity prescription derived from laboratory experiments. *Astron. Astrophys.*, **367**, 1087–1094.

Ida, S., G. Bryden, D.N.C. Lin, and H. Tanaka (2000) Orbital migration of Neptune and orbital distribution of trans-neptunian objects. *Astrophys. J.*, **534**, 428–445.

Ingersoll, A.P. (1990) Atmospheric dynamics of the outer planets. *Science*, **248**, 308–315.

Ingersoll, A.P. and D. Pollard (1982) Models of the interiors and atmospheres of Jupiter and Saturn: Scale analysis, anelastic equations, barotropic stability criterion. *Icarus*, **52**, 61–80.

Ingersoll, A.P., R.F. Beebe, J.L. Mitchell, G.W. Garneau, G.M. Yagi, and J.P. Müller (1981) Interaction of eddies and mean zonal flow on Jupiter as inferred from Voyager 1 and 2 images. *J. Geophys. Res.*, **86**, 8733–8743.

Ingersoll, A. P., R. F. Beebe, B. J. Conrath, and G. E. Hunt (1984) Structure and dynamics of Saturn's atmosphere. In T. Gehrels and M. S. Matthews (Eds.), *Saturn* (pp. 195–238). University of Arizona Press, Tucson, AZ.

Ingersoll, A.P., P.J. Gierasch, D. Banfield, A.R. Vasavada, and the Galileo Imaging Team (2000) Moist convection as an energy source for the large-scale motions in Jupiter's atmosphere. *Nature*, **403**, 630–632.

Ingersoll, A.P., T.E. Dowling, P.J. Gierasch, G.S. Orton, P.L. Read, A. Sánchez-Lavega, A.P. Showman, A.A. Simon-Miller, and A.R. Vasavada (2004) Dynamics of Jupiter's atmosphere. In F. Bagenal, W. McKinnon, and T. Dowling (Eds.), *Jupiter: The Planet, Satellites and Magnetosphere*, pp 105–128. Cambridge University Press, Cambridge, U.K.

Irion, R. (2002) California astronomers eye 30-meter telescope. *Science*, **298**, 1151–1152.

Iro, N., D. Gautier, F. Hersant, D. Bockelée-Morvan, and J.I. Lunine (2003) An interpretation of the nitrogen deficiency in comets. *Icarus*, **161**, 513.

Irwin, P.G.J. (1999) Cloud structure and composition of Jupiter's atmosphere. *Surveys in Geophysics*, **20**, 505–535.

Irwin, P.G.J. (2008) Detection methods and properties of known exoplanets. *Exoplanets: Detection, Formation, Properties, Habitability*, pp. 1–20. Springer/Praxis, Heidelberg, Germany/Chichester, U.K.

Irwin, P.G.J. and U. Dyudina (2002) The retrieval of cloud structure maps in the equatorial region of Jupiter using a principal component analysis of Galileo/NIMS data. *Icarus*, **156**, 52–63.

Irwin, P.G.J., S.B. Calcutt, F.W. Taylor, and A.L. Weir (1996) Calculated *k* coefficients for hydrogen- and self-broadened methane in the range 2000–9500 cm^{-1} from exponential sum fitting to band modelled spectra. *J. Geophys. Res.*, **101**, 26137–26154,

Irwin, P.G.J., A.L. Weir, S.E. Smith, F.W. Taylor, A.L. Lambert, S.B. Calcutt, P.J. Cameron-Smith, R.W. Carlson, K. Baines, G.S. Orton *et al.* (1998) Cloud structure and atmospheric composition of Jupiter retrieved from Galileo near-infrared mapping spectrometer real-time spectra. *J. Geophys. Res.*, **103**, 23001–23021.

Irwin, P.G.J., A.L. Weir, F.W. Taylor, S.B. Calcutt, and R.W. Carlson (2001) The origin of belt/zone contrasts in the atmosphere of Jupiter and their correlation with 5-micron opacity. *Icarus*, **149**, 397–415.

Irwin, P.G.J., P. Parrish, T. Fouchet, S.B. Calcutt, F.W. Taylor, A.A. Simon-Miller, and C.A. Nixon (2004) Retrievals of Jovian tropospheric phosphine from Cassini/CIRS. *Icarus*, **172**, 37–49.

Irwin, P.G.J., K. Sihra, N. Bowles, F.W. Taylor, and S.B. Calcutt (2005) Methane absorption in the atmosphere of Jupiter from 1800 and 9500 cm^{-1} and implications for vertical cloud structure. *Icarus*, **176**, 255–271.

Irwin, P.G.J., N.A. Teanby, and G. R. Davis (2007) Latitudinal variations in Uranus' vertical cloud structure from UKIRT UIST observations. *Astrophys. J.*, **665**, L71–L74.

Irwin, P.G.J., N. Teanby, R. de Kok, L. Fletcher, C. Howett, C. Tsang, C. Wilson, S. Calcutt, C. Nixon, and P. Parrish (2008). The NEMESIS planetary atmosphere radiative transfer and retrieval tool. *J. Quant. Spectrosc. Rad. Trans.*, **109**, 1136–1150.

IUPAC (International Union for Pure and Applied Chemistry) (1991) Isotopic compositions of the elements. *Pure Appl. Chem.*, **63**, 991–1002.

Jacquinet-Husson, N., N.A. Scott, A. Chédin, K. Garceran, R. Armante, A.A. Chursin, A. Barbe, M. Birk, L.R. Brown, C. Camy-Peyret *et al.* (2005) The 2003 edition of the GEISA/IASI spectroscopic database. *J. Quant. Spectrosc. Rad. Trans.*, **95**, 429–467.

James, J.F. and R.S. Stern (1969) *The Design of Optical Spectrometers*. Chapman & Hall, London.

Jewitt, D.C., H.E. Matthews, T. Owen, and R. Meier (1997) Measurement of $^{12}C/^{13}C$, $^{14}N/^{15}N$, $^{32}S/^{34}S$ in Comet Hale-Bopp (C/1995 O1). *Science*, **278**, 90–93

Jones, B.W. (1999) *Discovering the Solar System*. John Wiley & Sons, Chichester, U.K.

Jones, B.W. (2007) *Discovering the Solar System* (Second Edition). John Wiley & Sons, Chichester, U.K.

Kalogerakis, K.S., J. Marschall, A.U. Oza, P.A. Engel, R.T. Meharchand, and M.H. Wong (2008) The coating hypothesis for ammonia ice particles in Jupiter: Laboratory experiments and optical modelling. *Icarus*, **196**, 202–215.

Karkoschka, E. (1994) Spectrophotometry of the Jovian planets and Titan at 300- to 1000-nm wavelength: The methane spectrum. *Icarus*, **111**, 174–192.

Karkoschka, E. (1998a) Methane, ammonia, and temperature measurements of the Jovian planets and Titan from CCD-spectrophotometry. *Icarus*, **133**, 134–146.

Karkoschka, E. (1998b) Clouds of high contrast on Uranus. *Science*, **280**, 570–572.

Karkoschka, E. (2001) Uranus' apparent seasonal variability in 25 HST filters. *Icarus*, **151**, 84–92.

Karkoschka, E. and M.G. Tomasko (1992) Saturn's upper troposphere 1986–1989. *Icarus*, **97**, 161–181.

Karkoschka, E. and M.G. Tomasko (1993) Saturn's upper atmospheric hazes observed by the Hubble Space Telescope. *Icarus*, **10**, 428–441.

Karkoschka, E. and M. Tomasko (2005) Saturn's vertical and latitudinal cloud structure 1991–2004 from HST imaging in 30 filters. *Icarus*, **179**, 195–221.

Kerola, D.X., H.P. Larson, and M.G. Tomasko (1997) Analysis of the near-IR spectrum of Saturn: A comprehensive radiative transfer model of its middle and upper troposphere. *Icarus*, **127**, 190–212.

Kim, J.H., S.J. Kim, T.R. Geballe, S.S. Kim, and L.R. Brown (2006) High resolution spectroscopy of Saturn at 3 microns: CH_4, CH_3D, C_2H_2, C_2H_6, PH_3, clouds, and haze. *Icarus*, **185**, 476–486.

Klein, M.J. and M.D. Hofstadter (2006) Long-term variations in the microwave brightness temperature of the Uranus atmosphere. *Icarus*, **184**, 170–180.

Krane, K. (1996) *Modern Physics* (Second Edition). John Wiley & Sons, Chichester, U.K.

Kunde, V., R. Hanel, W. Maguire, D. Gautier, J.P. Baluteau, A. Marten, A. Chedin, N. Husson, and N. Scott (1982) The tropospheric gas composition of Jupiter's North Equatorial Belt (NH_3, PH_3, CH_3D, GeH_4, H_2O) and the Jovian D/H ratio. *Astrophys. J.*, **263**, 443–467.

Kunde, V.G., F.M. Flasar, D.E. Jennings, B. Bézard, D.F. Strobel, B.J. Conrath, C.A. Nixon, G.L. Bjoraker, P.N. Romani, R.K. Achterberg *et al.* (2004) Jupiter's atmospheric composition from the Cassini thermal infrared spectroscopy experiment. *Science*, **305**, 1582–1587.

Lacis, A.A. and V Oinas (1991) A description of the correlated-*k* distribution method for modelling nongray gaseous absorption, thermal emission, and multiple scattering in vertically inhomogeneous atmospheres. *J. Geophys. Res.*, **96**, 9027–9063.

LeBeau, R.P. and T.E. Dowling (1998) EPIC simulations of time-dependent three-dimensional vortices with application to Neptune's Great Dark Spot. *Icarus*, **132**, 239–265.

Lécuyer, C., Ph. Gillet, and F. Robert (1998) The hydrogen isotope composition of sea water and the global water cycle. *Chem. Geol.*, **145**, 249–261.

Lellouch, E., P. Romani, and J. Rosenqvist (1994) The vertical distribution and origin of HCN in Neptune's atmosphere. *Icarus*, **108**, 112–136.

Lellouch, E., H. Feuchtgruber, Th. de Graauw, B. Bézard and M. Griffin (1997) *ESA-SP*, **419**, 131. ESA, Noordwijk, The Netherlands

Lellouch, E., J. Crovisier, T. Lim, D. Bockelée-Morvan, K. Leech, M.S. Hanner, B. Altieri, B. Schmitt, F. Trotta, and H.U. Keller (1998) Evidence for water ice and estimate of dust production rate in comet Hale-Bopp at 2.9 AU from the Sun. *Astron. Astrophys.*, **339**, L9–L12.

Lellouch, E., B. Bézard, T. Fouchet, H. Feuchtgruber, T. Encrenaz, and T. de Graauw (2001) The deuterium abundance in Jupiter and Saturn from ISO/SWS observations. *Astron. Astrophys.*, **670**, 610–622.

Lellouch, E., B. Bézard, J.I. Moses, G.R. Davis, P. Drossart, H. Feuchtgruber, E.A. Bergin, R. Moreno, and T. Encrenaz (2002) The origin of water vapor and carbon dioxide in Jupiter's stratosphere. *Icarus*, **159**, 112–131.

Lellouch, E., R. Moreno, and G. Paubert (2005) A dual origin for Neptune's carbon monoxide. *Astron. Astrophys.*, **430**, L37–L40.

Leovy, C. (1986) Eddy processes in the general circulation of the Jovian atmospheres. *NASA GISS: The Jovian Atmospheres*, pp. 177–196.

Leovy, C.B., A.J. Friedson, and G.S. Orton (1991) The quasiquadrennial oscillation of Jupiter's equatorial stratosphere. *Nature*, **354**, 380–382.

Lewis, J.S. (1995) *Physics and Chemistry of the Solar System*. Academic Press, London.

Lewis, S.R. (1988) Long-lived eddies in the atmosphere of Jupiter, D.Phil. thesis, University of Oxford.

Li, X. and P.L. Read (2000) A mechanistic model of the quasi-quadrennial oscillation of Jupiter's stratosphere. *Planet. Space Sci.*, **48**, 637–669.

Limaye, S.S. (1986) Jupiter: New estimates of the mean zonal flow at the cloud level. *Icarus*, **65**, 335.

Lin, D.N.C. and J.C.B. Papaloizou (1986a) On the tidal interaction between protoplanets and the protoplanetary disk, III: Orbital migration of protoplanets. *Astrophys. J.*, **309**, 846.

Lin, D.N.C. and J.C.B. Papaloizou (1986b) On the tidal interaction between protoplanets and the primordial solar nebula, II: Self-consistent nonlinear interaction. *Astrophys. J.*, **307**, 395.

Lindal, G.F. (1992) The atmosphere of Neptune: An analysis of radio occultation data acquired with Voyager 2. *Astron. J.*, **103**, 967–982.

Lindal, G.F., G.E. Wood, G.S. Levy, J.D. Anderson, D.N. Sweetman, H.B. Hotz, B.J. Buckles, D.P. Holmes, P.E. Doms, V.R. Eshleman *et al.* (1981) The atmosphere of Jupiter: An analysis of the Voyager radio occultation measurements. *J. Geophys. Res.*, **86**, 8721–8727.

Lindal, G.F., D.N. Sweetnam, and V.R. Eshleman (1985) The atmosphere of Saturn: An analysis of the Voyager radio occultation measurements. *Astron. J.*, **90**, 1136–1146.

Lindal, G.F., J.R. Lyons, D.N. Sweetnam, V.R. Eshleman, D.P. Hinson, and G.L. Tyler (1987) The atmosphere of Uranus: Results of radio occultation measurements with Voyager 2. *J. Geophys. Res.*, **92**, 14987–15001.

Lindal, G.F. (1992) The atmosphere of Neptune: An analysis of radio occultation data acquired with Voyager 2. *Astron. J.*, **103**, 967–982.

Little, B., C.D. Anger, A.P. Ingersoll, A.R. Vasavada, D.A. Senske, H.H. Breneman, W.J. Borucki, and the Galileo SSI Team (1999) Galileo images of lightning on Jupiter. *Icarus*, **142**, 306–323.

Lockwood, G. and M. Jerzykiewicz (2006) Long-term atmospheric variability on Uranus and Neptune. *Icarus*, **180**, 442–452.

Lockwood, G. and D. Thompson (2002) Photometric variability of Neptune, 1972–2000. *Icarus*, **156**, 37–51.

Lockwood, G.W., D.T. Thompson, B.L. Lutz, and E.S Howell (1991) The brightness, albedo, and temporal variability of Neptune. *Astrophys. J.*, **368**, 287–297.

Lodders, K. and B. Fegley Jr, (1994) The origin of carbon monoxide in Neptune's atmosphere. *Icarus*, **112**, 368-375.

Lopez-Puertas, M. and F.W. Taylor (2001) *Non-LTE Radiative Transfer in the Atmosphere* (Third Edition). World Scientific, Singapore.

Magalhães, J.A., A.L. Weir, B.J. Conrath, P.J. Gierasch, and S.S. Leroy (1989) Slowly moving features on Jupiter. *Nature*, **337**, 444–447.

Magalhães, J.A., A.L. Weir, B.J. Conrath, P.J. Gierasch, and S.S. Leroy (1990) Zonal motion and structure in Jupiter's upper troposphere from Voyager infrared and imaging observations. *Icarus*, **88**, 39–72.

Mahaffy, P.R., H.B. Niemann, A. Alpert, S.K. Atreya, J. Demick, T.M. Donahue, D.N. Harpold, and T.C. Owen (2000) Noble gas abundance and isotope ratios in the atmosphere of Jupiter from the Galileo Probe Mass Spectrometer. *J. Geophys. Res.*, **105**, 15061–15072.

Malfait, K., C. Waelkens, J. Bouwman, A. de Koter, and L.B.F.M. Waters (1999) The ISO spectrum of the young star HD 142527. *Astron. Astrophys.*, **345**, 181.

Marcy, G.W. and R.P. Butler (1996) A planetary companion to 70 Virginis. *Astrophys. J. Lett.*, **464**, L147.

Marcy, G.W. and R.P. Butler (1998) Detection of extrasolar giant planets. *Ann. Rev. Astron. Astrophys.*, **36**, 57–98.

Marten, A., R. Courtin, D. Gautier, and A. Lacombe (1980) Ammonia vertical density profiles in Jupiter and Saturn from their radioelectric and infrared emissivities. *Icarus*, **41**, 410–422.

Marten, A., D. Gautier, T. Owen, D.B. Sanders, H.E. Matthews, S.K. Atreya, R.P.J. Tilanus, and J.R. Deane (1993) First observations of CO and HCN on Neptune and Uranus at millimeter wavelengths and their implications for atmospheric chemistry. *Astrophys. J.*, **406**, 285–297.

Marten, A., H.E. Matthews, T. Owen, R. Moreno, T. Hidayat, and Y. Biraud (2005) Improved constraints on Neptune's atmosphere from submillimetre wavelength observations. *Astron. Astrophys.*, **429**, 1097–1105.

Masset, F.S. and J.C.B. Papaloizou (2003) Runaway migration and the formation of Hot Jupiters. *Astrophys. J.*, **588**, 494

Massie, S.T. and D.M. Hunten (1982) Conversion of para- and ortho-hydrogen in the Jovian planets. *Icarus*, **49**, 213–226.

Mason, J.W. (2008) *Exoplanets: Detection, Formation, Properties, Habitability*. Springer/Praxis, Heidelberg, Germany/Chichester, U.K.

Matcheva, K.I. and D.F. Strobel (1999) Heating of Jupiter's thermosphere by dissipation of gravity waves due to molecular viscosity and heat conduction. *Icarus*, **140**, 328–340.

Matcheva, K.I., B.J. Conrath, P.J. Gierasch, and F.M. Flasar (2005) The cloud structure of the jovian atmosphere as seen by the Cassini/CIRS experiment. *Icarus*, **179**, 432–448.

Max, C.E., B.A. Macintosh, S.G. Gibbard, D.T. Gavel, H.G. Roe, I. de Pater, A.M. Ghez, D.S. Acton, O. Lai, P. Stomski *et al.* (2003) Cloud structures on Neptune observed with Keck telescope adaptive optics. *Astron. J.*, **125**, 364–375.

Mayer, L., T. Quinn, J. Wadsley, and J. Stadel (2002) Formation of giant planets by fragmentation of protoplanetary disks. *Science*, **298**, 1756–1759.

Mayor, M. and D. Queloz (1995) A Jupiter-mass companion to a solar-type star. *Nature*, **378**, 355.

McCaughrean, M.J and C.R. O'Dell (1996) Direct imaging of circumstellar disks in the Orion Nebula. *Astron. J.*, **111**, 1977.

McCaughrean, M.J., J.T. Rayner, and H. Zinnecker (1994) Discovery of a molecular hydrogen jet near IC 348. *Astrophys. J.*, **436**, L189-L192.

McCaughrean, M.J, H. Chen, J. Bally, E. Erickson, R. Thompson, M. Rieke, G. Schneider, S. Stolovy, and E. Young (1998) High-resolution near-infrared imaging of the Orion 114-426 silhouette disk. *Astrophys. J.*, **492**, 157.

Meier, R. and T. Owen (1999) Cometary deuterium. *Space Sci. Rev.*, **90**, 33–44.

Meier, R., T. Owen, H.E. Matthews, D.C. Jewitt, D. Bockelée-Morvan, N. Biver, J. Crovisier, and D. Gautier (1998) A determination of the HDO/H_2O ratio in Comet C/1995 O1 (Hale–Bopp). *Science*, **279**, 842.

Minnaert. M. (1941) The reciprocity principle in lunar photometry. *Astrophys. J.*, **93**, 403–410.

Mishchenko, M.I., L.D. Travis, and D.W. Mackowski (1996) T-matrix computations of light scattering by nonspherical particles: A review. *J. Quant. Spectrosc. Rad. Trans.*, **55**, 535–575.

Mishchenko, M.I., L.D. Travis, and A.A. Lacis (2006) *Multiple Scattering of Light by Particles: Radiative Transfer and Coherent Backscattering*. Cambridge University Press, Cambridge, U.K.

Mizuno, H. (1980) Formation of the giant planets. *Progress of Theoretical Physics*, **64**, 544–557.

Moreno, F. and J. Sedano (1997) Radiative balance and dynamics in the stratosphere of Jupiter: Results from a latitude-dependent aerosol heating model. *Icarus*, **130**, 36–48.

Moreno, R., A. Marten, H.E. Matthews, and Y. Biraud (2003) Long-term evolution of CO, CS and HCN in Jupiter after the impacts of comet Shoemaker-Levy 9. *Planet. Space Sci.*, **51**, 591–611.

Moses, J.I., B. Bézard, E. Lellouch, G.R. Gladstone, H. Feuchtgruber, and M.Allen (2000) Photochemistry of Saturn's atmosphere, I: Hydrocarbon chemistry and comparisons with ISO observations. *Icarus*, **143**, 244–298.

Moses, J.I., T. Fouchet, R.V. Yelle, A.J. Friedson, G.S. Orton, B. Bézard, P. Drossart, G.R. Gladstone, T. Kostiuk, and T.A. Livengood (2004) The stratosphere of Jupiter. In F. Bagenal, W. McKinnon, and T. Dowling (Eds.), *Jupiter: The Planet, Satellites and Magnetosphere*, pp. 129–157. Cambridge University Press, Cambridge, U.K.

Mousis, O., D. Gautier, D. Bockelée-Morvan, F. Robert, B. Dubrulle, and A. Drouart (2000) Constraints on the formation of comets from D/H ratios measured in H_2O and HCN. *Icarus*, **148**, 513–525.

Muñoz, O., F. Moreno, A. Molina, D. Grodent, J.C. Gérard, and V. Dols (2005) Study of the vertical structure of Saturn's atmosphere using HST/WFPC2 images. *Icarus*, **169**, 413–428.

Naylor, D.A., G.R. Davis, M.J. Griffin, T.A. Clark, D. Gautier, and A. Marten (1994) Broad-band spectroscopic detection of the CO $J = 3$–2 tropospheric absorption in the atmosphere of Neptune. *Astron. Astrophys.*, **291**, L51–L53.

Nellis, W.J. (2000) Metallization of fluid hydrogen at 140 GPa (1.4 Mbar): Implications for Jupiter. *Planet. Space Sci.*, **48**, 671–677.

Nellis, W.J., M. Ross, and N.C. Holmes (1995) Temperature measurements of shock-compressed liquid hydrogen: Implications for the interior of Jupiter. *Science*, **269**, 1249–1252.

Niemann, H.B., S.K. Atreya, G.R. Carignan, T.M. Donahue, J.A. Haberman, D.N. Harpold, R.E. Hartle, D.M. Hunten, W.T. Kasprzak, P.R. Mahaffy *et al.* (1996) The Galileo Probe Mass Spectrometer: Composition of Jupiter's atmosphere. *Science*, **272**, 846–849.

Niemann, H.B., S.K. Atreya, G.R. Carignan, T.M. Donahue, J.A. Haberman, D.N. Harpold, R.E. Hartle, D.M. Hunten, W.T. Kasprzak, P.R. Mahaffy *et al.* (1998) The composition of the Jovian atmosphere as determined by the Galileo probe mass spectrometer. *J. Geophys. Res.*, **103**, 22831–22845.

Nixon, C.A., P.G.J. Irwin, S.B. Calcutt, F.W. Taylor, and R.W. Carlson (2001) Atmospheric composition and cloud structure on Jovian 5-micron hotspots from analysis of Galileo NIMS measurements. *Icarus*, **150**, 48–68.

Nixon, C.A., R.K. Achterberg, B.J. Conrath, P.G.J. Irwin, N.A. Teanby, T. Fouchet, P.D. Parrish, P.N. Romani, M. Abbas, A. LeClair *et al.* (2007) Meridional variations of C_2H_2 and C_2H_6 in Jupiter's atmosphere from Cassini CIRS infrared spectra. *Icarus*, **188**, 47–71.

Noll, K.S. and H.P. Larson (1991) The spectrum of Saturn from 1990–2230 cm^{-1}: Abundances of AsH_3, CH_3D, CO, GeH_4, and PH_3. *Icarus*, **89**, 168–189.

Noll, K.S., R.F. Knacke, T.R. Geballe, and A.T. Tokunaga (1988) The origin and vertical distribution of carbon monoxide in Jupiter. *Astrophys. J.*, **324**, 1210–1218.

Noll, K.S., T.R. Geballe, and R.F. Knacke (1989) Arsine in Saturn and Jupiter. *Astrophys. J.*, **338**, L71–L74.

Noll, K.S., H.P. Larson, and T.R. Geballe (1990) The abundance of AsH_3 in Jupiter. *Icarus*, **83**, 494–499.

Noll, K.S., D. Gilmore, R.F. Knacke, M. Womack, C.A. Griffith, and G. Orton (1997) Carbon monoxide in Jupiter after Comet Shoemaker-Levy 9. *Icarus*, **126**, 324–335.

Ortiz, J.L., F. Moreno, and A. Molina (1993) Absolutely calibrated CCD images of Saturn at methane band and continuum wavelengths during its 1991 opposition. *J. Geophy. Res.*, **98**, 3053–3063.

Ortiz, J.L., F. Moreno, and A. Molina (1995) Saturn 1991–1993: Reflectivities and limb-darkening coefficients at methane bands and nearby continua–temporal changes. *Icarus*, **117**, 328–344.

Ortiz, J.L., F. Moreno, and A. Molina (1999) Saturn 1991–1993: Clouds and hazes. *Icarus*, **119**, 53–66.

Ortiz, J.L., G.S. Orton, A.J. Friedson, S.T. Stewart. B.M. Fisher, and J.R. Spencer (1998) Evolution and persistence of 5-μm hot spots at the Galileo Probe entry latitude. *J. Geophys. Res.*, **103**, 23051–23069.

Orton, G.S. and C.D. Kaminski (1989) An exploratory 5-micron spectrum of Uranus. *Icarus*, **77**, 109–117.

Orton, G.S. and P.A. Yanamandra-Fisher (2005) Saturn's temperature field from high-resolution middle-infrared imaging. *Science*, **307**, 696–698.

Orton, G. S., D.K. Aitken, C. Smith, P.F. Roche, J. Caldwell, and R. Snyder (1987) The spectra of Uranus and Neptune at 8–14 and 17–23 microns. *Icarus*, **70**, 1–12.

Orton, G.S., A.J. Friedson, J. Caldwell, H.B. Hammel, K.H. Baines, J.T. Bergstralh, T.Z. Martin, M.E. Malcom, R.A. West, W.F. Golisch *et al.* (1991) Thermal maps of Jupiter: Spatial organization and time dependence of stratospheric temperatures, 1980 to 1990. *Science*, **252**, 537–542.

Orton, G.S., J.H. Lacy, J.M. Achtermann, P. Parmar, and W.E. Blass (1992) Thermal ~spectroscopy of Neptune: The stratospheric temperature, hydrocarbon abundances, and isotopic ratios. *Icarus*, **100**, 541–555.

Orton, G.S., A.J. Friedson, P. Yanamandra-Fisher, J. Caldwell, H.B. Hammel, K.H. Baines, J.T. Bergstralh, T.Z. Martin, R.A. West, G.J. Veeder Jr. *et al.* (1994) Spatial organization and time dependence of Jupiter's tropospheric temperatures, 1980–1993. *Science*, **265**, 625–631.

Orton, G., E. Serabyn, and Y. Lee (2000, Corrigendum 2001) Vertical distribution of PH_3 in Saturn from observations of its 1–0 and 3–2 rotational lines. *Icarus*, **146**, 48–59. [Corrigendum, *Icarus*, **149**, 489-490, 2001.]

Orton, G.S., T. Encrenaz, C. Leyrat, R. Puetter, and A.J. Friedson (2007a) Evidence for methane escape and strong seasonal and dynamical perturbations of Neptune's atmospheric temperatures. *Astron. Astrophys.*, **473**, L5–L8, doi:10.1051/0004-6361:20078277.

Orton, G.S., M. Gustafsson, M. Burgdorf, and V. Meadows (2007b) Revised ab initio models for H_2–H_2 collision-induced absorption at low temperatures. *Icarus*, **189**, 544–549.

Orton, G.S., H. Hofstadter, C. Leyrat, and T. Encrenaz (2007c) Spatially resolved thermal imaging and spectroscopy of Uranus and Neptune. *Workshop on Planetary Atmospheres, November 6-7, Greenbelt, MD.* LPI Contribution No. 1376, 93–94.

Orton, G.S., P.A. Yanamandra-Fisher, B.M. Fisher, A.J. Friedson, P.D. Parrish, J.F. Nelson, A.S. Bauermeister, L.Fletcher, D.Y. Gezari, F. Varosi *et al.* (2008) Semi-annual oscillations in Saturn's low-latitude stratospheric temperatures. *Nature*, **453**, 196–199, doi: 10.1038/nature06897

Owen, T. and A. Bar-Nun (1995) Comets, impacts and atmospheres. *Icarus*, **116**, 215–226.

Owen, T. and T. Encrenaz (2003) Element abundances and isotope ratios in the giant planets and Titan. *Space Sci. Rev.*, **106**, 121–138.

Owen, T. and T. Encrenaz (2006). Compositional constraints on giant planet formation. *Planet. Space Sci.*, **54**, 1188–1196.

Owen, W.M., R.M. Vaughan, and S.P. Synnott (1991) Orbits of the six new satellites of Neptune. *Astron. J.*, **101**, 1511–1515.

Owen, T., P. Mahaffy, H.B. Niemann, S. Atreya, T. Donahue, A. Bar-Nun, and I. de Pater (1999) A low-temperature origin for the planetesimals that formed Jupiter. *Nature*, **402**, 269–270.

Owen, T., P.R. Mahaffy, H.B. Niemann, S. Atreya, and M. Wong (2001) Protosolar nitrogen. *Astrophys. J.*, **553**, L77–L79.

Papaloizou, J.C.B., R.P. Nelson, and M.D. Snellgrove (2004) The interaction of giant planets with a disc with MHD turbulence, III: Flow morphology and conditions for gap formation in local and global simulations. *Monthly Notices of the Royal Astronomical Society*, **350**, 829.

Pearl, J.C. and B.J. Conrath (1991) The albedo, effective temperature, and energy balance of Neptune, as determined from Voyager data. *J. Geophys. Res.*, **96**, 18921–18930.

Peek, B.M. (1958) *The Planet Jupiter*. Faber & Faber, London.

Penzias, A.A. and R. W. Wilson (1965). A measurement of excess antenna temperature at 4080 Mc/s. *Astrophys. J.*, **142**, 419–421.

Pérez-Hoyos, S. and A. Sánchez-Lavega (2006a) On the vertical wind shear of Saturn's equatorial jet at cloud level. *Icarus*, **180**, 161–175.

Pérez-Hoyos, S. and A. Sánchez-Lavega (2006b) Solar flux in Saturn's atmosphere: Penetration and heating rates in the aerosol and cloud layers. *Icarus*, **180**, 368–378.

Pérez-Hoyos, S., A. Sánchez-Lavega, R.G. French, and J.F. Rojas (2005) Saturn's cloud structure and temporal evolution from ten years of Hubble Space Telescope images (1994–2003). *Icarus*, **176**, 155–174.

Piccioni, G., P. Drossart, A. Sanchez-Lavega, R. Hueso, F.W. Taylor, C.F. Wilson, D. Grassi, L. Zasova, M. Moriconi, A. Adriani *et al.* (2007) South-polar features on Venus similar to those near the north pole. *Nature*, **450**, 637–640.

Pickett, M.K. and A.J. Lim (2004) Planet formation: The race is not to the swift. *Astron. Geophys.*, **45**, 1.12–1.17.

Pirraglia, J.A., B.J. Conrath, M.D. Allison, and P.J. Gierasch (1981) Thermal structure and dynamics of Saturn and Jupiter. *Nature*, **292**, 677–679.

Plass, G.N., G.W. Kattawar, and F.E. Catchings (1973) Matrix operator method of radiative transfer, 1: Rayleigh scattering. *Appl. Opt.*, **12**, 314–329.

Podolak, M. and M. Marley (1991) Interior model constraints on super-abundances of volatiles in the atmosphere of Neptune. *Bull. Am. Astron. Soc.*, **23**, 1164.

Podolak, M., J.I. Podolak, and M.S. Marley (2000) Further investigations of random models of Uranus and Neptune. *Planet. Space Sci.*, **48**, 143–151.

Pollack, J.B., K. Rages, S.K. Pope, M.G. Tomasko, P.N. Romani, and S.K. Atreya (1987) Nature of the stratospheric haze on Uranus: Evidence for condensed hydrocarbons. *J. Geophys. Res.*, **92**, 15037–15065.

Pollack, J.B., O. Hubickyj, P. Bodenheimer, J.J. Lissauer, M. Podolak, and Y. Greenzweig (1996) Formation of the giant planets by concurrent accretion of solids and gas. *Icarus*, **124**, 62–85.

Porco, C.C., R.A. West, A. McEwen, A.D. Del Genio, A.P. Ingersoll, P. Thomas, S. Squyres, L. Donas, C.D. Murray, T.V. Johnson *et al.* (2003) Cassini imaging science at Jupiter. *Science*, **299**, 1541–1547.

Porco, C.C., E. Baker, J. Barbara, K. Beurle, A. Brahic, J.A. Burns, S. Charnoz, N. Cooper, D.D. Dawson, A. Del Genio *et al.* (2005) Cassini imaging science: Initial results on Saturn's atmosphere. *Science*, **307**, 1243–1247.

Prangé, R., T. Fouchet, R. Courtin, J.E.P. Connerney, and J.C. McConnell (2006) Latitudinal variation of Saturn photochemistry deduced from spatially-resolved ultraviolet spectra. *Icarus*, **180**, 379–392.

Press, W.H., S.A. Teukolsky, W.T. Vetterling, and B.P. Flannery (1992) *Numerical Recipes in Fortran* (Second Edition). Cambridge University Press, Cambridge, U.K.

Pryor, W.R., R.A. West, K.E. Simmons, and M. Delitsky (1992) High-phase-angle observations of Neptune at 2650 and 7500 Å: Haze structure and particle properties. *Icarus*, **99**, 302–317.

Radousky, H.B., A.C. Mitchell, and W.J. Nellis (1990) Shock temperature measurements of planetary ices: NH_3, CH_4, and "synthetic Uranus". *J. Chem. Phys.*, **93**, 8235–8239.

Rae, A.I.M. (1985) *Quantum Mechanics*. Adam Hilger, Bristol, U.K.

Ragent, B., D.S. Colburn, K.A. Rages, T.C.D. Knight, P. Avrin, G.S. Orton, P.A. Yanamandra-Fisher, and G.W. Grams (1998) The clouds of Jupiter: Results of the Galileo Jupiter Mission Probe Nephelometer Experiment. *J. Geophys. Res.*, **103**, 22891–22909.

Rages, K., J.B. Pollack, M.G. Tomasko, and L.R. Doose (1991) Properties of scatterers in the troposphere and lower stratosphere of Uranus based on Voyager imaging data. *Icarus*, **89**, 359–376.

Rages, K., R. Beebe, and D. Senske (1999) Jovian stratospheric hazes: The high phase angle view from Galileo. *Icarus*, **139**, 211–226.

Rages, K.A., H.B. Hammel, and A.J. Friedson (2004) Evidence for temporal change at Uranus' south pole. *Icarus*, **172**, 548–554.

Read, P.L. (1986) Stable, baroclinic eddies on Jupiter and Saturn: A laboratory analogue and some observational tests. *Icarus*, **65**, 304–334.

Read, P.L. (2001) Transition to geostrophic turbulence in the laboratory, and as a paradigm in atmospheres and oceans. *Surveys in Geophysics,*, **22**, 265–317.

Read, P.L. and R. Hide (1983) Long-lived eddies in the laboratory and in the atmospheres of Jupiter and Saturn. *Nature*, **302**, 126–129.

Read, P.L. and R. Hide (1984) An isolated baroclinic eddy as laboratory analogue of the Great Red Spot. *Nature*, **308**, 45–48.

Read, P.L., Y.H. Yamazaki, S.R. Lewis, P.D. Williams, K. Miki-Yamazaki, J. Sommeria, H. Didelle, and A. Fincham (2004) Jupiter's and Saturn's convectively driven banded jets in the laboratory. *Geophys. Res. Lett.*, **31**, L22701.

Reuter, D.C., A.A. Simon-Miller, A. Lunsford, K.H. Baines, A.F. Cheng, D.E. Jennings, C.B. Olkin, J.R. Spencer, S.A. Stern, H.A. Weaver, and L.A. Young (2007) Jupiter cloud composition, stratification, convection, and wave motion: A View from New Horizons. *Science*, **318**, 223–225, doi: 10.1126/science.1147618.

Rhines, P.B. (1973) Observations of energy-containing oceanic eddies and theoretical models of waves and turbulence. *Bound.-Layer Meteor.*, **4**, 345–360.

Rhines, P.B. (1975) Waves and turbulence on a beta plane. *J. Fluid Mech.*, **69**, 417–443.

Rice, W.K.M. and P.J. Armitage (2003) On the formation timescale and core masses of gas giant planets. *Astrophys. J. Lett.*, **598**, 55.

Rice, W.K.M., P.J. Armitage, I.A. Bonnell, M.R. Bate, S.V. Jeffers, and S.G. Vine (2003) Substellar companions and isolated planetary-mass objects from protostellar disc fragmentation. *Monthly Notices of the Royal Astronomical Society*, **346**, L36.

Richard, D. and J.-P. Zahn (1999) Turbulence in differentially rotating flows. What can be learned from the Couette–Taylor experiment? *Astron. Astrophys.*, **347**, 734–738.

Rodgers, C.D. (2000) *Inverse Methods for Atmospheric Sounding: Theory and Practice*. World Scientific, Singapore.

Rogers, J.H. (1995) *The Giant Planet Jupiter*. Cambridge University Press, Cambridge, U.K.

Roos-Serote, M., and P.G.J. Irwin (2006) Scattering properties and location of the jovian 5-micron absorber from Galileo/NIMS limb-darkening observations. *J. Quant. Spec. Rad. Trans.*, **101**, 448–461.

Roos-Serote, M., P. Drossart, T. Encrenaz, E. Lellouch, R.W. Carlson, K. Baines, L. Kamp. R. Mehlman, G.S. Orton, S. Calcutt, P. Irwin, F. Taylor, and A. Weir (1998) Analysis of Jupiter NEB hotspots in the 4–5 μm range from Galileo/NIMS observations: Measurements of cloud opacity, water and ammonia. *J. Geophys. Res.*, **103**, 23023–23041.

Roos-Serote, M., A.R. Vasavada, L. Kamp, P. Drossart, P. Irwin, C. Nixon, and R.W. Carlson (2000) Proximate humid and dry regions in Jupiter's atmosphere indicate complex local meteorology. *Nature*, **405**, 158–160.

Rosenqvist, J., E. Lellouch, P.N. Romani, G. Paubert, and T. Encrenaz (1992) Millimeter-wave observations of Saturn, Uranus, and Neptune: CO and HCN on Neptune. *Astrophys. J.*, **392**, L99–L102.

Rothman, L.S., D. Jacquemart, A. Barbe, D. Chris Benner, M. Birk, L.R. Brown, M.R. Carleer, C. Chackerian, K. Chance, L.H. Coudert *et al.* (2005) The HITRAN 2004 molecular spectroscopic database. *J. Quant. Spectrosc. Rad. Trans.*, **96**, 139–204.

Russell, C.T. and J.G. Luhmann (1997) *Saturn: Magnetic Field and Magnetosphere. Encyclopedia of the Planetary Sciences*. Chapman & Hall, London.

Sada, P.V., G.H. McCabe, G.L. Bjoraker, D.E. Jennings, and D.C. Reuter (1996) ^{13}C–Ethane in the atmospheres of Jupiter and Saturn. *Astrophys. J.*, **472**, 903.

Sada, P.V., G.L. Bjoraker, D.E. Jennings, P.N. Romani, and G.H. McCabe (2005) Observations of C_2H_6 and C_2H_2 in the stratosphere of Saturn. *Icarus*, **173**, 499–507.

Salyk, C., A.P. Ingersoll, J. Lorre, A. Vasavada, and A.D. Del Genio (2006) Interaction between eddies and mean flow in Jupiter's atmosphere: Analysis of Cassini imaging data. *Icarus*, **185**, 430–442

Sánchez-Lavega, A. (1982) Motions in Saturn's atmosphere: Observations before the Voyager encounters. *Icarus*, **49**, 1–16.

Sánchez-Lavega, A. (2002) Observations of Saturn's Ribbon wave 14 years after its discovery. *Icarus*, **158**, 272–275.

Sánchez-Lavega, A. (2005) How long is the day on Saturn? *Science*, **307**, 1223–1224, doi: 10.1126/science.1104956.

Sánchez-Lavega, A., I. Miyazaki, D. Parker, P. Laques, and J. Lecacheux (1991) A disturbance in Jupiter's high-speed north temperate jet during 1990. *Icarus*, **94**, 92–97.

Sánchez-Lavega, A., J. Lecacheux, F. Colas, and P. Laques (1993) Ground-based observations of Saturn's North Polar Spot and Hexagon. *Science*, **260**, 329–332.

Sánchez-Lavega, A., J. Lecacheux, J.M. Gómez, F. Colas, P. Laques, K. Noll, D. Gilmore, I. Miyazaki, and D. Parker (1996) Large-scale storms in Saturn's atmosphere during 1994. *Science*, **271**, 631–634.

Sánchez-Lavega, A., J. Lecacheux, F. Colas, J.F. Rojas, and J.M. Gómez (1999) Discrete cloud activity in Saturn's equator during 1995, 1996 and 1997. *Planet. Space Sci.*, **47**, 1277–1283.

Sánchez-Lavega, A., J. Rojas, and P. Sada (2000) Saturn's zonal winds at cloud level. *Icarus*, **147**, 405–420.

Sánchez-Lavega, A., S. Pérez-Hoyos, J. R. Acarreta, and R. G. French (2002) No hexagonal wave around Saturn's Southern Pole. *Icarus*, **160**, 216–219, doi:10.1006/icar.2002.6947.

Sánchez-Lavega, A., S. Pérez-Hoyos, J.F. Rojas, R. Hueso, and R.G. French (2003) A strong decrease in Saturn's equatorial jet at cloud level. *Nature*, **423**, 623–625.

Sánchez-Lavega, A., R. Hueso, S. Pérez-Hoyos, J.F. Rojas, and R.G. French (2004) Saturn's cloud morphology and zonal winds before the Cassini encounter. *Icarus*, **170**, 519–523.

Sánchez-Lavega, A., R. Hueso, S. Pérez-Hoyos, and J.F. Rojas (2006) A strong vortex in Saturn's South Pole. *Icarus*, **184**, 524–531.

Sánchez-Lavega, A., G.S. Orton, R. Hueso, E. García-Melendo, S. Pérez-Hoyos, A. Simon-Miller, J.F. Rojas, J. M. Gómez, P.A. Yanamandra-Fisher, L. Fletcher *et al.* (2008) Depth of a strong jovian jet from a planetary-scale disturbance driven by storms. *Nature*, **451**, 437–440, doi: 10.1038/nature06533.

Saumon, D.S. and T. Guillot (2004) Shock compression of deuterium and the interiors of Jupiter and Saturn. *Astrophys. J.*, **609**, 1170–1180.

Schilling, G. (1999a) Submillimeter astronomy reaches new heights. *Science*, **283**, 1836.

Schilling, G. (1999b) Telescope builders think big—really big. *Science*, **284**, 1913–1915.

Schilling, G. (1999c) Lofty observatory gets boost. *Science*, **284**, 1915.

Seiff, A., D.B. Kirk, T.C.D. Knight, R.E. Young, J.D. Mihalov, L.A. Young, F.S. Milos, G. Schubert, R.C. Blanchard, and D. Atkinson (1998) Thermal structure of Jupiter's atmosphere near the edge of a 5-mm hot spot in the north equatorial belt. *J. Geophys. Res.*, **103**, 22857–22,889.

Shakura, N.I. and R.A. Sunyaev (1973) Black holes in binary systems: Observational appearance. *Astron. Astrophys.*, **24**, 337–355.

Showman, A.P. and I. de Pater (2005) Dynamical implications of Jupiter's tropospheric ammonia abundance. *Icarus*, **174**, 192–204.

Showman, A.P. and T.E. Dowling (2000) Nonlinear simulations of Jupiter's 5-micron hotspots. *Science*, **289**, 1737–1740.

Showman, A.P., P.J. Gierasch, and Y. Lian (2006) Deep zonal winds can result from shallow driving in a giant-planet atmosphere. *Icarus*, **182**, 513–526.

Shu, F.H., S. Tremaine, F.C. Adams, and P. Ruden (1990) Sling amplification and eccentric gravitational instabilities in gaseous disks. *Astrophys. J.*, **358**, 495–514.

Simon-Miller, A.A., B. Conrath, P.J. Gierasch, and R.F. Beebe (2000) A detection of water ice on Jupiter with Voyager IRIS. *Icarus*, **145**, 454–461.

Simon-Miller, A.A., D. Banfield, and P.J. Gierasch (2001) Color and the vertical structure in Jupiter's belts, zones, and weather systems. *Icarus*, **154**, 459–474.

Simon-Miller, A.A., P.J. Gierasch, R.F. Beebe, B. Conrath, F.M. Flasar, R.K. Achterberg, and the Cassini CIRS team (2002) New observational results concerning Jupiter's Great Red Spot. *Icarus*, **158**, 249–266.

Simon-Miller, A.A., N.J. Chanover, G.S. Orton, M. Sussman, I.G. Tsavaris, and E. Karkoschka (2006) Jupiter's White Oval turns red. *Icarus*, **185**, 558–562.

Smith, B.A., L.A. Soderblom, T.V. Johnson, A.P. Ingersoll, S.A. Collins, E.M. Shoemaker, G.E. Hunt, H. Masursky, M.H. Carr, M.E. Davies *et al.* (1979a) The Jupiter system through the eyes of Voyager 1. *Science*, **204**, 951–957, 960–972

Smith, B.A., L.A. Soderblom, R. Beebe, J. Boyce, G. Briggs, M. Carr, S.A. Collins, T.V. Johnson, A.F. Cook, G.E. Danielson *et al.* (1979b) The Galilean satellites and Jupiter: Voyager 2 imaging science results. *Science*, **206**, 927–950.

Smith, B.A., L. Soderblom, R. Beebe, J. Boyce, G. Briggs, A. Bunker, S.A. Collins, C.J. Hansen, T.V. Johnson, J.L. Mitchell *et al.* (1981) Encounter with Saturn: Voyager 1 imaging science results. *Science*, **212**, 163–191

Smith, B.A., L. Soderblom, R. Batson, P. Bridges, J. Inge, H. Masursky, E. Shoemaker, R. Beebe, J. Boyce, G. Briggs *et al.* (1982) A new look at the Saturn system: The Voyager 2 images. *Science*, **215**, 504–537.

Smith, B.A., L.A. Soderblom, R. Beebe, D. Bliss, R.H. Brown, S.A. Collins, J.M. Boyce, G.M. Briggs, A. Brahic, J.N. Cuzzi *et al.* (1986) Voyager 2 in the Uranian system: Imaging science results. *Science*, **233**, 43–64.

Smith, B.A., L.A. Soderblom, D. Banfield, C. Barnet, R.F. Beebe, A.T. Bazilevskii, K. Bollinger, J.M. Boyce, G.A. Briggs, and A. Brahic (1989) Voyager 2 at Neptune: Imaging science results. *Science*, **246**, 1422–1449.

Smith, R.A., F.E. Jones, and R.P. Chasmus (1968) *The Detection and Measurement of Infrared Radiation* (Second Edition). Oxford University Press, Oxford, U.K.

Sromovsky, L.A. and P.M. Fry (2005) Dynamics of cloud features on Uranus. *Icarus*, **179**, 459–484.

Sromovsky, L.A. and P.M. Fry (2007) Spatially resolved cloud structure on Uranus: Implications of near-IR adaptive optics imaging. *Icarus*, **192**, 527–557.

Sromovsky, L.A. and P.M. Fry (2008) The methane abundance and structure of Uranus' cloud bands inferred from spatially resolved 2006 Keck grism spectra. *Icarus*, **193**, 252–266.

Sromovsky, L.A., H.E. Revercomb, V.E. Suomi, S.S. Limaye, and R.J. Kraus (1982) Jovian winds from Voyager 2, Part II: Analysis of eddy transports. *J. Atmos. Sci.*, **39**, 1433–1445.

Sromovsky, L.A., H.E. Revercombe, R.J. Kraus, and V.E. Suomi (1983) Voyager 2 observations of Saturn's northern mid-latitude cloud features: Morphology, motions and evolution. *J. Geophys. Res.*, **88**, 8650–8666.

Sromovsky, L.A, S.S. Limaye, and P.M. Fry (1993) Dynamics of Neptune's major cloud features. *Icarus*, **105**, 110–141.

Sromovsky, L.A., A.D. Collard, P.M. Fry, G.S. Orton, M.T. Lemmon, M.G. Tomasko, and R.S. Freedman (1998) Galileo probe measurements of thermal and solar radiation fluxes in the Jovian atmosphere. *J. Geophys. Res.*, **103**, 22929–22977.

Sromovsky, L.A., J. Spencer, K. Baines, and P. Fry (2000) Ground-based observations of cloud features on Uranus. *Icarus*, **146**, 307–311

Sromovsky, L.A., P.M. Fry, K.H. Baines, S.S. Limaye, G.S. Orton, and T.E. Dowling (2001a) Coordinated 1996 HST and IRTF imaging of Neptune and Triton, I: Observations, navigation, and differential deconvolution. *Icarus*, **149**, 416–434.

Sromovsky, L.A., P.M. Fry, T.E. Dowling, K.H. Baines, and S.S. Limaye (2001b) Coordinated 1996 HST and IRTF imaging of Neptune and Triton, II: Implications of disk-integrated photometry. *Icarus*, **149**, 435–458.

Sromovsky, L.A., P.M. Fry, K.H. Baines, and T.E. Dowling (2001c) Coordinated 1996 HST and IRTF imaging of Neptune and Triton, III: Neptune's atmospheric circulation and cloud structure. *Icarus*, **149**, 459–488.

Sromovsky, L.A., P.M. Fry, T.E. Dowling, K.H. Baines, and S.S. Limaye (2001d) Neptune's atmospheric circulation and cloud morphology: Changes revealed by 1998 HST imaging. *Icarus*, **150**, 244–260.

Sromovsky, L.A., P.M. Fry, and K.H. Baines (2002) The unusual dynamics of northern dark spots on Neptune. *Icarus*, **156**, 16–36.

Sromovsky, L.A., P.M. Fry, S.S. Limaye, and K.H. Baines (2003) The nature of Neptune's increasing brightness: Evidence for a seasonal response. *Icarus*, **163**, 256–261.

Sromovsky, L.A., P.G.J. Irwin, and P.M. Fry (2006) Near-IR methane absorption in outer planet atmospheres: Improved models of temperature dependence and implications for Uranus cloud structure. *Icarus*, **182**, 577–593.

Sromovsky, L.A., P.M. Fry, H.B. Hammel, I. de Pater, K.A. Rages, and M.R. Showalter (2007) Dynamics, evolution, and structure of Uranus' brightest cloud feature. *Icarus*, **192**, 558–575.

Stam, D.M., D. Banfield, P.J. Gierasch, P.D. Nicholson, and K. Matthews (2001) Near-IR spectrophotometry of saturnian aerosols: Meridional and vertical distribution. *Icarus*, **152**, 407–422.

Stevenson, D. J. (1980) Saturn's luminosity and magnetism. *Science*, **208**, 746–748.

Stone, P.H. (1976) The meteorology of the Jovian atmosphere. In T. Gehrels (Ed.), *Jupiter*. University of Arizona Press, Tucson, AZ.

Stratman, P.W., A.P. Showman, T.E. Dowling, and L.A. Sromovsky (2001) EPIC simulations of bright companions to Neptune's great dark spots. *Icarus*, **151**, 275–285.

Taylor, F.W., S.B. Calcutt, P.G.J. Irwin, C.A. Nixon, P.L. Read, P.J.C. Smith, and T.J. Vellacott (1998) Investigation of Saturn's atmosphere by Cassini. *Planet. Space Sci.*, **46**, 1315–1324.

Taylor, F.W., S.K. Atreya, T. Encrenaz, D.M. Hunten, P.G.J. Irwin, and T.C. Owen (2004) The composition of the atmosphere of Jupiter. In F. Bagenal, W. McKinnon, and T. Dowling (Eds.), *Jupiter: The Planet, Satellites and Magnetosphere*, pp 59–78. Cambridge University Press, Cambridge, U.K.

Teanby, N.A., L.N. Fletcher, P.G.J. Irwin, T. Fouchet, and G.S. Orton (2006) New upper limits for hydrogen halides on Saturn derived from Cassini-CIRS data. *Icarus*, **185**, 466–475.

Tokunaga, A.T., S.C. Beck, T.R. Geballe, J.H. Lacy, and E. Serabyn (1981) The detection of HCN on Jupiter. *Icarus*, **48**, 283–289.

Toomre, A. (1964) On the gravitational stability of a disk of stars. *Astrophys. J.*, **139**, 1217.

Turcotte, S. and R.F. Wimmer-Schweingruber (2002) Possible in situ tests of the evolution of elemental and isotopic abundances in the solar convection zone. *J. Geophys. Res.*, **107**, SSH 5-1, CiteID 1442, doi: 10:1029/2002JA009418.

Turcotte, S., J. Richer, G. Michaud, C.A. Iglesias, and F.J. Rogers (1998) Consistent solar evolution model including diffusion and radiative acceleration effects. *Astrophys. J.*, **504**, 539.

Tyler, G.L., D.N. Sweetnam, J.D. Anderson, S.E. Borutzki, J.K. Campbell, V.R. Eshleman, D.L. Gresh, E.M. Gurrola, D.P. Hinson, N. Kawashima *et al.* (1989) Voyager radio science observations of Neptune and Triton. *Science*, **246**, 1466–1473.

Vallis, G.K. and M.E. Maltrud (1993) Generation of mean flows and jets on a beta-plane and over topography. *J. Phys. Oceanogr.*, **23**, 1346–1362.

Vanasse, G.A. (Ed.) (1983) *Spectrometric Techniques*, Vol. III. Academic Press, San Diego, CA.

Vasavada, A.R. and A.P Showman (2005) Jovian atmospheric dynamics: An update after Galileo and Cassini. *Rep. Prog. Phys.*, **68**, 1935–1996.

Vasavada, A.R., S.M. Hörst, M.R. Kennedy, A.P. Ingersoll, C.C. Porco, A.D. Del Genio, and R.A. West (2006) Cassini imaging of Saturn: Southern hemisphere winds and vortices. *J. Geophys. Res.*, **111**, E5004, doi: 10.1029/2005JE002563

Wallner, O., R. Flatscher, and K Ergenzinger (2006) Exo-zodi detection capability of the Ground-Based European Nulling Interferometry Experiment (GENIE) Instrument. *Appl. Opt.*, **45**, 4404–4410,

Ward, W.R. (1986) Density waves in the solar nebula: Differential Lindblad torque. *Icarus*, **67**, 164.

Ward, W.R. (1997) Protoplanet migration by nebula tides. *Icarus*, **126**, 261.

Ward, W. and D.P. Hamilton (2002) The obliquities of the giant planets. Paper presented at *Eurojove: Jupiter after Galileo and Cassini, Lisbon, June.*

Warwick, J.W., J.B. Pearce, D.R. Evans, T.D. Carr, J.J. Schauble, J.K. Alexander, M.L. Kaiser, M.D. Desch, B.M. Pedersen, A. Lecacheux *et al.* (1981) Planetary radio astronomy observations from Voyager 1 near Saturn. *Science*, **212**, 239–243.

Weast, R.C. (Ed.) (1975) *Handbook of Chemistry and Physics* (56th Edition), CRC Press, Cleveland, OH.

Weaver, H.A., K.H. Baines, A.A. Simon-Miller, A.F. Cheng, G.R. Gladstone, K.D. Retherford, H.B. Throop, J.M. Moore, J.R. Spencer, S.A. Stern, and the New Horizons Science Team (2007) *New Horizons Observations of Polar Lightning on Jupiter* (American Astronomical Society, DPS meeting #39, #1.05).

Weidenschilling, S. J. (1977) Aerodynamics of solid bodies in the solar nebula. *Monthly Notices of the Royal Astronomical Society*, **180**, 57–70.

Weir, S.T., A.C. Mitchell, and W.J. Nellis (1996) Metallization of fluid molecular hydrogen at 140 GPa (1.4 Mbar). *Phys. Rev. Lett.*, **76**, 1860–1863.

Weisstein, E.W. and E. Serabyn (1994) Detection of the 267 GHz $J = 1$–0 rotational transition of PH_3 in Saturn with a new Fourier transform spectrometer. *Icarus*, **109**, 367–381.

West, R.A. (1979a) Spatially resolved methane band photometry of Jupiter, II: Analysis of the SEB and STrZ reflectivity. *Icarus*, **38**, 34–53.

West, R.A. (1979b) Spatially resolved methane band photometry of Jupiter, I: Absolute reflectivity and CTL variations in the 6190, 7250 and 8900 Å bands. *Icarus*, **38**, 12–33.

West, R.A. (1999) *Atmospheres of the Giant Planets: Encyclopedia of the Solar System* (edited by P.R. Weissman, L.-A. McFadden, and T.V. Johnson). Academic Press, San Diego, CA.

West, R.A. and M.G. Tomasko (1980) Spatially resolved methane band photometry of Jupiter, III: Cloud vertical structures for several axisymmetric bands and the GRS. *Icarus*, **41**, 278–292

West, R.A., M.G. Tomasko, M.P. Wijensinghe, L.R. Doose, H.J. Reitsema, and S.M. Larson (1982) Spatially resolved methane band photometry of Saturn I: Absolute reflectivity and centre to limb variations in 6190, 7250-, and 8900-Å bands. *Icarus*, **51**, 51–64.

West, R.A., M. Sato, H. Hart, A.L. Lane, C.W. Hord, K.E. Simmons, L.W. Esposito, D.L. Coffeen, and R.B. Pomphrey (1983) Photometry and polarimetry of Saturn at 2640 and 7500 Å. *J. Geophys. Res.*, **88**, 8679–8697.

West, R.A., D.F. Strobel, and M.G. Tomasko (1986) Clouds, aerosols and photochemistry in the Jovian atmosphere. *Icarus*, **65**, 161–217.

West, R., K. Baines, and J. Pollack (1991) Clouds and aerosols in the Uranian atmosphere. In J. Bergstrahl, E.D. Miner, and M.S. Matthews (Eds.), *Uranus*. University of Arizona Press, Tucson, AZ.

West, R.A., A.J. Friedson, and J.K. Appleby (1992) Jovian large-scale stratospheric circulation. *Icarus*, **100**, 245–259.

West, R.A., K.H. Baines, A.J. Friedson, D. Banfield, B. Ragent, and F.W. Taylor (2004) Jovian clouds and hazes. In F. Bagenal, W. McKinnon, and T. Dowling (Eds.), *Jupiter: The Planet, Satellites and Magnetosphere*, pp 79–104. Cambridge University Press, Cambridge, U.K.

Westphal, J.A., W.A. Baum, A.P. Ingersoll, C.D. Barnet, E.M. De Jong, G.E. Danielson, and J. Caldwell (1992) Hubble Space Telescope observations of the 1990 equatorial disturbance on Saturn: Images, albedoes and limb-darkening. *Icarus*, **100**, 485–498.

Williams, G.P. (1978) Planetary circulations, 1: Barotropic representation of Jovian and terrestrial turbulence. *J. Atmos. Sci.*, **35**, 1399–1426.

Williams, G.P. (1979) Planetary circulations, 2: The Jovian quasi-geostrophic regime. *J. Atmos. Sci.*, **36**, 932–968.

Williams, G.P. (1985) Jovian and comparative atmospheric modelling. *Adv. Geophys.*, **28A**, 381–429.

Williams, G.P. (1996) Jovian dynamics, Part I: Vortex stability, structure and genesis. *J. Atmos. Sci.*, **53**, 2685–2734.

Williams, G.P. (2002) Jovian dynamics, Part II: The genesis and equilibration of vortex sets. *J. Atmos Sci.*, **59**, 1356–1370.

Williams, G.P. (2003a) Barotropic instability and equatorial superrotation. *J. Atmos. Sci.*, **60**, 2136–2152.

Williams, G.P. (2003b) Jovian dynamics, Part III: Multiple, migrating, and equatorial jets. *J. Atmos. Sci.*, **60**, 1270–1296.

Williams, G.P. and J.B. Robinson (1973) Dynamics of a convectively unstable atmosphere: Jupiter. *J. Atmos. Sci.*, **30**, 684–717.

Wong, M.H., G.L. Bjoraker, M.D. Smith, F. M. Flasar, and C.A. Nixon (2004a) Identification of the 10 μm ammonia ice feature on Jupiter. *Planet. Space Sci.*, **52**, 385–395.

Wong, M.H., P.R. Mahaffy, S.K. Atreya, H.B. Niemann, and T.C. Owen (2004b) Updated Galileo probe mass spectrometer measurements of carbon, oxygen, nitrogen, and sulfur on Jupiter. *Icarus*, **171**, 153–170.

Yamazaki, Y.H., D.R. Skeet, and P.L. Read (2004) A new general circulation model of Jupiter's atmosphere based on the UKMO unified model: Three-dimensional evolution of isolated vortices and zonal jets in mid-latitudes. *Planet. Space Sci.*, **52**, 423–445.

Yanamandra-Fisher, P.A., G.S. Orton, B.M. Fisher, and A. Sánchez-Lavega (2001) NOTE: Saturn's 5.2 μm cold spots: Unexpected cloud variability. *Icarus*, **150**, 189–193.

Young, A.T. (1985) What color is the solar system? *Sky & Telescope*, **69**, 399–403.

Young, R.E. (2003) The Galileo probe: How it has changed our understanding of Jupiter. *New Astron. Rev.*, **47**, 1–51.

Zarka, P. (1985) On detection of radio bursts associated with jovian and saturnian lightning. *Astron. Astrophys.*, **146**, L15–L18.

Yelle, R.V., J.C. McConnel, D.R. Strobel, and L.R. Doose (1989) The far ultraviolet reflection spectrum of Uranus: Results from the Voyager encounter. *Icarus*, **77**, 439–456.

Yelle, R.V., C. Griffith, and L.A. Young (2001) Structure of the Jovian stratosphere at the Galileo probe entry site. *Icarus*, **152**, 331–346

Young, L.A., R.V. Yelle, R. Young, A. Seiff, and D.B. Kirk (1997) Gravity waves in Jupiter's thermosphere. *Science*, **276**, 108–111.

Yung, Y.L. and W.B. DeMore (1999) *Photochemistry of Planetary Atmospheres*. Oxford University Press, Oxford, U.K.

von Zahn, U., D.M. Hunten, and G. Lehmacher (1998) Helium in Jupiter's atmosphere: Results from the Galileo probe Helium Interferometer Experiment. *J. Geophys. Res.*, **103**, 22815–22829.

Index

Printing: Mercedes-Druck, Berlin
Binding: Stein+Lehmann, Berlin